Methods of
Experimental Physics

VOLUME 14

VACUUM PHYSICS AND TECHNOLOGY

METHODS OF EXPERIMENTAL PHYSICS:

L. Marton and C. Marton, *Editors-in-Chief*

1. Classical Methods
   Edited by Immanuel Estermann
2. Electronic Methods, Second Edition (in two parts)
   Edited by E. Bleuler and R. O. Haxby
3. Molecular Physics, Second Edition (in two parts)
   Edited by Dudley Williams
4. Atomic and Electron Physics—Part A: Atomic Sources and Detectors, Part B: Free Atoms
   Edited by Vernon W. Hughes and Howard L. Schultz
5. Nuclear Physics (in two parts)
   Edited by Luke C. L. Yuan and Chien-Shiung Wu
6. Solid State Physics (in two parts)
   Edited by K. Lark-Horovitz and Vivian A. Johnson
7. Atomic and Electron Physics—Atomic Interactions (in two parts)
   Edited by Benjamin Bederson and Wade L. Fite
8. Problems and Solutions for Students
   Edited by L. Marton and W. F. Hornyak
9. Plasma Physics (in two parts)
   Edited by Hans R. Griem and Ralph H. Lovberg
10. Physical Principles of Far-Infrared Radiation
    L. C. Robinson
11. Solid State Physics
    Edited by R. V. Coleman
12. Astrophysics—Part A: Optical and Infrared
    Edited by N. Carleton
    Part B: Radio Telescopes, Part C: Radio Observations
    Edited by M. L. Meeks
13. Spectroscopy (in two parts)
    Edited by Dudley Williams
14. Vacuum Physics and Technology
    Edited by G. L. Weissler and R. W. Carlson
15. Quantum Electronics (in two parts)
    Edited by C. L. Tang

Volume 14

Vacuum Physics and Technology

Edited by

G. L. WEISSLER

Department of Physics
University of Southern California
University Park
Los Angeles, California

and

R. W. CARLSON

Department of Physics
University of Southern California
University Park
Los Angeles, California

Earth and Space Sciences Division
Jet Propulsion Laboratory
California Institute of Technology
Pasadena, California

1979

ACADEMIC PRESS

A Subsidiary of Harcourt Brace Jovanovich, Publishers

New York London Toronto Sydney San Francisco

ACADEMIC PRESS, INC.
111 Fifth Avenue, New York, New York 10003

United Kingdom Edition published by
ACADEMIC PRESS, INC. (LONDON) LTD.
24/28 Oval Road, London NW1 7DX

Library of Congress Cataloging in Publication Data
Main entry under title:

Vacuum physics and technology.

(Methods of experimental physics; v. 14)
Includes bibliographical references.
1. Vacuum. 2. Vacuum technology. I. Weissler, G. L. II. Carlson, Robert Warner, Date
III. Series.
QC166.V34 533'.5 79-17200
ISBN 0-12-475914-9 (v. 14)

PRINTED IN THE UNITED STATES OF AMERICA

79 80 81 82 9 8 7 6 5 4 3 2 1

CONTENTS

2. Measurement of Total Pressure in Vacuum Systems
by D. R. Denison

6. Metal Vacuum Systems and Components
by N. MILLERON and R. C. WOLGAST

16. Vacuum Deposition Techniques
by M. T. Thomas

CONTRIBUTORS

Numbers in parentheses indicate the pages on which the authors' contributions begin.

V. O. ALTEMOSE, *Technical Staff Division, Corporate Research Laboratories, Corning Glass Works, Corning, New York 14830* (313)

W. M. BRUBAKER, *1954 Highland Oaks Drive, Arcadia, California 91006* (81, 183, 216)

R. W. CARLSON, *Department of Physics, University of Southern California, University Park, Los Angeles, California 90007, and Earth and Space Sciences Division, Jet Propulsion Laboratory, California Institute of Technology, Pasadena, California 91103* (6, 11)

D. R. DENISON, *Perkin-Elmer, Ultek Division, P.O. Box 10920, Palo Alto, California 94303* (35)

Z. C. DOBROWOLSKI, *Kinney Vacuum Company, 495 Turnpike Street, Canton, Massachusetts 02021* (111, 468)

M. H. HABLANIAN, *Varian Vacuum Division, 121 Hartwell Avenue, Lexington, Massachusetts 02173* (101, 141, 180, 457, 465, 466, 472)

DAVID J. HARRA, *Varian Associates, Vacuum Division, 611 Hansen Way, Palo Alto, California 94303* (193)

LAWRENCE T. LAMONT, JR., *Varian Associates, Vacuum Research and Development, 611 Hansen Way, Palo Alto, California 94303* (231, 449, 477, 491, 505)

DAVID LICHTMAN, *Department of Physics, University of Wisconsin, Milwaukee, Wisconsin 53201* (25, 345)

N. MILLERON, *S*E*N VAC Services, Berkeley, California* (275, 425)

G. OSTERSTROM, *Sargent–Welch Company, Vacuum Products Division, 7300 North Linder Avenue, Skokie, Illinois 60077* (247)

Y. SHAPIRA*, *Department of Physics, University of Wisconsin, Milwaukee, Wisconsin 53201* (345)

M. T. THOMAS, *Battelle Pacific Northwest Laboratory, Battelle Memorial Institute, P.O. Box 999, Richland, Washington 99352* (521)

G. L. WEISSLER, *Department of Physics, University of Southern California, University Park, Los Angeles, California 90007* (1)

R. C. WOLGAST, *Department of Mechanical Engineering, Lawrence Berkeley Radiation Laboratory, 1 Cyclotron Road, Berkeley, California 94720* (275, 425)

* Present address: School of Engineering, Tel Aviv University, Ramat Aviv, Tel Aviv, Israel 69978.

FOREWORD

In this publication, "Methods of Experimental Physics," three volumes are not directed to specific fields of research: Classical Methods (Volume 1), Vacuum Physics and Technology (the present volume), and Quantum Electronics (the following Volume 15). Each of these has broad reference value, but the present volume is of importance to the widest range of research interests. Vacuums are required in nearly every physics laboratory and in many chemistry and engineering laboratories as well.

The general approach taken by the Editors should make this volume one of extreme value, and one that can be used in conjunction with almost every other volume of Methods.

To all involved with this presentation go our deepest thanks.

L. Marton
C. Marton

PREFACE

When we were asked by the Editor of "Methods of Experimental Physics," Dr. L. Marton, to compose one volume on the subject of vacuum physics and technology, we immediately thought of the influence on our own combined 60 years of experimental research of such path-breaking authors as John Strong in "Procedures of Experimental Physics" and Saul Dushman in "Scientific Foundations of Vacuum Technique."

Dushman mentions in his preface of 1949 that " . . . vacuum technique has emerged from the purely scientific environment of the research laboratory and has entered the engineering stage on a scale that could not have been conceived even as recently as two decades ago." Similarly, the decades since 1949 have seen an even more explosive expansion into industrial applications with an extremely strong market for a wide range of commercially produced vacuum equipment. This equipment, in its turn, has become available to the experimental physicist in his research. Unfortunately, the cost of some of these items in an era of both inflation and retrenchment in research funds forces many to either build vacuum devices themselves or at least select with great care from the many commercially available.

With this in mind, we contacted the contributors to this volume with the hope that their respective specialized knowledge would communicate to those less sophisticated in vacuum physics the basic scientific ideas and today's generally accepted methods of producing, maintaining, and measuring various degrees of vacua, as required for their specific efforts.

Obviously, the different contributors to this volume make impossible the uniformity of style and of viewpoint which single authors bring to their books. On the other hand, we hope that the diversity of outlook and of opinion of our authors will be fresh and stimulating to the reader and will provide him with a variety of useful opinions that can be applied to his unique experimental requirements. It is the intent of this volume to provide the individual experimentalist with, we hope, all the up-to-date information in this vast field, and we thank our authors for their efforts in presenting their particular views of and extensive experience in vacuum technology.

G. L. WEISSLER
R. W. CARLSON

1. INTRODUCTION

1.1. Survey*

Individuals concerned with the design, building, and maintenance of a vacuum system for purposes of performing certain tasks or certain physical measurements will initially have to decide on the approximate ultimate vacuum range they require, and this needs to be expressed in certain pressure units. In conformity with most texts on this subject, we use here the unit of Torr = 1 mm column height of a mercury (Hg) manometer, and we provide in Table I a set of factors to convert to other pressure units.

Once the ultimate vacuum range for his work has been established, the experimentalist will find some general information in Table II, which may help him to plan his apparatus. As an example of his evolving requirements, he may initially wish to measure spark-over potentials for spherical gaps in an atmosphere of room air. His first vacuum apparatus might only comprise a "rough" or "fore" vacuum system with an appropriate gage connected to a chamber containing the sphere-gap with high-voltage leads, similar to what is shown on the left side of Fig. 1. This would enable him to measure sparking potentials in air from 760 to about 1 Torr.

Certain inconsistencies in his data from one day to the next may convince him that the partial pressure of H_2O vapor may be responsible for these day-to-day variations. As a consequence, the investigator decides that he needs a high-vacuum system, similar to the diffusion pump system on the right side in Fig. 1, in which he not only can control the absence or presence of H_2O vapor but in which he can test separately the sparking potentials of the constituent gases of air, say N_2, O_2, and Ar.

While his data will now show far less statistical fluctuations, he may decide on the basis of his experimental evidence that extremely small quantities of polar molecules (such as H_2O by way of its capability of trapping a free electron to form a stable negative ion) may cause serious changes in the sparking potentials. He also suspects that concommitant changes may be due to different work functions or different surface structures of his sphere gaps. In order to outgas and otherwise control these surfaces and to be certain that even minute amounts of say H_2O vapor are ex-

---

* Chapter 1.1 is by G. L. Weissler.

ISBN 0-12-475914-9

TABLE I. Conversion Factors for Pressure Units[a]

Unit	Torr	Pascal	dyne cm^{-2}	bar	Atmosphere (standard)	Pound (force) $inch^{-2}$
1 Torr (0°C)[b]	1	1.3332×10^{2}	1.3332×10^{3}	1.3332×10^{-3}	1.3158×10^{-3}	1.9337×10^{-2}
1 Pascal (Newton m^{-2})	7.5006×10^{-3}	1	10	1.0×10^{-5}	9.8692×10^{-6}	1.4504×10^{-4}
1 dyne cm^{-2}	7.5006×10^{-4}	0.1	1	1.0×10^{-6}	9.8692×10^{-7}	1.4504×10^{-5}
1 bar	7.5006×10^{2}	1.0×10^{5}	1.0×10^{6}	1	0.98692	1.4504×10^{1}
1 Atmosphere (standard)	760	1.0133×10^{5}	1.0133×10^{6}	1.0133	1	1.4696×10^{1}
1 Pound (force) $inch^{-2}$	5.1715×10^{1}	6.8948×10^{3}	6.8948×10^{4}	6.8948×10^{-2}	6.8047×10^{-2}	1

[a] For example, to convert from torr to Pascal, multiply by 1.3332×10^{2}.
[b] A subunit is the micron = 10^{-3} Torr.

TABLE II. An Overview of Important Points for Different Pressure Ranges[a]

Pressure range (Torr) 10^2 1 10^{-2} 10^{-4} 10^{-6} 10^{-8} 10^{-10}	Type of vacuum	Mean free path[a] (cm)	Number[b] of molecules per cm^3	Type of pump	Type of pressure gauge	Time to form monolayer (sec)	Principal residual gases
←——→ -----	Rough or fore vacuum	$\sim 10^{-5}-10^{-1}$	$\sim 10^{19}-10^{15}$	Mech. oil-sealed, steam-ejector, or sorption pumps	Liq.-filled U-tube manometer, membrane compression (spiral) gauge		Air, H_2O, CO_2
-- ←——→ ----	Intermediate or booster vacuum	$\sim 10^{-2}-10^{1}$	$\sim 10^{16}-10^{13}$	Oil-ejector or oil-booster pumps	Thermo couple or alphatron gages		As above plus pump oils, hydrocarbons, and H_2
-- ←——→ ----------	High vacuum	1 cm to larger than vac. vessel	$\sim 10^{14}-10^{10}$	Oil or Hg diffusion pumps	Ordinary ionization gauge	$\sim 10^{-3}-10^{-1}$	Pump oils, hydrocarbons
----- ←——→	Ultrahigh vacuum	Much larger than vacuum vessel	less than 10^{10}	Ion, Ti-getter, cryo-, or rotary roots-type pumps	Bayard–Alpert type ion gauge	$\sim 1-10^{6}$	H_2 and CO, He, and other rare gases

[a] Saturated water-vapor pressures; at liquid N_2 (−190°C): less than 10^{-6} Torr; at dry ice temperature (−78°C): about 10^{-3} Torr; between +10 and +30°C: about 10–30 Torr. Saturated Hg-vapor pressures: at liquid N_2 or dry ice temp: less than 10^{-6} Torr; at or near room temperature: about 10^{-3} Torr. Saturated CO_2-vapor pressures: at liquid N_2 temperature (−190°C): about 10^{-7} Torr; at dry ice temperature (−78°C): about 793 Torr.

[b] From simple kinetic theory.

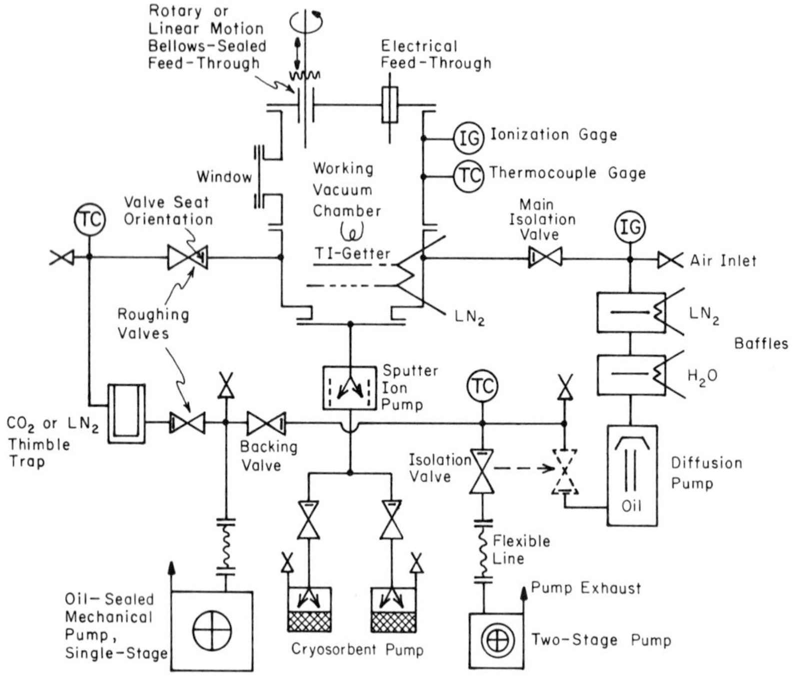

FIG. 1. A typical vacuum system using the standard symbols of the American Vacuum Society.* On the left is shown a so-called "roughing" system, while on the right a diffusion pump with a smaller capacity fore pump provides a high vacuum for the working chamber. For purposes of providing a more nearly complete list of symbols, an ultrahigh vacuum system has been added in the center, showing from the top a titanium getter cryopump, a sputter ionization pump, and two chemsorption pumps, which can be cooled to LN_2 temperatures.

cluded from the gas phase, our experimentalist decides on an ultrahigh vacuum system, shown in the center of Fig. 1, which in addition will be designed for bake-out at, say, 400°C (not indicated in Fig. 1).

In a similar sequence of using progressively better vacua, we find another good example in the measurements of the photoelectric effect from surfaces. Historically, Herz (1887) and Hallwachs (1888) used a very rough vacuum to establish the existence of the effect. DuBridge (ca. 1932) refined these measurements by using a high vacuum for his surfaces and later yet Spicer (in the 1960s and 1970s) among others used ultrahigh vacua for his studies of photoelectron energy distributions.

It is left to the reader to find his own examples as to the type of physical

* For graphic symbols in vacuum technology, see AVS Standard 7.1-1966, *J. Vac. Sci. Technol.* **4,** 139 (1967).

measurements which may require a rough, a high, and/or an ultrahigh vacuum.

A composite drawing of these three systems is presented in Fig. 1, primarily for the purpose of acquainting the reader with the usefulness of such a schematic sketch in the planning, construction, and maintenance of such systems, and in getting him used to the graphic symbols of components as approved by the American Vacuum Society, AVS Standard 7.1-1966.

Figure 1 includes a good cross section of these symbols, but by no means all of them. Here, a rough or fore vacuum system is shown on the left, which can also be used as a preliminary roughing system after the working chamber in the upper center has been opened to the atmosphere. The high-capacity single-stage (one circle) oil-sealed mechanical pump will reach a suitable fore vacuum in a few minutes, in order that an oil diffusion pump may take over to achieve a high vacuum. Its backing pump is usually a much smaller and therefore less costly mechanical oil-sealed pump with two stages (two circles in the symbol).

It is of course understood that this working chamber may be as small as a flashlight bulb or as large as several cubic meters, with corresponding differences in the sizes of the pumps.

The ultrahigh vacuum system shown in the center of Fig. 1 consists of a liquid N_2-cooled titanium (Ti) getter pump, in the drawing built into the working chamber, followed by a sputter ion pump, with the cryosorption pumps (one working, while the other is being reactivated) providing the initially required rough vacuum. If the experiment chamber is relatively small, say of the order of a few liters in volume, the central system in Fig. 1 by itself may be sufficient. If, however, the volume is larger or if frequent openings to atmospheric pressure (use dry nitrogen) are necessary, then a diffusion pump system (see the right side of Fig. 1) together with a roughing pump (as shown on the left side of Fig. 1) may also be necessary.

In the following chapters we hope to discuss in detail the many aspects of such systems as are sketched in Fig. 1.

1.2. Basic Equations*

1.2.1. Introduction

The necessity for creating and maintaining vacuum conditions is usually related to the need to reduce the number density of gaseous molecules or their surface collision rate. For laboratory vacuum systems, an exact statistical description of these quantities is quite complicated or impossible since one must deal with complex surface–gas reactions and anisotropic velocity distributions. Specific characteristics of the surfaces depend upon time and the prior history of the system, while temperatures are often ill defined. Nevertheless, it is often convenient to discuss vacua in terms of an ideal gas, using the parameter of pressure as an indicator of the quality of the vacuum. In many cases, the ideal gas laws lead to correct relationships, but it must be recognized that such an approximation is to be applied with caution, particularly for quantities which depend upon molecular collisions and lose significance when the mean free path is less than or comparable to the size of the vacuum chamber (e.g., viscosity, thermal conductivity).

In this chapter we summarize a few of the ideal gas laws which are applicable to vacuum systems. The transport of gases is treated in Chapter 1.3. Gas–surface interactions, which usually dominate the behavior of real vacuum systems, are treated in Chapters 1.4 and 5.1.

1.2.2. Kinetic Properties of Ideal Gases

The statistical physics of ideal gases is developed in many textbooks,[1–3] which should be referred to for more detail. The following summarizes only the most relevant portions of the theory.

1.2.2.1. Maxwell–Boltzmann Velocity Distribution. If there are n_i molecules of species i per unit volume, then the number dn_i with velocities in the range $\mathbf{v}$ to $\mathbf{v} + d\mathbf{v}$ is given by the Maxwell–Boltzmann distribution

[1] E. H. Kennard, "Kinetic Theory of Gases." McGraw-Hill, New York, 1938.
[2] R. D. Present, "Kinetic Theory of Gases." McGraw-Hill, New York, 1958.
[3] F. Reif, "Statistical and Thermal Physics." McGraw-Hill, New York, 1965.

* Chapter 1.2 is by R. W. Carlson.

ISBN 0-12-475914-9

$$dn_i = (n_i/\pi^{3/2}v_0^3)e^{-(v/v_0)^2}\, d^3\mathbf{v}, \tag{1.2.1}$$

where v_0 is the most probable speed. The differential volume element $d^3\mathbf{v}$ is related to the speed v and differential solid angle $d\Omega$ as

$$d^3\mathbf{v} = v^2\, dv\, d\Omega, \tag{1.2.2}$$

allowing the distribution law over all solid angles to be expressed in terms of the speed v as

$$dn_i = (4n_i/\pi^{1/2}v_0^3)e^{-(v/v_0)^2}v^2\, dv. \tag{1.2.3}$$

The most probable speed v_0 is found from kinetic theory to be

$$v_0 = (2kT/M_i)^{1/2}, \tag{1.2.4}$$

M_i being the molecular mass, k denoting Boltzmann's constant,† and T is the absolute temperature expressed in degrees K. One can also define two other characteristic speeds: the mean speed $\bar{v}$ and the root-mean-square speed v_{rms}, which are related according to

$$\bar{v} = \frac{2}{\sqrt{\pi}}\, v_0, \qquad v_{\mathrm{rms}} = \sqrt{\frac{3}{2}}\, v_0. \tag{1.2.5}$$

Illustrative values for v_0 are given in Table I. Note that these speeds are comparable to the speed of sound.

1.2.2.2. Impingement Rate. The flux Φ_i incident on a surface of unit area is

$$\Phi_i = \tfrac{1}{4}n_i\bar{v} = \frac{1}{2\pi^{1/2}}\, n_i v_0. \tag{1.2.6}$$

For normal air at 10^{-6} Torr and room temperature, this flux is 3.8×10^{14} molecules $\mathrm{sec}^{-1}\ \mathrm{cm}^{-2}$. Figure 1 shows the value as a function of partial pressure for various gases and temperatures.

It is of interest to estimate the time required to form a monolayer on a surface. Assuming an accommodation coefficient of unity (that is all of the molecule stick upon impact) and a separation d between surface molecules, then the monolayer time t_{mono} is

$$t_{\mathrm{mono}} \cong 2\pi^{1/2}/v_0 n_i d^2. \tag{1.2.7}$$

For $d = 3$ Å, monolayer times at room temperature are ~3 sec for air at 10^{-6} Torr, illustrating the necessity for ultrahigh vacuum in sensitive surface studies. The monolayer time as a function of pressure is illustrated in Fig. 1.

---

† $k = 1.38054 \times 10^{-16}$ erg deg^{-1} (CGS) or 1.38054×10^{-23} J deg^{-1} (MKS).

TABLE I. Most Probable Speed v_0 for Various Species (in 10^4 cm sec^{-1}).

Species	100 K	200 K	300 K
H_2	9.08	12.84	15.72
He	6.42	9.08	11.12
H_2O	3.03	4.29	5.25
N_2	2.43	3.44	4.21
CO	2.43	3.44	4.21
O_2	2.27	3.21	3.93
A	2.03	2.87	3.52
Electron	551	779	954
Average kinetic energy (eV)	0.0129	0.0258	0.0388

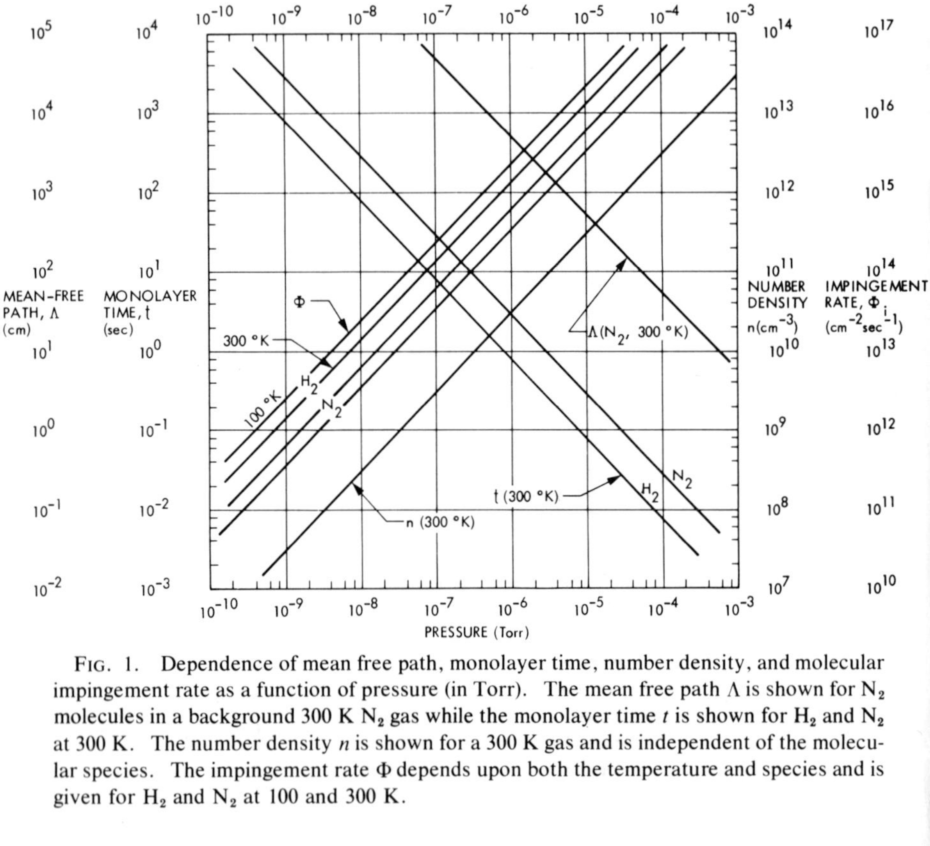

FIG. 1. Dependence of mean free path, monolayer time, number density, and molecular impingement rate as a function of pressure (in Torr). The mean free path Λ is shown for N_2 molecules in a background 300 K N_2 gas while the monolayer time t is shown for H_2 and N_2 at 300 K. The number density n is shown for a 300 K gas and is independent of the molecular species. The impingement rate Φ depends upon both the temperature and species and is given for H_2 and N_2 at 100 and 300 K.

1.2.2.3. Pressure. Computation of the momentum transfer to a surface by incident molecules shows that the partial pressure due to species i is

$$p_i = \tfrac{1}{3}n_i M_i v_{\mathrm{rms}}^2 = \tfrac{1}{2}n_i M_i v_0^{\,2} \tag{1.2.8}$$

or, from Eq. (1.2.4)

$$p_i = n_i kT. \tag{1.2.9}$$

The numerical relationship between pressure and number density is illustrated in Fig. 1. Obviously, for a multicomponent gas, the total pressure is the sum of the partial pressures. For equal volumes of gases at the same temperature and pressure, the number densities n are equal, as are the corresponding total number of molecules (Avogadro's law). At standard conditions (0°C, 760 Torr), the number density is given by Loschmidt's number ($n_0 = 2.6868 \times 10^{19}\ \mathrm{cm}^{-3}$). The volume occupied by one mole of a gas at standard conditions is $V_0 = 22.414 \times 10^3\ \mathrm{cm}^3$ and contains $N_A = n_0 V_0 = 6.02217 \times 10^{23}$ molecules (Avogadro's number).

Although pressure itself usually has no physical significance in vacuum systems, it is traditional to use this parameter as a measure of the degree of vacuum. The customary unit is presently the torr although one expects the SI unit of pressure (the *Pascal*) to become more common in the near future. Table I (Chapter 1.1) gives interrelationships between various pressure units now in usage.

1.2.2.4. Mean Free Path. One of the more important parameters for gases in vacuum systems is the mean free path—the average distance traveled before collision with another molecule. This quantity and its relationship to dimensions of the vacuum system is particularly important for determining flow properties in addition to thermal conductivity and viscosity. For a gas of single species i, the mean free path Λ_{ii} is

$$\Lambda_{ii} = (\sqrt{2}\, n_i \sigma_{ii})^{-1}, \tag{1.2.10}$$

where σ_{ii} is the effective molecular collision cross section. The $\sqrt{2}$ factor arises from the motion of both the target and projectile molecule. For unlike molecules, the mean free path of projectile i in a background gas of species j at the same temperature is

$$\Lambda_{ij} = \left[\left(1 + \frac{M_i}{M_j}\right)^{1/2} n_j \sigma_{ij}\right]^{-1}. \tag{1.2.11}$$

The cross section σ_{ij} is related to the molecular diameters D as

$$\sigma_{ij} = \tfrac{1}{4}\pi(D_i + D_j)^2. \tag{1.2.12}$$

Typical values for molecular diameters are given in Table II. For multicomponent gases, the inverse of the mean free path is the sum of the

TABLE II. Molecular Weights and Diameters

Species	Molecular weight	Diameter (10^{-8} cm)
H_2	2.016	2.74
He	4.002	2.18
H_2O	18.02	4.60
N_2	28.02	3.75
O_2	32.00	3.61
A	39.94	3.64

reciprocals of the individual path lengths. For further discussion of atomic and molecular collision phenomena and cross sections, the reader is referred to volumes by Hasted[4] and McDaniel.[5]

The mean free path is illustrated in Fig. 1 for an N_2 gas as a function of pressure. Note that the mean free path becomes comparable to the size of laboratory vacuum systems in the 10^{-4} Torr range. The mean time between collisions with other molecules (species j) is

$$\tau_i = \Lambda_{ij}/\bar{v}_i. \tag{1.2.13}$$

At low pressures, wall collisions are of course much more frequent.

[4] J. B. Hasted, "Physics of Atomic Collisions." Butterworths, London, 1964.
[5] E. W. McDaniel, "Collision Phenomena in Ionized Gases." Wiley, New York, 1964.

1.3. Molecular Transport*

1.3.1. Gaseous Flow

1.3.1.1. Flow Quantities. There are two quantities of concern in describing the flow characteristics of vacuum systems, the *throughput* and the corresponding *volumetric flow*. The throughput Q is closely related to the mass flow, while the volumetric flow S is the net volume of gas passing through a given area in unit time. As the number of molecules per unit volume is proportional to the pressure p [Eq. (1.2.9); 1.2.2.3], then the throughput† is proportional to both the pressure and volumetric flow rate and is usually defined as the product

$$Q = pS, \tag{1.3.1}$$

where p is in Torr, S is expressed in liter $\sec^{-1}$, with Q therefore given in Torr liters $\sec^{-1}$.‡

The volumetric flow rate is of fundamental importance in vacuum technology since most pumps and ducts exhibit constant volumetric flow rates. For vacuum pumps, this flow rate is termed the pumping speed and is customarily given in units§ of liters per second. The natural manner in which volumetric flow arises in vacuum systems is illustrated by the following ideal example. Consider a pumping port of area A connected to a vacuum vessel. For an ideal pump, all of the molecules impinging on the port area A will be transmitted into the pump and evacuated; this corresponds to a molecular rate of $\Phi A = \frac{1}{4} n\bar{v}A$ molecules $\sec^{-1}$ [cf. Eq. (1.2.6); Section 1.2.2.2]. Independent of the number density n (or corresponding pressure p) a constant volumetric flow rate or pumping speed rate of $S = \frac{1}{4}\bar{v}A$ is obtained. If $\bar{v}$ is expressed in cm $\sec^{-1}$ and A in cm^2, then the volumetric rate is obtained in $cm^3 \sec^{-1}$ (or 10^{-3} liter $\sec^{-1}$). For air (with mean molecular mass $M = 28.97$) at ambient temperature, the pumping speed per unit pump area (for an ideal pump) is

† These definitions apply to a single-component gas. For a multicomponent gas, throughput and volumetric flow apply to each species, when pressures are sufficiently low so that free molecular flow occurs.

‡ 1 Torr liter = 3.535×10^{19} molecules for 273 K temperature.

§ In the United States, mechanical forepumps are often rated in cubic feet per minute (CFM). 1 CFM = 0.4719 liter $\sec^{-1}$.

* Chapter 1.3 is by R. W. Carlson.

ISBN 0-12-475914-9

$\frac{1}{4}\bar{v} = 11.7 \times 10^3$ cm^3 sec^{-1} cm^{-2} or 11.7 liters sec^{-1} cm^{-2}. Values for other gases and temperatures are given in Table I.

1.3.1.2. Pressure Regimes. The physics of gas flow depends upon the relationship between the mean free path for molecular collisions [Eq. (1.2.10); Section 1.2.2.4] and characteristic dimensions of the duct. At low pressures wall collisions are more prevalent than collisions with other molecules; consequently, gaseous throughput is dependent upon the geometry of the duct and density differentials, but not upon the absolute density of the gas. This is termed the *free molecular flow* regime. At higher pressures, where collisions with other molecules are frequent, the flow of an individual molecule is determined by the neighboring molecules and their net motion, rather than by wall collisions. In this case, called *viscous flow,* the throughput depends upon the geometry of the conductor, the pressure gradient, *and* the absolute pressure of the gas. At these high pressures, the flow can be either laminar* or turbulent, with corresponding differences in flow properties.[1] Turbulent flow generally occurs at quite high pressure gradients and high gas velocities.

The pressure region p_t, which marks the gradual transition between viscous and free molecular flow, occurs when the mean free path Λ is comparable to the diameter d of the duct or orifice. For rough estimates, one can use the relationship

$$p_t(\mu) \sim 5/d,$$

where p_t is given in microns (10^{-3} Torr) and d is in centimeters.

1.3.1.3. Flow Relationships. In general, vacuum systems consist of complex geometries of traps, baffles, and ducts which connect the pump to the vacuum vessel. These interconnections impede the flow of gas and reduce the net pumping effect of the system. In order to estimate the resulting degree of vacuum, it is necessary to include the flow characteristics of such elements, since most vacua are ultimately determined by the continual release, flow, and subsequent pumping of gases.

In the case of steady flow, where there is no absorption or release of gases from the duct walls, then the throughput Q is constant. If measured at two points in the system $Q_1 = Q_2 = Q$, or

$$p_1 S_1 = p_2 S_2 = Q. \tag{1.3.2}$$

This relationship states that the pressure at any point in the vacuum system is inversely proportional to the corresponding volumetric flow (or effective pumping speed).

[1] S. Dushman, "Scientific Foundations of Vacuum Technique." Wiley, New York, 1949.

* Also called Poiseuille flow.

TABLE I. Volumetric Flow per Unit Aperture Area, $\frac{1}{4}\bar{v}$, for Various Gases and Temperatures[a]

Species	100 K	200 K	300 K
H_2	25.6	36.2	44.3
He	18.1	25.6	31.4
H_2O	8.55	12.1	14.8
N_2	6.86	9.70	11.9
CO	6.86	9.70	11.9
O_2	6.40	9.06	11.1
A	5.73	8.10	9.93
Air	6.74	9.54	11.7

[a] Units are liter sec^{-1} cm^{-2}.

For low pressures and Maxwellian distributions, Knudsen demonstrated that the throughput was related to pressure differences according to

$$Q = C(p_2 - p_1), \tag{1.3.3}$$

where C is termed the *conductance* between points 1 and 2 and is expressed in units of volume per unit time (e.g., liters per second). As noted above, in the free-molecular-flow regime, the throughput depends upon the pressure gradient and the duct geometry, but not upon the absolute pressure. Therefore, free-molecular-flow conductance values are independent of pressure. In the viscous flow region, the effective conductance is larger than the free-molecular-flow value and increases with pressure. A slight minimum is found in the transition region due to "slipping" of the gas at the walls.[1]

At low pressures, when conductances are independent of pressure, an often-used relationship for a series combination of individual ducts is

$$\frac{1}{C} = \frac{1}{C_1} + \frac{1}{C_2} + \cdots, \tag{1.3.4}$$

where C is the net conductance and C_i represent the separate elements. Obviously, this relationship loses physical meaning if the addition of a series conductance does not significantly alter the original trajectories of an individual molecule.* More accurate methods for determining the effects of series conductors are given in Section 1.3.2.4.

At the same level of approximation as Eq. (1.3.4), the net pumping speed S_2 at the entrance of a duct system with conductance C is

* For example, two closely spaced, identical, thin aperture plates would not reduce the conductance by the factor of two predicted by Eq. (1.3.4).

$$\frac{1}{S_2} = \frac{1}{S_1} + \frac{1}{C}, \tag{1.3.5}$$

where S_1 is the pumping speed of the pump or system to which the duct is connected.

For a parallel combination of ducts (or pumps), the net conductance (or pumping speed) is the sum of the individual contributions, i.e.,

$$C = C_1 + C_2 + \cdots . \tag{1.3.6}$$

1.3.1.4. Time Behavior. It is often of interest to estimate the temporal behavior during the pump down phase of a vacuum system. Consider a volume V being pumped with an effective pumping speed S. At any instant of time, the total number of molecules N in the volume is described by the product of the instantaneous pressure p and the volume V. The rate at which these molecules are removed is

$$\frac{d}{dt}(pV) = -pS, \tag{1.3.7}$$

which results in an exponential pressure dependence,

$$p(t) = p_0 e^{-t/\tau} \tag{1.3.8}$$

The pressure falls with a characteristic time constant

$$\tau = V/S. \tag{1.3.9}$$

This time constant is useful for estimates in a limited region of pressures. At high pressures, conductances and pumping speeds are pressure dependent, while at low pressures, leaks, outgassing, and surface effects determine the temporal behavior. These aspects are discussed in Chapters 1.4, 4.1, and 5.1.

1.3.2. Conductance Calculations*

1.3.2.1. Basic Considerations. The conductance of a duct is a measure of its ability to transport gas and is expressed in units of volume per unit time. For low pressures, intermolecular collisions are infrequent compared to wall collisions, so it is the latter which determine the gaseous flow characteristics through the geometry of the conductor.

There are two aspects which determine the conductance of a duct:

(1) the rate at which molecules enter the duct and

* Helpful inputs by Milleron are gratefully acknowledged.

(2) the probability that these molecules are transmitted through the system.

The first item depends upon the area of the entrance aperture while the latter is determined by the ensuing series of reflections from the walls which result in the molecule eventually being transmitted through the duct or reflected back into the original volume.

Consider first the case of a very thin aperture plate, for which the area A is more important in determining the conductance than any wall reflections. The volume of gas transmitted from one side of the aperture to the other side per unit time—the aperture conductance—is

$$C_A = \tfrac{1}{4}\bar{v}A = (2\sqrt{\pi})^{-1}v_0 A, \tag{1.3.10}$$

when the gas can be described by Maxwell–Boltzmann statistics. Conductance values depend upon the molecular mass and the kinetic temperature [Eq. (1.2.4); Section 1.2.2.1]. Numerical formulas are given in Table II, while Table I gives values of $\tfrac{1}{4}\bar{v}$ for various gases and temperatures.

The contrasting case in which wall collisions are more important than the conductance of the aperture is considered below. The more common intermediate case can be described with the general methods of Section 1.3.2.3.

1.3.2.2. Knudsen's Formulation. The conductance C_T of a long tube of length L, with variable cross-sectional area A, and perimeter H, was calculated by Knudsen[2] to be

$$C_T = \frac{4}{3}\left[\bar{v}\Big/\int_0^L \frac{H(l)}{A^2(l)}\,dl\right]. \tag{1.3.11}$$

This general result was obtained by assuming (1) that the length of the tube was much greater than the diameter, (2) that the direction of reflected molecules is independent of the incident direction, and (3) reflected molecules are distributed equally per unit solid angle, as in optical reflection from a Lambertian surface (cosine law).* Relationships derived from Eq. (1.3.11) are given in Table II for some simple geometries.

One of the consequences of assumption (1) above is that the effect produced by the entrance aperture is insignificant; the conductance value

[2] M. Knudsen, *Ann. Phys.* **28,** 75, 999 (1909); **35,** 389 (1911).

* Specular reflection will increase the conductance slightly, while rough walls will provide more backscattering and decrease the conductance. See Refs. 7 and 17.

TABLE II. Conductance Formulas for Simple Geometries and Free Molecular Flow

Geometry	Aperture conductance, C_A	Tube conductance, C_T	Net conductance, C
Thin aperture of area A	$\frac{1}{4}\bar{v}A$	—	C_A
Tube with uniform cross section, area A, perimeter H, length L	$\frac{1}{4}\bar{v}A$	$\frac{4}{3}\frac{A^2}{LH}\bar{v}$	$[1 + (\frac{3}{16}LH/A)]^{-1}C_A$
Cylindrical tube, radius R, length L	$\frac{1}{4}\pi\bar{v}R^2$	$\frac{2}{3}\pi\frac{R^3}{L}\bar{v}$	$[1 + \frac{3}{8}L/R]^{-1}C_A$
Rectangular duct, sides A and B, length L	$\frac{1}{4}\bar{v}A\cdot B$	$\frac{2}{3}\frac{A^2B^2}{(A+B)L}\bar{v}$	$\left[1 + \frac{3}{8}\frac{(A+B)L}{AB}\right]^{-1}C_A$
Elliptical duct, semiaxes A and B, length L	$\frac{1}{4}\pi\bar{v}A\cdot B$	$\frac{2\pi}{3}\frac{A^2B^2}{L[(A^2+B^2)/2]^{1/2}}\bar{v}$	$\left\{1 + \frac{3}{8}\frac{[(A^2+B^2)/2]^{1/2}L}{AB}\right\}^{-1}C_A$
Conical duct, radius, R_1 at one end, R_2 at the other, $\bar{R} = (R_1 + R_2)/2$, length L	—	$\frac{2\pi}{3}\frac{R_1^2R_2^2}{\bar{R}L}\bar{v}$	

given by Eq. (1.3.11) is for molecules which are well inside the tube, sufficiently removed from the entrance so that it is of no consequence. In an attempt to correct this deficiency, a rough approximation is to include the series conductance of the entrance aperture. For example, for a uniform tube of perimeter H, area A, and length L, the tube conductance from Eq. (1.3.11) is

$$C_T = \tfrac{4}{3}(A^2/LH)\bar{v}, \tag{1.3.12}$$

while the aperture conductance is [Eq. (1.3.10); Section 1.3.2.1]

$$C_A = \tfrac{1}{4}\bar{v}A. \tag{1.3.13}$$

Combining these conductances according to Eq. (1.3.4) (Section 1.3.1.3), we obtain a refined *approximation* for a uniform tube:

$$C = (1 + \tfrac{3}{16}LH/A)^{-1}\,\tfrac{1}{4}vA, \tag{1.3.14a}$$

$$= [1 + \tfrac{3}{16}(LH/A)]^{-1}\,C_A. \tag{1.3.14b}$$

Table II includes some entries with these approximate end corrections.

1.3.2.3. Clausing's Formulation. In the preceding paragraph, the net conductance of a tube was found to be approximately related to the conductance of the entrance aperture through the factor $[1 + \tfrac{3}{16}(LH/A)]^{-1}$. This factor can be interpreted as the probability that a molecule which is incident upon the aperture will be transmitted through the tube and leave the other end. In fact, it is convenient to discuss conductances in terms of the entrance aperture and the corresponding *probability of passage*—the *Clausing factor*. If $P_{1\to 2}$ represents the probability that a molecule incident on side 1 of a duct is transmitted through to side 2, then the conductance is

$$C = C_A P_{1\to 2} = \tfrac{1}{4}\bar{v}A_1 P_{1\to 2}. \tag{1.3.15}$$

Since conductance must be independent of direction,

$$A_1 P_{1\to 2} = A_2 P_{2\to 1}. \tag{1.3.16}$$

For vacuum pumps, the ratio of the pumping speed to the full aperture volumetric flow corresponds to this probability of passage and is called the *pumping efficiency*. The probability of passage factor is useful since it is independent of the size of the duct, but depends only on dimensionless ratios. Approximate values for some probabilities of passage are shown in Table II and are useful when errors of order 10% are permissible. Clausing[3] performed more refined calculations for some simple cases,

[3] P. Clausing, *Ann. Phys.* **12,** 961 (1932); *J. Vac. Sci. Technol.* **22,** 111 (1972) (English trans.).

while results for more complicated geometries have been obtained using Monte Carlo machine calculations,[4–8] Clausing's analytical method,[9,10] and variational methods.[11] Certain of these computations have also been verified experimentally.[5–7] Figures 1–4 show probability factors for a variety of geometries. Details of the method and calculations for other geometries can be found in the cited literature. Table III lists some of the geometries that have been investigated and the corresponding references.

1.3.2.4. Combining Conductances. Although the sum-of-reciprocals method for calculating series conductances receives widespread use, this approximation can often be in serious error, particularly for short ducts. An alternate method, due to Oatley,[18] makes use of the probabilities of passage (Clausing factors) to estimate the effect of series conductances.

For the two ducts shown in Fig. 5, we ascribe forward and reverse transmission probabilities $P_{1\to2}, P_{2\to1}$ and $P_{2\to3}, P_{3\to2}$ for the two conductors, respectively. The interface region is assumed to share a common area (i.e., no discontinuities). Multiple "reflections" occur across this interface, resulting in a net volumetric flow

$$\tfrac{1}{4}\bar{v}A_1P_{1\to3} = \tfrac{1}{4}\bar{v}A_1P_{1\to2}P_{2\to3}[1 + (1 - P_{2\to3})(1 - P_{2\to1}) + (1 - P_{2\to3})^2(1 - P_{2\to1})^2 + \cdots] \qquad (1.3.17)$$

or

$$P_{1\to3} = \frac{P_{1\to2}P_{2\to3}}{P_{2\to1} + P_{2\to3} - P_{2\to3}P_{2\to1}}, \qquad (1.3.18)$$

where $P_{1\to3}$ is the effective transmission probability for the combined ducts.

Recalling that $P_{2\to1} = (A_1/A_2)P_{1\to2}$ [Eq. (1.3.16); Section 1.3.2.3], then

$$P_{1\to3} = \frac{P_{1\to2}P_{2\to3}}{(A_1/A_2)P_{1\to2} + P_{2\to3} - (A_1/A_2)P_{1\to2}P_{2\to3}}. \qquad (1.3.19)$$

The net conductance is

[4] D. H. Davis, *J. Appl. Phys.* **31,** 1169 (1960).

[5] D. H. Davis, L. L. Levenson, and N. Milleron, *in* "Rarified Gas Dynamics" (L. Talbot, ed.), p. 99. Academic Press, New York, 1961.

[6] L. L. Levenson, N. Milleron, and D. H. Davis, *Trans. Natl. Symp. Vac. Technol. 7th* (1960). Pergamon, Oxford, 1961.

[7] D. H. Davis, L. L. Levenson, and N. Milleron, *J. Appl. Phys.* **35,** 529 (1964).

[8] D. Pinson and A. W. Peck, *Trans. Natl. Vac. Symp. 9th,* p. 407 (1962). Macmillan, New York, 1962.

[9] E. W. Balson, *J. Phys. Chem.* **65,** 1151 (1961).

[10] R. P. Iczkowski, J. L. Margrave, and S. M. Robinson, *J. Phys. Chem.* **67,** 229 (1963).

[11] A. S. Berman, *J. Appl. Phys.* **36,** 3365 (1965).

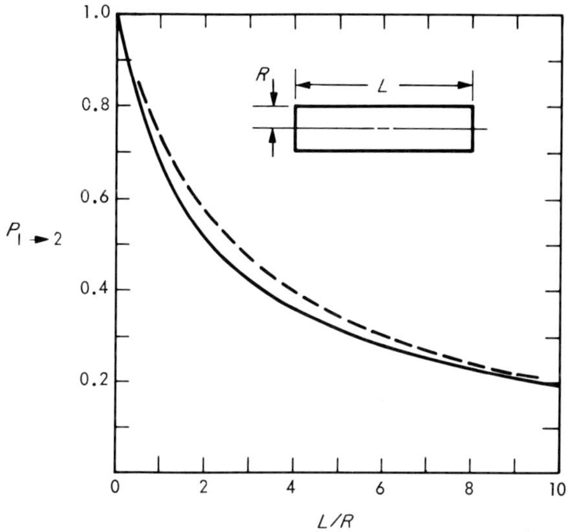

FIG. 1. Probability of passage for a cylindrical tube as a function of the ratio of the length to radius. The conductance is $C = \frac{1}{4}\pi\bar{v}R^2P_{1\to2}$. The solid curve is from Clausing's calculation, while the dashed curve corresponds to the *approximation* $P = 1/[1 + (3L/8R)]$ given in Eq. (1.3.14) and Table II.

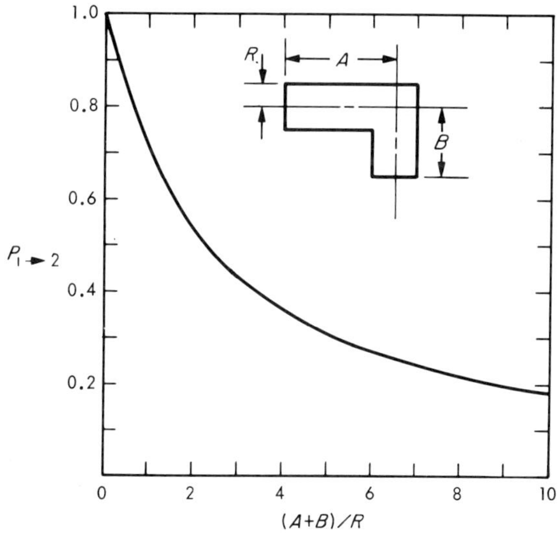

FIG. 2. Probability of passage for a cylindrical elbow. Note that the probability depends only on the ratio of the total central path length to the radius. See references cited in Table III for additional discussion and experimental verification.

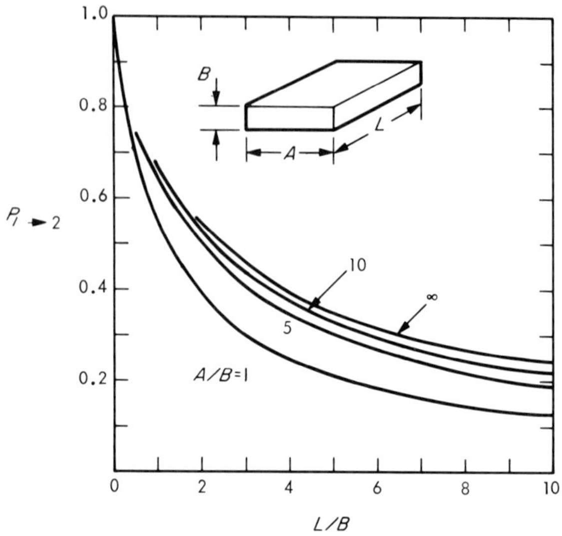

FIG. 3. Probability of passage for rectangular ducts calculated by Milleron.

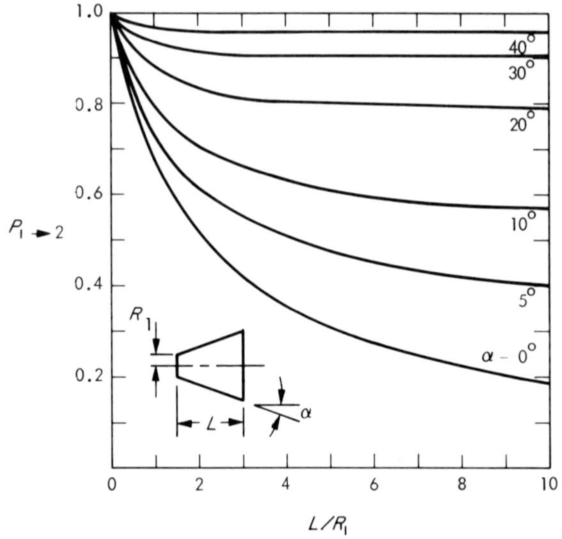

FIG. 4. Probability of passage for a conical duct. $P_{1\to2}$ refers to passage from the smaller to the larger end. The reverse probability can be found from Eq. (1.3.16).

TABLE III. Probabilities of Passage for Various Geometries

Geometry	References	Notes
Cylindrical tube	4, 6, 7, 12, 15, 16, 17	Experimental data in 6, Ref. 7 incorporates back reflection from walls. Reference 15 includes wall sorption. References 16 and 17 include specular reflection
Cylindrical elbow	4, 6	Experimental data in 6
Cylindrical annulus	4	
Cylindrical tube with restricted opening(s)	4, 5, 6	Experimental data in 5 and 6
Cylindrical tube with restricted opening(s) and internal baffle	4, 5, 6	Experimental data in 5 and 6
Baffled diffusion pump system	6	Includes experimental data
Conical tube	5, 8, 9, 10, 13, 14	Reference 14 includes specular reflection and surface diffusion
Louver	5, 6	Experimental data in 5 and 6
Chevron	5, 6	Experimental data in 5 and 6
Flat plates	11	Variational method
Bed of spheres	11	Variational method
Bulged elbows	6	Experimental measurements
Cubic elbows	6	Experimental measurements

$$C = \tfrac{1}{4}\bar{v}A_1P_{1\to3}, \tag{1.3.20}$$

while the reverse transmission probability must be

$$P_{3\to1} = (A_1/A_3)P_{1\to3}. \tag{1.3.21}$$

This method differs from the classical sum-of-reciprocals method through the probability product in the denominator of Eq. (1.3.18).

When there is a discontinuity in area, as shown in Fig. 6, Oatley[18]

[12] W. E. Demarcus and E. H. Hooper, *J. Chem. Phys.* **23,** 1344 (1955).

[13] L. Füstöss, *Vacuum* **22,** 111 (1972).

[14] J. W. Ward, M. V. Fraser, and R. L. Bivins, *J. Vac. Sci. Technol.* **9,** 1056 (1972).

[15] G. G. Smith and G. Lewin, *J. Vac. Sci. Technol.* **3,** 92 (1966).

[16] W. E. Demarcus, *in* "Rarified Gas Dynamics" (L. Talbot, ed.), p. 161. Academic Press, New York, 1961.

[17] D. Blechschmidt, *J. Vac. Sci. Technol.* **11,** 570 (1974).

[18] C. W. Oatley, *Br. J. Appl. Phys.* **8,** 15 (1957).

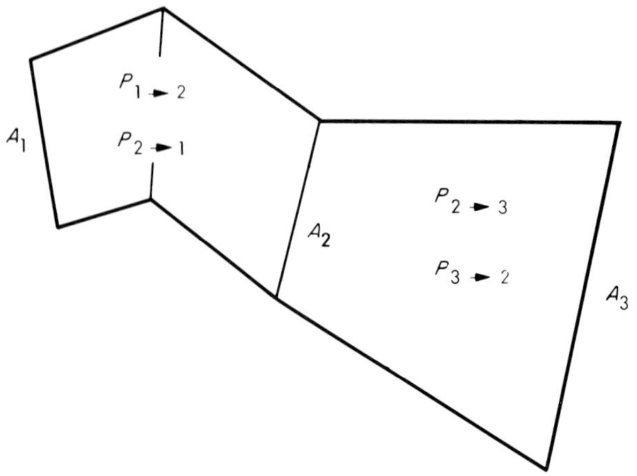

FIG. 5. Illustration of two ducts which combine to give a net probability of passage $P_{1\to3}$ as described by Eq. (1.3.18). If a discontinuity occurs at the interface, then one must resort to Eq. (1.3.23) (see Fig. 6).

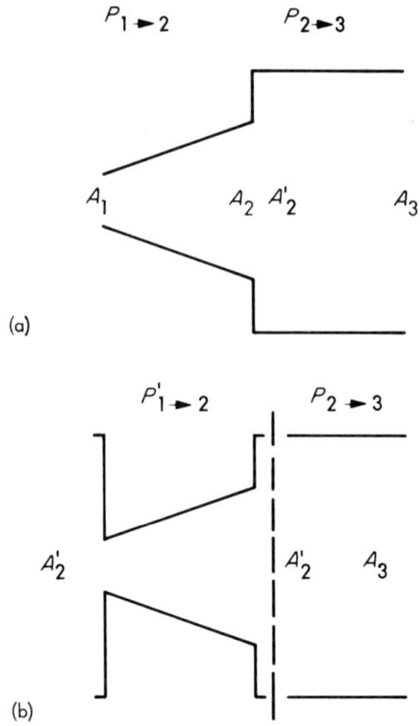

FIG. 6. Illustration of the method for mathematically combining two ducts when there is a discontinuity in area. The first duct in (a) is thought of as having entrance and exit apertures increased to A_2' (b), but with probabilities of passage reduced by the ratio of areas [Eq. (1.3.22)].

suggests the following procedure (cf. Refs. 19–21). Mathematically replace the first duct with one with increased area A_2' at the two ends (see Fig. 6). The transmission probabilities for this configuration are reduced by the area ratios, so

$$P'_{1\to 2} = (A_1/A_2')P_{1\to 2}, \tag{1.3.22a}$$

$$P'_{2\to 1} = \frac{A_2}{A_2'} P_{2\to 1} = \frac{A_2}{A_2'}\frac{A_1}{A_2} P_{1\to 2}. \tag{1.3.22b}$$

Using Eqs. (1.3.18) and (1.3.20), the net conductance is

$$C = \frac{1}{4}\bar{v}A_2' \frac{P_{1\to 2}P_{2\to 3}}{P_{1\to 2} + (A_2'/A_1)P_{2\to 3} - P_{1\to 2}P_{2\to 3}}. \tag{1.3.23}$$

1.3.2.5. Numerical Example. In order to illustrate the above procedures for conductance calculations, let us evaluate the combination of ducts and pump illustrated in Fig. 7. Assume that a 10-cm diameter pump

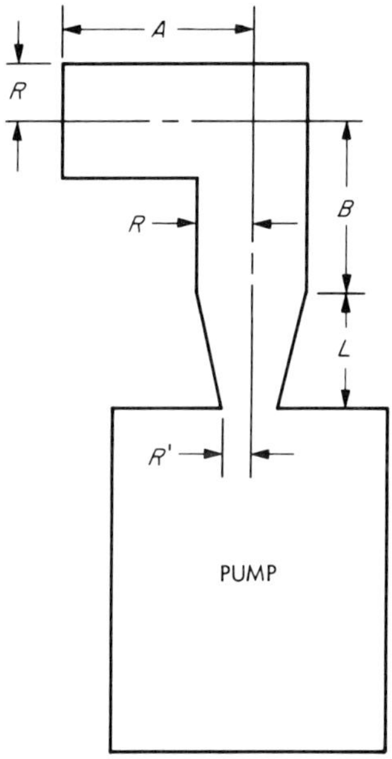

FIG. 7. Hypothetical vacuum system for numerical example. In this example we assume $R' = 5$ cm, $L = 5$ cm, $R = 10$ cm, $A + B = 30$ cm, and the pumping speed at the inlet to the pump is $S_{\text{pump}} = 200$ liters sec^{-1}. See text for example calculations.

[19] J. O. Ballance, *Trans. Int. Vac. Conf. 3rd,* Vol. 2, Part 1, p. 85. Pergamon, Oxford, 1967.

[20] B. B. Dayton, *J. Vac. Sci. Technol.* **9,** 243 (1972).

[21] L. Füstöss and G. Tóth, *J. Vac. Sci. Technol.* **9,** 1214 (1972).

with a pumping speed of 200 liters sec^{-1} is connected to a 5-cm long 45° conical duct and a 20-cm diameter cylindrical elbow with a central length $A + B = 30$ cm. If the air pressure inside the pump is 10^{-6} Torr; what is the pressure at the inlet to the system? We will first use Oatley's method and then the less accurate sum of the reciprocals method.

We first combine the elbow and conical duct. For the elbow $(A + B)/R = 3$, so $P_{1\to2} = 0.43$ (from Fig. 2). The probability of passage from the smaller end of the conical duct is ≈ 0.97, so $P_{2\to3} = (\pi 5^2/\pi 10^2) \times 0.97 = 0.24$. Since there is no discontinuity at the interface, Eq. (1.3.19) gives the forward transmission probability as $P = 0.43 \times 0.24/[1 \times 0.43 + 0.24 - 1 \times 0.43 \times 0.24] = 0.182$. We now combine this equivalent duct with the pump. For the pump aperture of $\pi \times 5^2 = 78.5$ cm^2 and 200 liters sec^{-1} pumping speed, the pumping probability is $200/(11.7 \times 78.5) = 0.22$. The total probability of a molecule at the inlet being pumped is $0.182 \times 0.22/[(\pi 10^2/\pi 5^2) \times 0.182 + 0.22 - (\pi 10^2/\pi 5^2) \times 0.182 \times 0.22] = 0.051$. The system pumping speed is $\pi 10^2 \times 0.051 \times 11.7 = 187$ liters sec^{-1}. As pressure is inversely proportional to pumping speed, the pressure at the inlet is $(200/187) \times 10^{-6} = 1.07 \times 10^{-6}$ Torr.

Let us repeat the calculation using classical conductance formulae (Table II) and the sum-of-cipricols method. For one half of the elbow, we find a conductance of $\pi 10^2 \times 11.7/(1 + 0.375 \times 1.5) = 2352$ liters sec^{-1}, while the conductance of the conical duct is calculated as 6535 liters sec^{-1}. The net pumping speed is the inverse of $1/2352 + 1/2352 + 1/6535 + 1/200$ or 167 liters sec^{-1}, which predicts a pressure of $(200/167) \times 10^{-6} = 1.20 \times 10^{-6}$ Torr.

1.4. Surface Physics*

1.4.1. Introduction

When one begins to work with vacuum systems, one very quickly becomes aware of the fact that gases from the walls of the system play an important part in determining the ultimate attainable pressure. The reason for this can be appreciated by considering the following simplified analysis.

To a very rough approximation one can consider a typical system to have the same number of square centimeters of internal surface area as there are cubic centimeters of internal volume. If we assume the surfaces to be covered with one monolayer of adsorbed gas (a reasonable approximation) and that we have N square centimeters of internal surface area, then there will be about $5 \times 10^{14} N$ molecules on the surface regions. Under the assumption made, the number of molecules in the gas phase is about $3.3 \times 10^{16} N$ molecules/Torr. One can then readily see that the number of adsorbed surface molecules corresponds to the number of gas phase molecules at a pressure of about 10^{-2} Torr.

If one now considers the approximation that in a typical system, during initial pumpdown, perhaps between 0.1% and 0.01% of the surface molecules are desorbing into the volume per second, one can understand why the rapid pumpdown curve begins to slow down at pressures in the range of 10^{-5}–10^{-6} Torr. At pressures below this range, molecules desorbing from the inner surfaces have a dominant effect on the achievable level of vacuum. In the remaining portion of this section we will review some of the general characteristics of adsorbed and absorbed gas molecules and the various processes which cause some of them to desorb into the volume of the system.

1.4.2. Adsorption

Adsorption involves the bonding to the surface of an atom or molecule that normally exists in the gas phase. This bonding is generally different for different combinations of molecules and substrates and may involve either condensation, physisorption, or chemisorption.

---

* Chapter 1.4 is by David Lichtman.

ISBN 0-12-475914-9

These terms can be understood by consideration of Fig. 1. Virtually all systems have a curve such as "A" with a relatively small binding energy well, denoted as E_p. An incoming atom or molecule can be trapped in this well and bound to the substrate with energy $\sim E_p$. Such a process is called *condensation or physisorption* and generally involves only Van der Waals forces.

For many systems there is a second potential energy curve such as "B" which involves a much deeper well E_c. Between the well of curve A and that of curve B there is often an energy barrier designated as E_a. Atoms or molecules which reach the well of curve A may overcome this barrier and drop into the well of curve B. The energy well of curve B involves a much stronger bond, may also involve dissociation of the incoming molecule, and leads to what is called *chemisorption*. This type of binding generally implies electron overlap or transfer between the adsorbed species and the substrate.

We will now briefly consider these types of adsorption separately. A detailed discussion of molecule–surface interactions and a list of many additional references is available in Ref. 1.

1.4.2.1. Condensation. Although the terms physisorption and con-

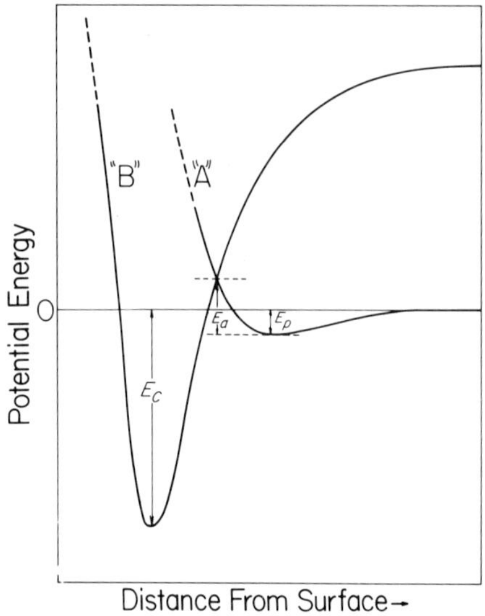

FIG. 1. Potential energy diagram illustrating possible physisorption and chemisorption states.

[1] P. A. Redhead, J. P. Hobson, and E. V. Kornelsen, "The Physical Basis of Ultrahigh Vacuum." Chapman & Hall, London, 1968.

densation are used interchangeably by many authors, I prefer to use them to designate two distinctly different phenomena. I will use the word condensation to indicate the process where an atom or molecule bonds to a substrate surface composed of like atoms or molecules. In this case, there is no reason why many layers of atoms or molecules could not build up. Therefore, a layer begins to develop of the liquid or solid phase of the particular element or material undergoing condensation.

Condensation, then, generally requires the maintenance of a surface temperature at which the gas in question can undergo a phase change to a liquid or a solid. Of special interest is the condensation coefficient, i.e., the probability that a molecule, arriving at a surface composed of its own species, will be trapped in the shallow minimum of curve A (Fig. 1). The value of the condensation coefficient will depend on the specie in question, the temperature of the surface, and the temperature of the incoming gas. Some typical values are shown in Table I. It should be noted that for sufficiently low temperatures the condensation coefficient generally approaches unity, that is, all the particles striking the surface remain at the surface.

1.4.2.2. Physisorption. Physisorption is generally used to describe the phenomena in which a gas atom or molecule is trapped in the shallow potential well (curve A) produced by a surface composed of some other material. The process is essentially similar to condensation but involves bonding of an atom or molecule to a surface composed of a dissimilar species.

The probability of bonding is strongly related to the particular combination of atoms or molecules involved, the temperature of the incoming gas specie, and, to a great extent, the temperature of the surface. Table II shows some experimental data on the probability for physisorption for

TABLE I. Condensation Coefficients of Common Gases at Room Temperature Impinging upon Their Own Frozen Deposits[a]

Surface temperature (K)	N_2	Ar	CO	CO_2
10	0.65	0.68	0.90	0.75
15	0.62	0.67	0.85	0.67
25	0.60	0.66	0.85	0.63

[a] Adapted from Ref. 1.

TABLE II. Physiosorption Probability of Some Common Gases on a Glass Surface[a]

	Surface temperature (°C)	
Gas	0	100
He	0.24	0.13
Ne	0.48	0.34
H_2	0.64	0.50
N_2	0.81	0.70
O_2	0.86	0.77
Ar	0.90	0.81

[a] Adapted from Ref. 1.

several gases at room temperature impinging on a glass surface held at two different temperatures.

The important difference between condensation and physisorption is that, with condensation, one has a buildup of many monolayers of the adsorbing specie, while with physisorption generally no more than one monolayer of adsorbed gas resides on the surface. Thus, in physisorption, when one monolayer is completed, additional incoming atoms or molecules would have to condense on a layer of their own kind and the surface temperature is usually not low enough for condensation.

1.4.2.3. Chemisorption. As an atom or molecule approaches a surface its potential energy can be represented as curve B in Fig. 1. It becomes strongly bound to the surface and we say that the particle has been chemisorbed. The probability that an arriving specie will become chemisorbed is called the sticking probability.

Several processes can occur in this situation. A specie may come in directly along the potential curve B and become chemisorbed in the deep potential well. Typical binding energies are of the order of several electron volts.

Another possibility is for the atom or molecule to first come in on curve A and become weakly bound in the shallow well. It may then have sufficient energy to overcome the potential barrier and drop into the deep well, thereby becoming chemisorbed.

A third possibility is for a molecule to approach the surface and arrive at the deep well either directly or via the shallow well. During such a process the molecule may dissociate, the individual atoms or fragments becoming strongly chemisorbed to the surface. This process is referred to as dissociative adsorption.

Chemisorption involves the bonding of a gas species to a substrate of some other material and almost always results in the accumulation of one monolayer of adsorbed gas. It is found experimentally that the sticking probability is a strong function of the nature of both the surface composition and the adsorbing gas particle, as well as the fraction of the surface already covered by the gas. Thus, initial sticking probabilities for many combinations are between 0.1 and 1, falling to much lower values as the surface is covered. Some typical sticking probability curves are shown in Figs. 2 and 3.

1.4.2.4. Surface Migration. In addition to the potential barrier which keeps an atom or molecule bound to the surface relative to the vacuum region, there also exists a potential barrier along the surface. If the barrier is relatively small, atoms or molecules may be able to migrate along the surface plane by hopping from one adsorption site to another. Of

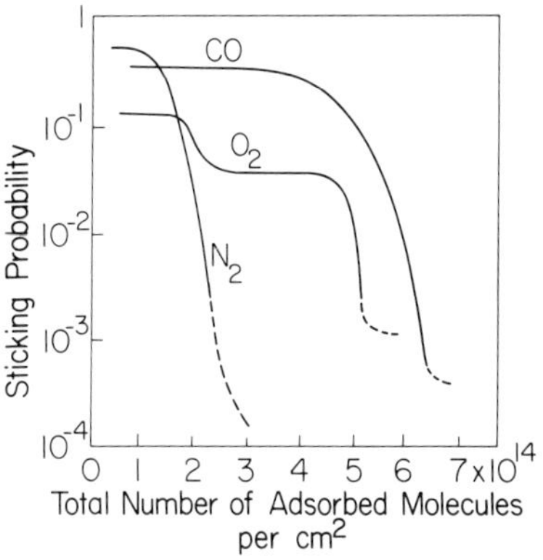

FIG. 2.

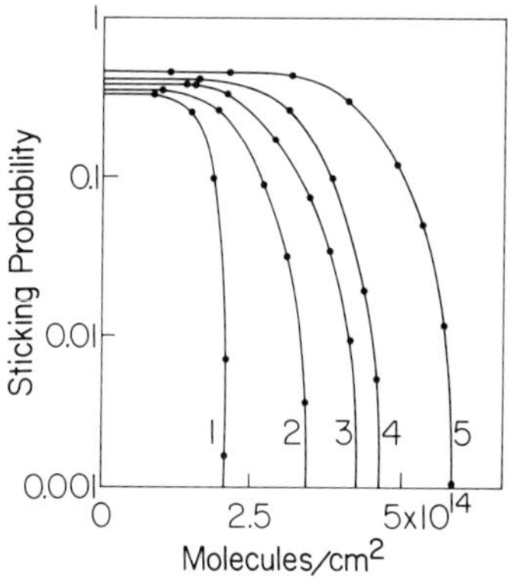

FIG. 3.

FIG. 2. Sticking probability as a function of surface coverage for various gases on single crystal tungsten at room temperature.

FIG. 3. Sticking probability for CO on single crystal tungsten as a function of coverage. Surface temperatures degrees K (1) 1100, (2) 900, (3) 700, (4) 500, (5) 300.

course, adjacent adsorption sites must be available and unoccupied for hopping to occur. Clearly, surface diffusion will depend very much on the surface temperature as well as the specific gas–surface combination involved.

1.4.2.5. Absorption. Atoms or molecules which arrive at a surface may also diffuse into the bulk of the surface material. This process is called absorption. In this case, two factors are important: the solubility of the gas in the solid and the diffusion constant of the gas moving in the solid. These processes are generally quite complicated. However, in practice few gases can diffuse readily through solids; typically only atoms such as hydrogen or helium are able to diffuse to any measureable extent. Further limitations to diffusion occur due to the fact that noble gases, such as helium, neon, etc., are insoluble in metals. For gases which are soluble in certain solids, both the solubility and the diffusion rate are generally very temperature dependent. The wide range of solubilities possible with different combinations and at different bulk temperatures is seen for some examples in Fig. 4. Similar variations in diffusion rates can be seen for some examples in Fig. 5.

The importance of bulk diffusion and absorption in the context of this discussion is that it frees surface sites to allow further adsorption from the gas phase. Of course one must remember that the process is reversible and the absorbed gas can become a serious source of outgassing under conditions that may cause absorbed gas particles to go to the surface where they can be desorbed.

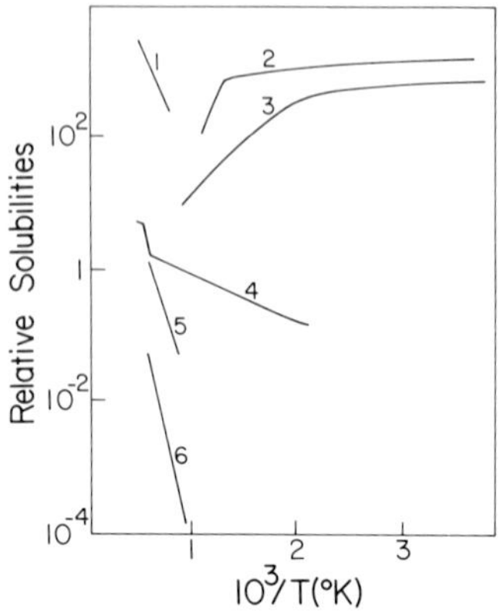

FIG. 4

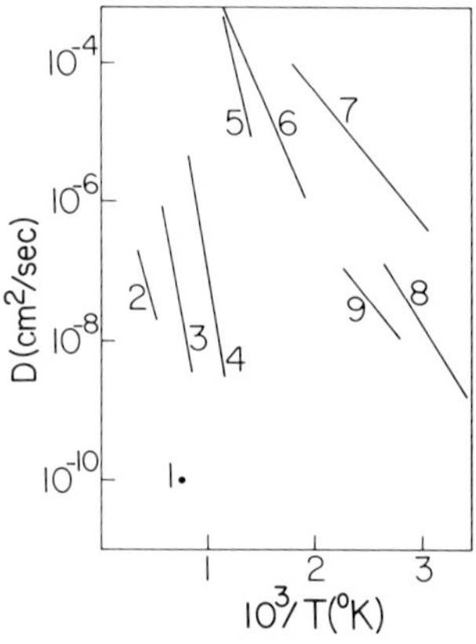

FIG. 5.

FIG. 4. Relative solubilities for various gas–metal combinations as a function of temperature. (1) H_2–Mo, (2) H_2–Zr, (3) H_2–V, (4) H_2–Ni, (5) N_2–Mo, (6) N_2–W.

FIG. 5. Diffusion constants for various gas–metal combinations as a function of temperature. (1) O_2–Fe, (2) H_2–W, (3) O_2–Ti, (4) O_2–Cu, (5) H_2–Al, (6) H_2–Cu, (7) H_2–Pd, (8) H_2–Fe, (9) H_2–Ni.

1.4.3. Desorption

The present discussion has concerned phenomena which remove gas species from the vacuum volume. This is generally, though not always, a desirable process since it helps reduce the gas phase pressure. The reverse process, i.e., gas leaving the solid surfaces and returning to the gas phase, is usually one which degrades the vacuum and is almost always the limiting factor in achieving high and ultrahigh vacuum. The energy necessary to cause desorption of adsorbed gases can be supplied to the adsorbed particle in a variety of ways. These energy sources can be thermal; particle beams such as electrons, ions, photons; ultrasonic waves; electric fields; etc. We will consider below those most frequently encountered.

1.4.3.1. Thermal Desorption. When atoms or molecules are adsorbed on a surface, a number of them will desorb per unit time generally according to the equation

$$dN/dt = \nu N e^{-E_d/kT} \quad \text{molecules/cm}^2\ \text{sec},$$

where N = number of adsorbed molecules/cm², ν is a rate constant, and E_d is the energy necessary for desorbing the specie in question. Values of E_d can vary from a few hundreths of an eV to several eV depending on the strength of the adsorption bond. See Table III. It is interesting to consider a situation for a species with a binding energy of about 0.90 eV. If

TABLE III. Some Reported Values of Binding or Desorption Energies for Various Gases on Various Substrates[a]

Adsorbent	Gas	Binding or desorption energy (eV)
Tungsten	CO (α, weakly bound)	0.85
	CO (β, strongly bound)	2.2–4.2
	H_2	1.66–2.2
Nickel	H_2	1.2
	CO	1.79
	O_2	4.55
Mo	H_2	1.7
Porous glass	Ar	0.11
Graphite	Ne	0.032
Iron	CO_2	2.9

[a] Adapted from data in Ref. 1.

one assumes a value of about 10^{13} sec^{-1} for ν (a value which seems to be experimentally valid for a large number of systems) one obtains for the ratio $dN/dt/N$ (the instantaneous fraction of the adsorbed monolayer desorbing per second) a value of about 2.5×10^{-3}. This value is the basis for the statements made in Section 1.4.1; namely, that gas desorbing from the wall is an important factor in determining the ultimate pressure in a vacuum system. Obviously, the desorption rate is a very strong function of both the binding energy E_d and the temperature of the surface. It is interesting to note that at room temperature the binding energy which separates rapidly desorbing species from those which experience long-term relative stability on the surface is about 0.9 eV.

For dissociated molecules which recombine on the surface before desorbing, the desorbing probability becomes proportional to N^2 and one finds a much smaller value for the rate constant ν.

For a particular gas species arriving at a surface from the gas phase, one may reach an equilibrium where the number arriving and sticking per unit time is equal to the number desorbing per unit time. One then observes what are called adsorption isotherms.

Thermal desorption is therefore seen to play a very important part in achieving high and ultrahigh vacuum. The strong effect of temperature can be appreciated by considering "the early morning effect." Most beginners in high vacuum work are surprised to note that, upon arriving at their system early in the morning, they observe a low value of pressure which, with nothing done to or on the system, increases to a higher value several hours later. The small temperature rise in the room (a few degrees) due to human activity and other warming processes during the day, e.g., sunlight, is enough to increase the desorption rate so that the equilibrium pressure in the system is seen to rise.

1.4.3.2. Electron Stimulated Desorption (ESD). The purpose of building a vacuum system is often to enable one to work with beams of particles, e.g., electrons, ions, and photons. These beams, at some time, will strike and interact with the surfaces in the system. This interaction may cause the desorption of gas atoms or molecules which will effect the pressure in the system.

Considerable work on ESD has led to the following general conclusions. Electrons interact with adsorbed molecular complexes with about the same cross sections that occur for electron–gas phase molecule interactions. Transitions to antibonding states provides sufficient kinetic energy to allow fragments of the surface particle to desorb. Desorbing fragments can be either charged or neutral. The process of ESD is very specific to the adsorbed particle–surface complex involved and the observed cross section for desorbing components can vary by many orders of magnitude depending on the system (e.g., it varies by a factor of 10^4 for two different phases of CO adsorbed on molybdenum). Some typical data is seen in Table IV.

ESD is sometimes used intentionally for cleaning of surfaces internal to a vacuum system.

1.4.3.3. Photodesorption. Photodesorption refers to the quantum process whereby a photon causes the desorption of a gas species. Thus, the effect of interest here is not related to thermal processes. The results of recent research indicates the following general observations. Photodesorption from metals is an extremely inefficient process and may not occur at all. Photodesorption from many semiconductors including metal

TABLE IV. Some Reported Electron Stimulated Desorption Parameters on Various Metals at 300 K[a]

Adsorbed molecule	Metal surface	Binding state	Desorbing species	Maximum cross section (cm^2)
O_2	W		O^+	1.5×10^{-20}
CO	W	Weak	O^+	3×10^{-20}
CO	W	Weak	CO^+	$1–2 \times 10^{-20}$
N_2	Mo		—	Not measureable
O_2	Mo	Weak	O^+	2×10^{-17}
CO	Mo	Strong	O^+	6×10^{-21}
CO	Ni	Weak	O^+	1×10^{-18}

[a] Adapted from Ref. 1.

oxides is a very efficient process with cross sections as high as 10^{-17} cm^2. The threshold energy for photodesorption corresponds to the band gap of the semiconductor. It should be noted that the surface of stainless steel is chrome oxide, a semiconductor, and therefore photodesorption from stainless steel is a strong effect. For many materials, the band gap is such as to put the threshold into the ultraviolet region, i.e., below $\lambda = 3500$ Å. Photodesorption gives rise only to neutral particle desorption and for most materials the primary desorbing species is CO_2. It has been found that this desorption is related to the impurity carbon content of the material of which the surface is composed.

UV radiation is sometimes used to clean vacuum surfaces. The process is still not as well understood as is, for example, ESD, but progress is being made.

1.4.3.4. Ion Induced Desorption. Although this process has not been studied in detail, it is expected that ion bombardment leads to the desorption of adsorbed gas species as neutrals as well as ions. The effect is in addition to the well-known sputtering process. As in the case of electrons and photons, ion bombardment is also used for surface cleaning.

1.4.4. Conclusion

This brief, rather simplified overview of some of the aspects of surface physics which are related to vacuum science and technology has attempted to show the importance of surface phenomena to the achievement and maintenance of high and ultrahigh vacuum. Many of the phenomena discussed will be considered further in later chapters dealing with specific aspects of vacuum technology.

2. MEASUREMENT OF TOTAL PRESSURE IN VACUUM SYSTEMS*

2.1. Introduction

2.1.1. General Remarks

The most fundamental of measurements relating to vacuum system parameters is that of the pressure of the gas or gases remaining within the vacuum vessel. Only at course or "rough" vacuum do there exist techniques for measurement of pressure directly as force per unit area. Over most of the pressure range from zero to atmospheric pressure, the pressure is inferred from other gas properties such as thermal conductivity or ionization rate by an electron stream. Literally hundreds of methods have been devised by researchers to measure the residual gas pressure but only a few are commonly used or find practical application. The discussion in this chapter is predicated on the assumption that the reader seeks practical information about pressure measurement and gauge characteristics. For this reason, the gauges described are those that are readily available and in general use. Special care is used to point out the gauge characteristics, their theory of operation, and precautions that should be exercised in obtaining valid data. Following the discussion on particular gauges is a section describing gauge calibration and the effects on gauge reading resulting from physical location in the vacuum system. The importance of this cannot be stressed too highly because improper gauge location can result in totally erroneous data even though extreme care is used in the operation of the gauge itself. Vacuum systems are rarely equilibrium systems but are dynamic or, at best, steady state systems and the gas flow characteristics must be considered.

2.1.2. Units

Historically, the units used in vacuum measurement have resulted from the adoption of the units used in meteorology. This has happened since the most convenient method of measuring variations in atmospheric pressure has been to fill a long tube, closed at one end, with some liquid (gen-

* Part 2 is by D. R. Denison.

ISBN 0-12-475914-9

erally mercury), and then invert the open end of the tube in a dish. The liquid in the tube will establish a height such that the force of the liquid column (weight of the liquid column) is balanced by the force on the liquid by the pressure of the atmosphere. The void that exists between the top of the liquid column and the closed end of the tube is a vacuum. In Fig. 1, such a column is depicted having a height h, a cross-sectional area A, and consisting of a liquid of density ρ. The force that the column exerts at its base is

$$F = mg = \rho h A g \tag{2.1.1}$$

and the pressure at the base of the column is

$$P = F/A = \rho h g, \tag{2.1.2}$$

where g is the acceleration due to gravity. For the liquid chosen, ρg is a constant and

$$P = Kh. \tag{2.1.3}$$

It is seen then that the pressure can be measured, relative to vacuum, as the equivalent of the height only of a liquid column and independent of its cross-sectional area. Such a device, using mercury as the liquid, is the

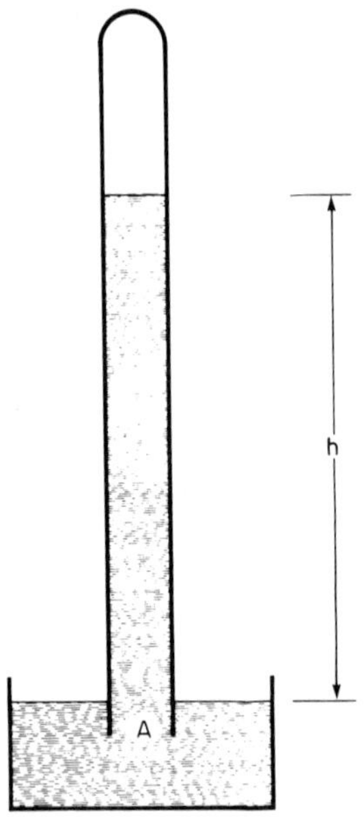

FIG. 1. Meteorological barometer demonstrates the measurement of pressure by the height of a liquid column.

barometer of meteorology and the atmospheric pressure can be recorded as so many millimeters of mercury (abbreviated mm Hg). It is a simple step to continue this tradition into vacuum measurement since the reference point is "zero" pressure. The mm Hg unit of pressure measurement was named the *Torr* in honor of the inventor of the barometer Torricelli (1608–1647). The standard atmosphere is defined to be 760 Torr or the atmospheric pressure required to balance a column of mercury 760 mm in height at 0°C with $g = 980.665\ \text{cm/sec}^2$. Additionally, another unit has continued in existence from the concept of mm Hg. One milli-Torr, 1×10^{-3} Torr, is equivalent to 10^{-6} meters of mercury and is commonly called the *micron*. Thus one will encounter reference to pressures listed interchangeably as milliTorr, Torr $\times\ 10^{-3}$, or microns.

At the time of this writing, the Torr is the most common usage in reporting vacuum pressure measurements but a move to make pressure units consistent with metric nomenclature has given rise to the use of the *millibar* and the *Pascal*. The *bar* is defined as 10^6 dynes/cm^2 and the standard atmosphere is *1013.3 mbar* so that the unit of pressure is in the cgs system of measurement. In the mks system, the unit of pressure is the newton/m^2 or *Pascal*. The standard atmosphere is 1.0133×10^5 *Pascal*. One Torr equals 133.3 Pascal or 1.333 mbar.

2.1.3. Pressure Ranges and Corresponding Measurement Techniques

Figure 2 shows a bar graph illustrating the useful range of the gauges described in this chapter. The bars indicate the usual range in which these gauges are used although it is possible to extend the range by special constructions. In the sections of this chapter dedicated to particular gauge types, these special constructions will be discussed.

No single gauge type has yet been developed that can span the full range from atmosphere to extreme high vacuum ($< 10^{-11}$ Torr).* In general, mechanical gauges are useful from atmosphere to 0.1 Torr, thermal conductivity gauges are useful from 1.0 to 10^{-3} Torr, and ionization gauges are useful from 10^{-3} to 10^{-12} or 10^{-13} Torr.

---

* Editorial comment: A proposal by L. Marton (*Bull. Acad. R. Belg.* **60,** 676–685 (1974) to use Rayleigh scattering of light for vacuum measurement may cover the range from 10^3 Torr to 10^{-12} Torr, and with special precautions possibly even to 10^{-15} Torr. No experiment was made to prove the validity of these calculations except for the upper atmosphere work reported in the paper, and it may not be easy to apply the principle to routine instruments which are the main concern of the present chapter. Nevertheless, if a light-scattering instrument is ever built, it could provide an absolute calibration of routine gauges, as its readings are linear and are derived from first principles.

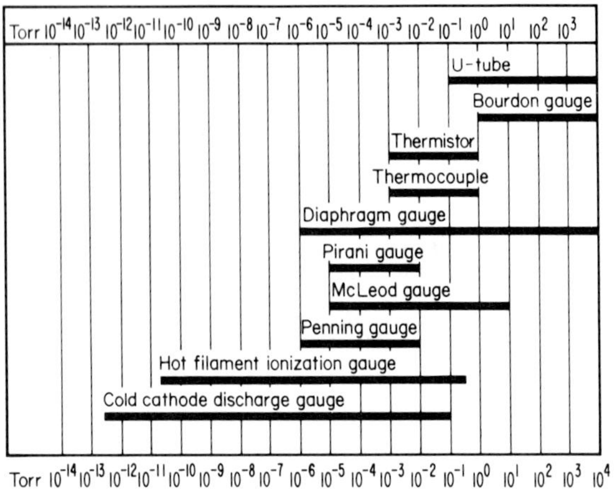

FIG. 2. The useful ranges of the gauges discussed in this chapter.

2.2. Manometers

2.2.1. Ideal Gas Laws

At low pressure, most gases behave very nearly as ideal gases. That is to say that the properties of the gas are describable with a theory that assumes that the gas molecules are negligible in size and exert no forces on each other except at the instant of impact in case of collision. The kinetic theory of gases is based on three hypotheses.

(i) Gases are made up of molecules and all molecules of a given substance are alike in terms of mass, size, shape, etc.

(ii) The gas molecules are in constant motion and collisions with each other and the walls are in accordance with the conservation laws of classical mechanics. Both momentum and energy are conserved.

(iii) Statistical methods may be used in describing the behavior of the gas.

At first this appears to be a gross oversimplification because it would apply at best to the noble gases which are nonreactive and atomic in form. In actual practice, however, we will see that gauge behavior for most gases indicates that the ideal gas laws are not a bad approximation to the truth. When particular gases have properties that result in gauge behavior different from that expected from the ideal gas laws they are discussed in the section for that gauge so that their behavior is understood. In general, deviations from the ideal gas laws result from characteristics peculiar to a given gauge, such as thermal decomposition or ionization cracking.

Let us consider an enclosed volume V containing N gas molecules in constant motion. If a molecule has mass m and collides with the wall with normal component to the velocity v_z, the change in momentum for an elastic impact is $2mv_z$. Consider the group of molecules with velocities such that they strike an area dS in a time interval dt. We will choose our coordinate system such that the z axis is normal to dS and we will choose dt short enough so that all molecules with a speed between v and $v + dv$ will strike the wall and not be scattered by molecular collisions. The number of molecules per unit volume n within the speed range and range of directions $d\omega$ is $nf(v)\, dv\, (d\omega/4\pi)$. The number of molecules that strike dS within time dt are then

$$nf(v)\, dv\, (d\omega/4\pi)v_z\, dt\, dS, \tag{2.2.1}$$

where $f(v)$ is the speed distribution function. The momentum imparted to the wall is then

$$nf(v)\, dv\, (d\omega/4\pi)2mv_z^2\, dt\, dS. \tag{2.2.2}$$

From classical mechanics, the pressure is given by the rate at which momentum is imparted to the wall per unit area. Thus

$$dP = nf(v)\, dv\, (d\omega/4\pi)2mv_z^2. \tag{2.2.3}$$

We can rewrite this in terms of the polar and azimuthal angles of impact θ and ϕ and recognizing that $v_z = v \cos\theta$ and $d\omega = \sin\theta\, d\theta\, d\phi$. Integration over the range of θ from 0 to $\pi/2$ and ϕ from 0 to 2π gives

$$dP = \tfrac{1}{3}nmv^2 f(v)\, dv. \tag{2.2.4}$$

Finally, integration over all possible molecular speeds gives

$$P = \tfrac{1}{3}nm \int_0^\infty v^2 f(v)\, dv = \tfrac{1}{3}nm\overline{v^2}, \tag{2.25}$$

where $\overline{v^2}$ is the mean square speed. Thus we see that for a gas at a constant temperature and, therefore, a constant mean square speed,

$$P = \tfrac{1}{3}m\overline{v^2}\, N/V \tag{2.2.6}$$

or

$$PV = \tfrac{1}{3}Nm\overline{v^2} = \text{constant}. \tag{2.2.7}$$

This is known as Boyle's law. From Boyle's law we see that the total kinetic energy in the volume V is

$$\tfrac{1}{2}m\overline{v^2} = \tfrac{3}{2}PV. \tag{2.2.8}$$

Empirically the temperature dependence of the ideal gas law is

$$PV = NkT, \tag{2.2.9}$$

where k is the Boltzmann universal constant. These two expressions for PV are reconciled if

$$\tfrac{3}{2}kT = \tfrac{1}{2}m\overline{v^2} \tag{2.2.10}$$

which relates the absolute temperature to the mean kinetic energy. An alternative form for (2.2.9) is

$$P = nkT, \tag{2.2.11}$$

which states that for constant volume and, therefore, constant molecular density n, the pressure is directly proportional to the absolute temperature. This is known as Charles' law. See Refs. 1 and 2 as suggested reading on kinetic theory and the gas laws.

2.2.2. Liquid Manometers

2.2.2.1. U-Tube Manometers. One of the simplest vacuum pressure gauges to construct and use is seen in Fig. 3. It consists of a length of transparent tubing, such as glass, bent into the form of a U. From its shape, the name is derived. Figure 3 shows three forms representing variations on the simple principle involved. Figure 3(a) is the most rudimentary form in which one leg of the U is sealed off. The liquid, usually mercury, is poured into the tube so that all air is displaced from the closed leg of the U. When the U tube is then placed as shown in the figure, the space over the mercury column is a "vacuum" containing only mercury vapor. Over the temperature range of 0–40°C, the vapor pressure of mercury is given by

$$\log P = 8.14 - 3243/T \quad \text{Torr}, \tag{2.2.12}$$

where the temperature is in degrees Kelvin. From this we see that even at 40° (313 K) the vapor pressure is only 6.1×10^{-3} Torr. In the case of the simple U-tube manometer, even this temperature is insufficient to cause an error that would be observable except by the use of more exact measurement of the column height. The vacuum established in the closed end of the U tube serves as the "reference" vacuum against which the unknown pressure will be measured. The open end of the U is con-

[1] R. D. Present, "Kinetic Theory of Gases," McGraw-Hill, New York, 1958.
[2] L. B. Loeb, "Kinetic Theory of Gases," 2nd ed. McGraw-Hill, New York, 1934.

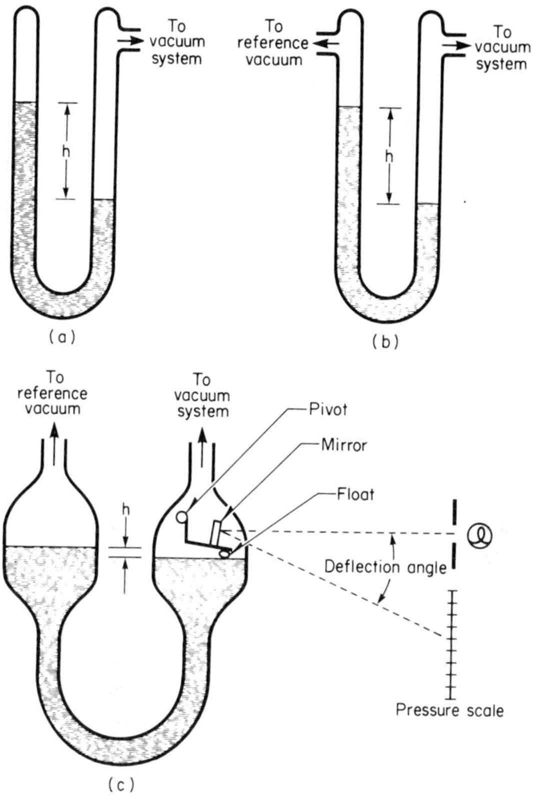

FIG. 3. The U-tube manometer used with a closed reference volume (a), a pumped reference (b), and optical lever amplification (c).

nected to the chamber in which the vacuum to be measured exists. A difference in mercury column height will then be established by the difference in gas pressure over the two columns. The pressure is determined by measuring the difference in column height, h in Fig. 3, and then calculated using Eq. (2.1.3). The constant in Eq. (2.1.3) is given by the product of the density of mercury, the acceleration due to gravity and a constant, to convert to pressure in Torr. For example, if ρ is in g/cm^3 and g in cm/sec^2, the scaling constant is 7.5×10^{-4} Torr/dyne cm^{-2}. The density of mercury is a function of temperature given by

$$\rho = 13.5955 - 0.0025T \quad \text{g/cm}^3 \tag{2.2.13}$$

over the range of 0–40°C, where T is in °C. At 20°C, Eq. (2.1.3) becomes, for h in centimeters,

$$P = 9.963h \quad \text{Torr}. \tag{2.2.14}$$

Note that the density is the only temperature coefficient we need consider. The expansion coefficient of the glass tubing plays no part since the area of the tube cancelled in the derivation of Eq. (2.1.3). The volumetric expansion of the mercury has been taken into account by the temperature dependence of its density. In a single tube manometer, known as a barometer, a correction must be applied for the capillary depression of the meniscus of the mercury in a glass tube but with the two column or U-tube manometer, the correction is applicable to both columns equally, and thus substracts out when determining the difference in height h.

The obvious improvement on the closed end U-tube manometer is to open the closed end and pump the "reference" side of the manometer, as in Fig. 3(b). This becomes important if a more sophisticated method of measuring column heights is utilized. The use of a traveling telescope on a precision travel screw can be used to measure pressures to 0.1 Torr. The pumped reference usefulness increases when one of the many modifications introduced over the years is used. Figure 3(c) shows one such modification to the U-tube manometer introduced by Schrader and Ryder.[3] A mirror mount is pivoted on supports attached to the walls of the U tube. The tube is enlarged to provide the necessary room for the mirror. The mirror is further attached to an arm and float which rides on the surface of the mercury. As the mercury level rises or falls, the mirror pivots through a small angle which is measured using an optical leveler. Pressure changes of as little as 1×10^{-3} Torr can be detected in this way. Carver[4] has described a more elaborate method of utilizing the optical lever to measure pressure changes to as low as 1×10^{-4} Torr. Glass floats have been used by Johnson and Harrison[5] and iron floats by Newbury and Utterback[6] to support mirror assemblies to obtain sensitivities of 10^{-3} Torr. Simon and Fehér[7] placed an electrode consisting of metal foil around the U-tube wall and electronically measured the capacitance between the mercury column and the external electrode. Sensitivities of the order of 10^{-2} Torr were obtained. This method has the advantage of not requiring the observer to read the manometer directly and pressure recording for a continuous record is facilitated.

Mercury is not the only liquid that may be used in a liquid manometer of the types discussed here. The sensitivity may be increased by noting from Eq. (2.1.3) that if the density of the liquid is reduced, the pressure to

[3] J. E. Schrader and H. M. Ryder, *Phys. Rev.* **13,** 321 (1919).
[4] E. K. Carver, *J. Am. Chem. Soc.* **45,** 59 (1923).
[5] M. C. Johnson and G. O. Harrison, *J. Sci. Instr.* **6,** 305 (1929).
[6] K. Newburg and C. L. Utterback, *Rev. Sci. Instr.* **3,** 593 (1932).
[7] A. Simon and F. Fehér, *Z. Elektrochem.* **35,** 162 (1929).

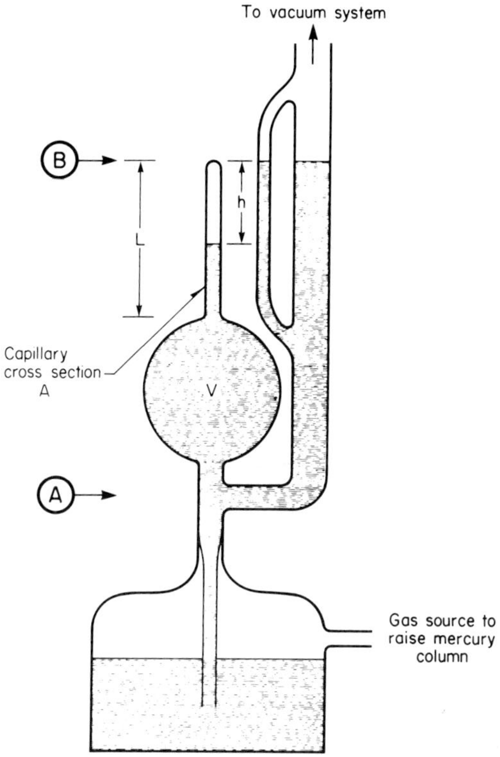

FIG. 4. A McLeod gauge utilizes Boyle's law to measure a low pressure.

cause a unit height column change is proportionally reduced. Burrows[8] has described a manometer using Apiezon oil. The author has used Dow Corning DC-704 diffusion pump oil with a density of 1.06 g/cm^3. At 20°C, the sensitivity is increased by a factor of 12.78. Visual readings of 0.1 Torr can be made and the optical lever techniques described earlier can be used to extend the useful range to 10^{-4} Torr or lower. The limitation of using a low vapor pressure oil as the manometer liquid lies in its tendency to wet the glass and its viscosity results in slow flow characteristics. For these reasons relatively slow pressure changes are observable and fast changes are either damped or have high error because of oil clinging to the manometer walls.

2.2.2.2. McLeod Gauge. A gauge described in 1874 by McLeod,[9] shown in Fig. 4, is basically a U tube with a twist. The gauge is constructed so that one leg of the U tube consists of an accurately calibrated

[8] A. Burrows, *J. Sci. Instr.* **20,** 21 (1943).

[9] H. McLeod, *Phil. Mag.* **48,** 110 (1874).

volume V and a length of precision bore tubing attached to it. The mercury level is lowered below point A so that the entire gauge is exhausted to the same pressure as the chamber to which the gauge is attached. The mercury is then raised so that the residual gases in the closed volume are trapped. The mercury continues to rise in both legs of the U tube and, in rising, compresses the trapped gas until it is finally compressed into the precision bore tubing. The mercury is raised until the column in the leg of the U tube connecting to the chamber being measured reaches point B as shown in Fig. 4. Some gauges commercially available have provision for self-leveling the mercury at level B. This provision is shown in Fig. 5. With the mercury in the open leg at level B, the difference in height between the two mercury columns, h in Fig. 4, is then a measure of pressure of the compressed gas within the precision bore tubing. Boyle's

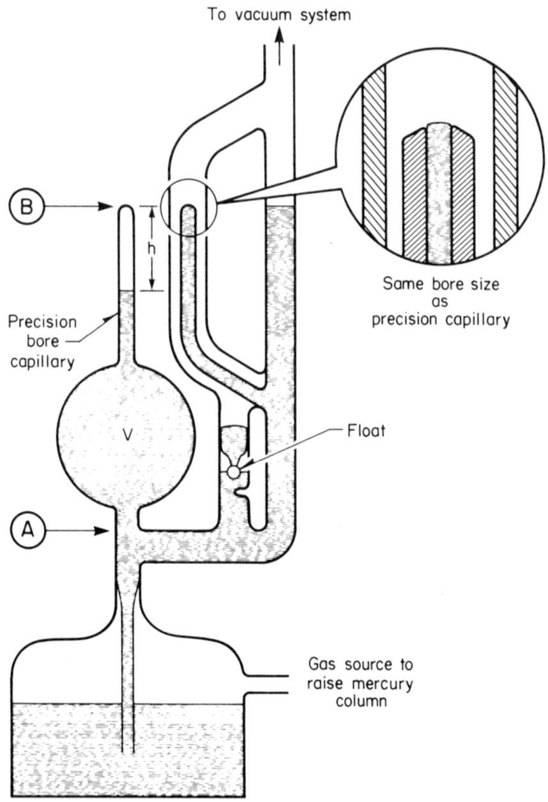

FIG. 5. A self-leveling device permits easier use of the McLeod gauge by increasing the repeatability of locating the reference side mercury column.

law, Eq. (2.2.7), can now be used to compute the pressure in the volume prior to compression which is equal to the chamber pressure to be measured. The volume of the gas after compression is hA, where A is the cross-sectional area of the precision bore tubing. The pressure after compression (assume the temperature to be 20°C) is given by Eq. (2.2.14) to be $9.963h$. The volume prior to compression is $V + LA$, where L is the total length of precision bore tubing. Boyle's law, PV = constant, then tells us that the unknown pressure is

$$P = 9.963Ah^2/(V + LA) = Ch^2. \tag{2.2.15}$$

The constant C is called the "gauge constant" and has the units of Torr per unit of length squared; e.g., Torr per cm². Note that the gauge constant has a temperature coefficient resulting from the variation of the density of mercury with temperature. For accurate measurements, corrections for temperature and meniscus must be applied.

Although Clark[10] has described a sealing technique for closing the end of the precision capillary in a manner as to leave the inside end as flat as possible, the end is rarely perfectly flat, and some error can be introduced at very low pressure. Rather than run the mercury close to the end of the capillary at low pressure, another method is to raise the mercury until the gas has been compressed into a known volume of the capillary. This takes some practice to stop the rising mercury at the chosen place. Let d be the length of capillary into which the gas is compressed and h is now the difference in mercury column height as shown in Fig. 6. The compressed gas volume $v = dA$ and the pressure is $9.963h$ as before. The pressure measurement is now

$$P = 9.963dAh/(V + LA) = C'h, \tag{2.2.16}$$

where C' is now a new gauge constant but the pressure is linearly related to the difference in mercury column heights. This method has the advantage of minimizing the error introduced by imperfections in the capillary bore. Needless to say, if the bulb volume V is very large compared to the capillary volume, Eqs. (2.2.15) and (2.2.16) become simpler.

The sensitivity of the McLeod gauge is improved by increasing the trapped volume V. The practical limit on this however is the weight of the mercury to be raised and the danger of breaking the gauge structure. The bore of the capillary may also be reduced to maximize the length for any given entrapped volume. If the bore diameter becomes smaller than about 0.5 mm, the mercury column tends to break when the mercury is

[10] R. J. Clark, *J. Sci. Instr.* **5**, 126 (1928).

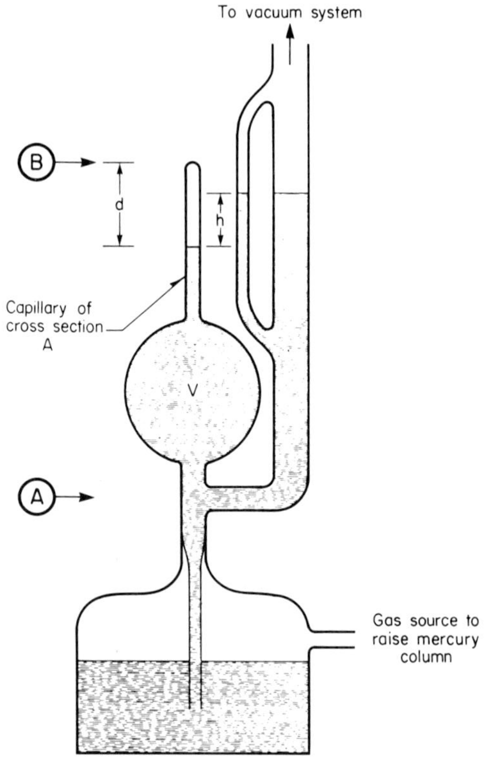

FIG. 6. The McLeod gauge used for lower pressure indication compresses the gas to a known volume.

lowered leaving a trapped portion in the capillary. This must be removed before another measurement may be made. Haase[11] has studied the effects of errors in McLeod gauge readings and concludes that two causes of error predominate:

(1) poor cleanliness of the inside surface of the capillary resulting in the mercury sticking to the surface;
(2) slight oxidation of the mercury surface when exposed to air results in a "scum" which adheres to the glass surfaces and distorts the normal raising and lowering of the mercury.

Haase also feels that one should only use vacuum distilled mercury to avoid dissolved gases and contaminated mercury. Condensable gases, such as water vapor and hydrocarbons from the pumping systems, should be kept out of the gauge. These gases are not accurately measured by the

[11] G. Haase, *Z. Tech. Phys.* **24,** 27 (1943); **24,** 53 (1943).

McLeod gauge because they condense as they are compressed during the measurement. At 20°C, water has a vapor pressure of 17.54 Torr, and thus will form a liquid phase if an attempt is made to compress it beyond this point. Coolidge[12] observed that condensation in the capillary occurs at lower pressure than the vapor pressure at the same observation temperature. In order to keep the condensable vapors out of the gauge and mercury out of the system being monitored, a liquid nitrogen trap is commonly used. A serious error can result from this, however, as shown by Ishii and Nakayama[13]. The liquid nitrogen cooled surface becomes an effective pump for mercury which streams from the McLeod gauge toward the trap. This stream of mercury vapor acts on the residual gases in the McLeod gauge much as a mercury diffusion pump. A differential pressure between the McLeod gauge and the chamber pressure being measured can result with the gauge reading as much as 200% too low.

Gaede[14] very carefully investigated the accuracy of the McLeod gauge and concluded that Boyle's law is accurately held even down to 1×10^{-4} Torr for nitrogen and hydrogen but errors are likely to arise with oxygen because of the formation of an oxide film on the mercury which then adheres to the capillary wall. He points out that careful heating of the capillary will remove this oxide layer. A new McLeod gauge will not measure accurately until the capillary is cleaned of water or other condensable vapors which adhere to the walls. This is accomplished by gentle heating, as with a hot air blower, while pumping on the gauge. This should only be done with the mercury lowered and the mercury should not be raised into the glass until it is fully cooled. The McLeod gauge should be used with great care to avoid breakage and spilling of the mercury.

2.2.3. Mechanical Manometers

2.2.3.1. Bourdon and Spiral Tube Gauges. Many attempts have been made to construct mechanical gauges requiring no electronics for pressure readout and responding to low pressures without gas type discrimination. One such attempt is a gauge consisting of a flat, tapered metal, glass, or quartz tube bent in the form of a spiral and closed on one end as in Fig. 7. It was invented by Bourdon, a 19th century inventor, and commonly referred to as a Bourdon gauge. As the pressure within the coiled tubing changes, the coil tightens or relaxes.

Almost everyone has encountered a Bourdon tube at a Halloween or

[12] A. S. Coolidge, *J. Am. Chem. Soc.* **49,** 708 (1927).
[13] H. Ishii and K. Nakayama, *Trans. Natl. Vac. Symp.* **8,** 519 (1961).
[14] W. Gaede, *Ann. Phys.* **41,** 289 (1913).

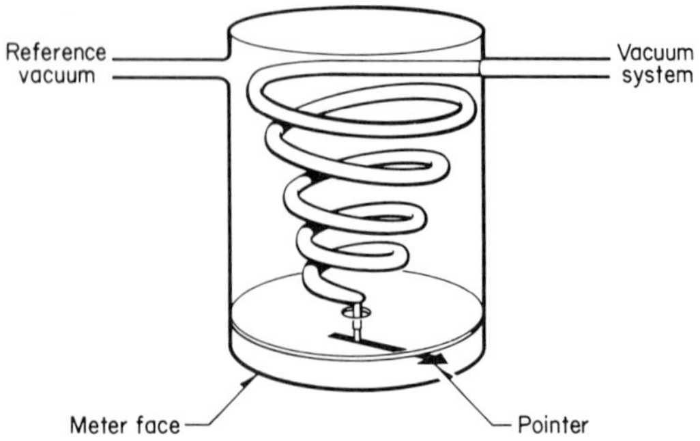

FIG. 7. The Bourdon gauge with a tapered quartz or metal tube which unwinds as the internal pressure increases relative to the eternal pressure.

New Year's party where they are constructed of paper and uncoil as the interior is pressurized by blowing in one end.

When used as a vacuum gauge, two methods of mounting the spiral are used. The more common version has a lightweight pointer attached to the closed end of the spiral perpendicular to the axis of the coil. As the spiral responds to pressure changes by coiling or uncoiling, the deflection is noted by the change in position of the pointer. The second method utilizes a mirror attached to the closed end of the spiral so that the mirror rotates as the spiral tightens or relaxes. A deflection of a light beam reflected off the mirror is then a measure of the pressure change. The spiral may be enclosed in a housing as shown in Fig. 7 so that the pressure external to the spiral may also be adjusted. In this way, a Bourdon gauge may be used as a differential pressure gauge. The Bourdon gauge is normally used as an indicator of crude or rough vacuum from atmosphere to about 10 Torr.

2.2.3.2. Aneroid Capsule Gauge. Another technique for mechanically measuring pressure is the deformation of a thin wall diaphragm. Wallace and Tiernan developed an absolute pressure gauge that utilizes the force on a sealed diaphragm to move a mechanical linkage and deflect a pointer on a calibrated front face. The capsule which consists of the closed diaphragm and support is evacuated prior to sealing and hence the name of aneroid capsule. The gauge sensitivity is a linkage design criterion and gauges are available with 0–50 mm Hg (Torr) full scale deflection and from 0 to 800 mm Hg full scale deflection. The practical limit to sensitivity is 0.2 Torr. The pressure sensitive capsule has mechanical stops so overpressure cannot cause damage.

The advantage of this design is simplicity and the readout is true pressure without respect to gas composition. A disadvantage is its lack of

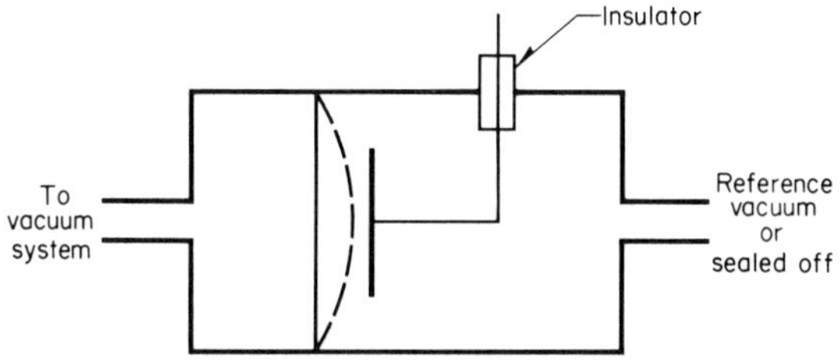

FIG. 8. A diaphragm gauge measures pressure by the capacitance change as the diaphragm deflects toward the fixed electrode.

electrical readout although the lack of auxiliary electrical equipment is an essential part of its simplicity.

2.2.3.3. Diaphragm Gauge. Another type of diaphragm gauge was reported by Alpert, Matland, and McCoubrey.[15] This device consists of a metal diaphragm stretched across a section of tubulation as shown in Fig. 8. A difference in pressure will cause the diaphragm to deflect toward or away from a fixed electrode mounted on one side. The capacitance between the fixed electrode and the diaphragm is measured and related to the differential pressure. The side of the gauge containing the fixed electrode is normally connected to a reference vacuum source that, essentially, is at zero pressure. For commercially available gauges this should be 10^{-6} Torr or better. An alternative is to evacuate and seal the reference side of the gauge. This has the advantage of simpler operation for a total pressure gauge. The deflection of a uniformly loaded circular membrane rigidly attached around its periphery[16] is

$$d = kPr^4/Et^3 \quad \text{m}, \tag{2.2.17}$$

where P is the differential pressure in Torr across the diaphragm, E is the elastic constant of the diaphragm material in newtons/m^2, and r and t are the diameter and thickness of the diaphragm in centimeters. The deflection of the diaphragm is linearly related to the differential pressure. The gauge is electrically a capacitor with a capacitance that varies with pressure. The capacitance of a parallel plate vacuum capacitor is

$$C = \epsilon_0 A/D \quad \text{F}, \tag{2.2.18}$$

where A is the area of the capacitor, ϵ_0 is the permittivity of free space, and D is the spacing between the fixed electrode and the diaphragm. The

[15] D. Alpert, C. G. Matland, and A. O. McCoubrey, *Rev. Sci. Instr.* **20,** 370 (1951).

[16] C. Carmichael (Ed.), "Kent's Mechanical Engineer's Handbook," pp. 8–32. Wiley, New York, 1950.

capacitance as a function of the diaphragm parameters is obtained by putting Eq. (2.2.17) into Eq. (2.2.18):

$$C = \frac{\epsilon_0 A}{(D - d)} = \frac{\epsilon_0 A E t^3}{(DEt^3 - Kr^4 P)} = \frac{K_1}{K_2 - K_3 P}. \tag{2.2.19}$$

Let us look at the magnitude of these constants with $A = 5 \times 10^{-4}$ m², $\epsilon_0 = 8.8 \times 10^{-12}$ F/m, $E = 1.52 \times 10^{11}$ n/m², $t = 2.5.\times 10^{-4}$ m, $D = 2.5 \times 10^{-3}$ m, $r = 0.025$ m, $K = 67$, then we find $K_1 = 1.1 \times 10^{-14}$, $K_2 = 6.8 \times 10^{-3}$, $K_3 = 2.8 \times 10^{-5}$ so that

$$C = (1.1 \times 10^{-11})/(6.3 - 0.028P) \quad \text{F}. \tag{2.2.20}$$

The point of this is to see that at low pressure where P is less than 1 Torr, the problem is to detect a small change in a small capacitance. Sophisticated modern electronics permits accuracy of better than 1% at 10^{-2} Torr but as Fig. 9 shows, the accuracy becomes less at lower pressure. Although this instrument is electronically sophisticated, its mechanical simplicity and compatibility with almost any vacuum environment make this gauge, often called a capacitance manometer, a most convenient gauge to use. Because it uses mechanical deflection resulting from a differential pressure across the diaphragm, the gauge reading is independent of gas type.

An interesting feature that emerges from Eq. (2.2.20) is that the first term of the denominator dominates and both it and the numerator contain the factor Et^3. Thus to a good approximation, the behavior of the gauge, in terms of sensitivity and accuracy, are independent of the diaphragm material or thickness.

This is the first gauge we have considered that is compatible with ultrahigh vacuum systems (pressure $< 10^{-8}$ Torr) in that it can withstand

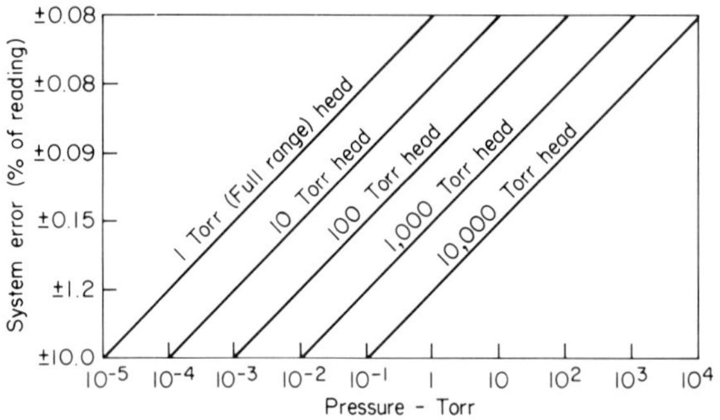

FIG. 9. The measurement accuracy vs pressure for diaphragm gauges of different full scale ranges. The units on the ordinate show that a lower limit to the error exists. Data provided courtesy of MKS Instruments, Burlington, Mass.

bakeout at high temperature. Even though it does not measure pressures in the ultrahigh vacuum region, the capacitance manometer diaphragm gauge is often used in monitoring the inlet of high purity gases into such systems and must be capable of being put in a very clean condition to avoid contamination of those gases.

Because the diaphragm gauge is truly a differential pressure gauge, it can be used as a gas flow monitor by flowing the gas through an orifice of known conductance C_0. If the two sides of the diaphragm are connected to the upper and lower pressure sides of this orifice, the difference in pressure $(P_1 - P_2)$ is displayed by the gauge and the gas flow Q is given by $Q = C_0(P_1 - P_2)$. In the same way, the differential feature can be used in gas mixing studies. Two gases can be filled to the same pressure in two chambers prior to mixing. Both chambers are pumped to low pressure then one gas is admitted to one of the chambers. The gauge is now referencing to low pressure and the absolute pressure is registered by the gauge. When the desired pressure is reached, gas is admitted to the second chamber until the gauge reads zero. The diaphragm is now in the null position and the pressures in the two chambers are equal. A like method may be used to fill to any two desired pressures.

2.3. Thermal Conductivity Gauges

2.3.1. Theory and Principles

If a surface is heated in a gaseous environment, energy can be lost in three ways. The first is by conduction loss through the support leads. Langmuir[17] studied these losses at length and developed a technique for computing the temperature distribution along heated wires mounted on fixed leads. These lead losses, although nonzero, can be minimized by proper choice of sizes and materials. The second energy loss mechanism is by thermal radiation given by

$$E = 5.67 \times 10^{-12} A(\epsilon_1 T_1^4 - \epsilon_0 T_0^4) \quad W, \tag{2.3.1}$$

where A is the radiating area at temperature T_1 and emissivity ϵ_1 and the surrounding walls have temperature T_0 and emissivity ϵ_0. The third mechanism is by thermal conduction through the gas. At high pressure, the gas acts like any other insulating material and the heat loss is determined by the surface area, the conductivity of the particular gas type, and the thermal gradient:

[17] I. Langmuir, *Phys. Rev.* **35,** 478 (1930).

$$Q = KA \frac{dT}{dx} \quad \text{cal/sec.} \tag{2.3.2}$$

Dickens[18] demonstrated that this relationship held over a wide pressure range. For air and CO_2, the conduction loss was constant within a few percent to as low as 11 Torr. Even for hydrogen, the thermal conduction loss was less than 3% lower at 130 Torr as at 760 Torr. Thus, over a wide pressure range above about 10 Torr, the thermal conduction loss from a heated surface is independent of pressure. At low pressure, however, where the distance from the hot to cold surfaces is less than the molecular mean free path, a different characteristic is observed. The reader is referred to a text on kinetic theory of gases for a deriviation of this theory. A particularly lucid account is provided in "A Treatise on Heat" by Saha and Srivastava.[19] In summary, analysis of the energy flow between two surfaces at temperatures T_1 and T_2, where $T_1 > T_2$, the energy transported between the surfaces is given by

$$E = \tfrac{1}{2}n\bar{v}k(T_1' - T_2') + \tfrac{1}{4}n\bar{v}(E_{r1} - E_{r2}), \tag{2.3.3}$$

where n is the molecular density, $\bar{v}$ is the mean speed of the gas molecules between the surfaces, T_1' and T_2' are the effective temperatures of the gas molecules leaving the surfaces of temperature T_1 and T_2, and E_{r1} and E_{r2} are the energies carried in rotational or vibrational modes in the case of polyatomic molecules. The first term represents energy carried by translational motion and the second term represents energy in the other degrees of freedom. The introduction of T_1' and T_2' keeps us aware that an incident molecule might not come fully to the temperature of the surface on which it impinges. Knudsen[20] introduced a quantity $\propto$ he called the accommodation coefficient which denotes the fractional extent to which incident molecules adjust or accommodate their mean energy toward the mean energy they would have if Eq. (2.2.10) defined the energy at the surface temperature T. If T_i is the effective temperature of the incident molecules, T_r is the temperature of the reflected molecules, and T_s is the surface temperature, then $\propto$ is defined as

$$\propto \equiv (T_\mathrm{r} - T_\mathrm{i})/(T_\mathrm{s} - T_\mathrm{i}). \tag{2.3.4}$$

If we assume that $\propto$ is the same for both the hot and cold surfaces,

$$T_1' - T_2' = \propto(T_1 - T_2') \quad \text{and} \quad T_2' - T_1' = \propto(T_2 - T_1')$$

or

[18] B. G. Dickins, *Proc. R. Soc. London* **A143,** 517 (1934).

[19] M. N. Saha and B. N. Srivastava, "A Treatise on Heat," p. 218. Indian Press, Calcutta, 1950.

[20] M. Knudsen, *Ann. Phys.* **35,** 593 (1911); **6,** 129 (1930).

$$T_1' - T_2' = [\alpha/(2 - \alpha)](T_1 - T_2). \tag{2.3.5}$$

We need to recall Eq. (2.2.11) which relates the gas pressure to the molecular density n and the translational and rotational/vibrational energies are related in the ratio of specific heats of constant pressure and constant volume, γ. When these are inserted in Eq. (2.3.3) with Eq. (2.3.5),

$$E = 3.64 \times 10^{-4} \frac{P}{2 - \alpha} \frac{\gamma + 1}{\gamma - 1} \frac{T_1 - T_2}{\sqrt{MT}} \quad \mathrm{W/cm^2},$$

where M is the molecular weight of the gas species, and T is given by

$$\sqrt{T} = 2\sqrt{T_1 T_2}/(\sqrt{T_1} + \sqrt{T_2}). \tag{2.3.6}$$

We see then that conduction of heat at low pressure is independent of the geometry and is proportional to the pressure although the heat conduction has mass dependence.

From the foregoing, some guidelines become apparent for constructing a gauge that would utilize thermal conductivity at low pressure as a measure of that pressure. To minimize the effects of radiative heat loss and conduction through the leads, the material chosen for the heated element should have low emissivity and should be operated at as low a temperature as practical. An accommodation coefficient near unity is desirable but normally one takes what one gets since other materials considerations dominate the choices, such as resistance to oxidation.

A thermal conductivity gauge is constructed using an element that is heated by an electric current. The heated element is contained within a vacuum tight housing with a port to connect it to the system where the pressure is to be monitored as in Fig. 10. Pirani[21] pointed out that such a gauge could be operated in three modes.

(1) One can apply a constant voltage across the heated element and observe the current as a function of pressure.

(2) One can maintain a constant current through the heated element and observe the resistance, as the ratio of the impressed voltage to current as a function of pressure.

(3) One can maintain constant resistance of the heated element and observe the required energy input as a function of pressure.

The choice of these depends on what configuration the gauge takes. Two basic types have evolved. These are the Pirani and thermistor gauges and the thermocouple gauge. In all the cases, the temperature of the surrounding housing, which provides the cold surface, plays an important role. Hale[22] constructed a second gauge as identical as possible to the

[21] M. Pirani, *Ver. Dtsch. Phys. Ges.* **8,** 686 (1906).

[22] C. F. Hale, *Trans. Am. Electrochem. Soc.* **20,** 243 (1911).

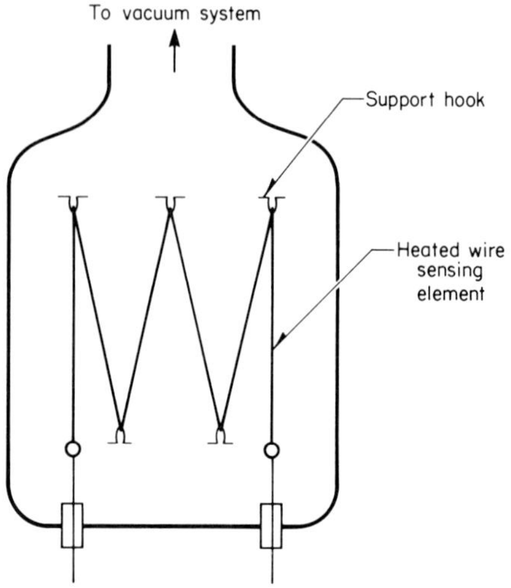

FIG. 10. The thermal conductivity gauge proposed by Pirani utilizes a long length of platinum wire as the heated element.

gauge to be used for pressure measurement and evacuated and sealed it for use as a compensation element in a Wheatstone bridge. Even with this compensation element, the gauge was not insensitive to variations in ambient temperature. If both the gauge and the compensation element were immersed in a constant temperature bath of 0°C the lower limit of sensitivity was about 10^{-5} Torr. Without temperature baths, the gauge will drift when first turned on until it reaches a thermal steady state with its surroundings. Care must be used to avoid locating the gauge where drafts can cause the housing to change temperature. Without constant temperature bath, commercially available gauges are useful from 1×10^{-3} to 1 Torr.

2.3.2. Pirani Gauge

The particular embodiment of the thermal conductivity gauge known as a Pirani gauge consists of a heated wire as the sensing element. The first gauges made by Pirani[21] and Hale[22] utilized a platinum wire operating at about 100–125°C. Ellett and Zabel[23] used nickel because it was readily flattened to provide a greater surface area for a given cross section and thereby increased the sensitivity. They showed that the sensitivity is proportional to the square root of the area of the wire. The wire is made as long as possible within the housing. Hale used 450 mm of 0.028-mm diameter platinum wire.

[23] A. Ellett and R. M. Zabel, *Phys. Rev.* **37,** 1102 (1931); **37,** 1112 (1931).

Figure 11 shows circuit configurations for operating a Pirani gauge in the three modes described in Section 2.3.1. The first method, that of constant voltage and observation of the current versus pressure, was the method chosen by Pirani. The operation is very simple but the sensitivity actually observed by Pirani was not as great as with the other two methods, about 10^{-4} Torr at the best.

Hale introduced the concept of the compensation element as one leg of a Wheatstone bridge [see Fig. 11(b)] and operated the gauge in a constant

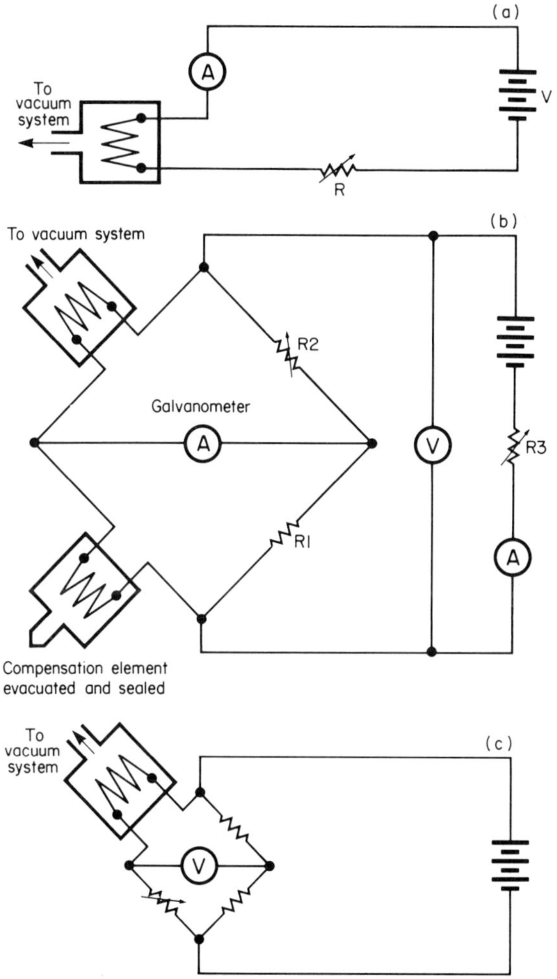

FIG. 11. Circuit configurations for operating a thermal conductivity gauge in constant voltage mode with current monitoring (a), constant current mode (b), or constant voltage with bridge unbalance monitoring (c).

current mode using the dual adjustments of the impressed voltage and the potentiometer to maintain the constant current. The change in resistance of the gauge wire resulting from the gas pressure is then measured from the resistance values. The more common method of operating the gauge was introduced by Campbell[24] who operated the gauge in a Wheatstone bridge but adjusted the impressed voltage to maintain the bridge in balance. In this way, the hot wire resistance and, hence, its temperature remained constant. This method has the advantage that the radiation and lead losses remain constant. Campbell found that if V_0 is the voltage impressed on the bridge at $P = 0$, and V is the observed voltage at pressure P, the gauge behavior follows

$$(V^2 - V_0^2)/V_0^2 = Kf(p), \tag{2.3.7}$$

where $Kf(p)$ is found to be closely proportional to P. Thus, the pressure is related to the square of the applied voltage. The constant K, varies with gas type. The gauge, operated in this way, was found to be linear with V^2 from 0 to about 0.15 Torr. Figure 12 shows typical operating curves for a gauge, operated as Campbell did, for various gases.

2.3.3. Thermistor Gauge

A thermistor is a semiconductor device which acts as a resistive element with a high, often negative temperature coefficient of resistivity. This replaces the hot wire of the conventional Pirani gauge. A compensation thermistor is commonly used in the opposite leg of the Wheatstone bridge but at atmosphere as a measure of T_0, the temperature of the gauge case. Otherwise, the gauge is operated either in the constant voltage mode or the constant temperature mode. If measurement no lower than 10^{-3} Torr is required, the constant voltage mode is acceptable with the bridge unbalance voltage being used as a measure of pressure. Figure 13 shows typical curves for this mode of operation. The bridge output is not linear with pressure but over a wide range, the pressure–voltage curve has reasonably constant curvature. Figure 14 shows a simple circuit that converts the signal to one that is linear with pressure to within a few percent. A McLeod gauge can be used to establish known pressures for setting the "gain" potentiometer. The gauge can be made linear within 10% over the range of 0–1 Torr (1000 mTorr or microns). A low pressure is established so that the "zero" potentiometer can be set for zero reading and then a pressure of a hundred milliTorr is set and the "gain" adjusted for the correct pressure readout. For example, the gain can be set so that milliTorr reads directly as tens of millivolts on a 10 V meter. Full scale of 10 V would correspond to 1000 mTorr (or microns).

[24] N. R. Campbell, *Proc. Phys. Soc. London* **33,** 287 (1921).

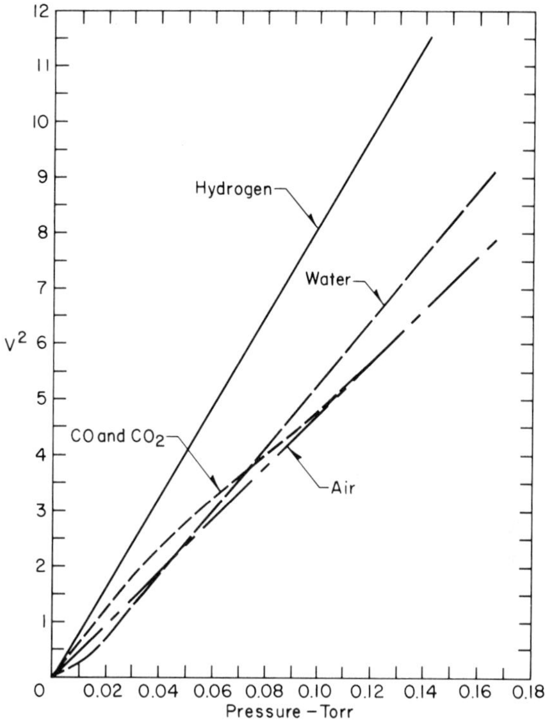

FIG. 12. Bridge output vs pressure for various gases utilizing a Pirani gauge operated at constant temperature.

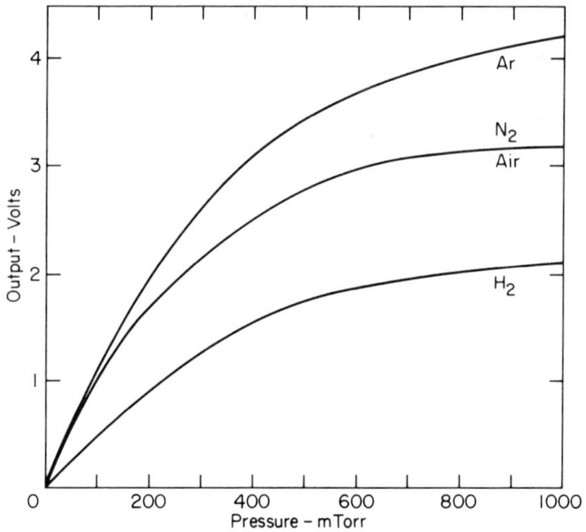

FIG. 13. Thermistor gauge output vs pressure for various gases operated with a circuit shown in Fig. 14.

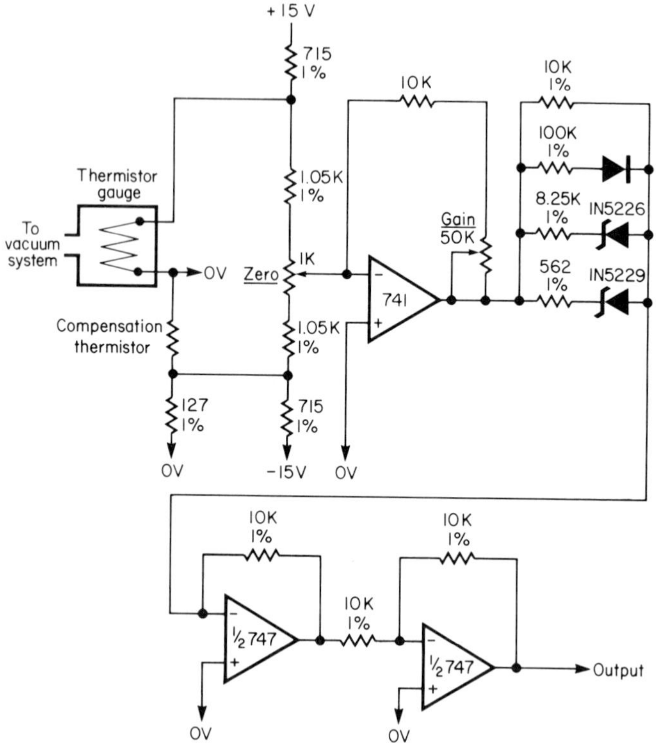

FIG. 14. An electronic circuit for use with a thermistor thermal conductivity gauge. The zener diodes produce a more linear output with pressure.

2.3.4. Thermocouple Gauge

Another variation on the principle of the thermal conductivity gauge is one that utilizes a thermocouple to measure the temperature of the heated element. The earliest use of this type of gauge was by Voege[25] who physically attached a fine wire thermocouple to a wire heated by a constant current. The thermocouple output was then used as a measure of the gas pressure. Most modern thermocouple gauges are still made in this fashion and operated in the constant current or constant voltage modes. Webber and Lane[26] designed a simple gauge consisting of one chromel and one alumel wire attached at their centers. A constant current source is applied between one chromel and one alumel lead and the thermocouple output is read on the other two leads.

[25] W. Voege, *Phys. Z.* **7**, 498 (1906).
[26] R. T. Webber and C. T. Lane, *Rev. Sci. Instr.* **17**, 308 (1946).

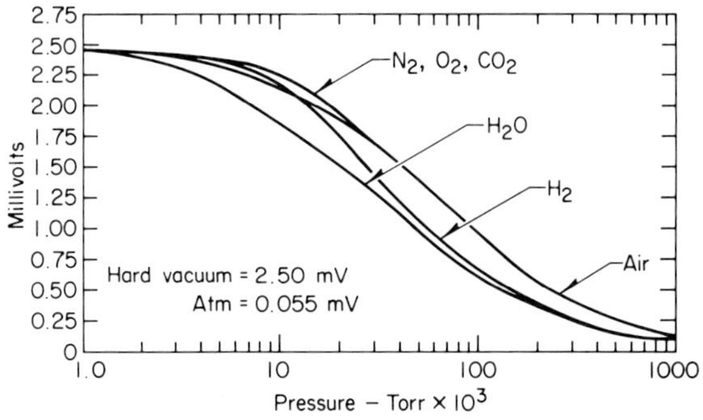

FIG. 15. Thermocouple gauge millivolt output vs pressure for various gases. Data provided courtesy of Teledyne Hastings–Raydist, Hampton, Virginia.

Commonly available thermocouple gauges have a sensitivity to about 10^{-3} Torr although Weisz and Westmeyer[27] were able, with extreme care and the use of a constant temperature bath, to extend the sensitivity to about 10^{-5} Torr. Figure 15 shows typical pressure versus thermocouple millivolt output curves for various gases.

2.3.5. Sensitivities of Thermal Conductivities for Various Gases

The thermal conduction of gas, as shown in Eq. (2.3.6), is inversely proportional to the square root of the molecular weight of the gas species conducting the heat. Unfortunately, life is rarely this simple. The matter is further complicated by the fact that the accommodation coefficient varies with gas type. Thus the gauge behavior as a function of gas type is difficult to predict. If a gauge is constructed, it must be calibrated for the various gas types for which it will be used. Commercial gauges read as though the gas were air. The pressure is then given in *air milli-Torr equivalent*. Unless the gas type is known and the conversion factors known, the actual pressure is *not* known. In the case of hydrogen or helium, the error would be substantial although these gases are rarely encountered without being aware of their introduction to a vacuum system. Table I lists relative sensitivity for various gases for a Bendix thermistor gauge, for a platinum wire Pirani gauge as built by Hale,[22] and for a Hastings Raydist thermocouple gauge. These must not be construed as being generally applicable but given here for illustration of the principles being discussed. The values are referenced to air with a relative sensitivity of 1.0.

[27] C. Weisz and H. Westmeyer, *Z. Instr.* **60,** 53 (1940).

TABLE I. Relative Sensitivities of Thermal Conductivity Gauges for Various Gases

Gas	Pirani	Thermistor	Thermocouple
Air	1.0	1.0	1.0
N_2	1.0	1.0	1.0
O_2	1.0	1.0	0.9
Ar	0.8	0.6	0.7
H_2	1.2	1.3	1.8
He	—	1.2	0.9
CO_2	1.0	1.0	1.0

2.4. Ionization Gauges

2.4.1. Theory and Principles

When electrons are accelerated through a potential difference V and their resultant kinetic energy, eV, is greater than some minimum value, collision of the electrons with gas molecules can result in the creation of positive ions. The minimum energy required is the binding energy of the outer electrons in the gas molecule. This energy, called the ionization potential, varies from 3.87 eV for cesium to 24.46 eV for helium. This leads us quickly to the conceptual design of a method for determining low pressure in a vacuum vessel. Consider a three electrode device where one electrode is a thermionic emitter of electrons, a second electrode is biased at a voltage positive with respect to the electron emitter and whose voltage is substantially higher than the ionization potential for the system gases, and a third electrode biased negative with respect to the thermionic emitter so that electrons cannot energetically reach it. The electrons emitted from the first electrode are accelerated to the second electrode and, in the process, collide with gas molecules creating ions from some fraction of the collision encounters. Some fraction of the ions thus created are collected on the third electrode and this ion current can be measured. The fraction of the ions collected on the third electrode is dependent on the electrode geometry. From Eq. (2.2.11) the pressure and the molecular density of the gas are related and if σ is the probability a gas molecule will be ionized in a collison by an electron. The ion current is related to pressure P by

$$i_+ = \sigma i_- L P, \tag{2.4.1}$$

where i_- is the electron current and L is the mean path length per electron. The parameter σ is not truly an ionization across section because of the spread of electron energies existing in a practical triode system and the collection efficiency is often not unity. For a given geometry, σ does represent the probability that any given electron will create an ion that will be collected on the third electrode. Some ions might be collected on the thermionic emitter if the geometry permits it. For a given geometry and given gas type (σ varies with gas species) σL is a constant and

$$i_+ = S\, i_-\, P \qquad (2.4.2)$$

or

$$S = i_+/i_-\, P. \qquad (2.4.3)$$

The proportionality constant is called the *gauge sensitivity* and has the units of Torr^{-1}. Sometimes i_- will be included in the definition of S in which case $S = i_+/P$ and has the units of amperes per Torr.

From Eq. (2.4.1), the gauge would be expected to give an ion current output that is linear with pressure. Over a wide range this is true but at high and very low pressure deviations from this behavior occur. Equation (2.4.1) holds so long as the mean free path for the electrons is long compared to the dimensions of the tube. When the pressure rises to where the mean free path becomes comparable to the path from emitter to anode, nonionizing inelastic collisions reduce the electron energy, and the gauge sensitivity suffers. Further, the free electron created in the ionizing collision is also collected at the anode and is indistinguishable from the emitted current. If the anode current is held constant, the apparent gauge sensitivity suffers. For a gauge with a sensitivity of 20 Torr^{-1}, this second electron source would equal the emitted current at 5×10^{-2} Torr. The apparent sensitivity would be decreased by half at 2.5×10^{-2} Torr. In 1947, Nottingham[28] showed that at the anode the collection of electrons at about 150 V energy results in the creation of a flux of soft X rays. The X rays arriving at the negative ion collector cause photoemission of electrons. This electron current leaving the ion collector is indistinguishable from ions arriving so a low pressure limit exists below which the gauge will not register a lower output. This limit is referred to as the *X-ray limit* of the gauge.

In our discussion we have used a thermionic electron emitter as the source of electrons. Different types of ionization gauges are characterized by the source of ionizing current and the geometric configuration of the gauge electrodes. The major variations are discussed in the following sections but it must be clearly stated that the most commonly used ionization gauge is the Bayard–Alpert gauge introduced in 1950.[29]

[28] W. B. Nottingham, *Ann. Conf. Phys. Elect.* **6,** (1947).

[29] R. T. Bayard and D. Alpert, *Rev. Sci. Instr.* **21,** 571 (1950).

2.4.2. Conventional Triode Hot Filament Gauges

The earliest ionization gauges were adaptations of commercial radio triode tubes. Buckley[30] reported a gauge in which the ion collector was the control grid interposed between the emitter and anode. In this way the negative grid potential served the dual purpose of ion collector and electron current control. Figure 16 shows this configuration. Despite the experience of Buckley, who obtained a sensitivity of 11 $Torr^{-1}$, the commercial gauges subsequently introduced operated with the electron collector as a grid interposed between the emitter and the ion collector. The rationale was increased sensitivity. Barkhausen and Kurz,[31] however, showed that high frequency oscillation of electrons through the grid could be maintained. When the grid is operated at a positive potential with respect to the emitter and the tube plate is negative with respect to the emitter, electrons can oscillate through the grid delivering high frequency energy into the external circuit. Electrons oscillating out of phase with the main space charge may derive sufficient energy to reach the negative ion collector causing highly erratic readings. These *Barkhausen oscillations*, as they came to be called, have caused many researchers considerable grief. Later developments, such as reported by Morse and Bowie,[32] showed that the geometry and operating parameters could be chosen to inhibit these oscillations.

The prototype of the commercial, conventional triode gauges was reported by Found and Dushman.[33] They determined that the linearity of ion and electron current predicted by Eq. (2.4.1) was true for a factor of more than 100 in the electron current. Figure 17 shows the effect of anode voltage on the collected ion current. The threshold of about 16 V for the argon used agrees with the published value of 15.68 for the ionization potential for argon. The ion current becomes essentially independent of the anode voltage above 150 V. The useful range of this gauge is from 10^{-3} Torr to the X-ray limit of 10^{-8} Torr. Schultz and Phelps[34] introduced a planar geometry gauge designed to perform at high pressure where ion collection efficiency results in falloff in gauge sensitivity. Figure 18 shows the configuration of this gauge. It consists of two parallel plates, one is the anode and one the ion collector, and a filament stretched midway between them. Because the gauge is meant to operate at high pressure, the filament is an iridium wire coated with thorium oxide to pro-

[30] O. E. Buckley, *Proc. Natl. Acad. Sci. U.S.A.* **2,** 263 (1916).
[31] H. Barkhausen and K. Kurz, *Phys. Z.* **21,** 1 (1920).
[32] R. S. Morse and R. M. Bowie, *Rev. Sci. Instr.* **11,** 91 (1940).
[33] C. G. Found and S. Dushman, *Phys. Rev.* **23,** 734 (1924).
[34] G. J. Schultz and A. V. Phelps, *Rev. Sci. Instr.* **28,** 1051 (1957).

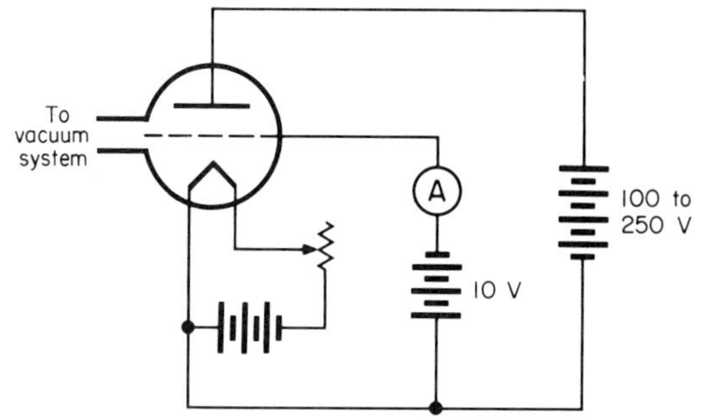

FIG. 16. The early ionization gauge of Buckley utilized a triode radio tube with the control grid also serving as the ion collector.

vide a low work function. Iridium is very oxidation resistant and so tolerates exposure to air at high pressure while hot, at least for short periods. The gap between the plates is only about 3 mm and the electrons pass directly from the filament to the anode plate across the 1.5 mm gap. The ion collector plate is as large as the anode and, therefore, efficient ion collection is assured. The very short electron path length results in a low gauge sensitivity and the useful range does not extend below 10^{-5} Torr but it extends as high as 1 Torr. The usable upper range is determined by the pressure at which even a 1.5 mm electron path becomes comparable to the mean free path and where the electrons to the anode are significantly originating within the ionization process. Table II is taken from the paper by Schulz and Phelps and gives the sensitivity for several gases as well as the pressure at which the gauge deviates 10% from linearity.

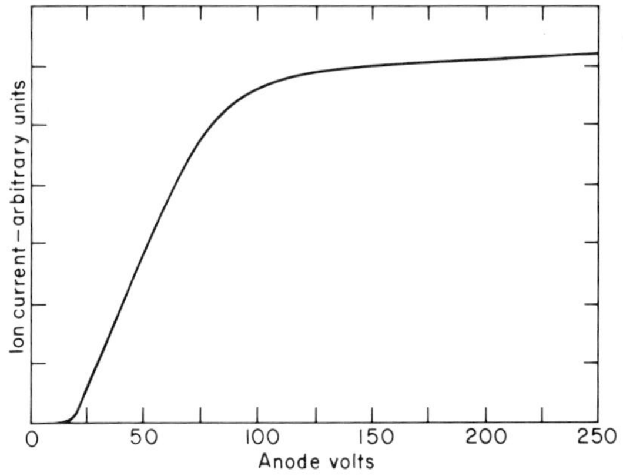

FIG. 17. Ion current vs anode (electron collector) voltage in a gauge where the anode is interposed between the electron emitter and ion collector.

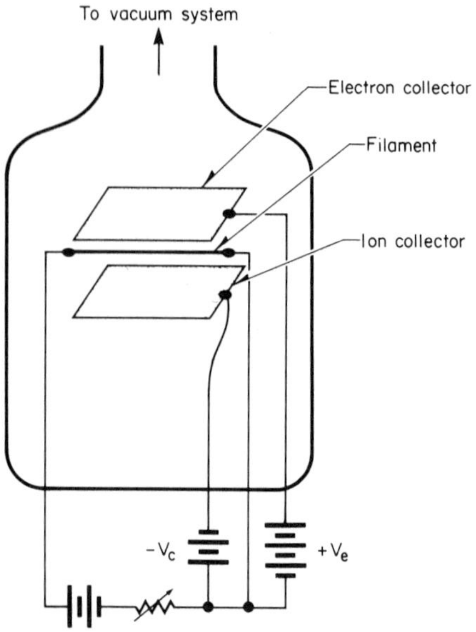

FIG. 18. A Schultz–Phelps gauge for use as high as 1 Torr pressure.

TABLE II. Sensitivity and Upper Limit to Linearity of a Schultz–Phelps Gauge

	Gas			
	N_2	He	H_2	CO
S, $Torr^{-1}$	0.6	0.09	0.43	0.63
P, Torr	0.6	6.0	0.6	0.6

2.4.3. Cold Cathode Discharge Gauge

An ionization gauge with no hot filament was reported in 1937 by Penning.[35] The mechanically simple gauge, shown diagramatically in Fig. 19 consists of a wire ring or short length of tubing centrally placed between two parallel plates with an equal space between the ring and each plate. A magnetic field is introduced perpendicular to the plates and ring. The two end plates are normally connected to ground through a microammeter, the high voltage is 1000–2000 V, and the magnetic induction, supplied by a permanent magnet, is a few hundred gauss. Penning operated his gauge at 2000 V and 370 G. When high voltage is applied to the anode, field emission, cosmic ray, or other energetic particle causes the dis-

[35] F. M. Penning, *Physica* **4**, 71 (1937).

charge to strike. For each ionizing collision, the ion travels to one of the end plates where it is registered by the meter, and the electron travels a very long oscillatory path, trapped in the magnetic field until it finally reaches the anode. Equation (2.4.1) still applies but here the electron current and ion current are equal in value in the steady state but the path length per electron is very long resulting in an adequately high sensitivity. Sensitivity here is more usefully given in amperes of ion current per Torr. Penning reports a sensitivity for air of about 1.0 ampere per Torr at 10^{-5} Torr decreasing to 0.7 A/Torr at 10^{-3} Torr. When the anode is a simple ring of wire, the gauge appears to have several discharge modes. At constant pressure, the pressure indication can fluctuate by as much as 10–20%. If the ring is replaced with a length of tubing, the gauge operation becomes much more stable. The lower usable limit to gauge operation is one of design. As Penning reported the gauge, the lower limit was about 10^{-5} Torr because of difficulty in maintaining the discharge. The same gauge, in a multiple Penning cell form, is usable to 10^{-11} Torr or lower as an ion pump if a stronger magnetic field is applied. The major disadvantage to this device is that its pumping speed is as much as 100 times that of a hot filament ionization gauge. It has found more application as a pump than as a gauge although in the range of 10^{-2}–10^{-5} Torr the gauge is used in industrial situations because of its simplicity and rugged construction.

The simple Penning gauge, often called the *Philips ionization gauge*, is limited at a low pressure limit of 10^{-5} Torr generally because of the pres-

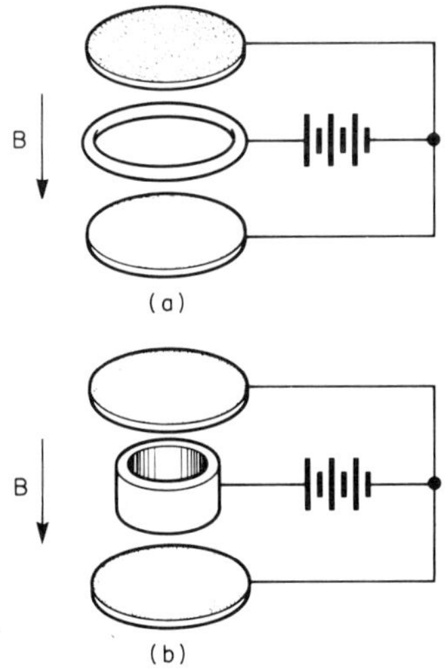

FIG. 19. Penning cold cathode discharge gauge using a ring anode (a) or tubular anode (b).

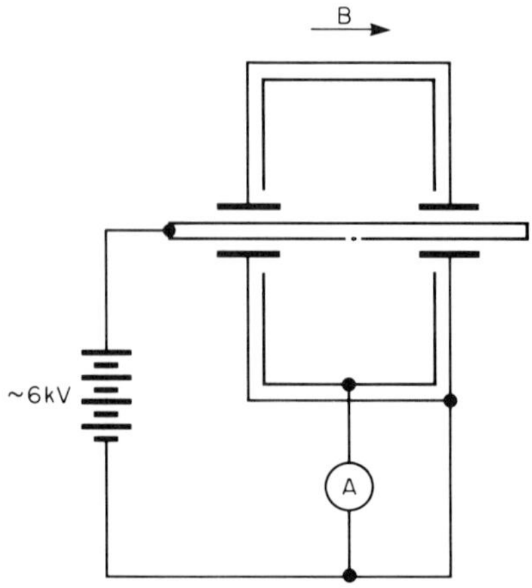

FIG. 20. The inverted magnetron gauge of Hobson and Redhead.

ence of field emission current which is indistinguishable from an ion current. Hobson and Redhead[36] developed a gauge that was designed to reduce the spurious currents to a very low value. Their gauge, shown in cross section in Fig. 20, utilizes crossed electric and magnetic fields and is referred to as an inverted magnetron gauge. Inverted, because a conventional magnetron has the cathode in the center with a surrounding anode. This geometry provides very efficient trapping for the electrons. The auxiliary cathode provides shielding for the ion collector and the stub cylinders that penetrate through the ion collector prevent field emission from the ion collector to the anode. Hobson and Redhead operated their gauge with a magnetic induction of 2000 G and an anode potential of 6000 V. They found that the ion current is a nonlinear function of the pressure having the form

$$i_+ = CP^n, \tag{2.4.4}$$

where C is a constant dependent on the anode voltage and n varies between 1.1 and 1.4 for various gauges. The parameter n, according to Hobson and Redhead, tends toward unity as the magnetic field is increased. Young and Hession[37] reported a Penning-type gauge operating at 2000 V and 1000 G which extended the useful range of this gauge to 10^{-12} Torr. In measurements with this type gauge, Lange[38] found discontinuities in the pressure–current relationship which result in a pressure dependent sensitivity.

[36] J. P. Hobson and P. A. Redhead, *Can. J. Phys.* **36,** 271 (1958).
[37] J. R. Young and F. P. Hession, *Trans. Natl. Vac. Symp.* **10,** 234 (1963).
[38] W. J. Lange, *J. Vac. Sci. Technol.* **3,** 338 (1966).

2.4.4. Bayard–Alpert Ionization Gauge

As described in Section 2.4.2, the triode ionization gauge was limited at low pressure to 10^{-8} Torr because of the X radiation resulting from electron bombardment of the anode. In 1950, Bayard and Alpert[29] reported a new configuration for the ionization gauge that substantially reduced the "X-ray current." They reasoned that the large area ion collector provided a high efficiency collector for the X rays. They could not eliminate the X rays but by turning the gauge inside out—filament on the outside of the grid and a fine wire ion collector in the center—the collection efficiency for the X rays could be made very small. Figure 21 diagramatically illustrates this configuration. When the ion collector is made from about 0.1 mm wire, the X-ray limit is reduced by almost three orders of magnitude to 2×10^{-11} Torr. The upper limit of this gauge's range results from an effect reported by Robinson.[39] Robinson investigated the nonlinear behavior of the Bayard–Alpert gauge at high pressure and concluded that at high pressure an ion space charge can develop around the collector wire thereby reducing the ion collection efficiency. The pressure at which this becomes important depends on the ion creation rate and, hence, the electron emission current. Nottingham and Torney[40] found that at a pressure of 2.4×10^{-3} Torr of nitrogen, the gauge could be in error by a factor of 6 at 1 mA emission, a factor of 2 at 100 μA, and read correctly at 10 μA or less. Many instances are reported in the literature of deviation from the linearity of ion current with pressure predicted by Eq. (2.4.1). A sampling of these are found in Refs. 39–50. The principle of operation of the Bayard–Alpert gauge is simple but the ratio of ion current to electron current is dependent on at least six variables. These are gas density (pressure), gas composition, electron path length, electron energy, ion collection efficiency, and electron collection efficiency. Unless all of these variables are controlled, any one of them can change and

[39] W. Robinson, *Vacuum* **6,** 21 (1956).
[40] W. B. Nottingham and F. L. Torney, *Trans. Natl. Vac. Symp.* **7,** 117 (1960).
[41] Y. Mizushima and Z. Oda, *Rev. Sci. Instr.* **30,** 1037 (1959).
[42] G. Schultz, *J. Appl. Phys.* **28,** 1149 (1957).
[43] R. E. Schlier, *J. Appl. Phys.* **29,** 1162 (1958).
[44] S. Aisenberg, *Ann. Conf. Phys. Elect.* **15** (1955).
[45] L. Riddiford, *J. Sci. Instr.* **28,** 375 (1951).
[46] G. Barnes, *Rev. Sci. Instr.* **31,** 1121 (1960).
[47] F. Baker, *Rev. Sci. Instr.* **31,** 911 (1960).
[48] H. Schwarz, *Z. Phys.* **117,** 23 (1921).
[49] D. Alpert, "Handbuch der Physik" (S. Flugge, ed.), p. 609. Springer-Verlag, Berlin and New York, 1958.
[50] P. A. Redhead, *Trans. Natl. Vac. Symp.* **7,** 108 (1960).

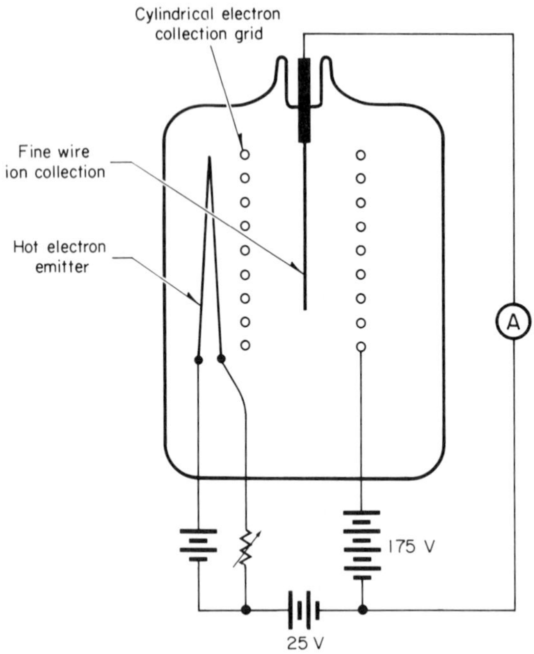

FIG. 21. The inverted triode hot filament gauge introduced by Bayard and Alpert.

cause an apparent gauge nonlinearity. Winters, Denison, and Bills[51] analysed the reported nonlinearities and traced most of them to lack of control of the gauge parameters. Some effects, such as reported by Robinson[39] and Nottingham[40], are fundamental to the nature of the gauge operation and are avoidable only by recognizing the limitations on the usable range or operating conditions.

In 1962, Ackley, Lothrup, and Wheeler[52] reported an anomalous behavior of the Bayard–Alpert gauge when operated at pressures below 10^{-7} Torr with gases containing oxygen. If the emission is decreased from the normal 4–10 mA current by 1 or 2 orders of magnitude, the ion current initially decreases by the same factor and then slowly increases by a factor of 10–100. If the emission is then increased again, the ion current is initially too high but decays to the original value. Outgassing the gauge causes the anomaly to temporarily disappear. The cause of this strange behavior is the adsorption of gas on the grid structure and subsequent desorption of the oxygen as ions by electron impact. At the higher

[51] H. F. Winters, D. R. Denison, and D. G. Bills, *Rev. Sci. Instr.* **33**, 520 (1962).

[52] J. W. Ackley, C. F. Lothrup, and W. R. Wheeler, *Trans. Natl. Vac. Symp.* **9**, 452 (1962).

4–10 mA current, the grid remained relatively gas free. Also the grid is warmer at the higher emission resulting in low gas coverage. As the emission is lowered, the grid cools, and the gas coverage increases. As it increases, the rate of oxygen ion desorption increases, and the anomalous current appears. The admonition is that the gauge should be operated at higher emission levels at very low pressure to avoid this problem.

Singleton[53] reported a similar effect when measuring the pressure of hydrogen gas. He attributed the anomalous pressure reading to an enhanced electron desorption of CO which occurs in the presence of hydrogen. Hickmott[54] found that atomic hydrogen formed at a hot filament can react with glass to produce CO, H_2O, and CH_4.

Also in 1962, Denison[55] reported that alkali metals desorbed from the hot filament can deposit on the grid and subsequently be electron impact desorbed as ions. The magnitude of the current observed from this cause is of the same order of magnitude as the X-ray current. The cause of the lower usable limit of the gauge may be more complex than just X rays.

For gases such as nitrogen, argon, and helium the gauge has been demonstrated to be linear as predicted by Eq. (2.4.1). Morrison[56] demonstrated the linearity with helium from 5×10^{-10} to 5×10^{-4} Torr within $\pm 2\%$.

Because of the creation of ionic species and the pyrolytic effects that can take place on hot filaments, it comes as no surprise to discover that the ionization gauge, in addition to measuring pressure, is also a pump. Bills and Carleton[57] studied the pumping effect on nitrogen and found that a clean, well degassed gauge operating at 10 mA emission pumped nitrogen at 0.5l/sec. Brennan and Fletcher[58] and Hickmott[59] observed that the pumping of hydrogen is independent of the emission current but depends directly on the filament temperature. The hydrogen is pumped by pyrolytic dissociation on the filament surface with the resulting highly reactive atomic hydrogen reacting with oxides on the system walls. Winters *et al.*[51] investigated the hydrogen pumping in two gauges; one with a tungsten filament and the other with a thoria coated iridium filament. As was predicted by the earlier work, the hydrogen pumping at 10 μA emission from the tungsten filament was the same as 10 mA emission from the thoria coated iridium filament since the two filaments in

[53] J. H. Singleton, *J. Vac. Sci. Technol.* **4,** 103 (1967).
[54] T. W. Hickmott, *J. Appl. Phys.* **31,** 128 (1960).
[55] D. R. Denison, *Trans. Natl. Vac. Symp.* **9,** 218 (1962).
[56] C. F. Morrison, *J. Vac. Sci. Technol.* **6,** 69 (1969).
[57] D. G. Bills and N. P. Carleton, *J. Appl. Phys.* **29,** 692 (1958).
[58] D. Brennan and P. Fletcher, *Proc. Roy. Soc.* (*London*) **A250,** 389 (1959).
[59] T. W. Hickmott, *J. Chem. Phys.* **32,** 810 (1960).

these conditions have essentially the same temperature. The pumping speed was measured at 0.035 l/sec although Bills[57] found as high as 5 l/sec for hydrogen in a gauge with a tungsten filament operating at 10 mA emission (1700°C filament temperature). In a gauge tube with a tungsten filament, oxygen is pumped by reaction with the tungsten surface and the subsequent evaporation of tungsten oxide. Winters *et al.* report that no pumping of oxygen was observed using a thoria coated iridium filament. CO and CO_2 can be produced by reaction of oxygen or water vapor with a filament containing carbon impurity.[60] Such carbon can be present either from impurity in the filament material or by contamination of the filament by pyrolitic decomposition of hydrocarbons which may be present in diffusion pumped vacuum systems. In general, pumping effects can be minimized by operating an ionization gauge at low emission levels to keep the filament temperature low and use of low work function filaments for even lower temperatures. In addition to thoria coated iridium, lanthanum hexaboride coated rhenium filaments have proved very effective.[61]

Knowledge of the pumping effects of the gauge lead one to the realization that proper use of the gauge requires that the conductance between the gauge and the system being monitored be as large as possible. Unless the gauge communicates through a high conductance, a clean gauge can register a pressure substantially lower than the system pressure or a "dirty" gauge with high outgassing can register a pressure substantially higher than the system pressure. To reduce this effect as much as possible, an ionization gauge structure can be mounted on a flange and inserted directly into the system volume. This is referred to as a "nude" gauge as opposed to a tubulated or appendage gauge.

2.4.5. Special Gauges for Ultravacuum and Extreme High Vacuum

The measurement of pressure in the 10^{-11} Torr range and lower presents special problems because of limiting effects such as the X-ray limit. One could even argue that total pressure measurement in these very low regions is of no particular value. However, undaunted, minds have turned to the solution of this problem.

2.4.5.1. Modulated Bayard–Alpert Gauge. The lower limit for pressure measurement with a Bayard–Alpert gauge results from the presence of an effective ion current to the collector which is independent of pressure. Redhead[62] reasoned that if a second "collector" were placed in the gauge and that it could be switched from grid potential to ion collector po-

[60] J. R. Young, *J. Appl. Phys.* **30,** 1671 (1959).
[61] J. M. Lafferty, *J. Appl. Phys.* **22,** 299 (1951).
[62] P. A. Redhead, *Rev. Sci. Instr.* **31,** 343 (1960).

tential, one could deduce what the real ion current is. Figure 22 illustrates the construction of this "modulation" gauge. If we let i_1 be the apparent ion current from the gauge when the second "collector" is at grid potential, it can be expressed as the sum of the true ion current i_1 and the X-ray current i_0:

$$i_1 = i_+ + i_0. \tag{2.4.5}$$

When the second "collector" or modulator is set at the same potential as the central gauge collector, the true ion current is divided between the two collectors, but the X-ray current is unchanged by this switching of potential. Thus, the new apparent collector current i_2 is

$$i_2 = \alpha i_+ + i_0, \tag{2.4.6}$$

where α is the fraction of the true ion current collected when the "modulator" is collecting ions. The two values i_1 and i_2 are the observables and solving the two equations for i_+ gives

$$i_+ = (i_1 - i_2)/(1 - \alpha). \tag{2.4.7}$$

The parameter α can be determined by operating the gauge at a high enough pressure that i_0 is negligible. In this way, the "modulation

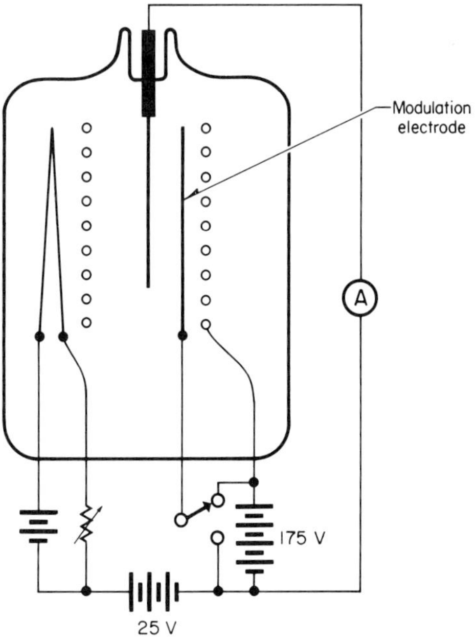

FIG. 22. The introduction of a "modulator" electrode in a Bayard–Alpert gauge permits the pressure range to be lowered by a decade to 10^{-12} Torr.

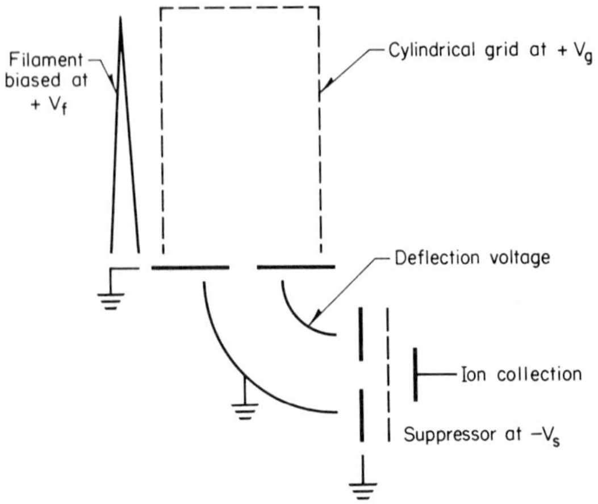

FIG. 23. The hidden collector gauge introduced by Helmer permits use to 10^{-13} Torr.

gauge'' can be used to extend the range of the Bayard–Alpert gauge by about one decade. When i_0 becomes large compared to i_+, the difference $i_1 - i_2$ becomes very small, and the measurement error limits the usefulness of the technique.

2.4.5.2. Helmer Gauge. If the collector is removed from a Bayard–Alpert gauge structure and an ion extraction potential is applied to one end of the ionization region, a gauge configured as in Fig. 23 can be obtained. Helmer and Hayward[63] added the deflection plates to completely isolate the ion collector from the source of X rays. The sensitivity is the same as a conventional Bayard–Alpert gauge but now the limit is only the practical ability to measure the current. The device presented by Helmer is usable to 10^{-13} Torr with a sensitivity of 17 Torr^{-1}.

This gauge has an additional benefit which results from the fact that the ion optics permit only gas phase ions to reach the collector. Ions desorbed from the electron collector, or grid, by electron impact have trajectories that preclude them from reaching the collector. This gauge, then, is relatively immune to the effect reported by Ackley, Lothrup, and Wheeler[52] and reported for hydrogen in the modulator gauge by Hobson and Earnshaw.[64]

2.4.5.3. Groszkowski Hidden Collector Gauge. In 1965, Groszkowski[65] reported the results of an investigation on the effect of collector position on Bayard–Alpert gauge sensitivity. He found that if end caps were added to the grid region and one cap had a hole as in Fig. 24, the col-

[63] J. G. Helmer and W. H. Hayward, *Rev. Sci. Instr.* **37**, 1952 (1966).

[64] J. D. Hobson and J. W. Earnshaw, *Proc. Int. Vac. Congr. 4th*, p. 619. Inst. Phys. and Phys. Soc., London, 1968.

[65] J. Groszkowski, *Bull. Acad. Pol. Ser. Sci. Technol.* **13**, 15 (1965).

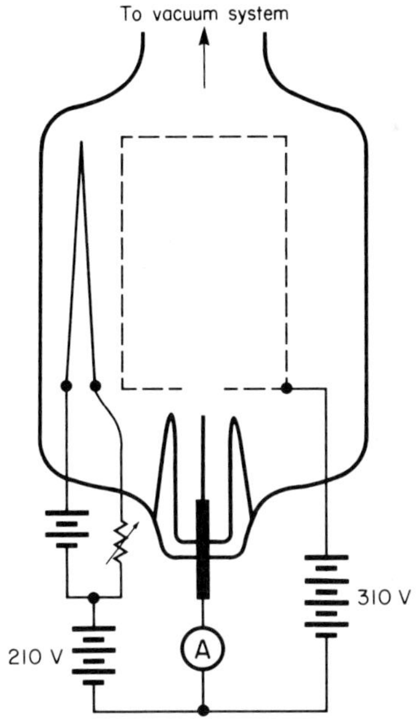

FIG. 24. A capped grid Bayard–Alpert gauge with "hidden" collector proposed by Groszkowski.

lector could be removed from the central region without reducing the sensitivity appreciably but reducing the X-ray effect by a factor of 600. The gauge sensitivity was shown to be constant to 10^{-12} Torr by Bernadet and Show.[66] The gauge, as constructed by Groszkowski, has a sensitivity of 50 $Torr^{-1}$ and is probably useful to 10^{-13} Torr. Like the Helmer gauge, the geometry is chosen so that electron desorption of ions from the grid structure does not affect the collected ion current.

2.4.6. Sensitivities of Ionization Gauges for Different Gases

The gauge sensitivity as defined by Eq. (2.4.2) must be linearly related, by Eq. (2.4.1), to the ionization probability σ and the mean path per electron L. The ionization probability depends on electron energy and gas type. The actual determination of sensitivity for particular gauges requires the accurate measurement of ion and electron currents at a *known* pressure. The problems associated with the establishment of these known calibration pressures are presented in Section 2.5. For the moment, assume this is possible. The sensitivity then can be calculated from the measured currents and pressure. Found and Dushman[33] mea-

[66] M. H. Bernadet and M. Show, *Le Vide* **134,** 80 (1968).

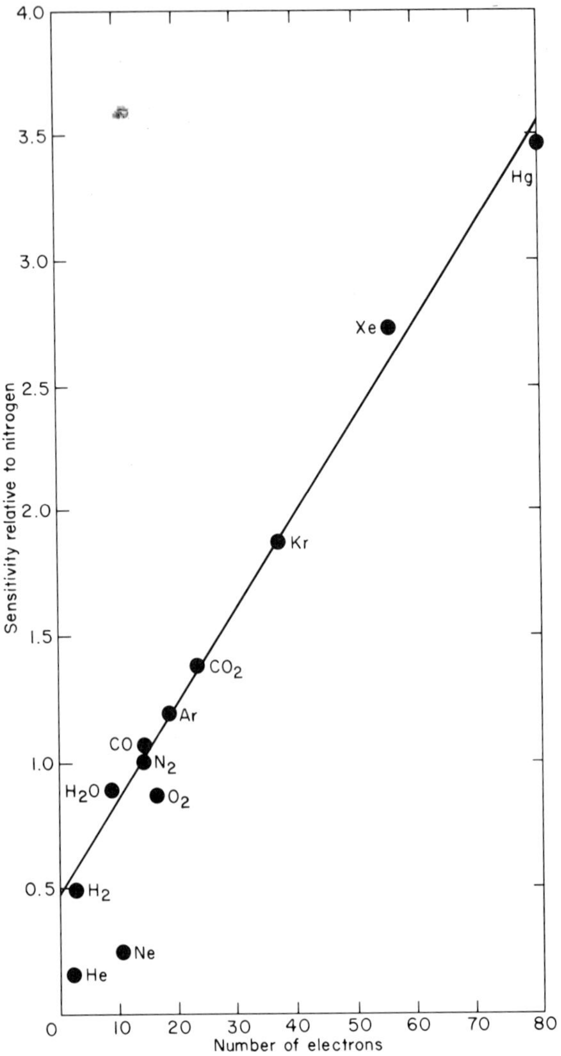

FIG. 25. Sensitivity of an ionization gauge relative to N_2 vs number of electrons in the atom or molecule shows the linearity seen by Found and Dushman.

sured the sensitivity of an ionization gauge of the type discussed in Section 2.4.2 utilizing a variety of gases. They determined that the empirical data suggested that the sensitivity was proportional to the number of electrons in the atom or molecule (see Fig. 25). No explanation has yet been offered for this phenomenon. Dushman and Young[67] repeated this work and arrived at the same conclusion. The agreement between this work and

[67] S. Dushman and A. H. Young, *Phys. Rev.* **68,** 278 (1945).

TABLE III. Sensitivity of Ionization Gauges for Various Gases Normalized to Nitrogen

Gas	N	S	S (calculated)
Ar	18	1.19	1.16
CO	14	1.07	1.01
CO_2	22	1.37	1.32
H_2	2	0.46	0.54
He	2	0.15	0.54
Hg	80	3.44	3.56
H_2O	8	0.89	0.77
Kr	36	1.86	1.86
N_2	14	1.00	1.01
Ne	10	0.24	0.85
O_2	16	0.84	1.08
Xe	54	2.73	2.55

that of Wagener and Johnson[68] for the commonly measured gases leads one to accept these as working sensitivities. Because of the variation in sensitivity between gauges from different manufacturers, Table III lists the sensitivity S normalized to that of nitrogen. All currently manufactured gauges are referenced to nitrogen and unless the output is adjusted to read directly for another gas, the pressure obtained is the *nitrogen pressure equivalent*. In other words, the gauge reads the pressure that *would* exist in the system if the same ionization current were being derived from pure nitrogen gas. For another gas, divide the reading by the ratio S of Table III. In Table III, the calculated value of the normalized sensitivity is given using the relationship $S = 0.038N + 0.465$. The agreement with this linear relationship, first seen by Found and Dushman, is clear. The two gases for which the relationship does not hold are helium and neon which give calculated values about three times the observed sensitivity. The higher error in the value for oxygen probably results from the difficulty in obtaining good data for oxygen. As was pointed out in Section 2.4.4, oxygen can react with the gauge filament impurities to form CO and CO_2. The difficulty in maintaining pure oxygen conditions within a gauge during measurement cannot be overemphasized.

2.5. Placement and Calibration of Gauges

2.5.1. Problems Associated with Gauge Placement

Almost every vacuum system of interest is a dynamic gas flow system and not a static system that permits pressure measurement in the usual

[68] S. Wagener and C. B. Johnson, *J. Sci. Instr.* **28**, 278 (1951).

Maxwellian sense. At best, a nude ionization gauge measures the gas molecular density in the vicinity of the gauge and tubulated gauges measure a "localized" pressure or the arrival rate of molecules at the entrance to the gauge structure.[69] In the case of thermal conductivity gauges operating at higher pressures, the gas flow may be in the viscous flow regime where measurement in a tube can result in errors arising from the Venturi effect. In all pressure ranges, the effects of gas beaming can result in measurement error. The proper placement of gauges is a major factor to be considered in the design of a vacuum system. The first step is to carefully consider what information is desired from the gauge and second is to ask if the gauge placement does, in fact, provide the required information. A few general rules are possible to guide in the design of gauging for a vacuum system and some examples are useful in illustrating the types of problems that can occur and be avoided.

(1) Even in large chambers, pressure gradients can occur in the vicinity of pumping ports. A gauge should not be located adjacent to such a port. Venema[70] showed that a chamber diameter at least three times the diameter of the pumping port would provide a volume that would exhibit a sufficiently random molecular motion to provide a Maxwellian pressure measurement. However, the gauge should be located as far from the pump port as is practical.

(2) In systems where gas flow characteristics are important, the gauge might be used as an arrival rate transducer, rather than a true "pressure" gauge, in the manner of Buhl and Trendelenburg.[69] It is important to recognize that the reading does not represent an equilibrium condition but a steady state arrival rate of molecules at a particular location on the system wall. This technique is very useful in determining flow characteristics at low pressure.

(3) Adequate baffling should be provided so that a gas inlet does not form a jet beaming gas into a gauge so that the effective pressure within the gauge is not representative of the system as a whole. This applies as well to the location of gauges where pump backstreaming can affect gauge operation. Blears[71] showed that hydrocarbons from a diffusion pump can result in erroneous pressure readings because an ionization gauge can "crack" one large molecule into many small ones and create a localized high pressure within the gauge. Backstreaming of oil into thermal conductivity gauges can change the surface accommodation coefficient and result in large errors.

[69] S. Buhl and E. A. Trendelenberg, *Vacuum* **15,** 231 (1965).

[70] A. Venema, *Vacuum* **4,** 272 (1954).

[71] J. Blears, *Nature* (*London*) **154,** 20 (1944).

(4) Care must be taken so that the gas to be measured is that with which the gauge interacts. The wall along the tubing connecting the gauge to the system can adsorb gas and pumping effects within the gauge can result in the gas composition within the gauge being greatly different than in the system and, as a result, a large error can result. For example well baked-out glass tubulation can adsorb large quantities of CO so that a few centimeters of 1 cm tubing can sustain a pressure difference of several decades pressure between gauge and system for several hours when CO is being used in the high vacuum region.

2.5.2. Gauge Calibration

2.5.2.1. Direct Comparison Methods. In the pressure range of 10^{-3} Torr to atmosphere, a gauge can be calibrated by direct comparison to a pressure measurement that results in an absolute pressure. The deflection of a Bourdon tube, for example, can be calibrated by direct comparison with a mercury U-tube gauge. A thermal conductivity gauge can be calibrated by direct comparison with a McLeod gauge for a wide range of gases. Even at low pressure, a gauge, once calibrated, may be used like a secondary standard and other gauges calibrated by direct comparison.

2.5.2.2. Pressure Extension Methods. For low pressure gauges, such as the ionization gauge, direct comparison techniques are no longer possible and other methods must be sought to establish a "known pressure" against which a gauge might be calibrated.

Two basically different techniques have been used to produce "known" pressures beginning with a pressure high enough to be measured by an absolute manometer or McLeod gauge. The first is a static technique which is basically a free gas expansion technique first proposed by Knudsen[72] in 1910. The second is a dynamic method first proposed in 1962 by Florescu[73] in which the pressure is divided across a series of orifices.

2.5.2.2.1. STATIC METHODS. Knudsen[72] proposed a method of generating pressures of known value by application of Eq. (2.2.7). By allowing isothermal expansion of a small volume of gas into a larger, evacuated volume, the new, lower pressure could be calculated. It is necessary that the larger volume be evacuated sufficiently well that the residual gas be a negligible contribution to the total gas after expansion. Multiple expansion cycles can generate a series of calibration points. If the initial volume V_0 contained gas at pressure P_0 after the nth expansion, the pressure is

[72] M. Knudsen, *Ann. Phys.* **31,** (1910).

[73] N. A. Florenscu, *Trans Natl. Vac. Symp.* **8,** 504 (1961).

$$P_n = P_0[V_0{}^n/(V_0 + V_1)^n], \tag{2.5.1}$$

where V_1 is the volume of the larger expansion volume. Two effects limit the lower range that this is useful. The volume V_1 must be evacuated each cycle and the residual pressure must be much less than P_n. This residual pressure is the first limitation and the second is the effect of adsorption or desorption of gas at the chamber walls. The quantity of gas adsorbed onto or desorbed off the walls during an expansion cycle must be small compared to the gas in V_1 after the expansion. Several variations on this method[74–81] have been proposed both to calibrate gauges and to determine unknown volumes. A range of general usefulness of atmosphere to about 10^{-4} Torr can be attained with this method although Meinke and Reich[82] reported using a method utilizing a four volume expansion method to obtain a pressure in the 10^{-7} Torr region with a total error of ±3%. This static, expansion method is particularly useful in the calibration of thermal conductivity gauges and other gauges whose useful range lies generally above 10^{-3} Torr. The technique is simple and apparatus easy to construct and use. An oil or mercury manometer may be used to calibrate the expansion volume by measuring the pressure before and after one expansion.

2.5.2.2.2. DYNAMIC METHODS. Figure 26 shows a pressure division scheme utilizing gas flow through a series of orifices separating vacuum chambers as proposed by Florescu. If the pump has sufficient speed to make P_3 negligible with respect to P_2, then

$$P_2 = P_1C_1/(C_1 + C_2). \tag{2.5.2}$$

By a proper selection of orifice sizes, the values P_2 of calibration interest will be generated by corresponding values P_1 in the higher pressure chamber that can be measured directly with a McLeod or other absolute gauge. This basic concept, although simple, is difficult to operate in practice. The orifice sizes are small enough that the chambers cannot be effectively pumped. Outgassing of the walls and pumping effects of the gauges being calibrated can contribute to substantial errors. Roehrig and Simons[83]

[74] A. Bobenrieth, *Le Vide* **1,** 62 (1964).
[75] G. C. Fryberg and J. H. Simons, *Rev. Sci. Instr.* **20.** 541 (1949).
[76] D. Alpert, *J. Appl. Phys.* **24,** 280 (1953).
[77] F. O. Smetana and C. T. Carley, Jr., *J. Vac. Sci. Technol.* **3,** 49 (1966).
[78] R. S. Barton and J. N. Chubb, *Vacuum* **15,** 113 (1965).
[79] C. Meinke and G. Reich, *Vak. Tech.* **11,** 86 (1962).
[80] C. B. Bichnell, *Trans. Natl. Vac. Symp.* **6,** 97 (1959).
[81] W. S. Kreisman, *Trans. Natl. Vac. Symp.* **7,** 75 (1960).
[82] C. Meinke and G. Reich, *J. Vac. Sci. Technol* **4,** 356 (1967).
[83] J. R. Roehrig and J. C. Simons, *Trans. Natl. Vac. Symp.* **8,** 511 (1961).

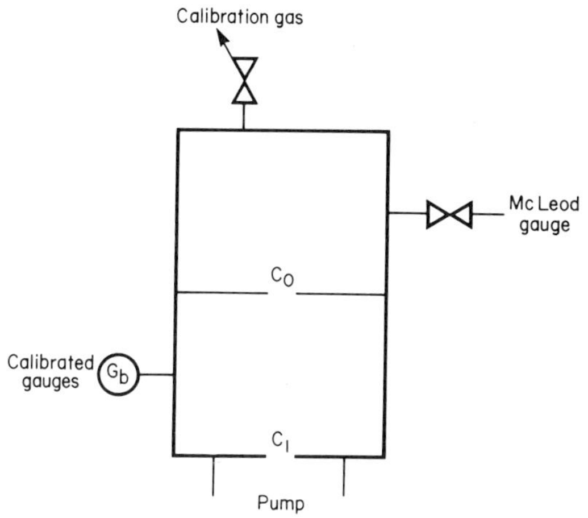

FIG. 26. A dynamic gauge calibration scheme as proposed by Florescu.

provided each stage with differential pumping so the gas at each stage was divided between its pump and the stage downstream. In this way, the orifice conductances could be enlarged and higher throughput of gas resulted. Gauge pumping and changing gas composition due to outgassing and desorption play a much smaller role in error production. Davis[84] extended the useful range of a pressure division apparatus by making one of the conductances stepwise adjustable. Morrison[85] introduced a new concept in the use of the pressure division technique. Figure 27 shows the apparatus diagrammatically. Two high conductance valves permit the transfer gauge G_t to be connected to either the upper or lower chambers. The gauges to be calibrated are G_b. The transfer gauge is used to stepwise increase the pressure at the calibrated gauges with the same pressures that exist in the higher pressure chamber in the previous step. To follow one cycle:

(a) open valve V_a and close valve V_b;

(b) set the gas flow so that the gauges G_b read about one decade higher pressure than the base pressure. This avoids problems with gas composition;

(c) read gauge G_t;

(d) close valve V_a and open valve V_b;

(e) increase the gas flow until G_t reads precisely the same value as when it was connected to chamber A. Record the reading on G_b;

[84] W. D. Davis, *Trans. Natl. Vac. Symp.* **9,** 253 (1962).
[85] C. F. Morrison, *J. Vac. Sci. Technol.* **4,** 246 (1966).

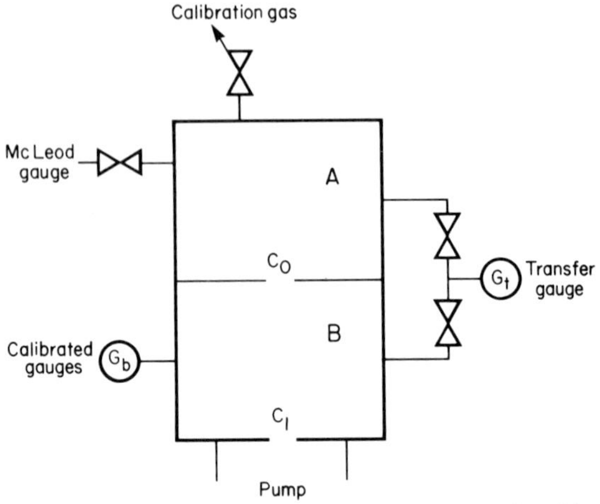

FIG. 27. The "transfer gauge" calibration apparatus introduced by Morrison.

(f) transfer G_t back to chamber A and read the pressure indication; and

(g) transfer G_t back to chamber B and repeat step (e), (f), ...

In this way a series of readings on G_b are accumulated all related to the previous reading by the fixed ratio U, where

$$U = (C_1 + C_2)/C_1. \tag{2.5.3}$$

When the upper chamber reaches a value that can be measured by an absolute method, all the other pressures may be computed from

$$P_n = P_r U^{n-r}. \tag{2.5.4}$$

The use of the "reference–transfer" method, as Morrison called it, creates a sequence of increasing pressures, $P_1, P_2, \ldots, P_m$, all related by the ratio U. The gauge G_t need not be a calibrated gauge because its only function is to establish the same pressure value in the two chambers. The only requirement for G_t is that it be repeatable and single valued.

The dynamic calibration methods may be used over a wide pressure range. Ionization gauges have been calibrated to 10^{-12} Torr using liquid helium pumping although most methods are usable to 10^{-9} Torr. The limit at low pressure results from the base pressure, being of unknown gas composition, causing apparent deviation from linear gauge behavior.

3. PARTIAL PRESSURE MEASUREMENT*

3.1. Mass Spectrometer†

3.1.1. Introduction to Partial Pressure Measurement

The use of mass spectrometers as partial pressure analyzers is crucial to our present-day technology. Without the knowledge of the density and identity of molecules in an evacuated vessel, it is difficult to comprehend the successful operation of the complex vacuum systems now in use. One of the first uses of the mass spectrometer in the attainment of ultrahigh vacuum conditions in sizable vacuum vessels was that of the helium leak detector. Because the concentration of helium in the earth's atmosphere is extremely low, about five parts per million, the use of helium as a probe gas with a mass spectrometer which responds only to helium provides a method for the detection of imperfections (leaks) in a vacuum vessel whose sensitivity exceeds that of other methods by orders of magnitude.

From the use of a mass spectrometer to locate imperfections (leaks) in the vacuum walls, it was an obvious step to employ a mass spectrometer to identify the species of the molecules which constitute the atmosphere of a vacuum system. This knowledge gives much insight as to the sources of these gases and of the steps which may be taken to greatly attenuate their density or to eliminate them from the system. One who has been accustomed to the luxury of having a partial pressure analyzer on each of his vacuum systems has said, "operating a vacuum system without a partial pressure analyzer is like trying to drive a car with all the windows frosted over."

Partial pressure analyzers are mass spectrometers. As such, they all consist of an ion source, a mass analyzer, and a detector–display device. Electrons from a thermally emissive filament are accelerated, focused, and driven through the ion source as a pencil or as a sheet. They are capable of ionizing molecules of all species. The probability that a given

† For general information on mass spectometry, see also Vol. 3 (second edition) Part 7, in this series, as well as Section 2.3.1 in Vol. 5.

* Part 3 is by W. M. Brubaker.

METHODS OF EXPERIMENTAL PHYSICS, VOL. 14

ISBN 0-12-475914-9

electron will produce an ion which subsequently can be caused to enter the mass analyzer is numerically nearly equal to the pressure in the ion source, in Torr.

In partial pressure analyzers, the pressure (density) of the gases in the ion source is essentially that of the pressure in the system. In some instances this may be extremely low. Under these conditions, the space charge of the ions is negligible and the space charge of the electrons dominates in the ion source. It is essential that conditions of gradients and potentials do not permit the electron space charge to form a potential minimum in which positive ions can be trapped. An analysis of the roles played by the space charge effects of electrons and ions in mass spectrometer ion sources is given by Brubaker.[1]

Ions formed by the ionizing electron beam are accelerated and focused into the mass analyzer. In all instances, the analyzer works best if the ion beam is homogenious in energy and in direction.

The different types of partial pressure analyzers are characterized by the operating principles of the mass analyzer systems. This topic is discussed in Section 3.1.3.1.

Detectors consist essentially of electronic current amplifiers which transform the very minute currents of the transmitted ion beams into signals which are capable of driving a display device. In order to gain sensitivity and speed of response, secondary emission multipliers are frequently used. This permits the continuous display of the spectrum on a cathode ray oscilloscope.

3.1.2. Requirements of Mass Spectrometers for Use as Partial Pressure Analyzers

3.1.2.1. Mass Range. In most cases, the composition of the residual gas in a vacuum vessel consists of gases with masses below 50 amu. It is normal to expect to find small amounts of air in the system at some times. It is a rare system that never has a leak. Only the noble gases krypton and xenon, and perhaps special local contaminents, contribute to the composition of the air above 50 amu and they are present at very low concentrations.

Information of the contribution of gases of molecular weight below 50 amu to the composition of the atmosphere in evacuated vessels is of prime importance. This mass range covers the usual gases found in the system. Additional information concerning portions of the mass spectrum above 50 amu may be of importance in very limited, special cases, but in general it is of little value.

[1] W. M. Brubaker, *J. Appl. Phys.* **26**, 1007 (1955).

3.1.2.2. Resolution. The mass resolving capability of partial pressure analyzers is modest by the standards of the more sophisticated instruments used for analytical purposes. As a partial pressure analyzer, a mass spectrometer must be able to clearly resolve ions of adjacent masses. That is, ions of a given mass should make a negligible contribution to the peak of an adjacent mass. This requirement means that the transmission of ions of a given peak must be attenuated by many orders of magnitude when the instrument is adjusted for the top of the peak which is 1 amu removed from the top of the given peak. This requirement of "unit" resolution is due to the complexity of the spectra of even simple molecules. For example, CO_2 makes a large contribution to mass 28 in addition to mass 44.

The above considerations imply that the expression of the resolution measured at the half peak height is a poor indicator of the adequacy of the mass resolving capabilities of the mass spectrometer. A much more meaningful measure of the adequacy of the resolving power is the contribution made by ions of a specific mass to the height of an adjacent peak because it is this number which gives an indication of the dynamic range of the pressures to which the instrument is responsive.

3.1.2.3. Sensitivity. The ability of a partial pressure analyzer to respond to and to identify gases which are present in extremely small concentrations is one of the most important aspects of the instrument. In order to analyze an atmosphere to include the 1% components, a partial pressure analyzer must operate at pressures two orders of magnitude lower than the total pressure gauge. Further, it is important that the instrument provide this measurement in a very short time interval, as there are many ionic species to be assessed.

A number of uniquely distinct factors influence the lowest partial pressure to which the instrument is responsive. First, a very small current must be measured in a very short time. The current through the mass analyzer consists of discrete charges which arrive at the collector (detector) sequentially and at a finite average rate. The statistical fluctuations in the arrival rate of these charges sets a definite limit to the magnitude of the current and to the related time interval required for measuring it. Additionally, there is the unavoidable signal which exists when there are no ions passing through the analyzer. This may arise from the Johnson noise generated in the current measuring resistor or it may be the dark current in a multiplier. Ion currents below this level are said to be "lost in the noise."

The above discussion emphasizes the importance of providing as large a signal of ions through the mass analyzer as is possible, consistent with the other constraints of resolution, etc. Sensitivity, expressed as amperes per torr, has been stressed in residual gas analyzers. In mass spectrometers used as partial pressure analyzers the sensitivity may range between 10^{-5} A/Torr to 10^{-2} A/Torr. If a mid-range value of 3.0×10^{-4} A/Torr is chosen and combined with a minimum current read by an electrometer of 10^{-15} A, the partial pressure associated with these choices is 3×10^{-12} Torr.

An electrometer which measures a current of 10^{-15} A is unavoidably slower than that which is normally acceptable for mass spectrometer applications. Fortunately, there are today "secondary electron multipliers" of various forms which readily perform a charge multiplication of several orders of magnitude (10^4–10^6). The use of such a multiplier solves the time of response problem.*

The multiplier provides no help for the uncertainty in the measurements due to the statistical nature of the transmitted ion current and it tends to aggravate the problem of the background current which is present when no ions are traversing the analyzer. Although lower counting rates are claimed, a rate of 1 count/sec is considered good for a system. An ion current signal of this magnitude is 1.6×10^{-19} A, which at an instrument sensitivity of 3×10^{-4} A/Torr corresponds to a pressure of 5×10^{-16} Torr. From these figures it appears that partial pressure instruments are available which can measure the lowest pressures attainable.

3.1.2.4. Linearity. Mass spectrometers used as partial pressure analyzers need not have the same degree of linearity as those units which are used for analytical purposes. This results from the fact that in the analysis of partial pressures, in most instances, it is not important to know these pressures with an accuracy of a few percent. However, it is very important to be able to observe with confidence small *changes* in the partial pressures of the various species of gases in the system. Thus, reproducibility or precision is of high importance.

As observed above, in Section 3.1.2.3 a high sensitivity of the instrument is very desirable because it lowers the minimum partial pressure to which the instrument is responsive and at any given pressure it reduces the statistical fluctuations of the output signal, making the observation times shorter. Thus, a higher sensitivity permits oscillographic presentation of the data at a lower pressure. If the demands on the linearity can be relaxed, the instrument can be designed to work at higher sensitivities, since one of the fundamental sources of nonlinearity is space charge in the ion source.

3.1.2.5. Cost. In partial pressure analyzers, as in any other instrument, the manufacturer who offers the most "performance" for the lowest cost sells the most units. Thus, it is unwise to build into a partial pressure an-

* See also Vol. 2 in this series, Section 11.1.3 (second edition).

alyzer refinements which are not essential for this particular application of mass spectrometers.

It is obvious that high sensitivity and a low response to background are performance factors of prime importance. Linearity and response to a mass range greater than that of prime interest are of lesser importance.

Because the information given by a mass spectrometer is of such a great value, relative to that provided by a total pressure monitor (ion gauge), it is important that the cost of partial pressure analyzers be kept low so that they can compete favorably with total pressure gauges in the marketplace. This is accomplished by recognizing which performance factors are important for this mass spectrometer usage and by relaxing the specifications (and costs) of those performance factors which are of much lesser importance for these measurements.

3.1.3. Types of Mass Spectrometers Used as Partial Pressure Analyzers

3.1.3.1. History. A great many types of mass analyzers have been used. When partial pressure analyzers first became available, any instrument which would present a mass spectrum and would indicate changes in the composition of the atmosphere of the "vacuum" was of great interest and was avidly sought by those who operated laboratory vacuum systems. During this period low cost was of prime importance. Questions concerning mass discrimination (varying sensitivity with mass) and linearity of response with pressure were seldom asked.

In the early days of the use of mass spectrometers as "residual gas analyzers," as they were called, system pressures seldom were below 10^{-8}, and frequently were higher than 10^{-6} Torr. Thus, the requirements of sensitivity were quite relaxed relative to today.

Prior to World War II all practical mass spectrometers employed magnetic sectors. In the early days of partial pressure analyzers, following the war, the magnetic sector dominated the field of mass analyzers used for this purpose. It was found that very small units, with a magnetic radius of 1 cm, deflection angle of 180°, completely immersed in a magnetic field, easily fulfilled the requirements of that day. However, as the pressures of interest dropped precipitously and the desire for oscillographic display increased, mass analyzers of more sophisticated designs became more popular.

During the period following the war many new types of mass analyzers were proposed. The vacuum requirements of the technology of the war had clearly indicated the great usefulness of mass spectrometers as leak detectors and as partial pressure analyzers. The technology developments made during the war provided vastly improved techniques, particularly in the precise measurement of short time intervals (radar).

The patent literature of the period following the war is rich with schemes for using pulsed and ac potentials for producing mass spectra. These are collected and described in a delightful little book by Blauth, "Dynamic Mass Spectrometers."[2] Although each of these schemes is capable of producing a mass spectrum, the marketplace gave its continued support only to those methods which do the job best and at lowest cost. In this treatise, space is given only to those devices which have enjoyed a good measure of success for a period of years. These include the time-of-flight, the omegatron, and the quadrupole. Historically, these were introduced in that order by Stephens, Hipple, and Paul.

3.1.3.2. Units. The "physics" of mass spectrometry is concerned with the acceleration of charged particles in electric and magnetic fields. For the classroom the familiar MKS system of units is encouraged. However, the meter is awkwardly large for the dimensions of most instruments, the kilogram is 27 orders of magnitude greater than the mass of ions of interest, and in one second even a 1 V nitrogen ion would go 2.5 kms.

Fortunately, for frequent or daily use there is another set of units which is much more suitable.[3] In these "mass spectrometer" units almost all quantities are expressed by numbers which lie between 1 and 1000, thus essentially eliminating the need to use powers of 10 in calculations. The symbols and their definitions in this system are:

V = volts as read on a laboratory voltmeter,
e = the elementary electronic charge is 0.965 which for all but the most critical calculations can be considered as unity,
m = mass, in atomic mass units (amu),
l = length, centimeters,
B = magnetic field strength, centi-Teslas (1 cT is 100 G),
E = electric field, volts per centimeter,
t = time, microseconds,
f = frequency, megahertz,
ω = angular frequency, radians per microsecond.

These units are used in this section with the electronic charge set to unity. They are useful for all kinematic calculations.

3.1.4. Discussion of Mass Spectrometer Types

3.1.4.1. Magnetic Mass Spectrometers. Magnetic mass spectrometers operate on the principle that charged particles, moving in planes

[2] E. W. Blauth, "Dynamic Mass Spectrometers." Elsevier, Amsterdam, 1966.
[3] W. M. Brubaker, *Ann. Conf. Mass Spectro. Allied Topics*, **14**, 528 (1966).

perpendicular to the direction of the magnetic field, travel in curved paths. The radius of curvature is proportional to the momentum of the particle and inversely proportional to the field strength. In its simplest concept the entire trajectory is immersed in the field. Many instruments have been built using this geometry. These are the so-called 180° instruments, as the maximum separation of ions of differing masses comes at the 180° point. Fortunately, instruments with sectors of much less than 180° work nearly as well. The increased instrument radius which is obtained by using a smaller sector angle has encouraged some designers to build instruments of this geometry.

The theory of operation of the magnetic sector is very well understood and has been developed to include the second and higher order effects. These are ray devices, and so have many characteristics of optical instruments, such as cameras. The focusing properties of magnetic sectors can be described in terms of principal planes and focal lengths, just as in light optics.

The magnetic sector alone is not a mass analyzer, because it has momentum, not mass dispersion. This is seen from the force equation for an ion of mass m (amu) moving through a magnetic field of intensity B (centitesla)

$$mv^2/R = Bv$$

or

$$mv = BR,$$

indicating that the sector is really a momentum spectrometer. An additional relation is required to make the sector a mass spectrometer. This is provided by the energy relation,

$$V = \tfrac{1}{2}mv^2,$$

where V is the potential difference, volts, through which the ions have been accelerated. Combining these two equations to eliminate v yields the "mass spectrometer equation,"

$$V = \tfrac{1}{2}B^2R^2/m.$$

The velocity with which ions leave the ion source and enter the magnetic analyzer has a small range of values, usually expressed as

$$v = v_0(1 + \beta).$$

Similarly, ions leave the source with a finite spread in the angular direction. The half-angle of the wedge of ionic trajectories is usually denoted by α.

All practical mass analyzers must accept ions whose angles and velocities differ from the fundamental values by the fractional values α and β, and bring them to a satisfactorily sharp (line) focus. The manner in which α influences the position of the ion beam in the focal plane is illustrated in Fig. 1. The motion of the beam from the zero position is related to α by

$$\Delta S_\alpha = 2R(1 - \cos\alpha) = R\alpha^2.$$

Since this expression contains no first-order term in α, the sector is said to be first-order focusing in α. A change in the velocity of the ion results in a change in the radius of curvature of the trajectory. For the 180° case the resulting incremental displacement of the beam at the focal plane

$$\Delta S = 2\Delta R.$$

Differentiation of the force equation yields

$$\Delta R = R\beta,$$

whence

$$\Delta S_\beta = 2R\beta.$$

These displacements of the ion beam by α and β must be small relative to the displacement of the beam by a change of the ionic mass by 1 amu. Differentiation of the mass spectrometer equation yields

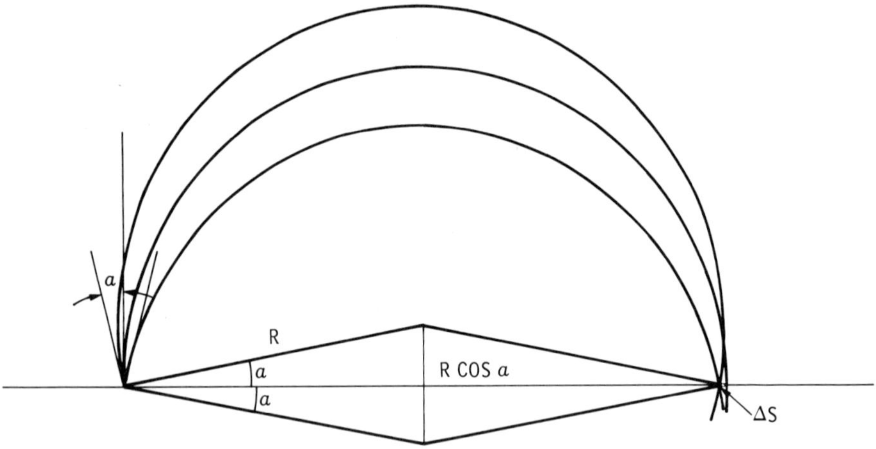

FIG. 1. Illustration of first-order directional focus in a totally immersed, 180° sector. Ions which leave the source from a line at directions $\pm\alpha$ relative to the central ray traverse angles of $180° \pm 2\alpha$ and come to a line on the focal plane displaced a distance ΔS from the central ray. ΔS is equal to $2R(I - \cos\alpha)$, or to $R\alpha^2$.

$$\Delta m/m = \tfrac{1}{2}(\Delta R/R),$$

whence

$$\Delta S_m = 2R(\Delta m/m),$$

which shows the separation between two beams of ions which differ by one atomic mass unit is

$$\Delta S \text{ (for } \Delta m = 1) = 2(R/m).$$

The above equation illustrates some of the dominant features of the magnetic sector instruments. First, the separation between adjacent peaks of given masses is proportional to the radius of the instrument. It is completely independent of all other operating parameters. This explains why the analytical instruments for high mass are so large and the partial pressure analyzers of limited mass range can be so small. Secondly, for a given instrument, it shows that the separation between adjacent peaks is inversely proportional to the masses of the ions. This accounts for the wide separation between peaks at the low mass range and the closeness of the peaks in the high mass range when the spectrum is scanned in a manner which allows equal times for all peaks to traverse the resolving slit.

Magnetic sectors have been made in a wide variety of sector angles. Angles of 180°, 90°, and 60° have been most frequently used. The analytical instruments are frequently made 180° or totally immersed. This avoids the passage of ions through fringe fields, which is of lesser importance for partial pressure analyzers. For first order, in a symmetrical geometry, the performance is independent of the angle of the magnetic sector. The magnification remains unity and the distance between peaks of adjacent masses remains the same for the same magnetic radius, of course.

3.1.4.2. Time of Flight. This was the first of the new mass analyzer types to be proposed after World War II.[4] It makes use of the pulsing techniques and the ability to measure extremely short time intervals which were developed during the war, particularly for radar. The instrument consists of a pulsed ion source, a field free drift tube, and a detector with high gain and exceptional time resolution.

In the pulsed ion source the ionizing electron beam is on for a fraction of a microsecond. Then the ions are accelerated through one or two field regions. It was found[5] that by delaying briefly the time between the ionization and the time of impressing the electric field in the ionizing region,

[4] W. E. Stephens, *Phys. Rev.* **69,** 691 (1946).

[5] W. C. Wiley and J. H. McLaren, *Rev. Sci. Inst.* **26,** 1150 (1955).

ions with initial velocities at the time of ionization could be made to transit the instrument in essentially the same time, thus coming to a focus in time. Another parameter which helps in adjusting the instrument is the use of two accelerating regions, with a constant voltage on the larger, second region and a much smaller voltage on the first region which is pulsed.

The field free drift space permits the ions of a given mass to come to a focus in time, as mentioned, but it also permits lighter ions to outdistance the slower, heavier ions to provide sequential arrival times at the collector. The time difference between the transit times for ions differing by one amu is a few tens of nanoseconds.

Because this instrument is capable of producing 20,000 complete mass scans each second, it is ideal for applications in which the composition of the sample is expected to change rapidly with time.

3.1.4.3. The Omegatron. The omegatron was first proposed and demonstrated by Sommer, Thomas, and Hipple.[6] It is a magnetic instrument, in which the ions make many full 360° turns. The ions are formed on the axis of the analyzer by a small pencil of electrons, collimated by the magnetic field. An alternating electric field, perpendicular to the magnetic field, imparts energy to the ions (see Fig. 2). Those ions whose natural or resonant frequency in the magnetic field matches the frequency of the alternating field traverse ever expanding orbits which are Archimedes spirals until they reach the ion collector. All other ions have trajectories of limited radius. They gain energy from the alternating field until the phase angle of their nearly circular trajectories has shifted through an angle of π. A further shift of phase causes the kinetic energy of the ions to be given back to the field and the trajectories become smaller in diameter.

Although it is straightforward, the derivation of the complete theory of the trajectories becomes tedious when done in sines and cosines. Berry[7] simplified the development considerably by using complex notation. By treating the pulsating field as a combination of equal vectors rotating in opposite directions, Brubaker and Perkins[8] obtained a very simple derivation of the trajectories. In a coordinate system rotating at the angular frequency of the applied radio frequency potentials, the trajectories become very simple. The ordered motion of the ions is simply the velocity E/B, as it is in all cases of crossed electric and magnetic fields. For resonant ions the trajectories are straight lines in the rotating coordinate system and Archimede's spirals in the laboratory system. When there is a slight

[6] H. Sommer, H. A. Thomas, and J. A. Hipple, *Phys. Rev.* **76,** 1877 (1949).
[7] C. E. Berry, *J. Appl. Phys.* **25,** 28 (1954).
[8] W. M. Brubaker and G. D. Perkins, *Rev. Sci. Inst.* **27,** 720 (1956).

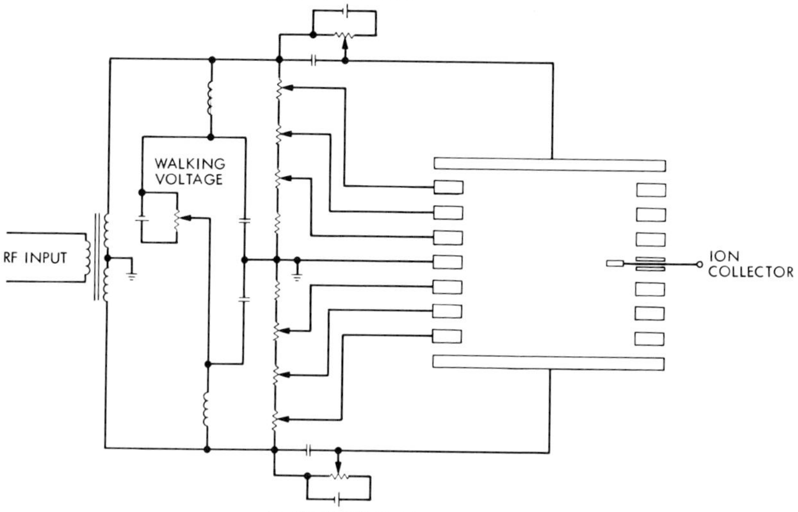

FIG. 2. The omegatron. A uniform electric field of strength $E \sin(\Omega t)$ is impressed in the volume reached by the circulating ions. The trapping voltages applied to the end electrodes provides axial gradients which prevents escape of ions as they reach the extremeties of the device.

difference between the ionic resonant frequency and the applied frequency the trajectory in the rotating system becomes a circle. The velocity at which the ions move along the arcs of the circles remains E/B. The angular frequencies at which the ions revolve around these circles are equal to the difference frequencies between the resonant and the applied frequencies. The diameters of these circles are the maximum distances at which these nonresonant ions proceed toward the collector. These distances are inversely proportional to the difference frequencies. Figure 3 depicts this description of the motions of the ions in the rotating coordinate systems.

Detailed development of the theory shows that the diameter of the circle traversed by a nonresonant ion is proportional to the level of the ac excitation and to the fractional deviation of the mass of the ion from resonance $\Delta m/m_0$, where m_0 is the mass of the resonant ion. Thus, when the scanning frequency departs from the resonant frequency of the ion in the magnetic field so that the diameter of the circle becomes less than the distance from the axis to the collector, transmission of ions of that mass ceases. Since $\Delta f/f = \Delta m/m$, this concept gives the half width of the transmitted peak.

The number of turns made by a resonant ion in reaching the collector

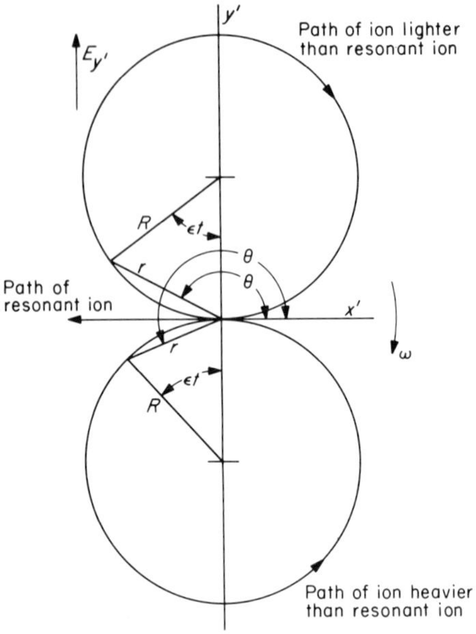

FIG. 3. Ion paths in the rotating coordinate system. The resonant ions move along the negative x' axis at a constant velocity, making Archimede's spirals in the laboratory coordinates. Nonresonant ions start out in the same direction, but travel in circles and return to the origin. In the laboratory coordinates they make spirals with ever smaller rotational increments until they reach their maximum distance, then spiral inward with ever increasing rotational increments.

follows directly from: radial velocity is E/B; distance to collector is r_0; transit time of resonant ion is r_0B/E; and number of turns is time x frequency or $(r_0B/E)f$. E is the intensity of the rotating vector, which is half that of the maximum field. A nonresonant ion which is at the edge of the peak travels a distance which is increased by the ratio of the circumference of the circle to the diameter $\pi/2$. The number of turns made by this ion is increased by the same ratio.

From the time of its introduction into the commercial market early in the 1950s until the emergence of the quadrupole mass filter, the omegatron enjoyed high popularity in the partial pressure analyzer field. Successful operation of the omegatron requires that the ions make a great many complete 360° turns in the magnetic field. During this time the ions are vulnerable to weak electric fields in which they have a drift velocity of E/B. Even though the disturbing electric fields which may result from surface and space charges are small and the resulting drift velocities are likewise small, the long transit times make the trajectories quite vulnerable to relatively small disturbing electric fields.

3.1.4.4. The Quadrupole. The quadrupole mass filter is quite unique as a mass analyzer. It is nonmagnetic and is excited by radio frequencies in the megaHertz range. It differs from all of the magnetic analyzers and from most of the other analyzers in that its operation is not dependent upon the ions having long, uninterrupted trajectories.

The use of a quadrupole structure as a mass separator was proposed by Paul and Steinwedel[9] in 1953. In 1955 Paul and Raether[10] published spectra, confirming the expectations. In 1958, Paul, Reinhard, and von Zahn[11] presented a more complete description, including theory and data. Because of the complexity of the theory and the difficulty in using it to describe the operation of the complete analyzer, most of the developments of the quadrupole have been made without reference to the solutions of the appropriate Mathieu equation. Brubaker[12] developed an approximate solution which gives a physical insight to the nature of the trajectories.

The quadrupole mass analyzer consists of four parallel rods. The theory assumes that the contours of these rods are hyperbolic, with the asymptotes being planes which cross at right angles and intersect each other on the instrument axis. These infinite planes make angles of 45° with the axes of the coordinate system. The rods are usually circular in cross section, with a radius of curvature that is about 15% larger than the smallest radius of curvature of the hyperbolae which they approximate.

Opposing rods are connected together electrically. The x and y axes pass through the centers of the round rods, as shown in Fig. 4. The potentials applied to the x rods are always mirrored in the y rod potentials, so the potential on the pair of intersecting planes that lie midway between the rods is always zero.

The rods are energized, as indicated in Fig. 4, by balanced ac and dc potentials. The ac potentials applied to the x and y rods are identical but with a 180° phase difference. The magnitude of the positive dc potential applied to the x rods is identical to that of the negative potential applied to the y rods. The dc potentials are about one-sixth of the peak values of the ac potentials. Mass scanning can be accomplished by a variation of the frequency of excitation, but it is usually done by varying the ac and dc excitation voltages in proportion.

Ions enter one end of the rod structure through an aperture, traverse the quadrupole, and exit through an aperture which is at least as large as

[9] W. Paul and H. Steinwedel, *Z. Naturforsch.* **8a,** 448 (1953).

[10] W. Paul and M. Raether, *Z. Phyz.* **140,** 262 (1955).

[11] W. Paul, H. P. Reinhard, and U. von Zahn, *Z. Phyz.* **152,** 143 (1948).

[12] W. M. Brubaker, "A Study of the Introduction of Ions into the Region of Strong Fields within A Quadrupole Mass Spectrometer," Final Rep. NASA, NASW-1298 (1967).

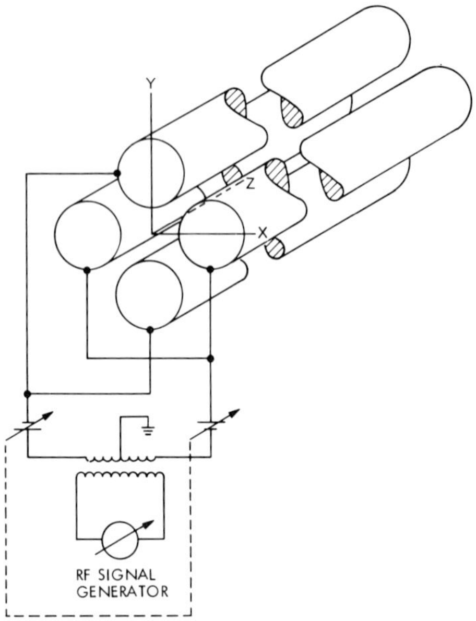

FIG. 4. Geometry and excitation schematic of quadrupole mass filter. Scanning is accomplished by varying either the excitation frequency or by varying all potentials in proportion.

the distance between opposing rods. As they emerge from the rods they are driven by a strong transverse electric field to an off-axis multiplier–detector. Placing the multiplier off-axis in an out-of-line-of-sight position from the ion source lowers the background signal due to photons from the ion source.

Unfortunately, the theory of operation of the quadrupole for ions in the quadrupole structure is quite complex. There is no first order, simplified description of the trajectories as there is for other instruments. However, it is easy to state the requirements for an ion to be transmitted through the quadrupole. The trajectories must be stable (bounded) for both the x and the y components of motion and the amplitudes of the trajectories must be less than the instrument radius.

When the rod surfaces are hyperbolic, the equipotential surfaces are all hyperbolic also. On the coordinate axes the potentials increase as the square of the distance from the instrument axis. The x component of the electric field is proportional to the perpendicular distance to the y axis, completely independent of the y position. The corresponding condition exists for the y component. The x and y components of motion are completely independent of each other. The trajectories for these two components of motion are distinct, as shown in Figs. 5(a) and (b).

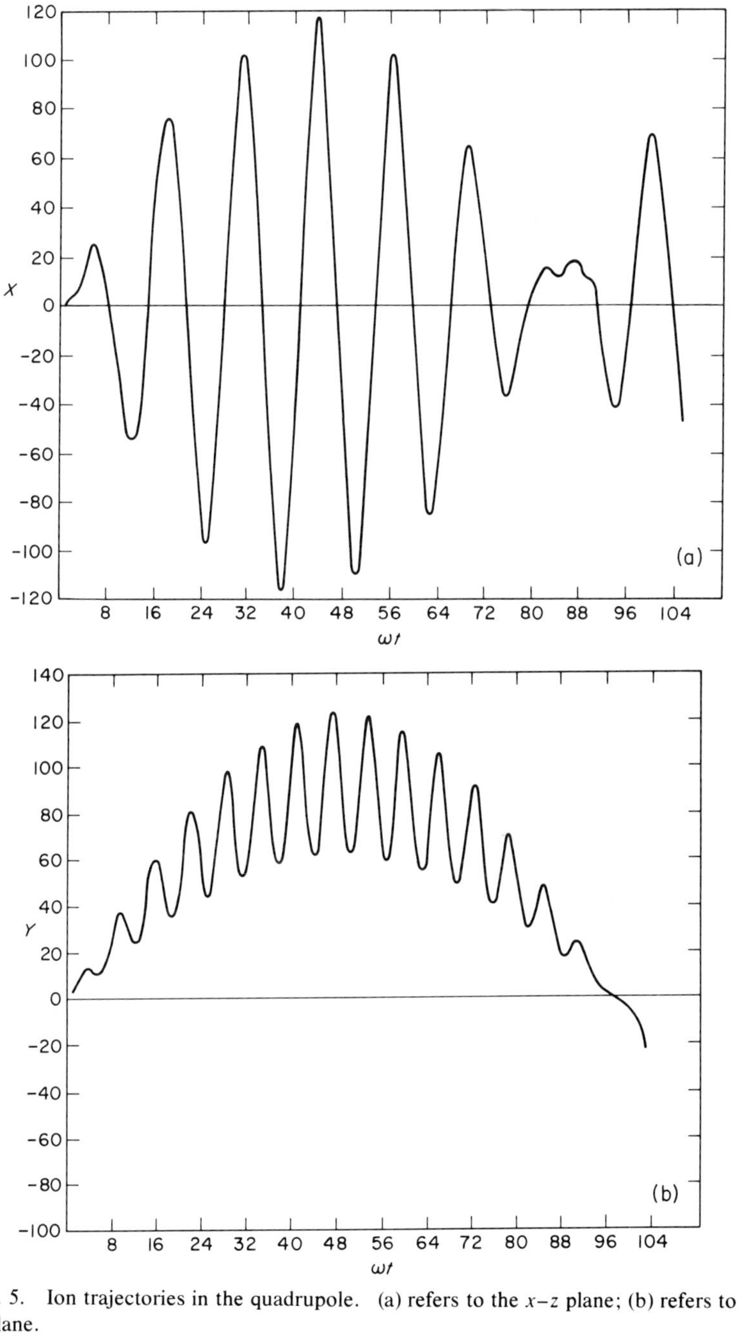

FIG. 5. Ion trajectories in the quadrupole. (a) refers to the $x-z$ plane; (b) refers to the $y-z$ plane.

The differential equations of motion are readily written, but there are no solutions in closed form. A simple and direct transformation of the variables brings the equations into the standard form of the Mathieu equation. The dimensionless variables are

$$a = 8V_{dc}/mr_0^2\omega^2 \quad \text{and} \quad q = 4V_{ac}/mr_0^2\omega^2.$$

The state of stability of the trajectories is determined uniquely by the values of a and q coordinates of the "working point" in the stability diagram of the solutions to the Mathieu equation. The plot of the stability regions in the a–q plane is presented in Fig. 6. In Fig. 7 the plot has been folded about the $q = 0$ axis in order to present on one chart the stability limits for both positive and negative values of a, representative of the dc potentials impressed on the x and on the y rods.

For a given level of excitation at a fixed frequency, the working points for ions of all masses fall on the scan line. When the scan line passes through the stability triangle, just below its apex, the working points for the range of values which lie within the triangle represent stable trajectories for ions with masses which correspond to this range of a, q values. These ions will be transmitted through the quadrupole if the amplitudes of their trajectories are less than r_0, the distance to the rods. Ions of heavier mass have working points nearer the origin and are lost to the y rods; ions of lighter mass have working points which lie beyond the triangle and are lost to the x rods. Adjusting the ratio of the dc excitation to the ac exitation changes the slope of the scan line and controls the width of the pass band for ionic transmission through the quadrupole. The incremental length, ΔS, of the scan line which lies within the triangle is related to the distance of the tip of the triangle from the origin S, as

$$\Delta S/S = \Delta m/m$$

to a good approximation.

3.1.4.5. The Monopole Mass Spectrometer. Electrically, the monopole is exactly a y quadrant of a quadrupole. It was first described and made to work by von Zahn.[13] It was noted in the discussion of the quadrupole that there is a pair of neutral axes on which the potential is always zero. These axes pass midway between adjacent rods on the quadrupole. Since the potentials on these planes are always zero, the potential and the field distributions are unchanged if conducting sheets are inserted on these planes. Although the fields are unchanged, the trajectories of the ions now have an additional restraint: they cannot cross these planes.

[13] U. von Zahn, *Rev. Sci. Inst.* **34,** 1 (1963).

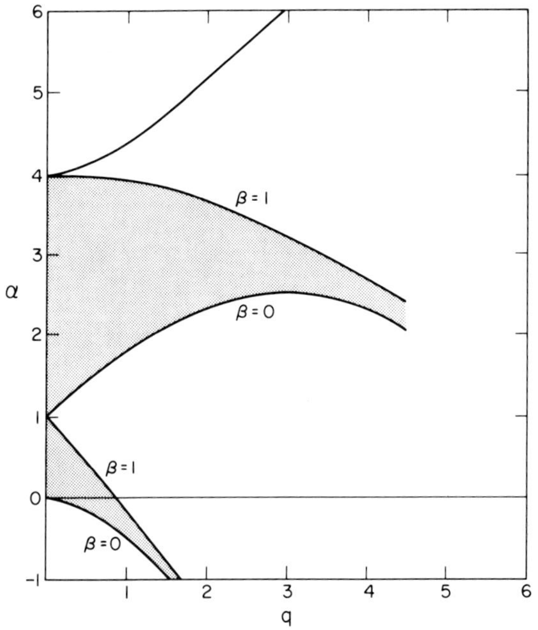

FIG. 6. Stability chart, showing alternate regions of stability and instability of the solutions of the Mathieu equation.

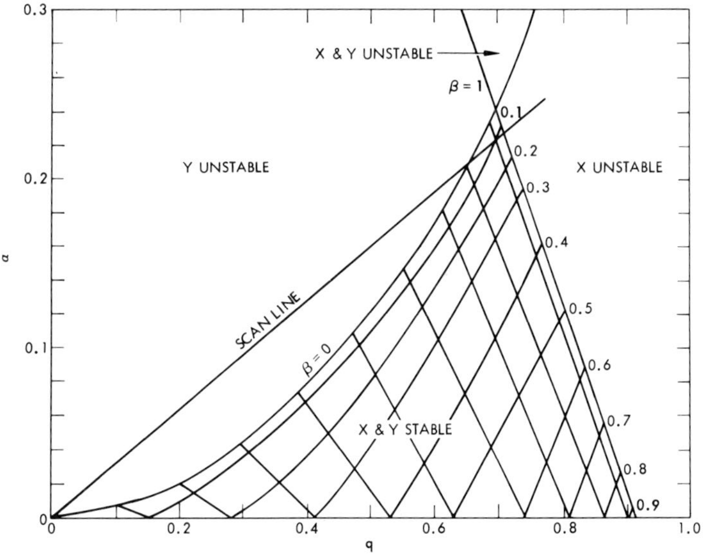

FIG. 7. Stability chart for small values of a and q, with the negative a values folded over the positive values.

The additional constraint imparts to the monopole some of the characteristics of the magnetic instruments. The trajectories now have a critical or minimum length between consecutive axial crossings. This criterion is similar to a focal length in the magnetic instrument.

In the stability diagram the monopole scan line passes well beneath the apex of the stability triangle, thus greatly easing the importance of a precise ratio of the dc to the ac excitation potentials. The portion of the stability diagram which is associated with ionic transmission becomes a very narrow band, separated a small distance from the y-stability limit. Thus the x stability is excellent and the ions are thereby constrained to within a narrow band in the symmetrical plane, and seldom come close to the V (grounded) electrode.

The monopole operates very much as a quadrupole, but has not enjoyed the popularity of the quadrupole.

3.1.5. Mass Spectra of the Atmosphere in Vacuum Systems

During the several decades of the widespread use of vacuum systems before the advent of partial pressure analyzers, our knowledge of the composition of the gases within the systems was essentially null. A paper before the American Vacuum Society in which movies of the cathode-ray tube display of the mass spectrum of the contained gases in a vacuum system was of intense interest. The audience was enthralled to see the immediate changes in composition which resulted from heating or cooling of various parts of the system.

The early partial pressure analyzers responded only to the dominant gases in the system. More recent instruments, of much higher sensitivity, may show a response at all but a very few of the mass numbers from 12 to 50 in an unbaked system. The sources of these species of molecules is, of course, the surfaces exposed to the system. Low volatile compounds on exposed surfaces provide enough molecules in the gas phase to provide trace spectra for very long times: days or weeks. Although it is very difficult to give complete and definite interpretations to these complex and unusual spectra, they are all due to the previous history of the materials and the surfaces of the components of the vacuum systems. Spectra of solvents used to clean the parts prior to assembly may be dominant when the instrument is first turned on after pump down. However, they usually pump out readily because of the high vapor pressure of these solvents.

The largest single peak in an unbaked, reasonably tight system is water vapor at mass 18 until the system has been pumped for days. This is true even in regions of low humidity. In a well baked system the water vapor peaks fade to insignificance. Oxygen is a very active gas and the mass 32 peak becomes very small unless oxygen is being supplied to the system, as by an air leak. Argon is almost the sole contributor to the mass 40 peak, so its presence is also an indication of an air leak.

The discussion in Chapter 5.1 shows how quickly the pressure in a system falls on pump down following venting to air, as far as the molecules in the space between surfaces (including the vacuum walls) are concerned. Unless driven off by heat, many molecules remain on surfaces for days. Water vapor is the worst offender in this respect, but the molecules of all gases seem to have finite, nonzero rest times on surfaces.

Other phenomena occur in ion sources to add to the complexity of the spectra and to complicate their interpretations. Many molecules are doubly ionized. Thus, argon, mass 40, gives a peak at mass 20. On the other hand, some ions attach themselves to neutral molecules to form heavier ions. An example is mass 3 in hydrogen. In a tight, well baked system, the spectra become simpler. Under these conditions the peak at mass 28 remains prominent. It is not nitrogen, but carbon monoxide. This is thought to be produced in the ion source by reactions on the hot filament. Oxygen, from either the gas or from water vapor, in contact with the hot filament, reacts with the carbon in the filament to produce CO. This peak (28) comes into prominence in the pressure range below about 10^{-8}.

The interpretation of spectra is complicated by the fact that more than one ionic species results from the electron bombardment of polyatomic molecules of any given species. For instance, the common molecule of methane, CH_4, fractures in nearly every possible manner when it is struck by an energetic electron. Besides hydrogen ions, there are generous contributions of ions of masses 12, 13, 14, 15, and 16, Carbon dioxide, mass 44, makes a large contribution at mass 28, which is the same as nitrogen and carbon monoxide.

In the application of a mass spectrometer to quantitative analytical work, the unraveling of the mixtures of gases which produce spectra in which the contributions to specific mass numbers may be due to several gaseous species, the classical technique is the use of the solutions of simultaneous, linear equations. Relatively simple computer systems perform these analyses quickly and efficiently. However, in the use of mass spectrometers as partial pressure analyzers, where the interest is usually in the spectra identification of the major components, this stratagem is seldom required. Additionally, for partial pressure analyses, there frequently is more interest in the manner in which a given spectrum duplicates or deviates from previously obtained spectra. In other words, is the system behaving normally or is there a deviation from the usual?

3.1.6. Practical Aspects

Although partial pressure analyzers are designed to be particularly effective at the lowest pressures, they are frequently used at the higher pressures, as at the beginning of a pump-down from atmospheric pressure. Under these conditions it is desirable to reduce the electron current to a very low level. Otherwise, the filament is more vulnerable to deterioration by any oxygen which may be present. Further, the surfaces which are bombarded by the nontransmitted ions receive less abuse. The impact of these ions in many instances leads to insulating layers. Continued incidence of ions results in surface charges which usually degrade the performance of the unit.

Aside from the possible deterioration of the performance of the instrument from operation at high pressures, the mere presence of the greater density of molecules interferes with the operation of the instrument because of ion–molecule collisions. Because of the unusually long trajectories of ions in an omegatron, this type of mass analyzer is the most vulnerable to operation at high pressures. It is interesting to note the manner in which high pressure operation influences the shape and height of the peaks. The quadrupole is most tolerant of ion–molecule collisions, because these collisions merely attenuate the peaks without broadening them in this instrument. On the other hand, the ray machines, which are the magnetic instruments, and to a lesser degree the omegatrons suffer peak broadening as well as attenuation of peak heights at high pressures.

If the instrument is "dirty," with insulating layers in sensitive regions which are struck by intense ion beams, proper operation may be impaired from this condition before ion–molecule collisions take their toll. In the ray machines pressure broadening of the peaks may become noticeable when the mean free path becomes as low as an order of magnitude longer than the trajectory. Quadrupoles can operate at a pressure an order of magnitude larger before appreciable attenuation of the peak heights occurs. These pressures are of the order of 10^{-4} and 10^{-3} Torr.

4. PRODUCTION OF HIGH VACUA

4.1. Overview and Formulation of General Requirements*

4.1.1. Gas Transport; Throughput

Compared to the experience near atmospheric pressure, molecular flow of gases is distinguished by very low mass flow, rather high volume flow, high pressure (or density) gradients, and a dependence on adsorption phenomena. The concept of conductance has been discussed in Section 1.3. Conductance and pumping speed are associated with description of gas flow in volumetric terms.

The actual amount of gas flowing through a duct or into a pump cannot be determined unless its density is also known. The more important quantity often is the mass flow expressed in units such as grams per second. Usually, the conversion from volume flow to mass flow can be easily made using the well known gas law relationships ($PV = RT$; $M = V \cdot d$). However, these relationships may lose their usual meaning when the gas approaches molecular beam conditions which may exist on occasions in high vacuum systems. It is not a simple matter to convert flow in grams per second into liters per second for a gas which does not have isotropic molecular velocity distribution (non-Maxwellian gas).

For this reason, great precision and accuracy is hardly important for considerations of design of most high vacuum systems.

Because mass flows are very low, it is customary to use throughput instead of mass flow. The assumption is made that heat transfer effects are negligible and the gas quickly accommodates to the temperature of the vessel or pumping ducts. When temperature is constant, the "mass flow" can be measured in units of throughput, Torr liters per second. Throughput then is simply the product of pressure and the volume flow of a gas ($Q = Sp$) for any given location (cross-sectional plane) in the flow passages.

Most analytical discussion of molecular flow found in textbooks on vacuum technology is associated with noncondensable gases and steady state

---

* Chapter 4.1 is by M. H. Hablanian.

METHODS OF EXPERIMENTAL PHYSICS, VOL. 14

ISBN 0-12-475914-9

flow condition. At ordinary temperatures, this may apply only to helium. Surface effects which govern the transient flow of condensable gases are often neglected. (A detailed discussion is presented by Lewin.[1]) Usually vapors of higher molecular weight are troublesome, but water vapor and even such "volatile" substances like acetone or alcohol are difficult to work with at room temperature. For example, if we attempt to measure the pumping speed of a diffusion pump for water vapor, the measurements of the flow rate have to be done by carefully considering the available surface area for evaporation from the liquid, heat transfer necessary for evaporation, the temperature of the tubes leading the vapor into the pump, etc. Even then it is difficult to separate the amount of water vapor actually pumped by the pump and that which is condensed inside the cold inlet areas. Pumping fluid vapor in a baked system may take a week to be noticed by an ionization gauge with a 4-cm long 2-cm diameter entrance tube; a mass spectrometer can retain a strong "memory" of acetone for two or three weeks if its ion source is 30 cm away from the pumps (2-cm diameter tube). In such cases, baking is indispensable.

In molecular flow environment, gas molecules rarely collide and, therefore, move independently of each other. Thus, there is no such thing as a true pressure gradient which would "guide" the molecules toward the region of lower pressure. An outgassing or evaporating molecule can move as easily away from the pump as well as into it. There exists a density (number of molecules) gradient toward the pump but, for condensable species, even this cannot be assured as temperature distribution may influence the flow more than the presence of a pump.

It is often assumed that residual collisions between gas (or vapor) molecules are completely negligible in high vacuum. This is not always true. Sometimes the remaining collisions may be the predominant method by which certain molecules pass through barriers such as cryogenic traps. Also, in some cases, introduction of one gas into the system may effect flow conditions of another. For example, in turbomolecular pumps, an improvement of density ratio for helium can be obtained by introducing a certain amount of argon into the pump.[2]

4.1.2. Pumping Speed in Liters per Second

It is customary in high vacuum technology to express gas flow in liters per second. This is convenient in basic computations if the temperature is assumed to be constant. There exist two distinct geometries in which

[1] G. Lewin, "Fundamentals of Vacuum Science and Technology." McGraw-Hill, New York, 1965.

[2] F. J. Schittko and S. Schmidt, *Vak. Tech.* **24,** 4 (1975).

the flow parameters are associated ($Q = pS$). One can be called orifice geometry (Knudsen geometry), the other pipe flow geometry. The two are not identical. In the case of the pipe flow, the pressure and speed are associated with the same cross-sectional plane; in the case of the orifice, the pressure is measured upstream from the orifice. The orifice geometry is shown in Fig. 1. The definition of orifice conductance C is derived from

$$Q = C(P_1 - P_2), \tag{4.1.1}$$

where P_1 and P_2 are upstream and downstream pressures. If P_2 is much smaller than P_1, we may speak of the orifice speed

$$S = C = Q/P_1 \quad \text{when} \quad P_2 \ll P_1. \tag{4.1.2}$$

Note that in this case, the P_1 is not associated with the vicinity of the orifice, but with an upstream location where the gas conditions may be considered to be isotropic.

It has been derived from kinetic theory of gases and confirmed by experiments that for air at room temperature, each square centimeter area of an orifice (having "zero" thickness) will have a pumping speed of approximately 11.6 liters/sec. When computing the number of molecules bombarding a small (compared to the chamber) unit area at a given pressure (measured far away), we can arrive at the same amount of 11.6 liters/sec. Note that this 11.6 liters/sec is a conversion from number of molecules and is associated with the remote, upstream density. The association of pumping speed with vacuum vessel evacuation is treated in Section 4.1.3.

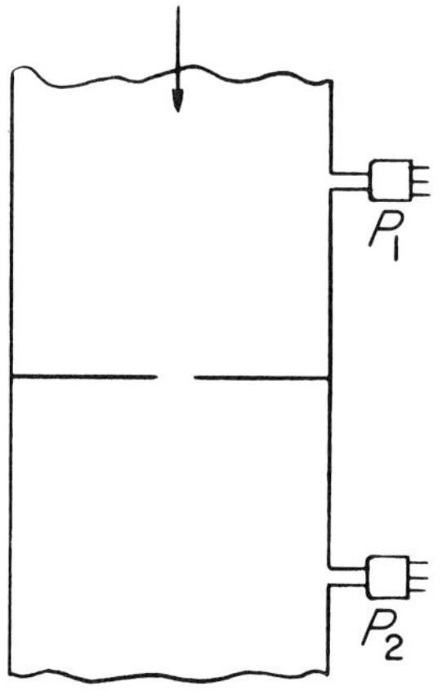

FIG. 1. Basic geometry defining the concept of orifice conductance.

The two geometries mentioned above occur in practice. Often the geometry represents a mixture of the two. For pipe flow and for a sequence of matched flow elements connected in series, the association of pumping speed at a cross section is more convenient. For a chamber with a small pumping port, the orifice association is more helpful. This duality of approaches is responsible for the difference between the pumping speed measurement standards (American Vacuum Society 4.1 and International Standard Organization 1607). The ISO standard tends toward the orifice treatment and gives 10–15% lower pumping speed values for a pump compared to AVS standard. The two can be related depending upon the ratio of pumping port to chamber diameter.[3]

The net pumping speed of the pumping system or the pumping stack can be determined by the use of the same standard measuring dome. To relate this measurement to the vacuum chamber in order to obtain an effective speed for the chamber, we should take into account a pressure difference existing between the center of the chamber and at the inlet into the pumping port. In practice, such considerations are not of great importance and simple approximate calculations can be made by adding conductances in series as long as no abrupt changes in geometry are present.

In cases where pronounced unisotropy exists, pumping speed consideration can become meaningless. Consider an imaginary molecular beam with parallel molecular trajectories (without collisions) entering directly into a pump. A gauge mounted on the wall in the vicinity of this beam may not know of its existence. In such cases, the meaning of $Q = pS$ is not obvious. Situations of this nature may arise in practice. See, for example, Section 5.3.

4.1.3. Pump-Down Time

One of the basic considerations in designing high vacuum systems is the time required to evacuate a vessel to a given pressure. Usually, this ranges from a few minutes to a few hours. However, in some cases, the desired time can be a few seconds or a few days. The prediction of time and pressure relationship by theoretical or experimental methods is very difficult because of the uncertainty associated with gas evolution rates from inner surfaces of the vessel (outgassing or virtual leak).

Gas evolution from a given material depends on temperature, surface finish, previous history of exposure to a variety of atmospheric conditions, and cleaning methods. Thus, even approximate prediction of evacuation time can be extremely difficult when outgassing becomes signifi-

[3] H. G. Nöller, *Vacuum* **13,** 539 (1963).

cant. Using well-known relationships applicable to higher pressures, the designer can predict a pressure–time curve from atmospheric pressure to about 10^{-2} Torr. From steady state (long time) outgassing data and from characteristics of pumping devices, the final (ultimate) pressures can sometimes be predicted. The region of greatest interest often is between these two points. The following discussion attempts to develop at least a qualitative appreciation of significant effects.

4.1.3.1. Constant Speed Case. In the pressure region between 760 and 10 Torr, the pumping speed of the fore-vacuum pump is usually nearly constant. A 10% reduction at 10 Torr compared to the speed near atmospheric pressure is common. Referring to Fig. 2 the evacuation process can be represented by the following relationship:

$$-V\,dp/dt = \mathrm{S}p, \tag{4.1.3}$$

where V is the volume of the vessel, dp/dt the rate of change of pressure with respect to time, and S the pumping speed. The physical meaning of Eq. (4.1.3) can be understood by noticing that the left side represents the amount of gas leaving the chamber (minus sign indicates pressure decrease), and the right side shows the gas entering the pumping duct. It is important to note that a solution of this differential equation will depend on the location where the pressure P is measured. Otherwise, to be valid, the equation must have the same pressure value on both sides of the equality sign. Thus, the equation can be used as an abstract definition of a pumping speed for the given chamber, pumping system, and gauge location.

The evacuation time from Eq. (4.1.3) for constant volume and speed is

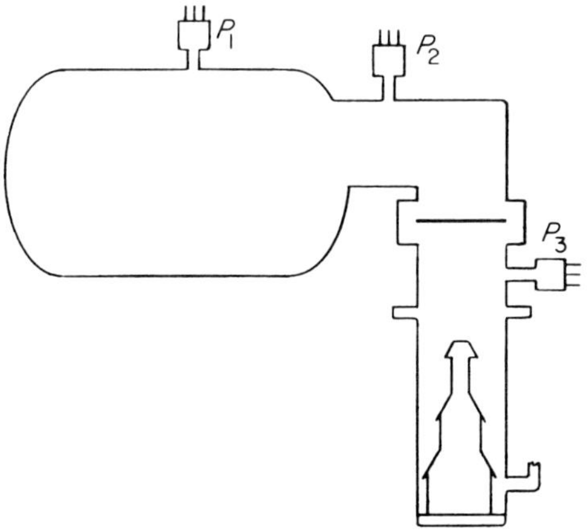

FIG. 2. A chamber with a pumping system showing three typical locations of vacuum gauges.

$$t = (V/S) \ln P_0/P, \tag{4.1.4}$$

where P_0 is the initial pressure and P the final pressure. This equation gives adequately accurate values after a few seconds from the start of evacuation and until approximately 10 Torr is reached. At lower pressures the outgassing becomes significant and the evacuation period is elongated. Generally, it is recommended to multiply the values obtained from Eq. (4.1.4) by about 1.5 between 10 and 0.5 Torr, by 2 between 0.5 and 5×10^{-2} Torr, and by 4 between 5×10^{-2} and 1×10^{-3} Torr.

For rapid pumping of small volumes (such as locks) the equation will not give accurate results due to complications of geometry, conductance and volume of pumping ducts, and transient effects within the pump itself. If the desired pumping time is less than 10 sec, experimental measurements may be advisable.

4.1.3.2. Constant Throughput Case. When constant throughput is used in Eq. (4.1.3) the solution becomes

$$t = (V/Q)(P_0 - P). \tag{4.1.5}$$

This is not an interesting case technologically, but it occurs in a narrow pressure region, 10^{-1}–10^{-3} Torr, in diffusion pump systems. This period is usually short, for example, less than 10 sec for a typical bell jar system.

4.1.3.3. In the Presence of a Leak. Equation (4.1.3) disregards a possibility of leaks and desorption gas loads. If $Q_\sim = Sp_\sim$ is a constant leak, we may write

$$-V\,dp/dt + Q_\sim = Sp, \tag{4.1.6}$$

with a corresponding solution

$$t = (V/S) \ln(P_0 - P_\sim)/(P - P_\sim). \tag{4.1.7}$$

The gas load after a long pumping time $Q_\sim$ may be due to an actual leak, outgassing of almost constant rate, or a finite permeability of the vessel walls.

4.1.3.4. In the Presence of Outgassing. For qualitative purposes, the outgassing rate of a surface in high vacuum can be represented as

$$Q = Q_0 \exp(-t/\tau), \tag{4.1.8}$$

where Q_0 is an initial outgassing rate, t the time, and τ is associated with the slope (decay) of the outgassing function relative to time and is assumed to be a constant. If necessary for proper matching of experimental outgassing curves, two superimposed exponential terms can be used, such as $Ae^{-at} + Be^{-bt}$. Q is the total outgassing rate and it can be expressed as qA, where A is the area of the outgassing surface.

The general differential equation describing the evacuation of a vessel can be then written as follows:

$$-V\,dp/dt + Q_{\sim} + Q_0 \exp(-t/\tau) = Sp. \qquad (4.1.9)$$

The expression for the evacuation time becomes

$$t = (V/S) \ln \frac{(P_0 - P_{\sim}) - Q_0/(S - V/\tau)}{(P_1 - P_{\sim}) - [Q_0/(S - V/\tau)] \exp(-t/\tau)}.$$

The expression for pressure decay relative to time will be

$$P = (P_0 - P_{\sim}) \exp(-St/V) + [Q_0/(S - V/\tau)][\exp(-t/\tau) - \exp(-St/V)] + P_{\sim}. \qquad (4.1.10)$$

The first parts of the last two equations can be recognized as solutions when outgassing is disregarded.

In the high vacuum region there does not exist a simple inverse relationship between pumping speed and time of evacuation which is obtained at higher pressures. Using Eq. (4.1.10) it can be demonstrated that for a common bell jar system, the pumping speed may have to be increased six or seven times in order to cut the time in half. However, when process outgassing occurs, the larger pump will maintain a lower process pressure inversely proportional to speed.

4.1.3.5. Condensable Species. The outgassing rates of various materials can vary by many orders of magnitude. Condensable species such as water may exist in ordinary surfaces in the amounts of the order of 10 to 100 monolayers. The behavior of such films is highly dependent on temperature and binding energies involved. Outgassing rate of many substances may be considered to vary exponentially with temperature.[2] The most disturbing species are those with desorption energies from 15 to 25 kcal/mole. Below that value, the pumping proceeds rapidly. Above 25 kcal/mole, the presence of the material in the vacuum space is likely to be below 10^{-11} Torr. The ordinary conductance relationships cannot be applied to gases in the temperature regimes in which they are condensable (see Section 4.1.1).

4.1.4. Ultimate Pressure

The ultimate pressure for a high vacuum pump in a system is the pressure established after a sufficient time after which further reductions of pressure will be negligible.

4.1.4.1. Summation of Partial Pressure. The total ultimate pressure consists of partial pressures of various gases present in the system. In positive displacement pumps such as fore-vacuum pumps, the pumping speed for various gases is approximately the same. However, under molecular flow conditions in a diffusion pump each species of gas is pumped with a different pumping speed and the ultimate pressure is given by the sum of the individual partial pressures obtained by dividing the gas load (throughput) of the gas by its pumping speed. In addition, the ultimate pressure comprises also the component arising in the pump such as pumping fluid or lubricant vapors. Generally, the ultimate pressure of a pump is obtained when an equilibrium is established between the amount of gas flowing into the pump and the amount flowing in reverse.

4.1.4.2. Pump Limit versus System Limit. A clear distinction should be made between an ultimate pressure of a pump and that of a system. System gas loads often are orders of magnitude higher than the internal pump gas evolution so that the system may never reach the ultimate pressure capability of the pump. Normally, pump performance data are given for a minimum volume and surface area at the inlet of the pump. In addition to the amount of gas loads, the conductance of the inlet ducts will also influence the system ultimate pressure.

4.1.4.3. Evacuation versus Process Pressure. Ultimate pressure in a system is rarely an end in itself. If certain work is to be performed inside the vessel after evacuation, it usually results in additional gas evolution. Therefore, if a process or an experiment is to be performed at a given pressure, the ultimate system pressure should be lower than that. A convenient and very approximate rule for work which produces low gas evolution is to specify ultimate system pressure ten times lower than the desired process environment.

4.1.5. Fore-Vacuum and High Vacuum Pumping

High vacuum technology deals with a very wide range of pressure or particle density conditions. Usually, the process of evacuation begins at atmospheric pressure and then proceeds to high or ultrahigh vacuum. No single pumping device can be expected to function efficiently at all pressure conditions. In addition to multistage pumping system, the type of pumping mechanism employed is different at atmosphere and at high vacuum. Even if, in principle, a given pumping method could be used throughout the entire pressure range, such an attempt would be impractical in regard to size, weight, or cost of equipment as illustrated in Fig. 3.

The most common pumping arrangement for production of high vacuum consists of a positive displacement mechanical pump for initial evacuation followed by a vapor–stream pump (usually called diffusion pump). Often parallel and series arrangements are used. For initial evacuation, the mechanical pumps are used alone and, for obtaining high vacuum, the two types of pumps are connected in series. In such a system,

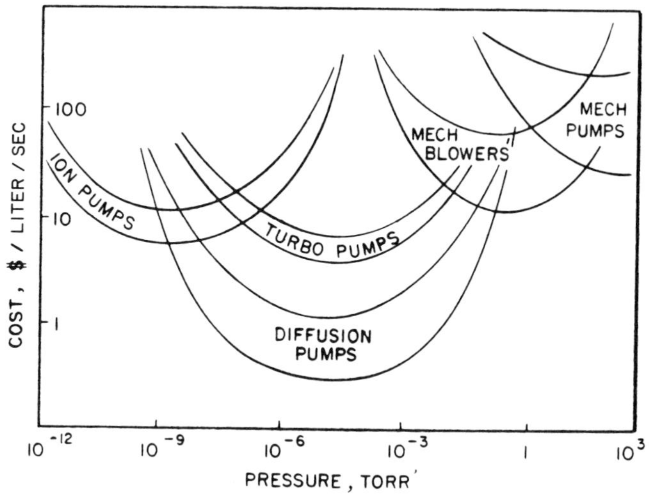

FIG. 3. A qualitative illustration of the pressure ranges in which the use of various pumps is most economical.

the gas enters the pumping train at high vacuum and is exhausted at atmospheric pressure by the last pump. In some cases, the device may be evacuated and sealed terminating the pumping process; in others, the pumping is continuously applied to compensate the gas evolution in the vacuum chamber.

In sorption pumps, and particularly in ion-getter pumps, the pumped gas is not exhausted to atmosphere. This has an obvious advantage of isolation from high pressure environment and the disadvantage of limitation in gas load capacity or the necessity for periodic regeneration.

The basic performance of pumps and compressors can be associated with flow and pressure factors. With appropriate allowances for size, power, pumped fluid characteristics, etc., all such devices generally behave according to the pressure–flow graph shown in Fig. 4.

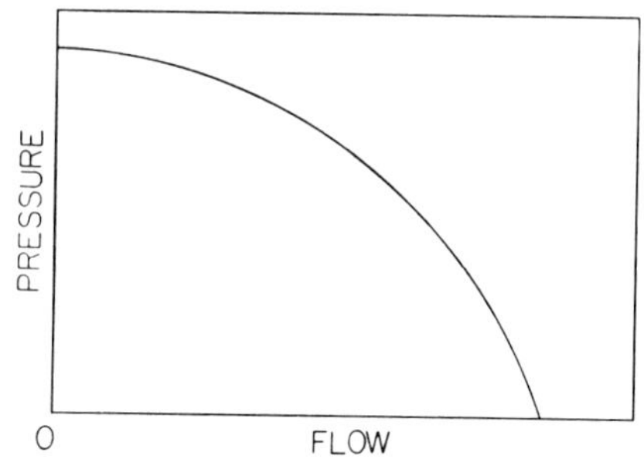

FIG. 4. The general qualitative relationship between pressure and flow for most compressors or pumps.

For high vacuum work, both high volume flow rate and high pressure ratio (inlet to discharge) are necessary. The flow rate is associated with the size of the pumping device. Thus, for a given size, the pressure ratio needs to be increased as much as possible. Both, mechanical vacuum pumps and diffusion pumps (and also turbo-molecular pumps) produce pressure ratios of over a million to one. This can be compared to industrial plant air compressors, automobile engines, or aircraft compressors not exceeding compression ratio of ten to one.

Obviously, high vacuum pumps require very special designs and the familiarity with their design, construction, operation, and maintenance is an important ingredient of success in production and use of high vacuum environment.

4.2. Fore-Vacuum Pumps*

Pumps which can produce the necessary fore-pressure for other pumps not capable of discharging to atmosphere are called fore-vacuum pumps. Some of these can be operated in two modes when discharging to atmosphere or when discharging into reduced fore-pressure.

4.2.1. Overview

Pumping devices which convey gases from a low pressure region to atmospheric pressure can be grouped into three major classes:

(a) gas transport is accomplished by one or more stages of compression;

(b) gas is entrained and transported by momentum transfer;

(c) gas is immobilized on large surfaces at low temperatures and is later thermally regenerated.

Mechanical vacuum pumps are representative of the first group of pumps while ejectors and sorption pumps are representative of the second and third group (Fig. 1). Some general sources for information on mechanical pumps are given in References 1–6.

Operating Pressure Range. Although all the pump types shown in Fig. 1 can discharge against atmospheric pressure, their effective operating range varies depending on the operating principle (Figs. 2 and 3). When sealing fluid is not used, it is impractical to maintain a high compression, mechanical clearances cannot be made sufficiently small to prevent significant reexpansion or reverse flow. Absence of sealants makes such pumps eminently suitable for operation near atmospheric

[1] C. M. VanAtta, "Vacuum Science and Engineering." McGraw-Hill, New York, 1965.

[2] R. Jaeckel, "Kleinste Drücke, Ihre Messung und Erzeugung." Springer-Verlag, Berlin and New York, 1950.

[3] K. Diels and R. Jaeckel, "Leybold Vacuum Handbook." Pergamon, Oxford, 1966.

[4] A. Guthrie, "Vacuum Technology." Wiley, New York, 1963.

[5] "Displacement Compressors, Vacuum Pumps and Blowers," **ANSI PTC-9** (1974), Am. Soc. Mech. Eng., New York.

[6] "Compressors and Exhausters," **ANSI PTC-10** (1975), Am. Soc. Mech. Eng., New York.

---

* Chapter 4.2 is by Z. C. Dobrowolski.

ISBN 0-12-475914-9

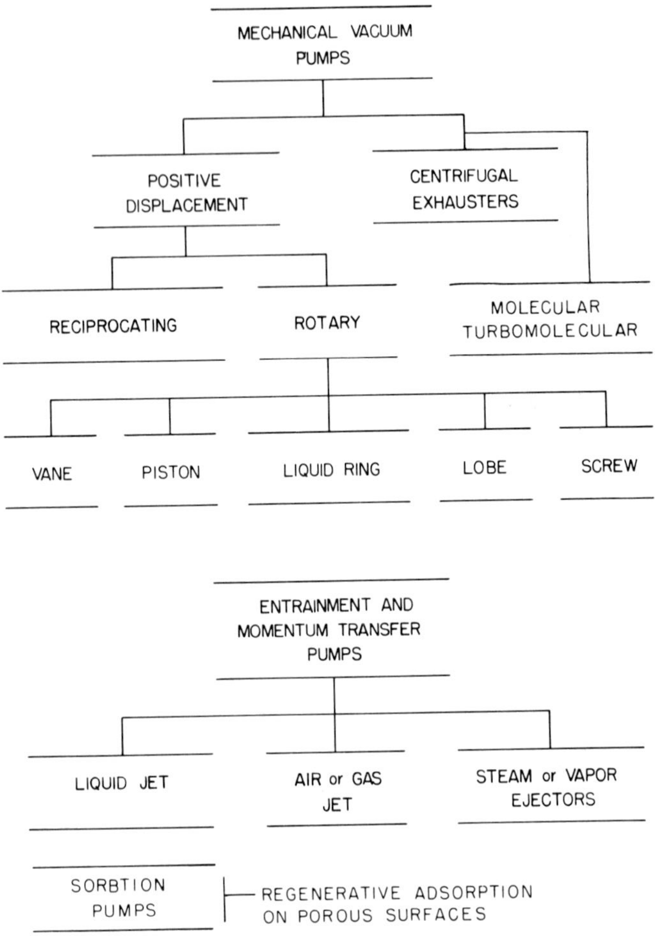

FIG. 1. Vacuum pump types which can discharge against atmospheric pressure and can produce a significant pressure reduction at suction.

pressure. Heat of compression, which is difficult to dissipate, imposes a restriction on the low pressure operation of some dry mechanical pumps, particularly the lobe and dry screw compressors. These pumps operate best at pressures above 300 Torr. Exploitation of self-lubricating characteristic of materials such as carbon, graphitized cast iron, and high temperature synthetics can extend the operating range of dry vane pumps to 150 Torr. The same pressure range can be served efficiently also by liquid ring pumps, reciprocating piston compressors, and, less efficiently, by ejectors. An even more extended pressure range to about 30 Torr is served by compound liquid ring, valveless multivane, two-stage reciprocating piston, liquid jet, and two stage steam ejector pumps. As the

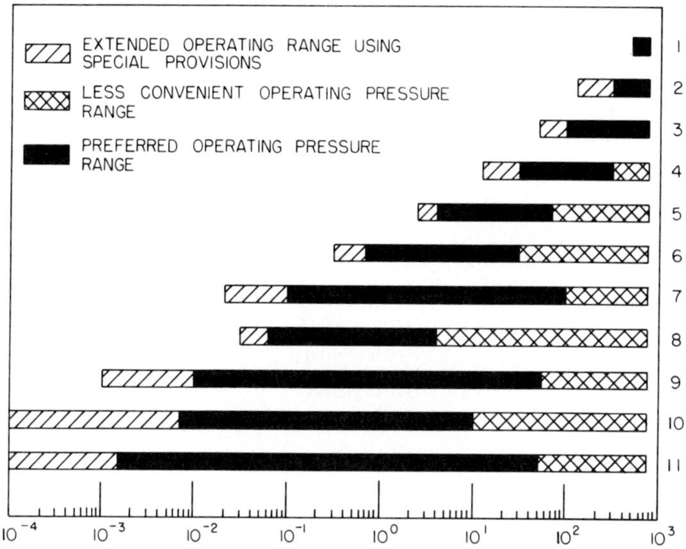

FIG. 2. Operating pressure ranges of different fore-vacuum pumps and pumping chains. The numbers represent: (1) centrifugal exhausters; (2) dry lobe screw or vane pumps; (3) single stage reciprocating piston, liquid ring, steam ejector, two stage water sealed lobe pumps; (4) oil flooded screw, two stage steam ejector, liquid jet, compound liquid ring, valveless oil sealed multivane pumps; (5) air ejector or steam ejector backed by liquid ring or liquid jet pump, three stage steam ejector, oil sealed compound liquid ring pump, two stage valveless multivane pumps; (6) dry lobe pump backed by liquid ring or liquid jet, four stage steam ejector; (7) oil sealed valved single stage rotary piston or vane pump, five stage steam ejector; (8) lobe pump, air ejector, liquid ring pump in series; (9) lobe pump, lobe pump, liquid ring pump, six stage steam ejector; (10) compound rotary pump, sorption pump(s); (11) lobe pump backed by single stage or compound rotary pump.

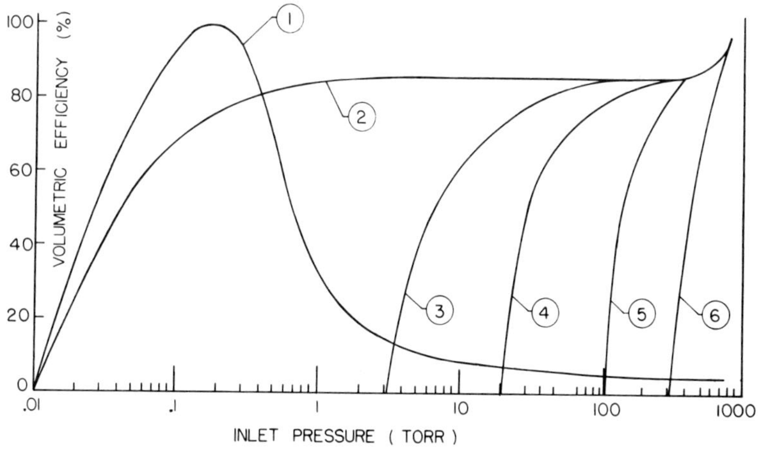

FIG. 3. Volumetric efficiency of major classes of vacuum pumps. The numbers signify: (1) five stage steam ejector (assumed volume efficiency of 100% at design point); (2) rotary oil sealed vacuum pumps; (3) liquid sealed valveless pumps; (4) liquid ring, reciprocating valveless pumps; (5) dry vane pumps; (6) dry lobe pumps.

operating pressure is lowered, the relatively simple pumps can no longer cover the extended operating pressure range and it is necessary to resort to rotary oil sealed vacuum pumps, multistaging, and the selection of low vapor pressure sealants. Useful operating pressure range of a single stage pump extends to 0.1 Torr limited primarily by oil contamination effects. The same considerations limit the low pressure operation of compound oil sealed pumps to slightly below 10 mTorr. Oil vapor pressure of about 1 mTorr prevents exploitation of lobe pumps backed by compound oil sealed pumps (at intrinsically available pumping speed) to 10^{-6} Torr range. Pumping below 1 mTorr is more efficiently performed by other pumping devices such as diffusion pumps, ion pumps, etc., and in terms of mechanical pumping, the range is best served by turbomolecular pumps.

When handling small volumes of gas, sorption pumps used singly or sequentially can cover a pressure range similar to that of compound oil sealed mechanical pumps, without the presence of oil or oil vapors.

Because of the broad pressure range covered, ruggedness, simplicity, reliability, and convenience, oil sealed rotary vacuum pumps are most commonly used in experimental laboratory work. Most of the following subject matter is devoted to the discussion of their characteristics.

4.2.2. Rotary Oil Sealed Pumps

By convention, this description is reserved for lubricated positive displacement pumps discharging to atmosphere through discharge valves and capable of producing a base pressure of at least 0.1 Torr.

4.2.2.1. Principle of Operation. There are two types of pumps commonly used: the rotary piston and rotary vane; however, their pumping principle is the same. As the rotor of the pump turns on its axis, a gas free space is created between the rotor and the cylinder wall (Fig. 4). Gas enters this pocket equalizing the pressure between the newly created space and the space upstream. This pressure difference is responsible for the transport of gas into the inlet of the pump. While gas free space is created at inlet, gas trapped in the discharge pocket is compressed and transported toward the discharge valve and is expelled into atmosphere.

The ratio between discharge and inlet pressures is called the compression ratio (R = discharge pressure/inlet pressure). Discharge pressure is normally constant at around 1000 Torr and represents the back pressure due to atmosphere and spring tension of the discharge valve. The highest compression ratio is established at no flow conditions and can have a limiting value of 100,000 for a single stage pump with an ultimate pressure of 0.01 Torr.

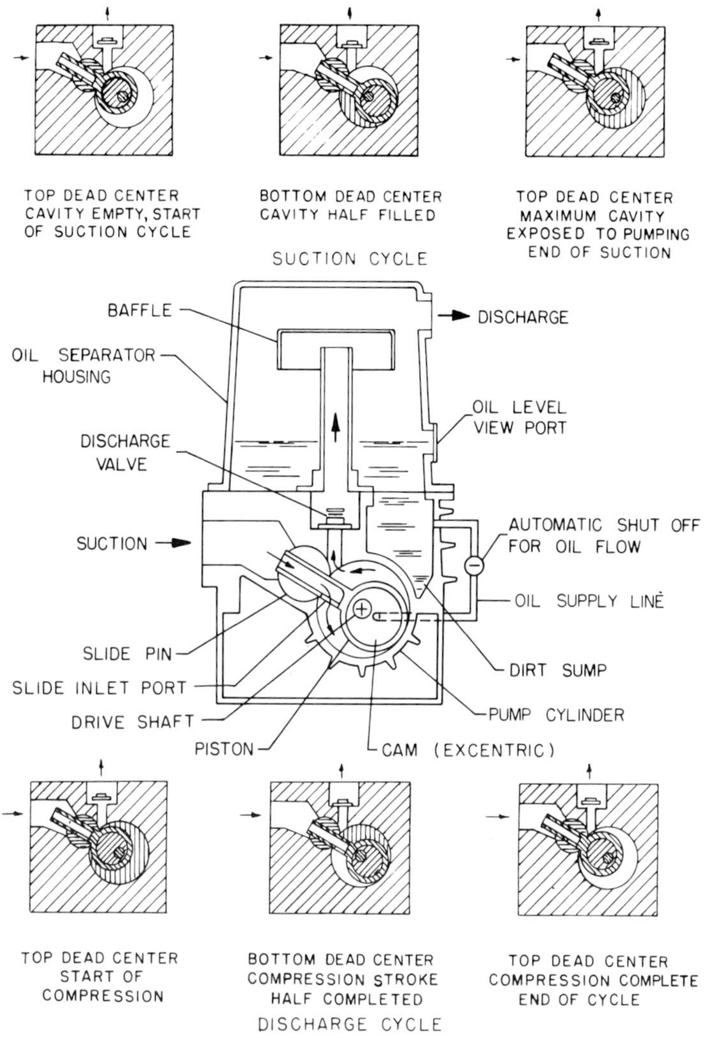

FIG. 4. Elements of rotary piston oil sealed vacuum pump; principle of operation.

Such an extraordinary compression ratio is possible because the discharge valve prevents atmospheric air from entering the compression pocket and the sealing fluid acts as a hydraulic valve lifter (when the gas pressure is low), in addition to completely filling the void between the discharge valve and the top dead center position of the rotor.

4.2.2.2. Ultimate Pressure. The basic pumping mechanism is not the limiting factor in pressure reduction; that limitation is imposed by air solubility in the lubricant. Oil discharged through the exhaust valve returns to the oil separator and then reenters the pump due to pressure difference between atmosphere and the working pressure inside the pump or by means of an oil feed pump. The returning oil is saturated with air which means that it brings back with it 10–15% by volume of dissolved air at standard conditions. Thus, the ultimate pressure (partial pressure of air) of a single stage pump is typically 5 mTorr. Small pumps (10 liters/sec) have relatively large oil flow rates because mechanical precision of small parts is difficult to control and manufacturing variations have to be compensated for by a more generous oil flow. In addition, oil flow orifices must have sensible dimensions to prevent clogging. The net result is that small pumps exhibit an ultimate pressure typically slightly higher than 10 mTorr. The ultimate pressure is stated in terms of partial pressure of air (McLeod gauge).

Ultimate pressure based on total pressure measurement introduces too many variables to be used as a universal yardstick. It introduces as parameters vapor pressure of oil (hence also pump temperature), oil contamination level, and gauge response to the gas mixture. Good lubricating vacuum pump oils have a vapor pressure of less than 10^{-5} Torr at room temperature. However, that low vapor pressure is not retained once the pump is set into operation. Local thermomechanical effects are responsible for a vapor pressure increase by at least 2 orders of magnitude. The total pressure of slow operating, well cooled pumps is at best 10 mTorr, making the apparent vapor pressure equal to 5 mTorr. On the other extreme, high speed air cooled pumps can exhibit a vapor pressure of 50 mTorr, while the partial pressure of air is 5 mTorr also. Diffusion pump oils (even those possessing good lubricating properties) do not exhibit low vapor pressure when used in rotary pumps. This is also true of less commonly used lubricants such as synthetic esters, polyglycols, fluorocarbons, triaryl or tricrescyl phosphates, and lubricating silicones.

Since air solubility limits the low pressure performance of single stage pumps, it follows that elimination of air from the lubricating oil would extend it.

Efficient outgassing is provided by connecting two pumps in series within the same housing (Fig. 5). The first stage is lubricated by an independent oil storage reservoir maintained in an outgassed (air free) condition by the atmospheric stage. The oil level in the high vacuum reservoir is maintained by oil splash produced by the second stage, while the oil circulation rate of the first stage is maintained by the available hydrostatic head which rarely exceeds a few inches of oil. In consequence, the oil circulating rate is very low and outgassing effects no longer limit the low pressure performance. Partial pressure of air at inlet of compound pumps is often in the 10^{-5} Torr range while ultimate total pressure is about

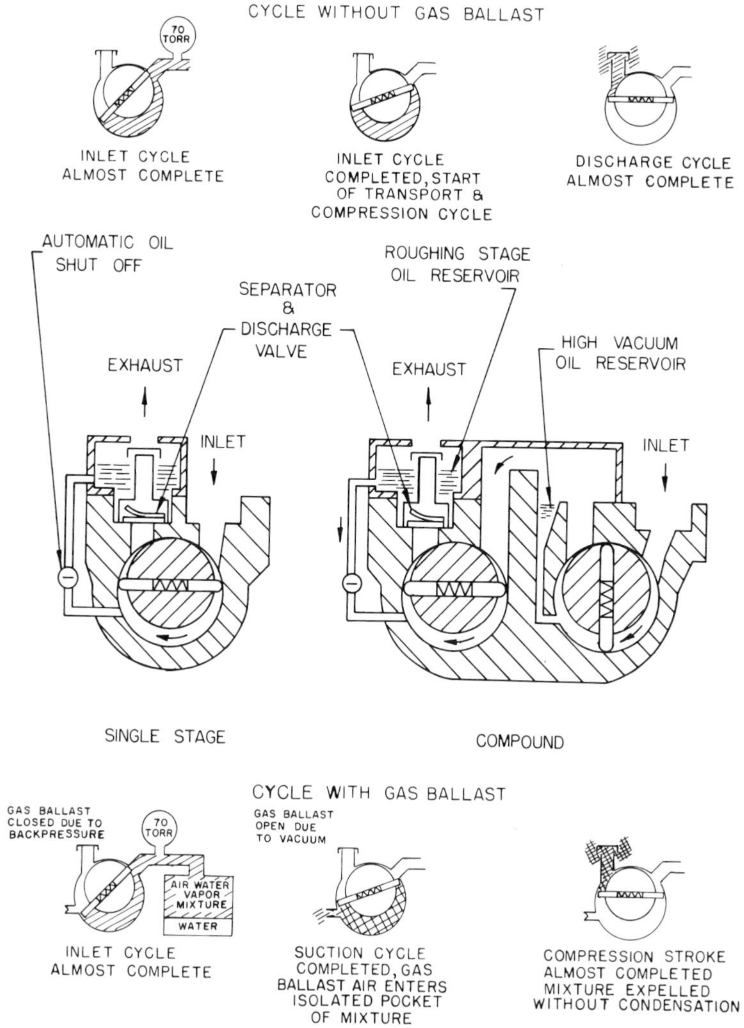

FIG. 5. Elements of rotary oil sealed vane vacuum pump; principle of operation of a single stage and compound (series) arrangement.

1 mTorr depending on the operating temperature of the pump and the effect of its mechanism on the stability of the lubricant. The pumping effect of the atmospheric stage helps to maintain the high vacuum stage oil free of light fractions.

4.2.2.3. Pumping Speed. As in the case of other gas pumps, the rate of gas transported by a vacuum pump at inlet conditions is its pumping speed. Displacement is obtained from pump geometry (swept volume

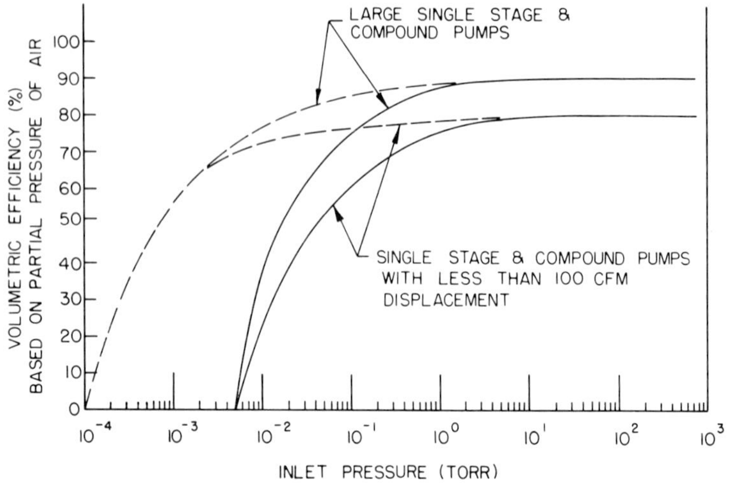

FIG. 6. Volumetric efficiency of small and large rotary oil sealed pumps (see Section 4.2.2.7.1 for gas ballast).

multiplied by rotational speed) and pumping speed is measured. The volumetric efficiency of vacuum pumps is typically 80% for pumps with displacement of less than 50 liters/sec and 90% for larger pumps (Fig. 6).

Pumping speed near ultimate pressure may vary between different pump designs. For single stage pumps it is sensitive to oil flow rates and operating temperature, and for compound pumps also to the staging ratio. The staging ratio is the ratio of the displacement of the high vacuum stage to the displacement of the atmospheric stage. The pumping speed is the same for all permanent gases and vapors which have solubility rates in oil similar to those of air and which do not condense during the compression cycle. In other words, pumping speed is independent of gas density. With compound pumps the operating pressure range below 1 mTorr is rarely exploited, consequently performance curves are often terminated at 0.1 mTorr.

4.2.2.4. Horsepower. The compressive work performed by the vacuum pump can be approximated by an ideal adiabatic compression cycle, where

$$\text{work} = \frac{k}{k-1} PD\left(\frac{P_a^{(k-1)/k}}{P} - 1\right),$$

where k is the ratio of specific heat (C_p/C_v), P the inlet pressure, D the displacement, and P_a the discharge pressure (approximate atmospheric). The maximum for that expression occurs at about 250 Torr and no significant compressive work is done at inlet pressures below 1 Torr (Fig. 7).

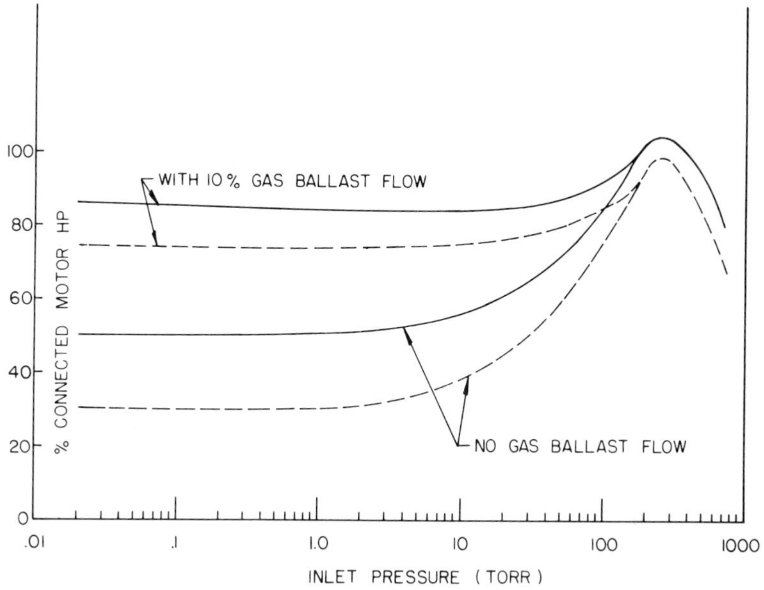

FIG. 7. Typical horsepower requirements of single stage and compound rotary oil sealed vacuum pumps. Dashed line, compound; solid line, single stage.

Use of gas ballast results in increased low pressure horsepower. At 10–15% gas ballast flow, power required approaches the motor rating. Since horsepower curves are not often published, these can be constructed by calculating the theoretical work done on compressing gas and adding it to an assumed friction effort. It is reasonable to assume that 50% of the rating of the recommended pump motor represents friction losses for single stage pumps and 30% for compound pumps. Oil viscosity can make a difference, but at normal operating conditions the oil film temperature is high and viscosity contribution is not significant. On the other hand, it is difficult to start up pumps at temperatures below 5°C unless heated or when using low temperature lubricants.

4.2.2.5. Single Stage and Compound Pumps. The basic difference in the method of lubrication of compound and single stage pumps is responsible for the difference in needs these pumps serve. The fractionating effect available in a compound pump, rather than its inherent ability to produce low ultimate pressure, makes it more suitable for low pressure work (below 1 Torr).

Contaminants, products of oil degradation, end up in the atmospheric stage oil and do not greatly affect the performance of the first stage. This means satisfactory operation between oil changes can be quite long. Application of gas ballast with its beneficial effect on oil cleanliness increases the partial pressure of air at inlet by a few millitorr only, at the same time reducing the operating noise. Even without gas ballast, the noise level is normally lower than that of single stage pumps. Hydraulic noise produced in the atmospheric stage with its high fluid flow rate is absent in the high vacuum stage. The low oil flow rate which makes the compound pump so attractive for low pressure work makes it unattractive for sustained operation at higher pressures.

With large mass flow of gas the high vacuum stage reservoir becomes depleted of oil and the backsplash does not work against a vigorous gas flow. When prolonged operation at high pressures (above about 10 Torr) becomes necessary, provisions for positive oil lubrication should be made.

The lubricating system of single stage pumps is based on the differential pressure principle. It works adequately well at pressures below 500–600 Torr. Backflow of gas through the oil passages interrupts oil flow at higher inlet pressures and continuous operation above 500 Torr is possible only with positive oil feed. Single stage pumps are normally equipped with oil gas separators which take into account the large mass flow rates which may have to be handled (Fig. 4).

Ample oil storage is provided not only to reduce the necessity for frequent oil changes but to provide sedimentation space for particulate contaminants and for decanting of separable liquid.

4.2.2.6. Comparison of Vane Pumps and Rotary Piston Pumps. True vane pumps are dynamically balanced by virtue of design. The center of rotation of the pump rotor is at the center of the shaft and the vanes have sufficiently small mass not to affect the vibrational characteristics significantly. During one revolution (two-vane pump), the pumping volume is swept twice which allows compact dimensions. The vane tip pressure on the cylinder wall tends to be high, particularly with high speed pumps and friction effects are significant. A critical dimension is the rotor to stator seal which separates the discharge port from the suction port. Differential pressure across this seal is high and it is necessary to keep that clearance as tight as possible. This means in practice tighter than 0.025 mm, which is not easy mechanically. In order to facilitate repair, some pumps are dowelled together after initial test to make subsequent handling easier. It is a common practice for undoweled pumps to be repaired at the source of manufacture. Spring loaded vanes which provide a positive contact with the cylinder wall help to hold some degree of vacuum on stopping. Oil transfer from the atmospheric stage to the high vacuum stage is sluggish which makes pressure recovery in rapid cycling applications rather slow. Oil immersed pumps, especially when stopped under vacuum, can become flooded with oil. To prevent this, various mechanical oil shutoff devices are used. The immersion of small vane pumps in an oil bath, used primarily to provide a vacuum seal between joints provides excellent noise damping. Larger pumps, which would become too bulky with this practice, are noticeably more noisy. Because there are no balancing limitations, high staging ratio pumps are possible and displacement penalty for compounding is small. In consequence, compound pumps predominate but the price of high staging ratio is poor vapor handling and a rather noisy operation at high gas ballast flow necessary to partly compensate for high interstage compression.

The rotary piston pump is a very rugged machine. The piston actuated by an eccentric cam, rolls along the cylinder wall pushing ahead of it a large slug of oil which makes for a large sealing area insensitive to mechanical damage. Except for the slide pin, the mechanism is self-compensating for wear. The mechanism at the same time is inherently unbalanced. The vibration level can be controlled reasonably well depending on the balancing method used. Piston pumps do not use oil immersion sealing and oil flow control is provided by automatic valves in the easily accessible oil passages. The vacuum holding characteristics are poor because the stationary contact between the piston and the cylinder is weak. The noise of pumps operating without gas ballast is generally slightly higher than that of comparable size vane pumps (particularly those immersed in oil). On the other hand, when operating with gas ballast, rotary piston pumps are quieter. The internal pump mechanism produces a vigorous internal oil splash and compound pump recovery times for rapid cycling are excellent. However, prolonged operation of compound piston pumps at high pressures is also not desirable without the necessary modifications.

The most outstanding characteristics of the rotary piston pump are its insensitivity to mechanical damage and ease of repair by unskilled personnel. The major disadvantage is inherently high vibration level, particularly if an unbalanced version of the pump is used.

4.2.2.7. Water Vapor Pumping. Large volume vapor pumping is performed most efficiently by condensing. Depending on the operating pressure, the condensing is done by direct contact with a water spray or by surface condensers cooled by a refrigerant. In these cases, the vacuum pumps are used only to remove noncondensible gases. On small scale installations, oil sealed vacuum pumps serve both as water vapor and gas pumps. The techniques available for water vapor handling apply generally to handling of other vapors as well, particularly those not soluble in pump oil.

4.2.2.7.1. GAS BALLAST. The most useful method of water vapor handling is the use of gas ballast. This consists of admitting a sufficient quantity of air into the pump during the compression stroke to prevent condensation. The addition of the gas changes the composition of the compressed mixture, thereby keeping the partial pressure of water vapor below the condensation point. When gas ballast is not used water vapor condenses inside the pumps and forms an oil–water emulsion. On recirculation, the water evaporates at pump inlet, diminishing the pumping speed. When the oil becomes fully saturated no further water vapor is pumped. The higher the operating temperature of the pump, the smaller the quantity of gas ballast needed. The rate of air at standard conditions admitted for the purpose of gas ballast is expressed as a percentage of displacement

$$\%\mathrm{GB} = \frac{\text{gas flow rate into gas ballast}\quad (\text{liters/sec at NTP})}{\text{pump displacement}\quad (\text{liters/sec})} \times 100.$$

The price one pays for this convenience is deterioration of inlet pressure. In compound pumps, only the atmospheric stage is gas ballasted so that inlet pressure deteriorates to only 20 mTorr even when using 10% gas ballast. With such a small effect, it is unprofitable to control the flow to match the load requirements and operation with fixed gas ballast flow is a most common practice. The effect of gas ballast flow on a single stage pump is more pronounced and gas ballast flow regulation to the maximum tolerable level is the general practice (Fig. 8).

Most pumps are provided with a full gas ballast flow rate such that at the operating temperature of the pump, water vapor can be handled at room temperature (20 Torr) without condensation inside the pump. While handling water vapor even with an adequate gas ballast flow, the oil contamination level increases linearly with operating pressure; roughly 10% of the operating pressure for single stage pumps and 1% for compound pumps. For example, the ultimate pressure of a compound pump handling water vapor at 20 Torr will be 1% of 20 Torr or 0.2 Torr (Fig. 9).

Contaminated oil can be cleaned by the use of gas ballast but the rate of water removal is very much slower than calculated since the removal does not occur at constant pressure and depends on heat input. Single stage pumps with their large oil storage are sometimes operated without gas ballast during low pressure processing and are cleaned up between cycles if the operation is not continuous.

4.2.2.7.2. Decanting. This method of water handling is based on demulsification properties of oil. It is most often applied to single stage pumps. Water condensed during compression mixes with the lubricant until the oil can no longer hold any more water and it separates out. The incoming water rate becomes equal to the water rejection rate. The sepa-

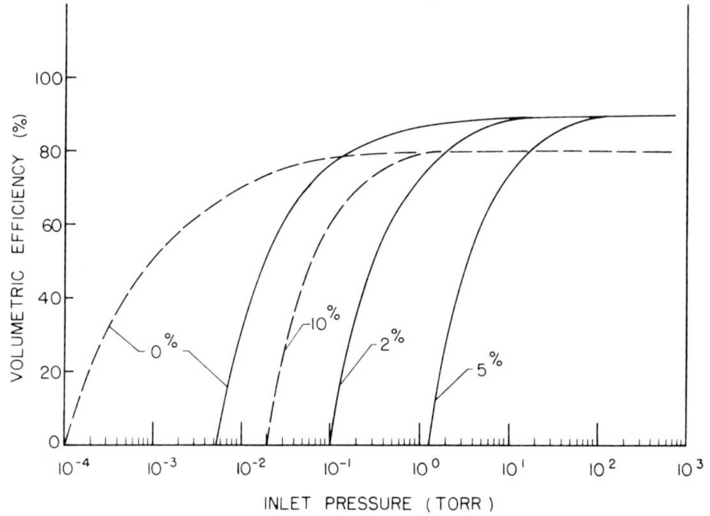

FIG. 8. Effect of gas ballast flow on the pumping speed of single stage and compound oil sealed vacuum pumps. Dashed line, compound; straight line, single stage.

rated water can be decanted either automatically or manually as needed. Depending on the oil properties and the character of the installation (inhibiting effects), this method can work very efficiently which means the water handling rate is as calculated from pump speed or not at all if a stable emulsion is formed. Small gas ballast flow and warm pump operation enhance this water handling mechanism. Presence of rust and other contaminants have an inhibiting effect. Decanting is usually employed

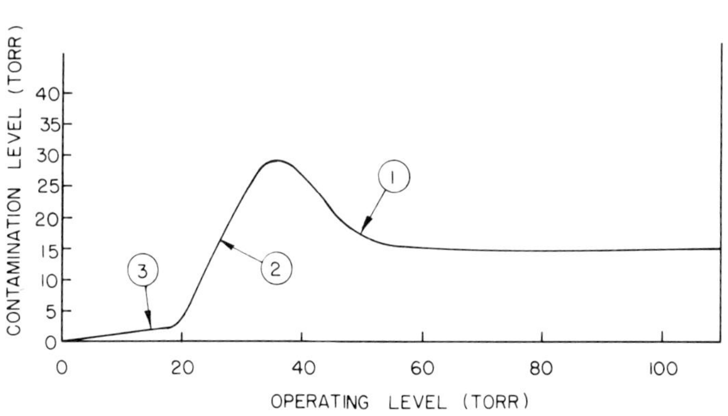

FIG. 9. Oil contamination level when handling water vapor as a function of method used. Numbers indicate: (1) self-induced boil off or gas ballast assisted decanting range; (2) outside gas ballast range contamination level approaches operational level water removal by decanting and gas ballast; (3) gas ballast range.

when water removal per se is of secondary importance, but the process itself occurs in the presence of water, as is the case, for example, with cold water degassing.

4.2.2.7.3. AIR STRIPPING. Oil drying involves blowing dry (compressed) air over the discharge valves. The dry air is quite efficient in removing water from the finely divided oil emulsion spray existing in the discharge pipe. It is particularly applicable to small single stage mechanical pumps handling water at pressures below 5 Torr since, unlike gas ballast, it does not affect the pressure at the inlet of the pump. The quantity of air involved is roughly 5% of pump displacement for operation at below 5 Torr and can be reduced with operating pressure. The hotter the pump, the drier and hotter the stripping air, the more efficient the oil conditioning. This method is wasted on compound pumps.

4.2.2.7.4. BOIL OFF. If the pump temperature can be kept above about 105°C (220°F) water vapor will pass through the pump without condensing. This is accomplished with thermostatically controlled oil heaters. Hot operation occurs spontaneously when the oil contamination level is sufficiently high (inlet pressure above 26 Torr for high speed pumps, 45 Torr for low speed pumps). Adiabatic vapor recompression generates enough heat to support self-generated boil off. Extension of the boil off method consists of separator pumping using a condensate stripper pump in a vacuum still arrangement.

A schematic diagram of the method is shown in Fig. 10. During initial pumpdown, air is expelled through the blow off valve, as well as through the smaller backing (liquid ring) pump. When the throughput is reduced, the blow off valve closes and oil circulation is maintained by the positive oil feed pump.

Heat generated by friction losses and compression keeps the incoming vapor in a superheated state and the distillation pressure is maintained in the separator by the backing pump. The heat exchanger in the sealant line regulates the operating temperature at a desired level.

4.2.2.7.5. OIL RECLAMATION. If the pump oil is circulated through an oil reclaiming device before returning, the pump performance is not affected. The oil is dried outside the pump most often by means of vacuum distillation or absorption. Partial reclamation is adequate with small water loads (inlet pressures below 1 Torr). Full reclamation is necessary when handling oil soluble fluids; this is often inconvenient and expensive but sometimes unavoidable.

4.2.2.8. Contamination. Troublesome contaminants consist of particulate matter and vapors or liquids.

The pump mechanism is not particularly sensitive to small mechanical damage on account of the inherent self-compensating clearance takeup.

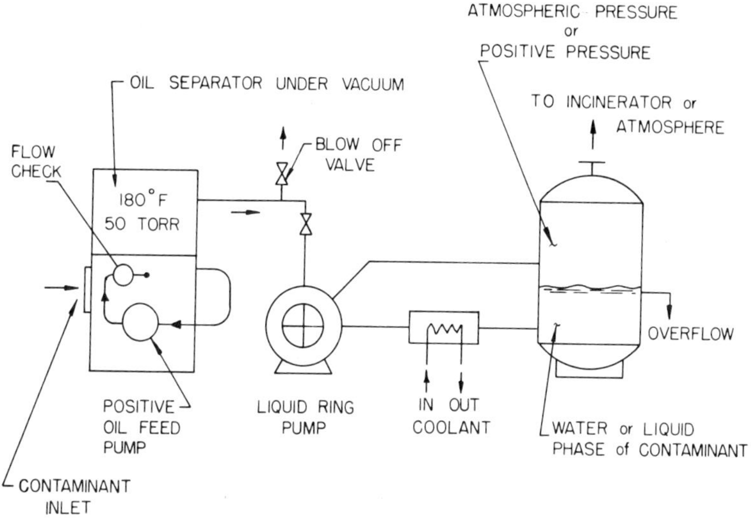

FIG. 10. Vacuum still arrangement of an oil sealed pump for the purpose of oil conditioning and reclamation of contaminant.

Sleeve bearings preferred in most designs are much less sensitive to dirt than are ball bearings. Once the particles separate from oil by gravity, oil storage spaces keep the sediment out of circulation until removed by cleaning. Oil filters can be used but this practice is rarely worthwhile. If the contamination rate is high, necessary filter changes become too frequent and it is simpler to rely on oil changes. With a dust load outside the manageable range it is better to intercept the particles before ingestion into the pump. Dry filters or wet oil impingement traps are most helpful. Selected protection method will depend on volume handled, particle size, and permissible pressure drop. In case of abrasive particles, pumps are sometimes additionally protected by hard surfacing of the working components.

Exposure to low vapor pressure substances is not unusual. These condense in the pump and mix with the lubricant either diluting it or forming varnishes and gumming deposits. Gas ballasting becomes impractical and it would be best to intercept the vapors before arrival at the pump. In many instances, this is not prudent because the pump operates over such broad pressure range that an efficient trap can not be designed or, because of the small quantities of contaminants involved, the maintenance of the trap presents a problem. Most often such situations are handled by periodic oil changes.

Corrosive contaminants, particularly when present in small amounts, are manageable. The lubricant provides a buffer between most metal surfaces and the corrodant. If, in addition, water condensation can be prevented, corrosive damage becomes quite tolerable. Dry gas blanketing can be helpful in preventing atmospheric condensation in the separator. It consists of maintaining the gas separator at a slight positive pressure of dry gas, so as to isolate the pump oil from the effects of ambient humidity.

Soluble vapors, that means substances which dissolve in oil and reappear out of solution at pump inlet, are handled with gas ballast but very inefficiently. Compound pumps are best in these situations provided oil viscosity in the atmospheric stage is preserved at a workable level. Oil conditioning by reclaimers or by vacuum distillation using contamination insensitive backing pump can work well (Fig. 10).

4.2.2.9. Installation. The connected vacuum system and the associated process cycle can have contaminating effects on the pump but an incorrectly installed pump can create problems for the vacuum system as well. Many of the potential difficulties can be avoided by correct installation procedures (Fig. 11).

All vacuum pumps should be vibration isolated. The isolation method depends on the characteristics of the pump mechanism and the power input. Unbalanced pumps, that means pumps producing a net external turning moment, must be firmly restrained. The degree of restraint can consist of simple bolting down to mounting on massive pillow blocks. Balanced pumps, which represent the bulk of modern vacuum pumps, should be vibration isolated at inlet, discharge, and from the floor. With small pumps, inlet and discharge piping can be flexible enough and have

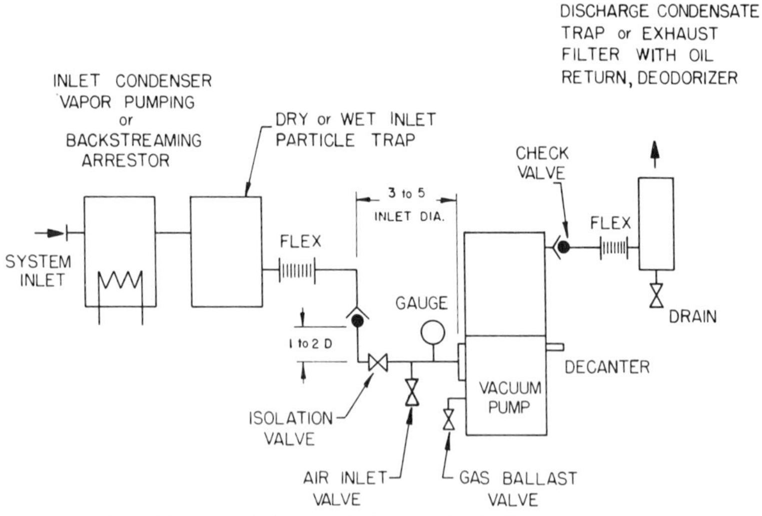

FIG. 11. Installation method of mechanical vacuum pumps.

sufficient inertia to produce the necessary isolation effect. Elastomer tubing or flexible bellows prevent not only propagation of vibration but provide noise isolation as well. Mounting isolation is provided most efficiently by metal or elastomer springs and when selected correctly the pumps can operate on any support capable of supporting their weight.

Design of inlet connection should meet the following objectives: to provide adequate conductance; to prevent pump fluid from entering the process chamber; and to prevent the entrance of particulates into the pump.

Since all oil sealed mechanical vacuum pumps produce oil splash at the pump inlet unless designed with internal splash baffling, the inlet manifolding must be designed to keep oil from traveling upstream of the pump. With small pumps the problem is not too serious but it grows with pump size. In an antisplash piping arrangement it is the length of the horizontal section of manifolding more than the height of the vertical riser that provides the necessary baffling effect. A simple particle trap can be provided by an oil pool in a tee connection at pump inlet.

The three principal problems to be overcome in the design of the discharge manifold are oil loss, return of condensate to the pump, and oil mist in the discharge gases. Although an efficient pump separator will limit the oil loss rate to about 20 cm^3 for each cubic meter of gas (NTP) pumped, not all separators can restrict the oil loss to such a low rate. Some separators which are effective at steady state operation do not prevent excessive oil loss during sudden pressure excursions. A discharge manifold of ample diameter can reduce oil loss, it can also prevent the return of liquid contaminants to the oil separator. Condensation takes place when vapors expelled with the hot exhaust gases condense on the walls of the exhaust manifold and run down to the pump. A drainable condensate trap, usually a tee with a sight gauge, will prevent it. Even the most efficient separator will not remove oil smoke from the pump's exhaust. Elimination of oil mist can be accomplished only by use of exhaust filters. In-line filters are generally preferred and these can be located where convenient and can be connected for outdoor venting.

On multipump installations it is often more economical to exhaust into one central system. The filtering capacity is based on the total gas load to be handled. On such installations, electrostatic precipitators or fume scrubbers are often useful as they exert no backpressure and are relatively maintenance free.

Any well designed vacuum system should include an isolation valve and an air admittance valve. When a vacuum pump is stopped, it is a source of leakage. Closing the isolation valve prior to stopping the pump will prevent a pressure rise in the chamber. Isolating the pump from the process chamber permits an independent check of the base pressure of the pump. The purpose of the air inlet valve is to break the vacuum at the pump inlet just before or at the time the pump is stopped. If this is not done, the pump may flood with oil, making subsequent starting difficult or causing damage because of high hydraulic loads. A gas ballast valve can be used for the purpose of pump venting provided the pump is isolated from the vacuum system.

In case of drive failure oil shut-off devices, such as normally open solenoid valves, will not close and it is possible to transfer the pump oil into the vacuum system. To prevent this, an inlet check valve allowing unrestricted gas flow but preventing reverse liquid flow can be used. If this is not desirable (pressure drop), a discharge check value which closes when the pump stops can be used. The pump will turn in reverse a few revolutions and the pressure between inlet on suction will equalize, preventing oil transfer. Zero speed switch can also be used to actuate the isolation and air admittance valves.

4.2.3. Lobe Pumps

These pumps are known also as "Roots" boosters or mechanical vacuum boosters when backed by another vacuum pump and used at pressures below 100 Torr. It is the mechanical vacuum booster range which is the subject of discussion. Mechanical boosters extend the operating pressure range of oil sealed pumps toward lower pressure and higher pumping speed very efficiently. They can also reduce the effect of oil contamination.

4.2.3.1. Principle of Operation. A lobe pump is a positive displacement pump with two symmetrical rotors rotating in opposite directions within the pump housing. The rotors are synchronized by a gear drive so that these move past one another and within close proximity to the casing walls without touching.

This allows operation at high rotational speeds with little friction. In consequence, the pumps provide high displacement for a given size at a reasonably low investment and operating cost. The volumetric efficiency is the difference between forward flow due to the sweeping action of the rotors and reverse flow through the tight pump clearances from discharge to suction. The reverse flow volume depends on Reynolds number and characteristic gas flow properties associated with it in the turbulent and laminar flow range. For a fixed staging ratio (ratio of pumping speed of the booster and pumping speed of the backing pump) reverse flow losses reach characteristic minimum in the molecular flow range where the volumetric efficiency is typically 80–93% for staging ratios of 10:1 or less

(Fig. 12). Such high staging ratios, with the resultant high pumping speed amplification with respect to the backing pump, are possible only at low operating pressures, usually below 1 Torr. Operation at higher pressures is limited by the temperature of compressed gas. This relationship is given as

$$\Delta T = \{T_1 \times TRC \times [(k - 1)/k][(P_2/P_1) - 1]\}/\eta_v,$$

where ΔT is the gas temperature rise, T the absolute gas temperature at inlet, k the C_p/C_v ratio of specific heat (heat capacity ratio), η_v the volumetric efficiency, P_2 the discharge pressure, P_1 the suction pressure, ΔT

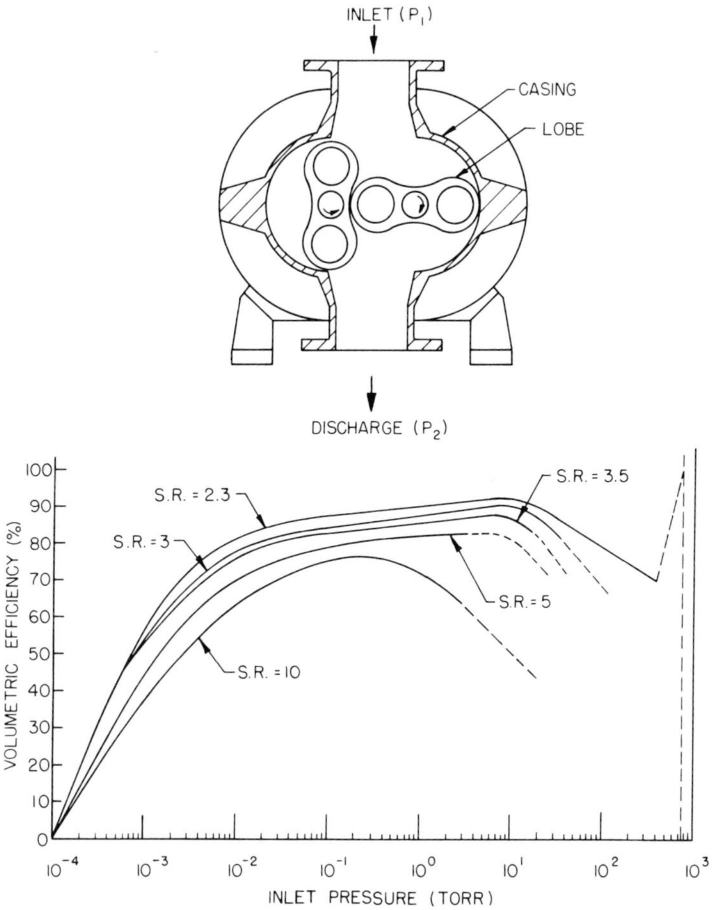

FIG. 12. Typical volumetric efficiency of lobe pumps with tip gear velocity of 3000–4500 fpm, displacement between 100 and 1500 liters/sec, pumping air at different staging ratios. S.R. is the staging ratio.

max 120°C, and TRC the temperature rise coefficient (≤ 1). That coefficient is obtained empirically and represents heat dissipating characteristics of the pump body (Fig. 13).

When gas is heated, pump rotors expand until all end clearances are taken up which results in binding and possible mechanical damage. For this reason the maximum permissible temperature rise should not be exceeded. In addition, discharge temperature cannot exceed 235°C because of thermal compatability of materials of construction.

The maximum pressure ratio permissible at any given inlet pressure which does not result in overheating is called the maximum staging ratio (Fig. 14) and corresponding inlet pressure, the maximum operating pressure. The temperature rise coefficient varies with pressure as shown in Fig. 13. Since it decreases sharply with reduced pressure, overheating does not occur below 0.6 Torr. The effect of hot gas generated by the mechanical booster on the backing pump must be taken into account. At low mass flow rates, load on the backing pump is not significant. With high throughput, compression ratio has to be based on temperature corrected gas volume, while the heat load carried into the backing pump must be accounted for in the heat balance of that pump. Excessive heat is usually dissipated in interstage gas coolers.

When the maximum compression ratio characteristics are known, it is possible to calculate the ultimate pressure of the mechanical booster for any given or assumed backing pressure. The booster produces a maximum compression ratio (P_2/P_1) when the net gas transport is zero as shown in Fig. 15 (no gas admitted at inlet). This value is obtained by measuring the inlet pressure corresponding to a regulator controlled interstage pressure, the booster acting as a one way valve. This operation is also subject to the temperature limitations of maximum permissible staging ratio (Fig. 14).

The horsepower requirements of a mechanical booster can be represented by a rectangular work diagram ($VP_2 - VP_1$) since no internal compression takes place. The operating horsepower can then be expressed as

$$\text{BHP} = 0.000179 \times D \times \Delta P + \text{FHP},$$

where D is the theoretical displacement in liters per second and $\Delta P = P_2 - P_1$ the pressure difference in torr. The numerical value of frictional horsepower can be taken as 60% of the timing gear diameter expressed in inches. In compressor work, lobe pump sizes are designated in terms of gear diameter and rotor length as identifying dimensional parameters.

The open clearances of the pump make the performance characteristics very pressure dependent and these are best summed up in graphical form.

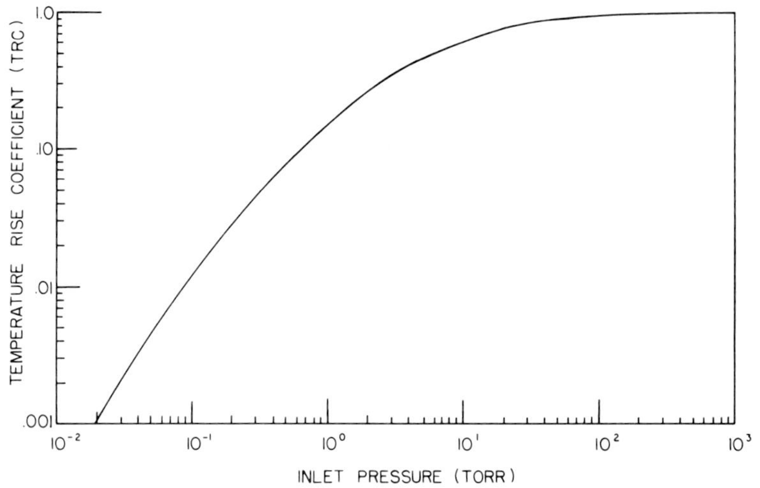

FIG. 13. Dependence of discharge gas temperature on temperature dissipating properties of lobe pumps operating with limitations as in Fig. 12.

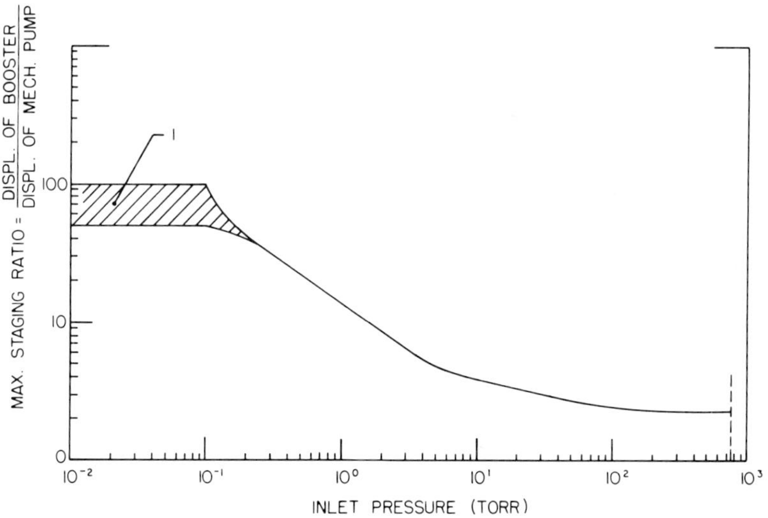

FIG. 14. Highest permissible operating pressure for continuous operation with limitations as in Fig. 12. Maximum staging ratio equals the displacement of booster/displacement of mechanical pump. The shaded area (1) is limited by clearances in the booster (compression ratio limited).

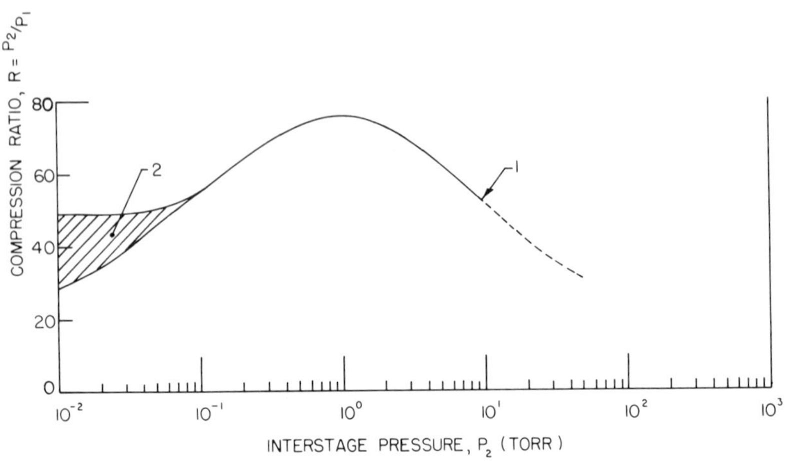

FIG. 15. Maximum compression available from a mechanical booster operating with limitations as in Fig. 12. (1) Continuous operation against 10 Torr or higher is not possible if mechanical booster is operating at blank off; (2) compression ratio below 0.100 Torr is primarily dependent on surface finish.

Volumetric efficiency versus pressure graph shown is valid for air and for gear tip velocities in the 3000–4500 ft/min range (Fig. 12).

Corrections for viscosity and molecular weight for fluid other than air have to be made in the appropriate flow regimes. The heat inertia of the pump rotors is exploited frequently in pumpdown applications where the allowable operating time period between the startup pressure and continuous operating pressure limit is kept short enough to prevent overheating. This period can be estimated based on the assumption that average calculated gas compression horsepower is absorbed by the rotors and that the average rotor temperature should not be more than 120°C. Cast iron is the usual material of construction for mechanical boosters.

4.2.3.2. Basic Performance. Mechanical design characteristics are an important consideration in pump selection. Pumps which use internal differential pressure seals and employ only one mechanical drive shaft seal have a much better service life but cannot be used where the pump lubricant and drive are affected by process gases and vapors. This is also true of canned motor (seal-less) pumps. In the case of corrosive or particle carrying gas loads, internal seals must be used. Pumps equipped with internal seals can be injected with liquids for the purpose of heat dissipation, sealing, or cleaning action, provided the rotors are of a nonpocketing design. When backing into a high forepressure such as presented by water sealed pumps, efficient boosters will produce a compression ratio outside the limits permissible for continuous operation. In such instances, the gas temperature is high but the heat load quite low and can be dissipated readily by either internal rotor cooling or by making the pump inefficient (reduced rotational speed) so that the compression ratio is reduced.

These methods may allow operation into forepressures as high as 40 Torr with zero flow at inlet. With low friction losses, the cooling requirements are very modest even when using large units. On the other hand, when using mechanical booster trains, interstage gas coolers are frequently necessary.

4.2.3.3. System Effects. Frequently, pump rotors present an optically dry appearance on inspection; this is particularly true of differential seal pumps. The absence of an oil film does not mean absence of oil vapor; it simply means that one deals with unsaturated vapor pressure of oil (with wet rotors saturated vapor pressure is the rule). Oil is responsible for the total pressure not being significantly below 1 mTorr (0.5 mTorr is typical) even with a very low partial pressure of air. Backstreaming can be intercepted with cold traps quite efficiently when necessary. Process stream effects on the mechanical booster can be prevented by the use of internal seals so that the bearing lubricant is isolated. Most unsaturated vapors and gases have no effect on the pump metals (cast iron) and the treatment of contaminants is associated with the backing pump only. A similar situation exists with respect to fine particulate matter which also ends up in the backing pump. Vapor trapping, except for the purpose of backstreaming, is most conveniently executed in the interstage. Particle trapping is done with inlet filters if the pressure drop is acceptable and with wet oil traps if the permissible pressure drop is very small.

Mechanical boosters are adaptations of low or medium pressure compressors to vacuum work. At low operating pressure differentials the work done is only a fraction of that for which the pumps were designed and the service life can be excellent. The bulk of mechanical failures which do occur are associated directly with seal malfunctions. These result in loss of adequate lubricant level and subsequent damage to the drive components. Use of differential seals reduces that hazard and the maintenance of one externally accessible mechanical seal is quite simple. Damage due to overheating because of excessive compression ratio can be prevented by the use of thermal sensors, while automatic sequencing is conveniently provided by pressure switches. (See Fig. 16.)

4.2.4. Liquid Piston Pumps

The liquid piston pumps (also known as liquid ring pumps) are positive displacement pumps. An excentrically located pinwheel rotor traps

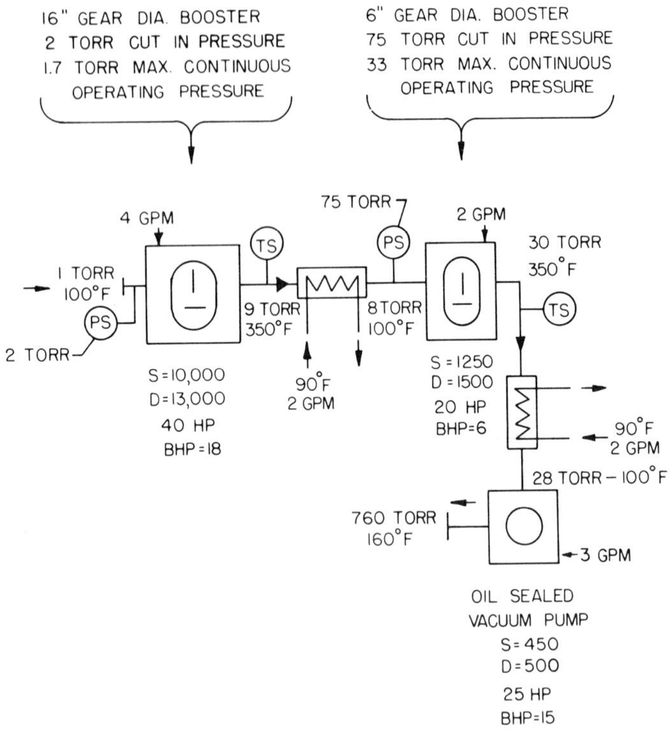

FIG. 16. Mechanical booster train selected for most efficient operation at 1 Torr or below. PS is a pressure switch set as indicated; TS is a temperature switch set at 375°F gas temperature; S the pumping speed (CFM); D the free air displacement (CFM). Total BHP is 39 at design point of 1 Torr; total connected power is 85 HP; and total cooling water needed is 35 GPM at 90°F.

pockets of gas in the space limited by the blades, the end walls of the cylinder in which it rotates, and the inner surface of the liquid ring as the blades enter into it on the compression stroke. Inlet and outlet ports are strategically located at the flat ends of the pump cylinder. The inlet port allows gas access to the liquid free space, while the the outlet port is located toward the top dead center in such a manner as to allow gas and liquid discharge once the gas pocket has entered the ring of liquid. Angular location and size of the discharge port (Fig. 17) determine the region of most efficient operation, both from the point of view of volumetric efficiency and operating power. Since such exhaust port location is fixed, a broader range of efficient operation is provided by the use of discharge valves.

An even more universal pressure range adaptation is exhibited by a compound liquid ring pump, particularly for pressures below $\frac{1}{2}$ atm. The

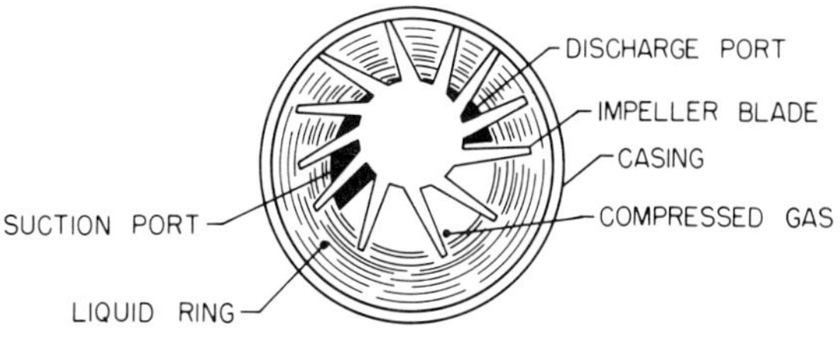

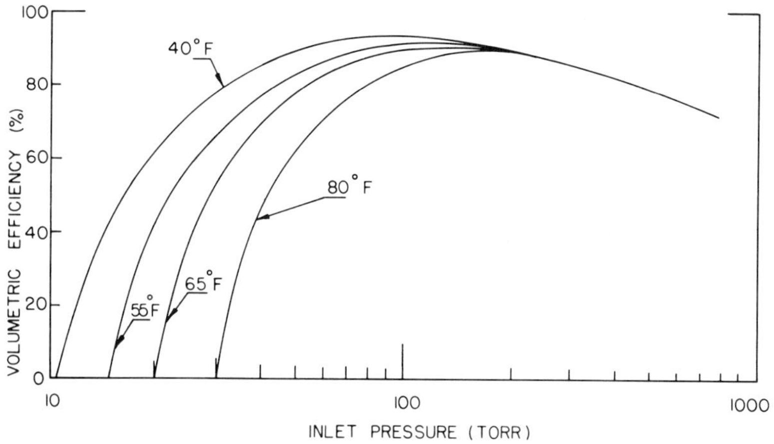

FIG. 17. Volumetric efficiency of compound liquid ring pump sealed with water at different temperatures; assumes 10°F temperature rise on completion of the compression cycle.

arrangement is similar to that of a compound rotary oil sealed pump except that a sealant is injected into the first stage and is discharged after sealing the atmospheric stage on the way out. In case of water, gas–liquid separation occurs at the drain or more often in a gas–liquid separator. Unless limited by operating pressure, part of the water is returned to the pump and makeup water for temperature control is fed into the pump. Both, the open ended system with once-through-water or partial water recovery arrangement, can be converted into a full recovery system with the recirculating fluid cooled through heat exchangers. This can represent a closed loop system with respect to external environment. While valveless single stage pumps are compression ratio limited to a pressure reduction of 10:1, compound pumps are either vapor pressure or gas solubility limited in their low end performance.

In order to exhaust to atmospheric pressure, the liquid ring vacuum pump must be able to produce a liquid pressure equal to or higher than atmospheric. This establishes the minimum rotational speed, while maximum speed is determined by more than linear increase in friction losses. Liquid ring pumps are quite inefficient and as much as 90% of the connected horsepower is dissipated in heat. Since hydraulic work represents the more significant portion of the effort of operating the pump, horsepower input is almost constant over the operating pressure range and very nearly synonymous with the rating of the drive.

The major advantage of a liquid ring pump is that the sealing fluid need not possess any lubricating characteristics and there are no critical clearances involved in the pump mechanism.

Using low vapor pressure sealants, the pump can be used at pressures down to several torr and a similar extension into low pressures operation is available by staging it with an air ejector. The resulting combination is simple but inefficient. A more efficient combination results from backing a mechanical booster with a liquid ring pump provided it is operated within its design parameters with respect to the permissible temperature rise, the interstage pressure associated with it, and staging ratio limitations. When used for backing of steam ejectors, the condensing effects of the liquid ring pump (direct contact condensers) are utilized. The condensing rate is limited by the ability to dissipate the imposed heat load. Cavitation can be a serious limitation when operating near the vapor pressure of the sealing fluid and is avoidable by the injection of air or inert gas. This is more practical with compound pumps because gas injection can be introduced into the interstage without a pronounced effect on suction pressure. Corrosion, if not anticipated with the selection of materials of construction or protection by surface coatings, can be catastrophic. Plating out of solids because of water hardness can produce binding, even with the generous clearances of the pump.

4.2.5. Ejectors

The operating medium (motive fluid) which can be vapor, gas, or liquid under pressure, enters the ejector inlet through the ejector nozzle. The ejector nozzle converts the pressure head of the motive fluid into a high velocity stream as it emerges from the nozzle into the suction chamber. Pumping action occurs when the fluids present in the suction chamber are entrained by the motive fluid, acquire some of its velocity, and are carried into the diffuser.

The velocity of the mixture is recovered to a pressure greater than the suction pressure but lower than the motive pressure. The diffuser pressure must be equal or higher than the backing pressure for stable operation.

The capacity of steam ejector is a direct function of the weight of the motive fluid used and its performance is best characterized by a graph representing the weight ratio of gas pumped to motive fluid required as a function of operating pressure (Fig. 18). Pump design is termed critical when the fluid velocity in the diffuser is sonic or noncritical if subsonic. Commonly used critical design ejectors are very sensitive to changes in operating parameters which are motive pressure, discharge pressure, and suction pressure. This sensitivity to a change in operating conditions is amplified for multistage installations. When the inlet pressure is raised, the throughput of the system increases and discharge pressure of each stage is also increased. A limit is reached when the discharge pressure in one of the interstage regions equals the maximum which the steam can attain in the diffuser during compression. If the flow is increased further, that jet collapses with the resulting flow of discharge gas and steam back up through the system into the connected process vessel. Consequences of such a blow back can be unpleasant as can be backstreaming when operating at zero throughput.

Interstage condensing is used for economy reasons whenever possible. Because of prevailing cooling water temperatures, condensing is possible only between third and second, second and first (atmospheric) stages. This implies that for systems having more than three stages, the motive fluid of each stage must be added to the process load. Each successive stage has to handle an ever increasing flow of gas and vapors. Despite this, steam ejectors are quite efficient precisely at low pressures but are subject to size limitations. Minimum mass flow rate limitations (orifice size) of the nozzle make it impractical to produce steam ejectors with capacities of less than 5000 liters/sec at 10 mTorr, 5000 liters/sec at 0.1 Torr, and a 50 liters/sec ejector at 1 Torr may not be quite practical.

Simplicity, reliability, and wide choice of materials of construction make the steam jet important, particularly in processing industries. On the other hand, high operating cost at pressures above 20 Torr and environmental incompatability in some applications can make its use difficult in spite of its superb simplicity.

Unlike the vapor ejector, which represents a single point design that means it has an optimum for a single set of conditions, the liquid jet performs very much like a positive displacement pump. This is not unexpected in view of the density difference between liquid and gas or vapor phase. The faucet aspirator, with a pumping speed between 0.1 and 0.01 liters/sec and operating range limited by vapor pressure of water, represents one end of a scale, while self-contained recirculating units with a 40 liters/sec cooling water flow and 50 liters/sec gas pumping capacity represent the other extreme. Primary applications of liquid jets involve pumping of condensible loads since large circulating coolant flow rates

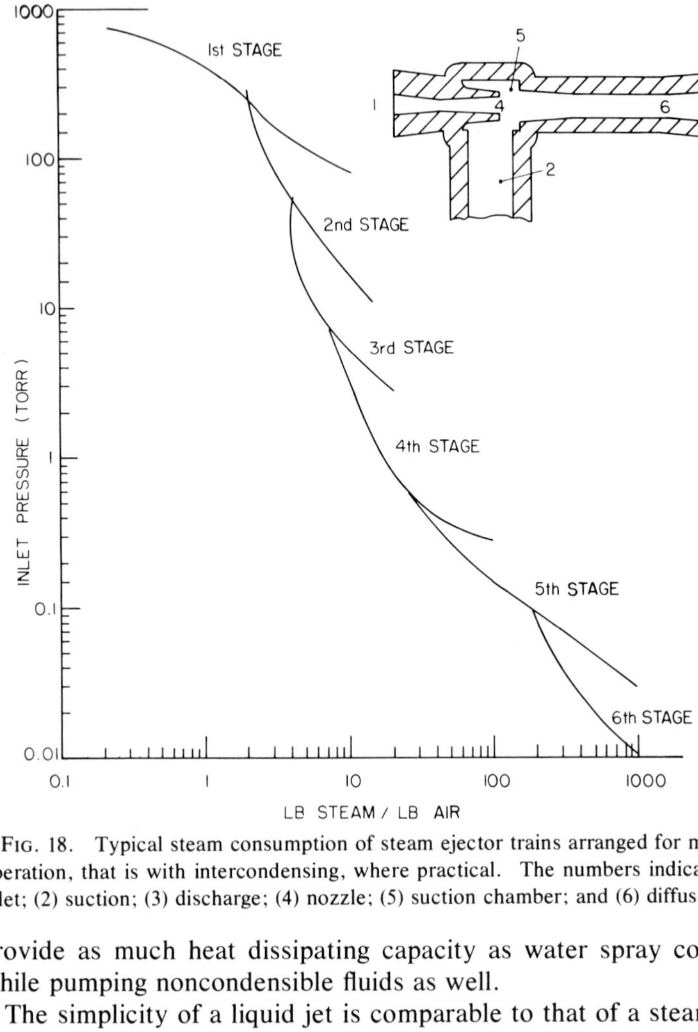

FIG. 18. Typical steam consumption of steam ejector trains arranged for most efficient operation, that is with intercondensing, where practical. The numbers indicate (1) steam inlet; (2) suction; (3) discharge; (4) nozzle; (5) suction chamber; and (6) diffuser.

provide as much heat dissipating capacity as water spray condensers, while pumping noncondensible fluids as well.

The simplicity of a liquid jet is comparable to that of a steam ejector. In self-contained units, the complications of the circulating pump, piping, and separator introduce difficulties in terms of available selection of materials.

4.2.6. Sorption Pumps

Substances possessing large surface area per unit of weight can be used to sorb gases. If, in addition, these surfaces have pore sizes similar to the molecular diameter of the species sorbed, significant gas pumping is possible. Such properties are exhibited by activated charcoal, activated alumina (Al_2O_3), and synthetic molecular sieves, such as alkali aluminosilicates (zeolite).

Artificial zeolites are produced with best porosity control and are most often used for sorption pumping. Zeolite 13X exhibits an average pore diameter of 10 Å, with a surface area of 1000 m^2/gm. The pore diameter is compatible with the most commonly pumped gases and vapors but not with hydrogen, neon, and helium which are only weakly sorbed. The adsorption of gases depends not only on the temperature of the sorbent (temperature isotherm) but also on the pressure above the sorbing surface. Pumping speed decreases with coverage so that it is not constant with time. Ultimate pressure depends on the composition of gases present at the start of the pumping process. Since oxygen, nitrogen, CO_2, hydrocarbons, water vapor, and argon are pumped efficiently, the residual pressure is determined by residual light gases (H_2, He, Ne).

With sorbent at a temperature of liquid nitrogen, an ultimate pressure of about 10 mTorr can be obtained and, if light gases are absent, base pressure of 1 mTorr can reasonably be reached. Adsorption isotherms of the sorbent material for different gases are often available from the supplier. Zeolite 13X (Figs. 19 and 20). when at liquid nitrogen temperature, will sorb nitrogen gas at a rate of about 200 Torr liters/g at a pressure between atmospheric and 10^{-2} Torr. The sorption rate decreases sharply at lower

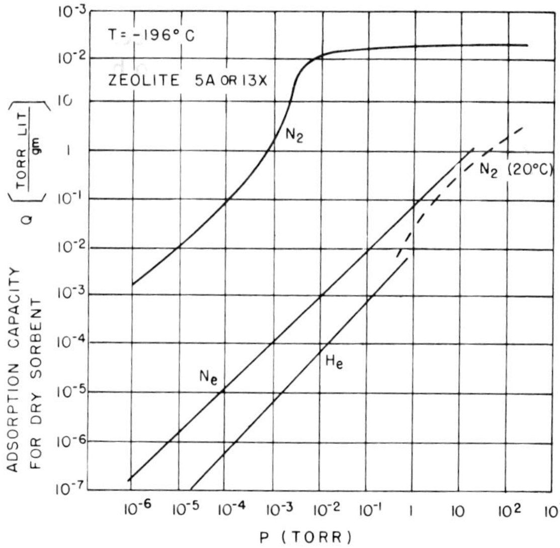

FIG. 19. Approximate adsorption isotherms for molecular sieves desorbed at temperatures between 200 and 350°C.

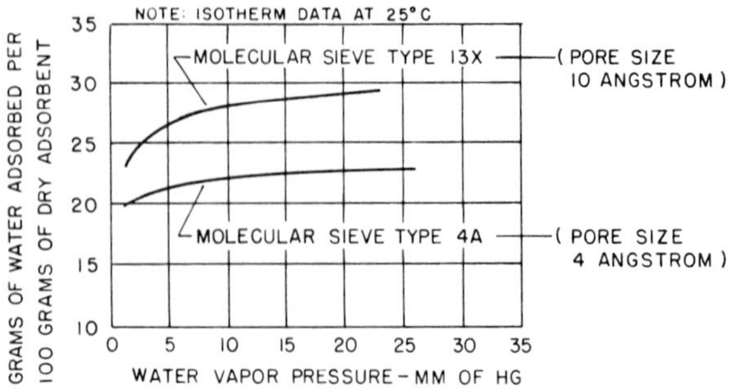

FIG. 20. Adsorption isotherms for dry molecular sieves for water vapor.

inlet pressure. Desorption is performed at room temperature except when handling moist gases in which case 200°C bakeout is advisable because of the high binding energies involved. Small vessels are evacuated by just one pump, while larger vessels can be sequentially pumped by several sorption pumps. When large pumps are contemplated, poor heat conductivity of the sorbent has to be taken into account. Where sizable chambers or frequent cycling are involved, it is more practical to remove the bulk of the gas present by another pump and then finish roughing with a sorption pump. In such a fashion, desorption during processing becomes unnecessary. Where presence of hydrocarbons has to be excluded even as a potential hazard, preroughing is accomplished by oil free pumps such as air or steam ejectors, aspirators, liquid ring pumps, dry piston pumps, or unlubricated multivane pumps.

4.3. Diffusion Pumps*

High vacuum pumping systems usually include at least one diffusion pump and one mechanical pump.

Mechanical vacuum pumps are used to remove about 99.99% of the air from the chamber (rough pumping). The remaining air, down to any residual pressure from 10^{-3} to 10^{-9} Torr, is removed by the diffusion pump discharging into the mechanical pump.

Diffusion pumps are normally used when constant high speeds for all gases are desired for long periods of time without attention.

Diffusion pumps cannot discharge directly into the atmosphere. A mechanical pump is required to reduce the pressure in the vacuum system to the correct operating range. This operation is commonly termed rough pumping or roughing. After suitable operating pressure conditions are reached, the diffusion pump can take over. The mechanical pump is now used to maintain proper discharge pressure conditions for the diffusion pump at the foreline connection. This operation is called backing or forepumping.

Diffusion pumps are essentially vapor ejectors specialized for high vacuum applications. In the past, the gas–vapor diffusion aspects and the vapor condensation have been overemphasized giving the established name for the pump (occasionally also called condensation pump).

The original designs go back to 1915. The basic design form was stabilized approximately 10 years later or 50 years ago. Modern pumps direct the vapor stream at high velocity in the pumping direction. Pumped gas entrainment into this stream, in principle, is not very different from steam ejectors or other vapor pumps or compressors. The original pumping fluid was mercury. Oil-like substances were first used in 1928.

In the following sections, the discussion is devoted mainly to oil diffusion pumps. Accessory devices such as baffles and traps are considered only when directly relevant (see Section 1.3.2 and Chapter 6.5).

4.3.1. Basic Pumping Mechanism

The typical diffusion pump consists of a vertical, usually cylindrical, body fitted with a flanged inlet for attachment to the system to be evac-

* Chapter 4.3 is by M. H. Hablanian.

METHODS OF EXPERIMENTAL PHYSICS, VOL. 14

ISBN 0-12-475914-9

uated. The bottom of the cylinder is closed, forming the boiler, which is fitted with a heater. The upper two thirds of the body is surrounded by cooling coils. And outlet duct (or foreline) is provided at the side of the lower pump body for discharge of the pumped gases and vapors to the mechanical forepump. The sectional view in Fig. 1 illustrates a schematic arrangement of a single stage diffusion pump.

A jet-forming structure (chimney) is located within the pump body. This consists of a concentric cylinder partially capped and fitted with flared ends to form jets through which the pump fluid vapors can emerge at high velocity and in a predetermined direction. There are no moving mechanical parts.

In operation, the working fluid in the boiler of the pump is heated by means of the electric element clamped to the lower body and a vapor stream is created. This vapor rises in the chimneys of the jet structure and is emitted through the annular jet in a downward and outward direction against the water-cooled wall of the pump body.

Gas molecules arriving at the pump inlet are entrained in the stream of working fluid vapor and are given a downward momentum. The vapor stream normally flows at supersonic velocities. The gas–vapor mixture travels downward toward the foreline. When the oil vapor constituents

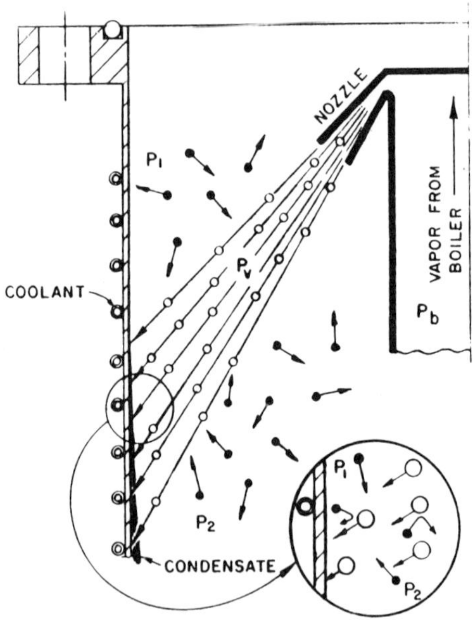

FIG. 1. Schematic arrangement of a single stage diffusion pump. Open circles, vapor jet molecules; closed circles, gas molecules.

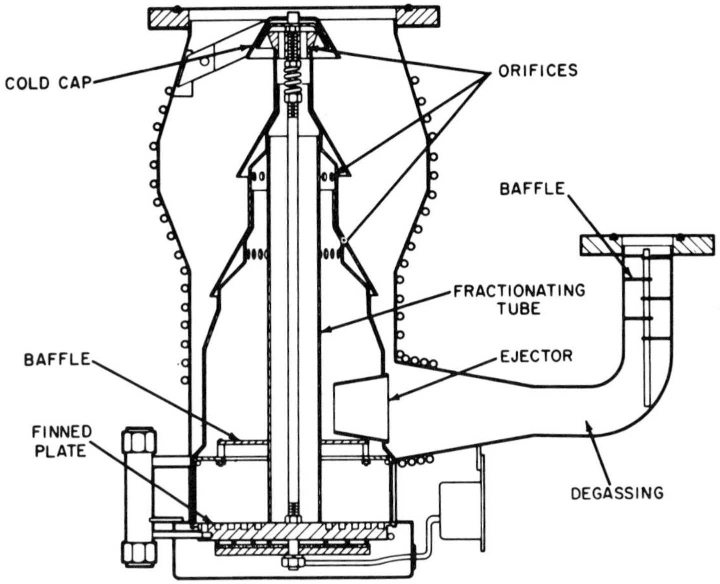

FIG. 2. Multistage diffusion pump.

of such a jet stream strike the water-cooled wall of the pump body, they are condensed and returned to the boiler in liquid form. The entrained gas molecules continue their path toward the exit, where they are removed by the mechanical forepump.

The condensed oil vapors return to the boiler. Heat is once more added and the oil revaporized to maintain the vapor flow to the jet assembly and the continuity of pumping mechanism. A sectional view of a typical multistage diffusion pump is shown in Fig. 2.

The pumping action in a diffusion pump results from collisions between vapor and gas molecules. It is more difficult for gas molecules to cross the vapor stream in the counterflow direction. Thus a pressure (or molecular density) difference is created across the vapor stream. The pressure ratio created by the vapor stream can be approximately expressed by

$$P_2/P_1 = \exp(\rho VL/D),$$

where ρ is vapor density, V its velocity, and L the width of the stream. D is related to the diffusion coefficient and depends on molecular weights and diameters of the vapor and gas[1]:

[1] R. Jaeckel, "Kleinste Drücke." Springer-Verlag, Berlin and New York, 1950.

$$D = \frac{3}{8(2\pi)^{1/2}} \left(RT \frac{M_1 + M_2}{M_1 M_2}\right)^{0.5} \left(\frac{\sigma_1 + \sigma_2}{2}\right)^{-2},$$

where subscript 1 refers to the pumped gas and 2 to the pumping fluid. From this it may be appreciated that the pressure ratio is much lower for lighter gases.

4.3.1.1. Operating Range. The pressure range for the use of diffusion pumps is between 10^{-10} and 10^{-1} Torr. Without the assistance of cryogenic pumping and without baking, the lowest inlet pressures conveniently achieved are in the 10^{-8} Torr range. At the high pressure end, the steady state pressure (at the pump inlet) normally should not exceed about 1×10^{-3} Torr for current pump designs. With the aid of cryogenic pumping, liquid nitrogen cooled traps, for example, inlet pressures near 1×10^{-10} Torr range can be obtained. The significance of ultrahigh vacuum capabilities of diffusion pumps is in the reduction of system contamination to a low enough level so that other sources of contamination become predominant. For many ultrahigh vacuum applications, ion pumps are often preferable and diffusion pumps are usually chosen for their higher mass flow or high gas load capacity. Baking and subsequent cooling of pumped chambers and inlet ducts can also extend the pressure range because it can reduce outgassing as well as produce sorption pumping effects.

4.3.1.2. Number of Stages. The number of pumping stages or vapor nozzles depends on particular performance specifications. A single stage pump would have conflicting requirements of high pumping speed and high compression ratio. Generally, the reasons for multistage pump designs are analogous to the reasons for having gear shift trains or multistage electronic amplifiers. Normally, the first stage at the inlet has high pumping speed and low compression ratio and the last (discharge) stage vice versa. Small pumps often have three stages and large ones five or even six. The initial stages have annular nozzles, the discharge stage sometimes has a circular nozzle and is called an ejector. There are no principle differences in such variations of geometry, although certain advantages are gained by one or the other choice (see Section 4.3.3.2, for example). Sometimes, to obtain certain performance effects, two diffusion pumps can be used connected in series. This has an effect of increasing the number of compression stages and it allows the use of different pumping fluids in the two pumps.

4.2.1.3. Vapor and Gas Distributions. As noted in Section 4.3.1.4, in regard to their basic pumping action, diffusion pumps or high vacuum vapor pumps are related to oil or steam ejectors. The potential energy of elevated pressure inside the jet assembly (boiler pressure) is converted to the kinetic energy of high velocity vapor stream or jet after it passes through a nozzle. The gas is pumped by the jet by momentum transfer in the direction of pumping. In another way, the pumping action can be related to the molecular pumps. Instead of a solid surface moving at high speed, we have a jet of vapor which imparts the necessary collisions. Because the pumping fluids used in diffusion pumps are easily condensable at room temperature, a multistage nozzle–condenser system can be fitted in a compact space (a steam ejector of equivalent volumetric capacity may fill a small room).

The interaction of pumping vapor and the pumped gas can be illustrated experimentally by finding the density distribution of both species in the pumping region. The gas density distribution obtained by traversing the pumping region with an ionization gauge probe is shown in Fig. 3 and the distribution of vapor arriving at the pump wall in Fig. 4. The pattern of diffusion of pumped air into the vapor jet, its relative absence in the core of the jet near the nozzle exit, and the gradual compression of the gas can be deduced from Fig. 3. The repetition of the process in the second stage is also evident.

4.3.1.4. Basic Performance Curve. The pumping performance of a diffusion pump is usually displayed in the form of a plot of pumping speed versus inlet pressure, as shown in Fig. 5. The graph consists of four distinct sections. To the left, the speed is seen to decrease near the limit of obtainable vacuum. The constant speed section results from the constant gas arrival rate at molecular flow conditions and a constant capture efficiency of the vapor jets. The part marked "overload" is a constant throughput section which indicates that the maximum mass flow capacity of the pump has been reached. In the last section, at the right, the performance is highly influenced by the size of the mechanical backing pump.

The performance of diffusion pumps is fundamentally similar to any other pump, compressor, blower, ejector, and similar devices. The essential elements of flow and pressure relationships are analogous.

4.3.2. Pumping Speed

4.3.2.1. System Speed and Pump Speed. The pumping speed measured by the AVS Standard refers to the inlet plane of the pump. Ducts connecting a pump to a chamber, and baffles and traps produce an impedance to flow resulting in a pressure difference or pressure drop. Under molecular flow conditions, it is common for a baffle or a trap to have a conductance numerically equivalent to the speed of the pump. Thus, the net speed at the pumping port in the chamber can be easily half or a third of the pump speed.

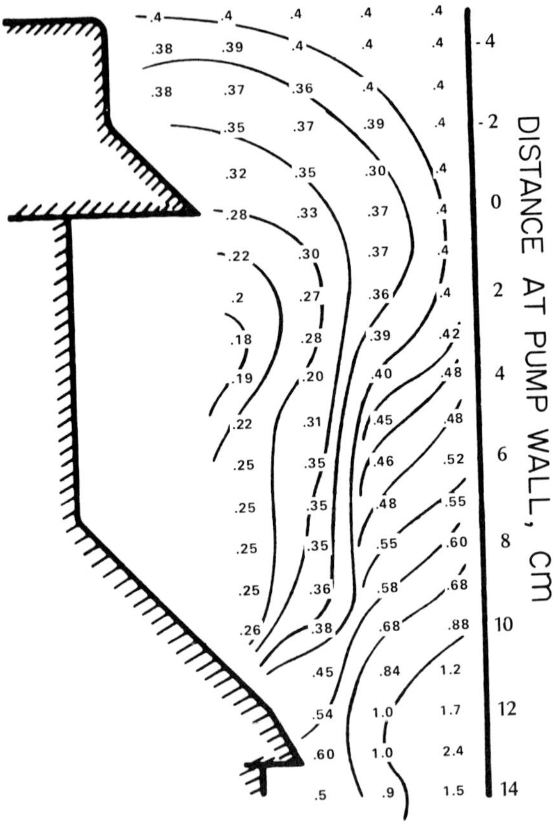

FIG. 3. Gas density distribution obtained by traversing the pumping region with an ionization gauge probe.

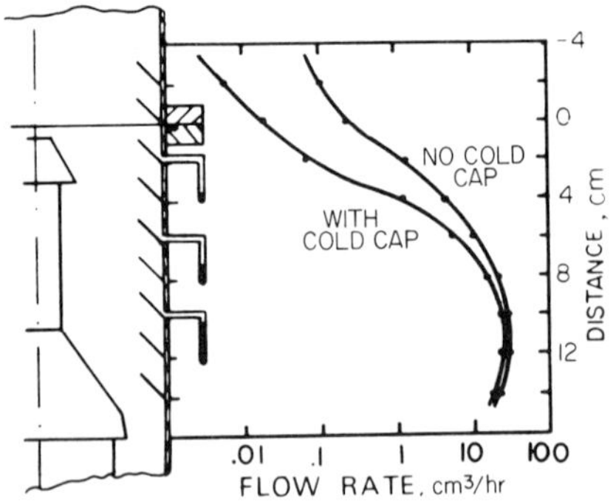

FIG. 4. Distribution of vapor arriving at the pump wall.

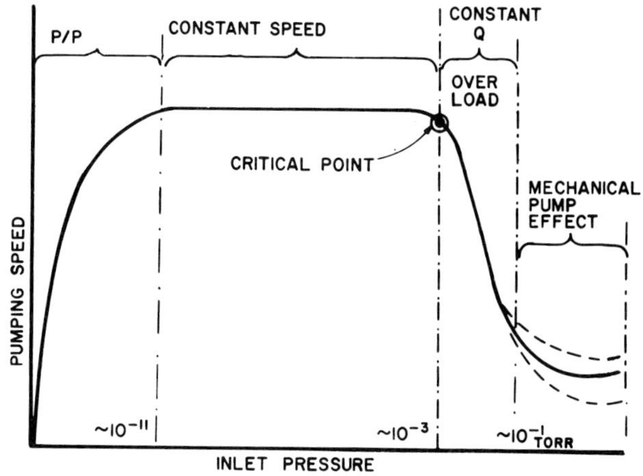

FIG. 5. Pumping performance of a diffusion pump displayed in the form of a plot of pumping speed versus inlet pressure.

In steady state flow, throughput remains constant so that the maximum throughput capacity of the pump is not affected by baffles, orifices, etc. The lower net speed at the chamber results, of course, in higher pressure for any given gas load. Because of outgassing, the gas load in high vacuum systems is never zero, and the ultimate pressure in the vacuum chamber, therefore, is always higher than the ultimate pressure of the pump itself.

4.3.2.2. **Speed Efficiency—Capture Probability.** The speed is determined by the American Vacuum Society Standard as

$$S = Q/(P - P_0),$$

where Q is the flow rate (throughput) and P_0 the ultimate pressure prior to the experiment. Figure 6 shows the standard recommended test setup.

The pumping speed of diffusion pumps is nearly proportional to the inlet area, but the larger pumps are somewhat more efficient as shown in Fig. 7. The entrance geometries of small and large pumps are not strictly similar. Thus, it is possible that the largest pumps can be constructed with a speed efficiency of 50% referred to the inlet plane rather than the conical surface where pumping action occurs. This can become an important consideration in systems where the desired pumping speed is so high that there is simply not enough wall space available for attaching additional pumps.

4.3.2.3. **Speed and Throughput.** Traditional emphasis on the pumping speed versus the inlet pressure relationship has caused some misconceptions about the pressure range in which diffusion pumps can be

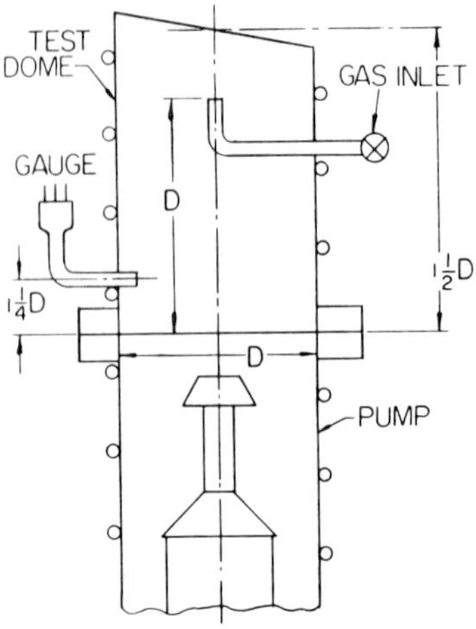

FIG. 6. Standard recommended test setup.

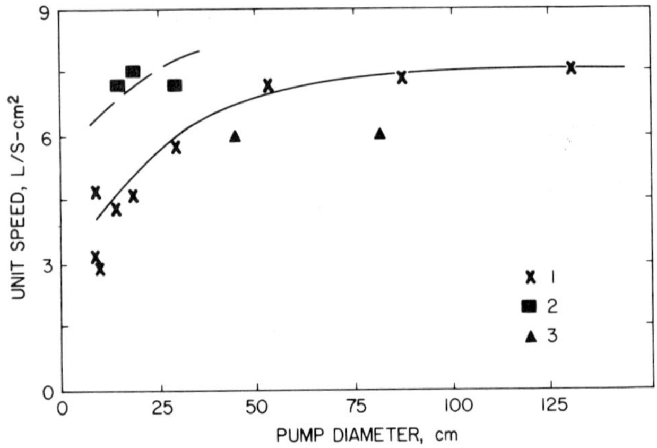

FIG. 7. Efficiency of larger pumps. (1) Conventional pumps, (2) bulged body pumps, (3) high throughput pumps.

used. Thus, it is often assumed that the diffusion pumps are unstable at system pressures above 10^{-3} Torr, but, for example, turbomolecular pumps are stable up to 2×10^{-2} Torr. Such judgments should not be made without regard to the pump size and system gas load. This is illustrated in Fig. 8 which shows pumping speed curves of common diffusion and turbomolecular pumps. In this particular comparison the smaller turbomolecular pump can be seen not to have any advantage over the larger diffusion pump in regard to pumping speed, gas load, and throughput relationships.

It is much simpler to see these relationships by looking at a throughput

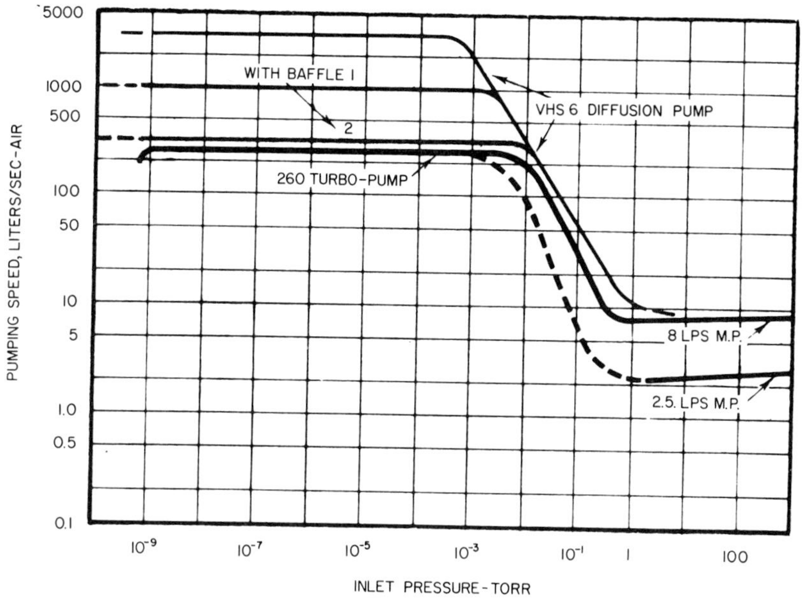

FIG. 8. Pumping speed curves of common diffusion and turbomolecular pumps.

versus inlet pressure graph, Fig. 9. Any combination of throughput and pressure included in the shaded region can be chosen for operation, provided that the pumped gas admittance is restricted (throttled) at inlet pressures above approximately 1×10^{-3} Torr. The pressure stability region can be extended even to 10^{-1} Torr (dashed diagonal line in Fig. 9) if the pumping speed is reduced. It also depends to some extent on the size of the backing pump (as indicated in Fig. 5).

4.3.2.4. Size Effects. Diffusion pumps are made from 5 cm (2 in.) to 120 cm (48 in.) inlet flange sizes. An obvious difference between the smaller and larger pumps is the distance which the pumping fluid (oil vapor) must travel from the nozzle to the condensing surface or pump wall.

It is apparent that the density of oil vapor by the time it arrives at the pump wall is lower in a large pump compared to a small one. This is the reason why the pumping speed plateau in the case of a 5-cm diffusion pump may be extended up to the pressure of 3 microns, while for a 120 cm (48 in.) pump it may be only 3.10^{-4} Torr. This is a significant difference of an order of magnitude in the steady state of stable operation region and must be considered in system design.

Special measures are necessary to improve the higher pressure operation of large pumps (relatively higher heat input, higher number of stages).

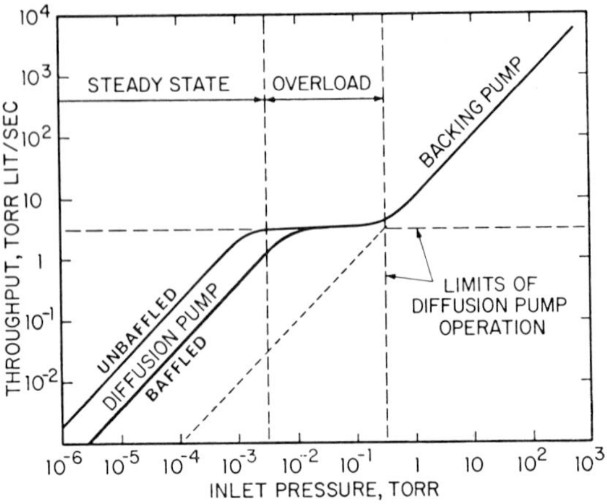

FIG. 9. Throughput vs inlet pressure graph.

It may be noted that the small and large pumps are not geometrically similar. The boiler pressure in small and in large pumps is approximately the same because we are limited in maximum temperature of the oil to avoid thermal breakdown. Therefore, the vapor density at the nozzle exit is nearly the same for all pumps. However, the vapor expands in both axial and radial directions and we may assume that the density is inversely proportional to the square of the distance from the nozzle. Thus, near the pump wall the jet is rare enough to be less effective in pumping gas molecules at higher pressures.

4.3.2.5. Speed for Various Gases. The pumping speed must be considered in relation to the partial pressure of each gas species. When pumping speed is measured, the values obtained near the ultimate pressure of the measurement system become meaningless. The total pressure values cannot be used for obtaining speed due to the uncertainties of gas composition and the condition of the gauge. The composite picture should look as shown in Fig. 10. Each gas has its separate speed and, what is more important, separate ultimate pressure. The normally measured "blank-off" is due to pump fluid vapor, cracked fractions, and perhaps water vapor remaining in the system. Connecting such "ultimate" pressure point by a smooth curve to the horizontal section of the air curve does not serve a functional purpose and can be misleading.

Although the arrival rate, expressed in liters per second, should be inversely proportional to the square root of molecular weight, the lighter gases are not pumped with the same efficiency as air (or nitrogen) and pumping speed values for different gases do not peak at the same power setting. Despite expectations, helium and hydrogen speeds can be much

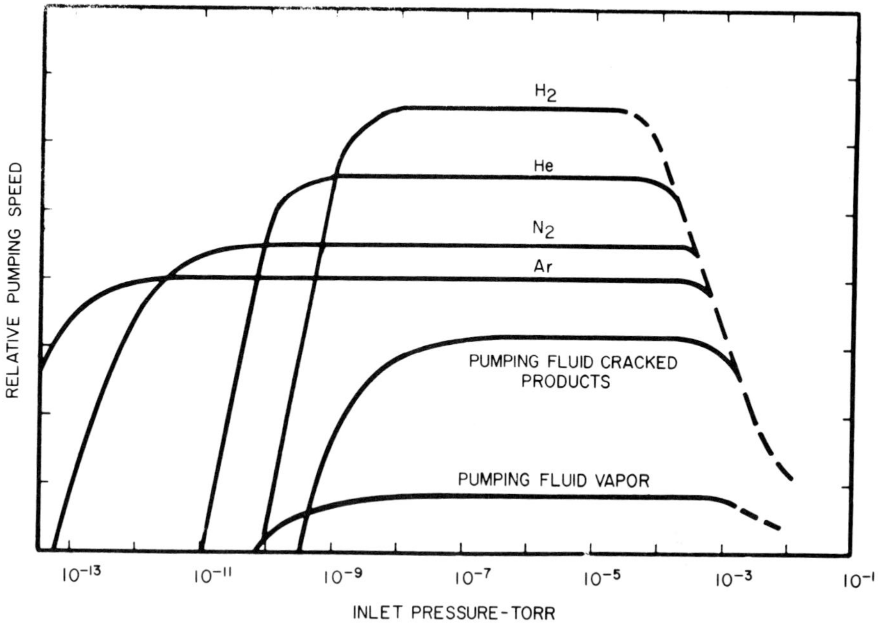

FIG. 10. Pumping speeds for various gases which may be present in a test chamber.

lower than air for poorly designed pumps. When actual values are necessary, separate measurements must be made. Usually the gases present in vacuum systems such as hydrogen, helium, water vapor, carbon monoxide and dioxide, nitrogen, and argon are pumped at approximately the same speed. It is common to see helium speed about 20% and hydrogen speed about 30% higher than air. The impedance of baffles and traps is lower for lighter gases compared to air. Thus, the net system speed values for lighter gases are relatively higher than those obtained for the same baffle with air. Figure 11 shows a typical set of results for helium and hydrogen in a large pump.

4.3.3. Throughput

4.3.3.1. Maximum Throughput. Referring to Section 4.3.2.3, it may be observed that maximum throughput is often the important aspect of diffusion pump performance rather than pumping speed. The value of the maximum throughput determines the amount of power required to operate a given diffusion pump. Dimensionally, throughput and power are equivalent. For pumps of current designs using modern pumping fluids, approximately 1 kW of power is required to obtain a maximum throughput of 1.2 Torr liters/sec.

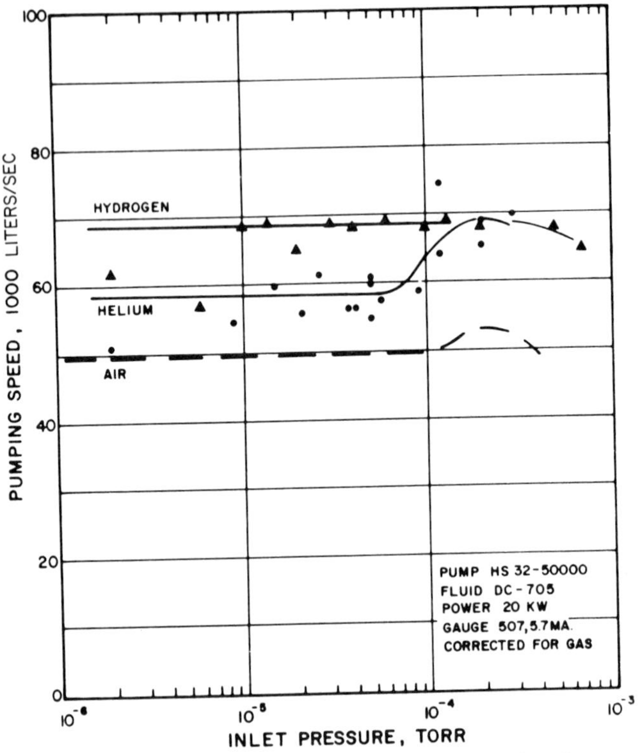

FIG. 11. Typical set of results for helium and hydrogen in a large pump.

For systems which remain under vacuum for long periods of time, the maximum throughput is of little value. In such cases, provisions can be made to reduce the power after the initial evacuation or to use lower power heaters.

Operation of the pump at half power is possible without changes in pump design. Special designs can be made for low power (low throughput) operation with appropriate attention to corresponding reduction of forepressure tolerance (see Section 4.3.4.4).

For rapid frequent evacuation and for high gas load application, the value of maximum throughput determines the choice of pump size. Knowing this value and the inlet pressure at which it is reached (the point where the speed suddenly decreases), it is possible to reconstruct both the speed versus pressure and throughput versus pressure characteristics. The former is constant below and the latter above this prominent pressure point.

4.3.3.2. Pressure versus Throughput Curve. If the graph in Fig. 9 is replotted interchanging the axis, a more customary arrangement of dependent and independent variables is obtained, Fig. 12. Then it will be simpler to view the maximum throughput and the pressure stability limits and the concept of overload when those limits are exceeded. The more convenient way of looking at this is to keep in mind that for a given system gas load presented to the pump there is a resulting inlet pressure. This will help in selecting the required pump size and in distinguishing between the requirements of evacuating a chamber and maintaining a desired operation pressure at a given process gas load.

A common example of tradeoff between pumping speed and pump inlet pressure range is given by the net pumping speed curve when the pump is used with baffles and traps. The constant speed region is extended to higher pump inlet pressures at a cost of some reduction of net speed. Another example is given by a pump designed specifically for sputtering work. It is based on a 25-cm diffusion pump design with a maximum throughput of about 7 Torr liters/sec, it is half as tall as a conventional pump and its steady state, constant speed region has been increased tenfold to 2×10^{-2} Torr with a corresponding reduction of pumping speed to near 400 liters/sec. The pump includes a closely spaced chevron baffle to prevent backstreaming under shorter mean free path conditions produced by higher pressure operation (Fig. 13).

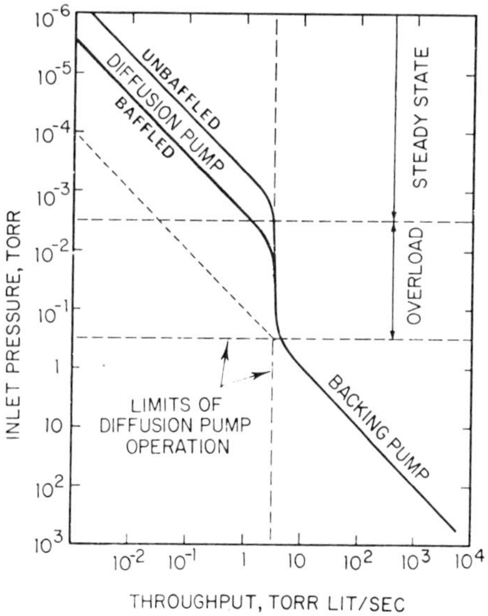

FIG. 12. Throughput versus input pressure.

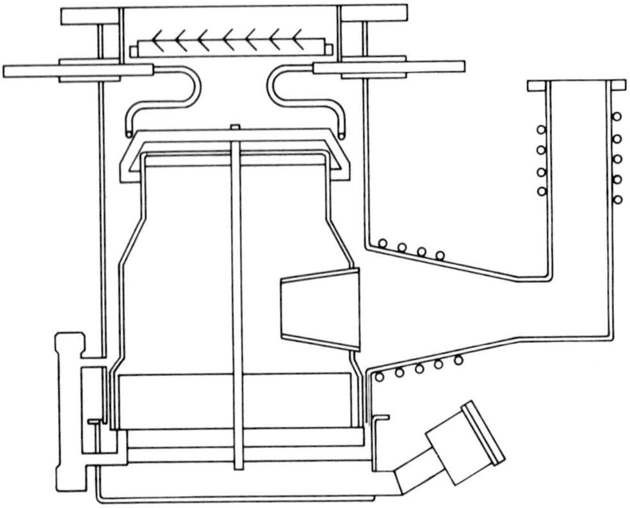

FIG. 13. Higher pressure sputtering pump.

These pressure and speed tradeoffs can be made easily as long as the maximum throughput is not exceeded.

4.3.4. Tolerable Forepressure

4.3.4.1. Discharge Pressure. A diffusion pump is designed for high vacuum application. Thus, its boiler pressure usually is 1–1.5 Torr. This implies that the maximum pressure the pump can be expected to produce is 1.5 Torr. In addition, diffusion pump working fluids cannot be boiled at high pressures because the resulting high temperatures will decompose the fluid at an unacceptable rate. Thus, diffusion pumps must always be backed by another pump which produces a pressure of generally less than 0.5 Torr at the discharge of the diffusion pump.

Tolerable forepressure of a diffusion pump is the maximum permissible pressure at the foreline.

Let it be said that the expected pumping action of the diffusion pump collapses when the tolerable forepressure is exceeded without attempting to describe what this collapse exactly entails.[2] Essentially, the vapor of the discharge stage of the pump does not have sufficient energy and density to provide a barrier for the air in the foreline when its pressure exceeds a certain value (usually near 0.5 Torr). Then this air will flow across the pump in the wrong direction carrying with it the pumping fluid vapor.

[2] H. G. Nöller, *Vacuum* **5,** 59 (1955).

Modern diffusion pumps have a boiler pressure of about 1.5 Torr. Approximately half of this initial pressure is recovered in the form of tolerable forepressure. When the tolerable forepressure is measured, a conventional nonultrahigh vacuum system usually does not reveal the dependence between the discharge and the inlet pressures. This is because the maximum pressure ratio for air is usually not exceeded unless the experiment is conducted under ultrahigh vacuum inlet pressures.

4.3.4.2. Backing Pump Requirements. To select an appropriate backing pump for a given diffusion pump, several questions must be considered. First, what is the size of the initial roughing pump and whether it is to be used for both roughing and backing? Second, is the backing pump expected to perform at the maximum throughput of the diffusion pump? Third, what is the tolerable forepressure of the diffusion pump? Also, what is the volume of the foreline ducts (or a special reservoir which could be used in the line)? The nominal pumping speed of the required backing pump for full load condition is obtained as follows:

$$S = Q_{\text{max}}/(\text{TFP}),$$

where Q_{max} is the maximum throughput of the diffusion pump and TFP stands for tolerable forepressure. Aside of consideration of safety factors and conductances of forelines, it should be noted that mechanical pumps often have reduced speed when their inlet pressure is equal to the tolerable forepressure of the diffusion pump. A safety factor of 2 may be required to make sure that the tolerable forepressure value is never exceeded even if mechanical pump and diffusion pump are not operating at their best. A numerical example follows. Assume a pump with a maximum throughput of 4 Torr liters/sec and tolerable forepressure (maximum permissible discharge pressure) of 0.5 Torr at full load (at maximum throughput). The required backing pump speed then becomes

$$S = \frac{4 \text{ Torr liters}}{\text{sec } 0.5 \text{ Torr}} = 8 \text{ liters/sec} = 17 \text{ CFM}.$$

A good choice for the backing pump would be a nominal 14 liters/sec (30 CFM) pump, provided that the conductance between the two pumps are not severely limited (see Section 4.3.4.3).

4.3.4.3. Safety Factors. The most important rule of diffusion pump operation is: Do not exceed tolerable forepressure. In other words, in an operating pump, the maximum permissible discharge pressure should not be exceeded under any circumstances. Observance of this most basic requirement will eliminate most of the difficulties encountered with diffusion pumps, especially problems with noticeable backstreaming of the pumping vapor into the vacuum system. High vacuum systems should be designed with interlocks, fail-safe valve arrangement, or clearly marked instructions to preclude the possibility of exceeding the tolerable fore-pressure. As in most engineering considerations, a safety factor should be included in establishing the maximum permissible discharge pressure. A factor of 2 is a good general recommendation. As much as 25% reduction of tolerable forepressure can be expected near maximum throughput operation (full load) and some reduction can be expected from low heater power. The selection of the mechanical backing pump must be made with these points in mind. Also, it should be noted that the desired discharge pressure must be obtained near the diffusion pump to avoid errors caused by limited conductance of forelines and the reduction of mechanical pump speed at lower pressures.

4.3.4.4. Tolerable Forepressure and Pressure Ratio. Tolerable forepressure (maximum permissible discharge pressure) and the maximum pressure ratio achievable with a given pump should not be confused. Tolerable forepressure is directly related to the boiler pressure, while the pressure ratio has a logarithmic dependence on vapor density in the jets. This distinction is indicated in Fig. 14. For air the maximum pressure ratio is usually so high that it cannot be measured under normal circumstances. Only when inlet pressure is less than 1×10^{-10} or 1×10^{-11} Torr,

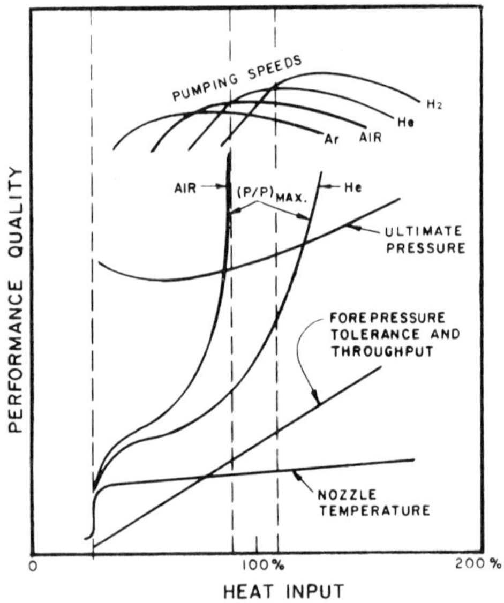

FIG. 14. Relation of forepressure to boiler pressure and the logarithmic dependence of pressure ratio on vapor density in jets.

a dependence between discharge and inlet pressure may be observed. However, for hydrogen and helium, the dependence can be observed in the high vacuum range.

The effects of boiler pressure variation on pump performance are summarized, qualitatively, in Fig. 14. Due to the strong dependency indicated in Fig. 14, clearly, small changes in vapor jet density may cause large variations in the maximum pressure ratio. In applications where stable pressure is required with lighter gases, this may require attention. Ordinary pumps often display inlet pressure variation exceeding 5% for helium. This is unacceptable for some applications, such as highly sensitive mass spectrometer leak detectors and analytical mass spectrometers where helium is used either as a tracer or a carrier gas. Thus, special pump designs having very stable boiler pressure and stable pumping action may be required.

In regard to pumping speed, it may be observed that the curves for helium and hydrogen can have a rather steep slope near the nominal heat input value (100%). Therefore, if variations of vapor density occur, they may result in noticeable pumping speed variations for these gases.

4.3.4.5. Tolerable Forepressure for Various Gases. The tolerable forepressure for various gases is approximately the same as far as the complete collapse of vapor jet (last stage) is concerned. However, in regard to the appearance of lighter gases at the inlet of the pump, when they are introduced in the foreline, the maximum pressure ratio effect becomes noticeable.

The results of pressure ratio measurements are shown in Fig. 15. The data were obtained by introducing helium into the foreline of the pump and observing the resulting effect on the inlet pressure. The discharge pressure was measured with a radioactive high-pressure ionization gauge (Alphatron) and the inlet pressure with hot-filament ionization gauge, both corrected for helium. The upper curve represents an older 5-cm pump with a 400-W heater. The comparison between the two pumps indicates a thousandfold improvement in maximum pressure ratio which can be sustained across the pump. This should permit the use of smaller mechanical backing pumps and elimination of small oil-booster pumps sometimes used in instruments which depend on the maximum pressure ratio for their performance.

For the older pump in Fig. 15, it is difficult to establish a concept of forepressure tolerance for helium, at inlet pressures below 10^{-6} Torr. The newer pump shows a behavior which can be interpreted as if a forepressure tolerance of 0.6 Torr for helium is present. This pump had a forepressure tolerance for air of about same magnitude.

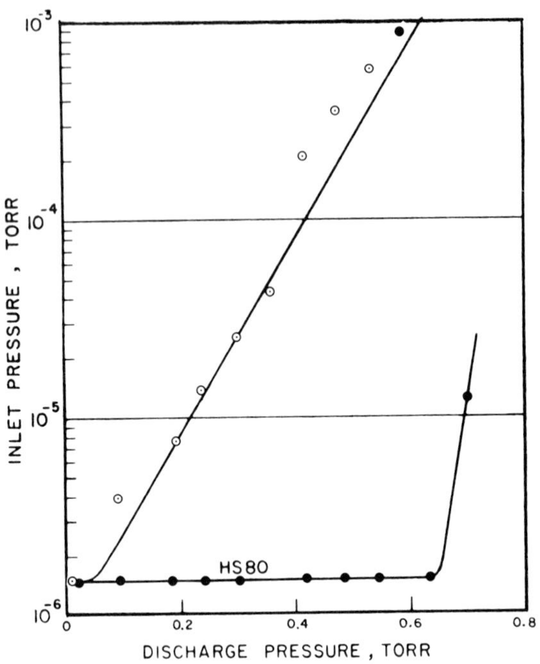

FIG. 15. Results of pressure ratio measurements.

4.3.5. Ultimate Pressure

4.3.5.1. Pressure Ratio Concept. Referring to Section 4.3.4.4, two distinct observations can be made regarding the relationship between the ultimate pressure of a pump. Ultimate pressure may be considered to be a gas load limit or a pressure ratio limit. Both are of significance in practice, the latter usually only with light gases. The pumping action of the vapor jet does not cease at any pressure, however low. The ultimate pressure of the pump depends on the ratio of pumped versus back-diffused molecules, plus the ratio of gas load to the pumping speed. In addition, the pump itself can contribute a gas load either through backstreaming of pump fluid vapor and its cracked fractions and the outgassing from its parts. Thus, in practice, the observed total ultimate pressure is a composite consisting of several elements. In practice, the commonly observed first limit is due to the pumping fluid, although with the best fluid some system degassing (baking) may be necessary to observe this limit when it is below 1×10^{-8} Torr.

When liquid nitrogen traps are employed, the limiting pressure is usually given by various gas loads in the system even if the system consists only of the measuring dome and a pressure gauge. Below 5×10^{-9} Torr thorough baking is usually required.

For helium and hydrogen a true pressure ratio limitation can be observed in most pumps (see Section 4.3.4.4), although special designs can be made to improve the limit beyond observable level.

4.3.5.2. Baffle and Trap Effects. Room temperature baffles do not affect the ultimate pressure except by shielding a gauge from a direct entry of pumping fluid vapor into the gauge. Water cooled baffles suppress the rate of re-evaporation of condensed or intercepted fluid thereby reducing the density of vapor in the space between the baffle and the trap. For substances such as diffusion pump fluids each 20°C temperature difference near room temperature will account for about an order of magnitude change in vapor pressure and hence the rate of evaporation. The lowered vapor density will reduce the possibility of intercollisions and consequent bypass through the trap without touching a refrigerated surface.

Cryogenic or refrigerated traps have basically two effects. They act as barriers for the flow of condensable vapors from pump to system but they also act as cryopumps for condensable vapors emanating from the system. The latter may be the primary effect on the ultimate pressure in many cases. In the high vacuum region and for unbaked systems using modern low vapor pressure pumping fluids the reduction of pressure (when traps are cooled) is primarily due to water vapor pumping. In unbaked systems after the initial evacuation, water may constitute 90% of the remaining gases and cooling of the trap simply increases the pumping speed for water vapor (usually factor of 2 or 3).

In rapid cycle leak detection apparatus, for example, the trap serves to protect the mass spectrometer tube from contamination from test samples rather than from diffusion pump backstreaming.

4.3.5.3. Pressure Ratio for Lighter Gases. As noted previously, the pressure ratio can be sufficiently small for the light gases to reveal the dependence of inlet pressure on the discharge pressure. Measurements of pressure ratio for various gases have been reported as follows: hydrogen 3×10^2–2×10^6, helium 10^3–2×10^6, neon 1 or 2×10^8, CO and argon 10^7, oxygen and krypton 3–5×10^7, and hydrocarbons (nC_2H_3) 7×10^8. In modern pumps the helium pressure ratio is closer to 10^7 and it can be increased even as high as 10^{10} by doubling the heat input. In practice, even an ion gauge operated in the foreline can produce sufficient hydrogen to cause an increase of the inlet pressure. Exactly the same occurrence has been observed in a turbomolecular pump system. The supply of hydrogen apparently comes from the mechanical backing pump oil. Further discussion will be found in Sections 4.3.4.4 and 4.3.4.5.

As far as ultimate pressure is concerned, hydrogen can be a substantial part of the residual gas composition due to its presence in metals, in the pumping fluid, and in water vapor. This can be an important consideration for ultrahigh vacuum work where some diffusion pumps may need a second pump in series (see Section 4.3.5.6). Same consideration may apply to helium in leak detectors, mass spectrometers, molecular beam experiments, etc.

4.3.5.4. Pumping Fluid Selection. A variety of "organic" liquids have been used as motive fluids in diffusion pumps. Criteria for the selection of the fluid are low vapor pressure at room temperature, good thermal stability, chemical inertness, nontoxicity, high surface tension to minimize creep, high flash and fire points, reasonable viscosity at ambient temperature, low heat of vaporization, and, of course, low cost. A list of the presently popular fluids and their properties is given in Table I. The selection of the fluid should be made giving due consideration to its operational stability in the pump boiler. Many of the fluids presently used have been developed within the last 20 years. Until 1960 most fluids used had vapor pressures (at 20°C) of 10^{-7}–10^{-8} Torr and the ultimate pressure without the use of cryogenic trapping was limited to this range. A breakthrough in the selection of fluids was made when Hickman[3] reported the use of five-ring polyphenyl ethers consisting of chains of phenyl groups interbonded by oxygen. This fluid offered exceptional thermal and chemical stability and enabled reaching ultimate pressures of 10^{-9} Torr (approaching its vapor pressure at ambient temperatures) with only water cooled baffles.

Operational characteristics of another low vapor pressure silicone fluid (DC-705) were discussed by Crawley *et al.*[4] Ultimate pressures of 10^{-9} Torr with water cooled baffles and 10^{-10} Torr with baffle at -20°C were reported by Oikawa and Mikami,[5] who also gave ultimate pressures of 10^{-9} Torr in metal pumps with water cooled baffles.

Whatever fluid is used, its vapor may pervade the pumped system, dependent on its vapor pressure, the pump design, and the type of trapping used. The vapor can be broken down by the presence of hot filaments and bombardment by charged particles. The polymerization of silicone fluids resulting from bombardment by charged particles may cause

[3] K. C. D. Hickman, *Trans. Am. Vac. Soc.* p. 307. Pergamon, New York, 1961.

[4] D. J. Crawley, E. D. Tolmie, and A. R. Huntress, *Trans. Am. Vac. Soc.* p. 67. AIP for AVS, New York, 1967.

[5] H. Okamoto and Y. Murakami, *Abst. Am. Vac. Soc. Symp.* p. 67. AIP for AVS, New York, 1967.

TABLE I. Properties of Diffusion Pump Fluids

Trade name	Chemical name	Molecular weight	Vapor pressure at 20°C (Torr)	Flash point (°C)	Viscosity 20°C at (centistokes)	Surface tension (dynes/cm)
Octoil	Diethyl hexyl phthalate	391	10^{-7}	196	75	
DC-704	Tetraphenyl tetramethyl trisiloxane	484	10^{-8}	216	47	30.5
Apiezon C	Paraffinic hydrocarbon	574	4×10^{-9}	265	295	30.5
DC-705	Pentaphenyl trimethyl trisiloxane	546	5×10^{-10} (25°C)	243	170 (25°C)	>30.5
Santovac 5	Mixed five-ring polyphenyl ether	447	1.3×10^{-9} (25°C)	288	2500 (25°C)	49.9

an insulating film to be produced on electrode surfaces, changing the characteristics of the electronic instrumentation. Octoil or polyphenyl ethers are usually recommended to eliminate this problem in applications where mass spectrometers and other electron optical devices are used.

4.3.5.5. Residual Gas Analysis. The errors and uncertainties associated with tubulated ionization gauges are perhaps even more serious with the mass spectrometer residual gas analyzers. This is particularly true in regard to condensable vapors of high molecular weight, so that it is extremely difficult to correlate the ion currents indicated by the mass spectrometer with the rate of backstreaming through the baffle.

Generally, the residual gas analysis of a trapped diffusion pump system cannot be reliably performed with spectrometers having poorer detectability than 10^{-9} Torr. When the spectrometer tube and other parts of the system are baked, the results can be very misleading. It may take weeks before equilibrium conditions are established.

Qualitative measurement can be made, however, with no more difficulty than the ultrahigh pressure measurements made with total pressure gauges such as an ionization gauge.

A typical mass spectrum from a baffled diffusion pump system, unbaked, operating in the ultrahigh vacuum region, with DC-705 motive fluid, is shown in Fig. 16. The mass numbers 16, 19, and 35 are unusually high owing to the characteristic properties of the spectrometer tube and the hydrogen peak is not shown. The peaks 50, 51, 52, 77, and 78 are characteristic for the pumping field. Crawley[4] also gives the residual gas analysis of diffusion pump system using silicone DC-705 fluid under both

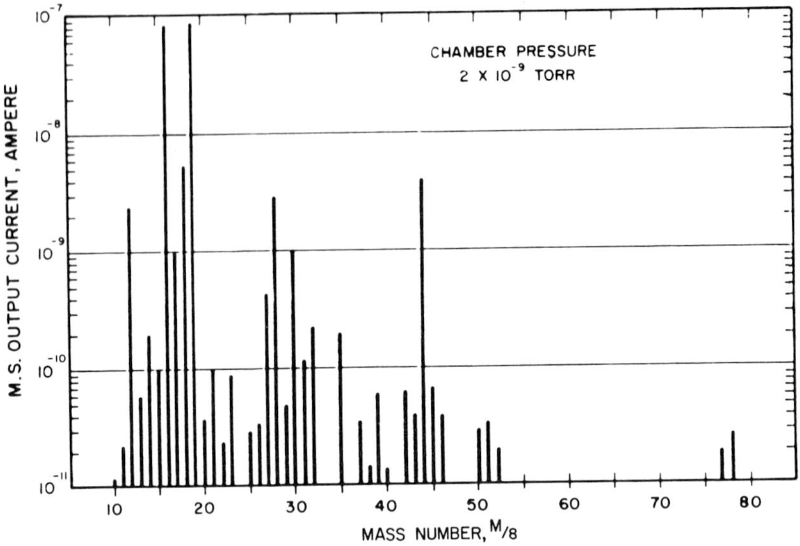

FIG. 16. Typical mass spectrum from a baffled diffusion pump. Chamber pressure is 2×10^{-9} Torr.

unbaked and baked conditions. Wood and Roenigk[6] and Cleaver and Fiveash[7] provide the mass spectra from which the significant identifying peaks can be chosen for the detection of the pumping fluid.

An example of practical value of residual gas measurements is given below.

In a new pump, the inner boiler compartment was omitted because of a vertical heater arrangement. The ultimate pressure was adversely affected. Normally with DC-705 or Santovac-5 pumping fluids, an ultimate pressure below 1×10^{-8} Torr is expected with a test dome where the ionization gauge is essentially baffled, but untrapped and unbaked. In this case the pressure was over 5×10^{-8} Torr. Residual gas analysis revealed that a substantial part of this was due to hydrocarbon vapors.

The residual gas measurements were made with a mass spectrometer (Varian, MAT 111) by simply substituting the new pump for the pump normally used at the spectrometer ion source section. The results are shown in Figs. 17 and 18. The only difference between the two arrangements from which these mass spectra were taken consists of a change in foreline design. The total pressure was measured by an ionization gauge which is part of the spectrometer. The reduction of system pressure with

[6] G. M. Wood, Jr. and R. J. Roenigk, Jr., *J. Vac. Sci. Technol.* **6**, 871 (1969).
[7] J. S. Cleaver and P. N. Fiveash, *Vacuum* **20**, 49 (1970).

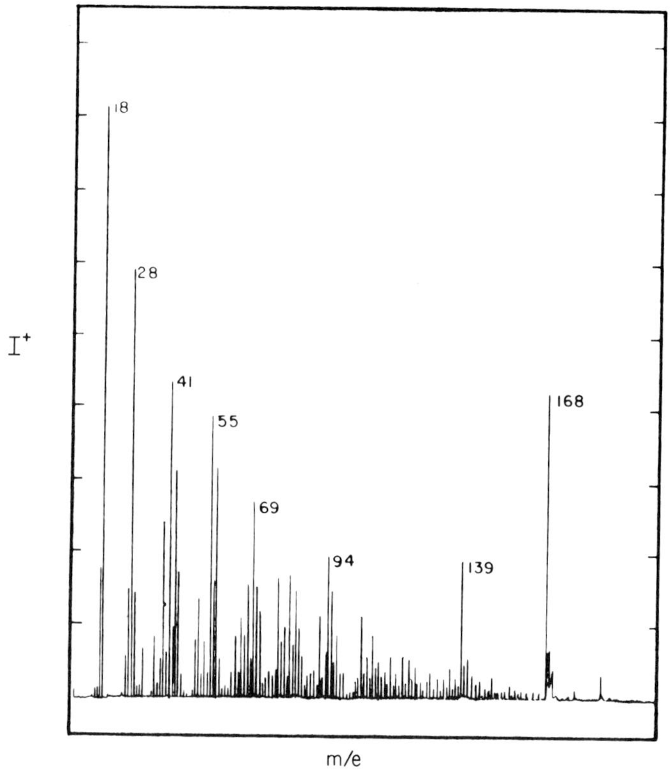

FIG. 17. Residual gas measurements made with a mass spectrometer by substituting a new pump for the pump normally used at the spectrometer ion source section.

an improved foreline (Fig. 18) is evident but the reduction in hydrocarbon peaks associated with pumping fluid (for example, mass peaks 139 and 168) is more pronounced, representing almost 500-fold improvement. The spectrometer sensitivity in Fig. 18 is about 10 times higher than in Fig. 17. The change in foreline design was primarily associated with the temperature of the baffle inserted into the vertical section of the foreline. It should be noted that this increase in temperature cannot be normally achieved by simply using higher heat input to the pump.

4.3.5.6. Foreline Booster Pumps. As noted in Section 4.3.5.3, diffusion pumps may have a noticeable limitation in the maximum pressure ratio for lighter gases. This may require the use of two pumps in series. Sometimes small booster diffusion pumps are used in series with the main pump. In addition to hydrogen and helium pumping, they also provide a barrier between the mechanical pump and the main pump, thus preventing or reducing the possibility of the mechanical pump oil vapors get-

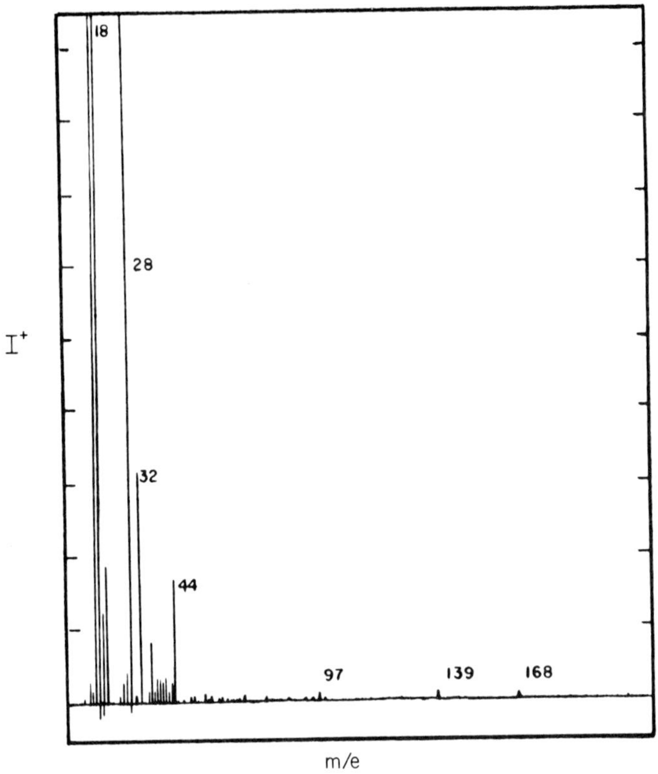

FIG. 18. Same as Fig. 17, with a resulting reduction of system pressure due to an improved foreline.

ting into the main diffusion pump. This practice was common ten years ago. Modern pumps do not show significant effects with the use of booster pumps. This is especially true for pumps which have a horizontal ejector stage as the last stage, facing the foreline. The ejector provides a more efficient arrangement compared to annular nozzles and it also produces a barrier for migration of mechanical pump vapor into the boiler of diffusion pumps.

Huber and Trendelenburg[8] describe a system using Octoil-S reaching ultimate pressures of low 10^{-9} Torr without the use of refrigeration. They used a booster diffusion pump for purification of the pump fluid by keeping the backstreaming mechanical pump oil from reaching the diffusion pump boiler. The booster also prevented the back diffusion of light

[8] W. K. Huber and E. A. Trendelenburg, *Trans. Am. Vac. Soc.* p. 146. Pergamon, New York, 1959.

gases and removed the cracked products of the pumping fluid from the forevacuum region. Modern diffusion pumps achieve the same results with the use of a side ejector stage.

4.3.5.7. Fluid Breakdown and Purification. The ultimate pressure in a vacuum system consists of the partial pressures of various gases. Components arising from gas evolution from sources other than the diffusion pump (including the inlet gasket) are not discussed here. From the components arising in the pump, the major constituent is the vapor of the motive fluid, the products of its decomposition in the boiler, the contaminant vapors from other system parts, and back diffused gases. In a system without the use of cryogenic trapping, the ultimate pressure due to the vapor pressure of the fluid is the minimum achievable. The other components of the total pressure can be reduced by optimizing the pump design.

Degassing of the condensed fluid can be achieved by controlling the temperature of the pump wall near the boiler high enough to permit fluid degassing before it enters the boiler.[9] A significant improvement of ultimate pressure can be obtained. This is particularly important in pumps which do not have separated boiler compartments producing vapor for inner stages (called fractionation). In pumps which have a horizontal ejector stage in the foreline, in addition to increasing the forepressure tolerance, the ejector stage maintains low pressures in the region under the lower annular stage. With this arrangement, the volatile components of the returning condensate are more easily removed to the forevacuum region. Pumps with a strong side ejector stage are very little influenced by the addition of a booster diffusion for the improvement of ultimate pressures even in the 10^{-10} Torr range.[10]

With the use of cryogenic traps and modern fluids, such as DC-705 or Santovac-5, ultimate pressures of 10^{-10} Torr and below can be achieved in diffusion pumped systems where contaminants from sources other than the diffusion pump are controlled.

An increase of heat input will usually increase the pressure ratio obtainable with a pump. However, depending on pump design, higher power can either improve or worsen the ultimate pressure. This depends on the ratio between fluid thermal breakdown rate and the purification ability of the particular pump. Small pumps present, generally, a more challenging problem in this respect because all distances are smaller and the control of fluid circulation and the temperature distribution is more difficult than in larger pumps.

[9] H. G. Nöller, G. Reich, and W. Bachler, *Trans. Am. Vac. Soc.* p. 6. Pergamon, New York, 1957.

[10] B. D. Power, "High Vacuum Pumping Equipment." Van Nostrand-Reinhold, Princeton, New Jersey, 1966.

4.3.6. Backstreaming

4.3.6.1. Backstreaming Definitions. Any transportation of the pumping fluid into the vacuum system may be called backstreaming. The possibility of this back flow is perhaps the most undesirable characteristic of diffusion pumps. It is in this area that most misunderstanding and misinformation exists. Pump manufacturers usually report the backstreaming rate at the inlet plane of an unbaffled pump. System designers are concerned about the steady state backstreaming above liquid nitrogen traps. The users are often troubled by inadvertent high-pressure air inrush into the discharge end of the pump and the breakup of the frozen liquid films on the liquid nitrogen traps. The quantities of backstreaming fluid involved in the above cases can range over many orders of magnitude.

As far as the pump itself is concerned, there can be several sources of backstreaming:

(1) the overdivergent flow of vapor from the rim of the upper nozzle;
(2) poorly sealed penetrations at the top nozzle cap;
(3) intercollision of vapor molecules in the upper layer of the vapor stream from the top nozzle;
(4) collisions between gas and vapor molecules, particularly at high gas loads (10^{-3}–10^{-4} Torr region);
(5) boiling of the returning condensate just before the entry into the boiler (between the jet assembly and the pump wall) which sends fluid droplets upward through the vapor jet; and
(6) evaporation of condensed fluid from the pump wall. Most of these items can be corrected by good design.

4.3.6.2. Primary and Secondary Backstreaming. Somewhat arbitrarily, all the items listed above which can be stopped or intercepted by a room temperature baffle may be called primary backstreaming. The re-evaporation of the pumping fluid from the baffle and the passage through the baffle may be called secondary backstreaming. The primary backstreaming can be effectively controlled by the use of cold caps (see Section 4.3.7.3).

In systems with liquid nitrogen traps (barring accidents and high gas-load operation) the backstreaming level can be controlled at such a low level that contaminants from sources other than the diffusion pump will predominate.[11] Properly operated and protected diffusion pump systems can be considered to be free of contamination from the pumping fluid for most applications.

[11] M. H. Hablanian, *J. Vac. Sci. Technol.* **6**, 265 (1969).

4.3.6.3. Speed and Backstreaming. The backstreaming rates measured by the American Vacuum Society Standard and given in pump specifications refer to the inlet plane of the pump. It is important to understand that this rate quickly diminishes at some distance from the pump inlet. At a distance from inlet equivalent to two pump diameters, the rate is reduced typically 50 times (Fig. 19). A 90° bend in the inlet duct will act as a baffle and reduce backstreaming to the level of the natural evaporation rate of the fluid at the ambient temperature. Without cryogenic traps or other similar devices, this rate cannot be reduced much further.

Figure 20 shows the qualitative relationship between conflicting requirements of minimum backstreaming and maximum retained pumping speed with baffles at ambient temperature. If a series of elements which reduce backstreaming is introduced above the pump, the backstreaming rate approaches the evaporation rate of the fluid. It may be observed that in many applications, complete opaque baffles are redundant. With efficient cold caps (Section 4.3.7.3) and low vapor pressure pumping fluids a system can be operated for a year or more with continuous liquid nitrogen cooling of the trap. The fluid will be returned into the pump and the buildup on the trap will not be detrimental.

In normal practice, an optimum design for a diffusion pump and trap combination can have nearly 40% net pumping speed and reduce backstreaming to less than 1×10^{-10} g/cm² min (at the inlet plane of the trap). Values of this magnitude have been measured.[12]

4.3.6.4. Surface Migration. Some pumping fluids may have a tendency to spread on metal surfaces as a thin film. This spreading is discussed in the literature dealing with lubrication of small instruments.[13] The spreading may be noticed sometimes as wetness on rough (sand blasted) surfaces. Modern low vapor pressure pumping fluids having surface tensions above 30 dynes/cm do not spread on ordinary metal surfaces. They are so-called "autophobic" liquids because they do not spread on their own monolayers covering a metal surface. There remains, however, the possibility of molecular surface diffusion sometimes referred to as "creep" in conjunction with diffusion pump backstreaming. The possibility of surface diffusion as a mechanism of backstreaming may be estimated as follows.

Referring to Section 4.1.3.5, the residence time associated with surface diffusion steps is lower compared to the period between evaporative events. The values of heat of adsorption for surface diffusion are rarely

[12] G. Rettinghaus and W. K. Huber, *Vacuum* **24,** 249 (1974).

[13] F. M. Fowkes, "Contact Angle," *Adv. Chem. Ser. 43*, Am. Chem. Soc. (1964).

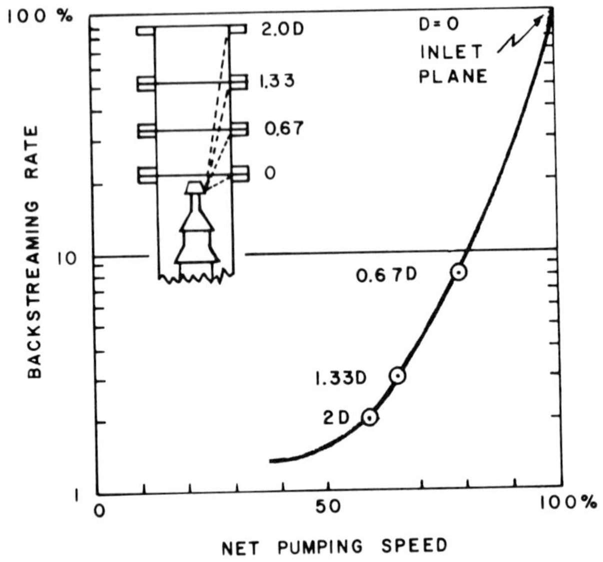

FIG. 19. Reduction of backstreaming rate.

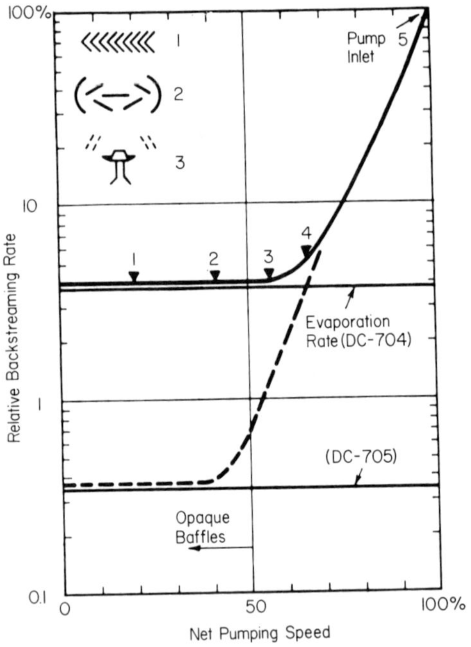

FIG. 20. Qualitative relationship between conflicting requirements of minimum backstreaming and maximum retaining pumping speed with baffles at ambient temperature.

found in literature, but they range from 0.55 to 1.0 compared to heat of adsorption for evaporation. A sojourn time ratio between evaporation and surface diffusion can be used to estimate the importance of surface migration. Surface diffusion for a 1-cm gap is significant when the sojourn time ratio is above 10^{15}.[14,15] In the case of pumping fluid, such as DC-705, the heat of adsorption values may be taken to be 28 and 16 kcal/mole, respectively. This gives a sojourn time ratio of only 10^5. The residence time corresponding to 16 kcal/mole is approximately 0.1 sec at room temperature. The distance between a pump (20-cm diam) and the chamber is approximately 50 cm. Assuming a single step distance to be about 10 Å, a molecule has to make about 10^9 steps traveling in the same direction to cross a baffle or a trap wall. Multiplying 0.1 sec per step times 10^9 steps, we obtain 10^8 sec or 3 yr.

It may be concluded that surface diffusion is not likely to be a significant backstreaming mechanism on surfaces at room temperature or below for pumping fluids with vapor pressures in the 10^{-9} Torr range and high surface tension.

The practical significance of creep barriers is not in arresting surface diffusion (and spreading) but in stopping the by-pass possibility in the narrow space between cold outer parts of the trap and the warm wall. The geometry of well designed trap should provide opportunity for re-evaporated molecules to go toward the cold surface. When this is done and when "autophobic" fluids are used, the major remaining steady state backstreaming mechanism is due to collisions and possibly thermal breakdown resulting in lighter fragments which are not fully absorbed on the cold trap surfaces.

4.3.6.5. "Accidental" Backstreaming. It is important to remember that malfunction and misoperation can destroy intentions of most intelligent designs. The most common causes of gross backstreaming are: "accidental" exposure of discharge side of the diffusion pump to pressure higher than the tolerable forepressure, high inlet pressure exceeding maximum throughput capacity for long periods of time, incorrect start up, and incorrect bakeout procedures.

Providing proper interlocks and protection in system design is as important as training of operating personnel. The worst possibility is to air release the system through the foreline when the diffusion pump is operating. Fact acting fail-safe valves in the foreline should be consid-

[14] J. H. de Boer, "The Dynamical Character of Adsorption," p. 226. Oxford, London and New York, 1953.

[15] P. A. Redhead, J. P. Hobson, and E. V. Kornelsen, "The Physical Basis of Ultrahigh Vacuum," p. 81. Chapman & Hall, London, 1968.

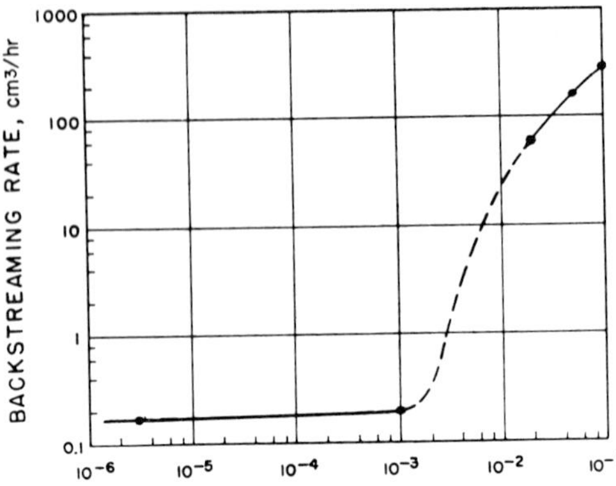

FIG. 21. Measurements of backstreaming rate at various inlet pressures.

ered whenever the equipment is left unattended. The signal for closure and power cutoff can be obtained from a gauge with a fast time response.

4.3.6.6. Backstreaming Rate versus Pressure. From lowest pressures to about 1×10^{-4} Torr, the backstreaming rate appears to be independent of pressure indicating that oil–gas collisions are not significant in this region. Between 10^{-4} and 10^{-3} Torr, a slight increase may be noticed.[16] Above the critical pressure point when the maximum throughput is reached (pumping speed begins to decline), the backstreaming rate may rise markedly.

Most measurements of backstreaming rate are conducted at the blank-off operation of the pump. One of the important questions is what happens during operation at higher inlet pressures and particularly during startup of the pump. Measurements of the backstreaming rate at various inlet pressures made under such undesirable conditions, are plotted in Fig. 21 and it can be seen that the rate does not change significantly as long as the inlet pressure is below about 10^{-3} Torr. This is the point where most diffusion pumps have an abrupt reduction in speed indicating that the top jet essentially stops pumping.

The sudden increase of backstreaming at inlet pressures above 10^{-3} Torr points out that a diffusion pump should not be operated in this range unless the condition lasts a very short time.

It is common practice in vacuum system operation to open the high vacuum valve after the chamber has been rough pumped to about 10^{-1} Torr exposing the diffusion pump to this pressure. In well proportioned

[16] D. H. Holkeboer, D. W. Jones, F. Pagano, and D. J. Santeler, "Vacuum Engineering." Technical Publishers, Boston, 1967.

systems this condition lasts only a few seconds. Extended operation at high pressure can direct unacceptably high amounts of oil into the vacuum system.

The tests shown in Fig. 21 were obtained with a collecting surface in the immediate vicinity of the top nozzle. If the measurements were made at the inlet plane as specified by the American Vacuum Society Standard, the rate would have been 10 or 100 times lower (see Fig. 19).

The higher pressure conditions exist at the inlet to the pump briefly after the high vacuum valve is opened. The associated backstreaming peak normally lasts only a few seconds unless the diffusion pump is severely overloaded.

The same condition will exist when the pump is initially heated. In this case, the duration is a few minutes and the amplitude a few times higher than the steady state backstreaming at lower pressures. The generalized picture is shown in Fig. 22. At pressures higher than 0.5 Torr, depending on diameter and length of ducts, the pumping fluid vapor may be swept back into the pump. This can be utilized for reduction of initial backstreaming and prevention of mechanical pump backstreaming by arranging a flow of air (or other gas) through a leak valve. However, such complications are not necessary for normal operation.

4.3.6.7. Mechanical Pump Effects. Another possibility for hydrocarbon contamination arises from the roughing pump. Roughing pump should not be left connected to the system for long periods of time. After initial evacuation, the roughing valve should be closed and pumping switched to the diffusion pump.

In systems where only a mechanical pump is used for maintaining vacuum conditions in the 10^{-2}–10^{-1} Torr range, sometimes it is advisable to pump periodically rather than continuously. A valve and a gauge are needed to start and stop pumping whenever necessary. This prevents continuous backdiffusion of lubricating oil and, in addition, keeps the oil cold reducing its vapor pressure.

4.3.6.8. Pumping Fluid Loss. The pumping fluid level in a well de-

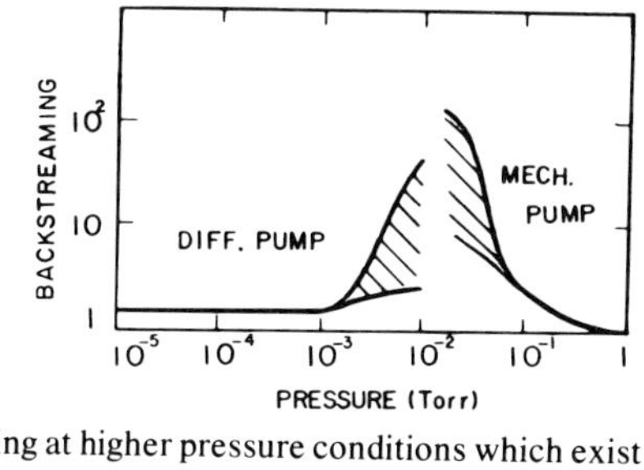

FIG. 22. Backstreaming at higher pressure conditions which exist at the inlet to the pump.

signed pump need not be precisely controlled. Generally, 30% above and below normal level should be tolerable without noticeable effects. When the level is too low, the boiling process may pass from nucleate boiling to partial film boiling which leads to overheating of the boiler heating surface. If this condition is continued for an extended period of time, particularly for large pumps, it can cause distortions of the boiler plate. This, in turn, may expose the center of the boiler plate above the liquid level and lead to further overheating. The resulting poor contact between heaters and the boiler plate may also overheat the heating elements and cause their failure.

If the liquid level is too high, the boiling process may foam the fluid and raise its level as high as the foreline opening.

Excluding normal backstreaming, the fluid may be lost out of the pump in several ways: prolonged operation above or near maximum throughput; accidental high pressure and high velocity air flow through the pump in either direction; and evaporation of higher vapor pressure fluids due to incorrect temperature distribution.

With relatively low gas loads, modern pump fluids, and correct system design and operation, diffusion pumps can be operated for many years without adding or changing pumping fluid. Operation exceeding ten years has been reported in the case of a particle accelerator.[17] Large pumps usually have means of monitoring the fluid level. To reduce fluid loss, some pumps have built-in foreline baffles.

4.3.6.9. The "Herrick Effect." An unusual mechanism of gross backstreaming has been observed in diffusion pump systems with liquid nitrogen traps. It occurs sometimes during or following the charging of the trap with liquid nitrogen and can be recognized by the appearance of small droplets upstream from the trap and by accompanying pressure fluctuations. The amount of pumping fluid which may reach the base plate of a typical bell jar vacuum system due to this mechanism may be few orders of magnitude greater than the normally expected steady state backstreaming rate. The phenomenon is produced by fracture of the frozen pumping fluid film due to the unequal temperature expansion coefficients between it and the metal surfaces on which it has been previously deposited. The elastic energy stored in the film is sufficient to impart high velocity to the fragments resulting from the fracture. This phenomenon has been observed in many traps of conventional design. However, once it is recognized, the design can be improved by avoiding certain geometric configurations which tend to accumulate heavy fluid films and produce

[17] J. Moenich and R. Trcka, *Trans. Am. Vac. Soc.* and *Int. Congr. Vac. Sci. Technol.*, p. 1133. Pergamon, New York, 1961.

highly stressed films which are likely to fracture. Also the qualities of the trapping surfaces can be modified to minimize the tendency of the film fracture.

Liquid nitrogen traps generally should not be placed too close to unprotected diffusion pumps. Ambient or water-cooled baffles, partial baffles or efficient cold caps which essentially remove primary backstreaming, should be used whenever possible. This is to reduce the accumulation of the pumping fluid on cryogenic surfaces. Traps which have removable internal structures should be cleaned periodically. The frequency will depend on the vapor pressure of the pumping fluid and possibly the degree of accumulation of either condensables such as water vapor, assuming that the trap is continuously kept refrigerated. The latter is a good practice under some conditions (such as humidity controlled rooms). A thin film (less than 10^{-3}-mm thickness) is not likely to fracture. In this connection, traps which have long liquid nitrogen holding time are particularly recommended.

If the trap has to be desorbed occasionally, such as during weekends, the restarting procedure needs some attention. It is a good practice to keep the high vacuum valve closed during and after the trap filling periods. In this manner, the fractured frozen film particles and products of desorption can be prevented from reaching the vacuum chamber. The valve should be opened a certain time after recooling the trap to allow time for repumping of condensable matter.

With the combination of the design and procedural remedies mentioned, the difficulties associated with the described phenomenon can be avoided. Figure 23 shows a record of a typical pressure variation due to this effect.

4.3.7. Other Performance Aspects

4.3.7.1. Design Features. Manufacturers of diffusion pumps traditionally report in their specifications only pumping speed, ultimate pressure, forepressure tolerance (measured at the foreline), throughput, and sometimes the backstreaming rate (referred to the inlet plane of the pump). There exist American Vacuum Society standards with the help of which most of these properties can be checked. However, there are other performance aspects listed below which can be used as guides in the selection of pumps for particular applications. (Most of these characteristics can be improved in case of special need, but not without sacrifice of others.)

Pumping Speed per Unit Inlet Area. This can be called speed efficiency and it is not likely to exceed 50% compared to a hypothetical pump

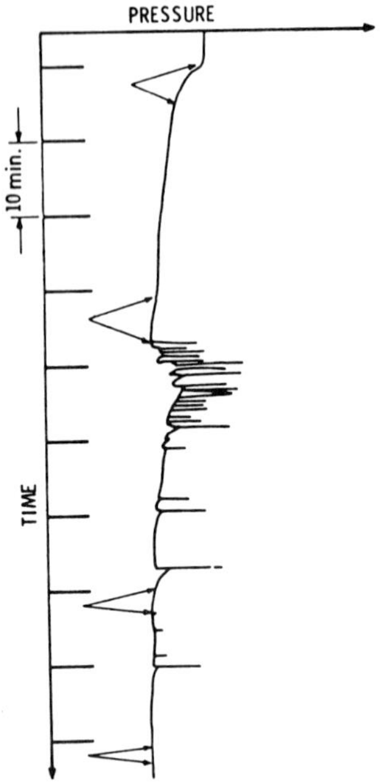

FIG. 23. Record of typical pressure variation when trap is desorbed. Arrows show trap filling periods.

with 100% capture probability for molecules which cross the inlet plane into the pump.

POWER REQUIRED TO OBTAIN A GIVEN MAXIMUM THROUGHPUT WITHOUT OVERLOADING AND WITHOUT OVERSIZED FOREPUMPS. Nearly 1 kW per 1.2 Torr liters/sec is used in common designs.

RATIO OF MAXIMUM THROUGHPUT AND FOREPRESSURE TOLERANCE. This determines the minimum required forepump speed which must be provided at the discharge of the diffusion pump (at its full load).

MAXIMUM PRESSURE RATIO FOR LIGHT GASES. Of particular interest are helium and hydrogen. This pressure ratio can be sufficiently small to reveal the dependence of inlet pressure on the discharge pressure. Diffusion pumps vary widely in this regard; reported figures range from 300 to 2,000,000 for hydrogen and from 1000 to 10,000,000 for helium.

PUMPING SPEED FOR HELIUM AND HYDROGEN. Compared to the speed for air, the ratio should be near 1.2. Note that the baffle conductances are greater for lighter gases. Hence, the net speed of a given pumping system for helium and hydrogen will have a greater relative value compared to air.

RATIO OF EVAPORATION RATE OF PUMPING FLUID AT AMBIENT TEM-

PERATURE AND THE ACTUAL PUMP BACKSTREAMING WITHOUT TRAPS. This ratio is approaching unity in some modern pumps which have efficient cold caps surrounding the top nozzle.

RATIO OF VAPOR PRESSURE OF WORKING FLUID AT AMBIENT TEMPERATURE AND ULTIMATE PRESSURE OBTAINED BY PUMP WITHOUT CRYOGENIC TRAPS. The target for this ratio should also be near unity. To achieve this, the pump must have low pumping fluid breakdown level and a high degree of fluid purification.

RATIO OF FOREPRESSURE TOLERANCE AND BOILER PRESSURE. This is usually about 0.5. The significance of this figure is in keeping the fluid temperature as low as possible to reduce thermal breakdown, while keeping the forepressure tolerance as high as possible.

RATIO OF PUMP DIAMETER AND HEIGHT. The height is normally minimized for the sake of compactness but some performance improvements could be realized if pumps were allowed to be taller.

PRESSURE STABILITY IN CONSTANT SPEED REGION. This can be expressed as a percent variation referred to an average value. Pressure instability is more common in smaller pumps and with lighter gases. It is probably associated with fluctuations of boiler pressure which can occur in a boiler with a relatively small volume for a given heat transfer area and power.

SENSITIVITY TO HEAT INPUT FLUCTUATIONS. In addition to variations of heater power, cooling water flow rate and temperature may have some significance. Modern pumps are not seriously affected by heat input variations, see Sections 4.3.4.4 and 12.4.1.

4.3.7.2. Boiler Design. Diffusion pumps of conventional design usually require about 1 kW power for a maximum throughput of 1.2 Torr liters/sec. This requirement, together with the necessity for prevention of pumping fluid breakdown, focuses attention on heat transfer conditions in the boiler.

The low degree of fluid breakdown and self-purification qualities of the pump can be judged by the degree with which the ultimate pressure of the pump follows the expected vapor pressure of the pumping fluid at room temperature. Figure 24 illustrates the performance of various fluids in a diffusion pump which must have a given boiler pressure (for example, 0.5 Torr). In modern diffusion pumps, the pressure of the oil vapor inside the jet assembly may be about 1.0 Torr. From Fig. 24 it can be seen that above the temperature range of 250°C, the performance of the pump deteriorates, as far as ultimate pressure is concerned. Generally, it is useless to employ fluids of very low vapor pressure because of resulting excessive boiler temperatures. However, one of the requirements for a diffusion pump fluid is low vapor pressure at the ambient temperature of

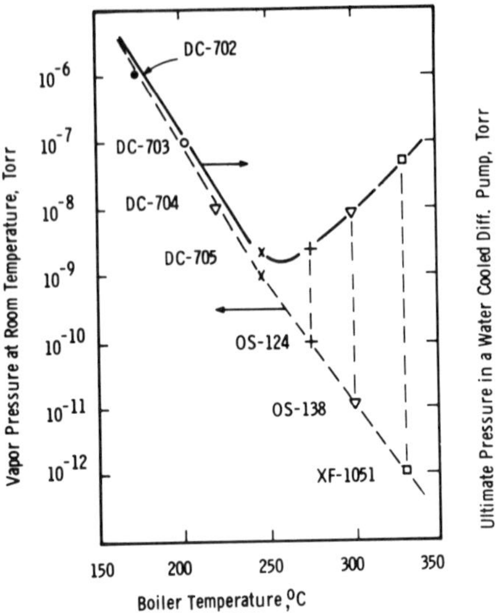

FIG. 24. Illustration of performance of various fluids in a diffusion pump with a given boiler pressure.

the vacuum system so that ultimate pressure without traps and the migration of oil is sufficiently low. Satisfactory design must reconcile the above conflicting requirements.

When using extremely low vapor pressure fluids (for example, Santovac-5) it is sometimes necessary to adjust water-cooling rate or power to obtain optimum performance. Some pumps may have to operate with more effective heat input, some with less depending whether they are below or above the valley of the curve in Fig. 24.

4.3.7.3. Cold Caps. Cold caps surrounding the top nozzle of the diffusion pump reduce backstreaming rates 50 or more times.[18–22] In medium-sized pumps, they are usually made of copper and cooled by radiation or conduction through supports contacting the water-cooled pump wall. This allows the fluid drops from the cold cap to fall into the

[18] B. D. Power and D. J. Crawley, *Vacuum* **4,** 415 (1954) (publ. 1957).

[19] S. A. Vekshinsky, M. I. Menshikov, and I. S. Rabinovich, *Proc. Int. Congr. Vac. Tech. 1st,* p. 63. Pergamon, London, 1958.

[20] N. Milleron and L. Levenson, *Trans. Am. Vac. Soc.,* p. 213. Pergamon, New York, 1960.

[21] M. H. Hablanian and H. A. Steinherz, *Trans. Am. Vac. Soc.* p. 333. Pergamon, New York, 1961.

[22] M. H. Hablanian and A. A. Landfors, *Trans. Am. Vac. Soc.,* p. 555. Pergamon, New York, 1960.

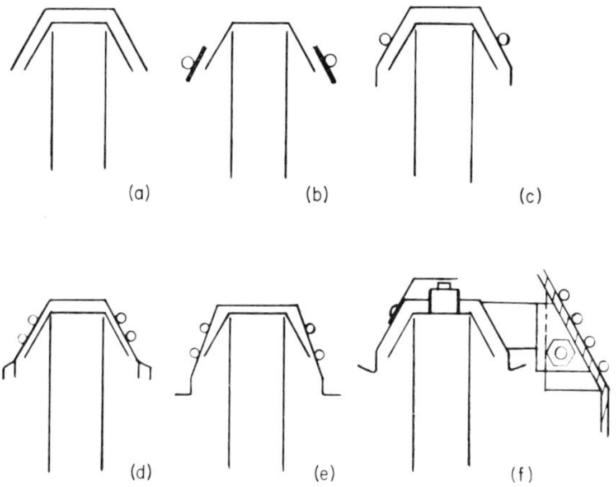

FIG. 25. Evolution of cold cap design.

pump, preventing evaporation into the regions outside the pump. Figure 25 shows the evolution of cold cap design.

Figure 26 shows the dependence of backstreaming on the temperature of the cold cap. The difference between the two curves is due to the degree of interception of the overdivergent portion of the vapor jet emanating from the top nozzle. It can be seen that temperatures below approximately 80°C are adequate for efficient condensation on the cold cap, while above 105°C the cold cap essentially ceases to function properly.

To ensure proper operation of cold caps, they must be thermally isolated from the hot top nozzle parts (jet assembly). In addition, it should be concentric with the nozzle and have adequate distance from it to prevent accumulation of pumping fluid of high viscosity between the hot and cold parts.

4.3.7.4. Pressure Stability. Pressure instabilities occasionally seen in diffusion pump systems can originate in the pump as well as from sources outside the pump. The sources outside the pump are (1) gas bubbles from the elastomer seals; (2) liquid dripping from the baffle; (3) high foreline pressure, light gases; (4) throughput overload; (5) trap defrosting; and (6) explosive breakup of a frozen layer in the trap. The sources inside the pump are (1) eruptive or unstable boiling; (2) boiling outside the jet assembly; (3) low pressure ratio, light gases; (4) liquid droplets in the nozzles; (5) cold top nozzle; and (6) leaks near the boiler.

Unstable boiling is one of the early recognized causes.[23] This produces

[23] K. Hickman, *J. Vac. Sci. Technol.* **9**, 960 (1972).

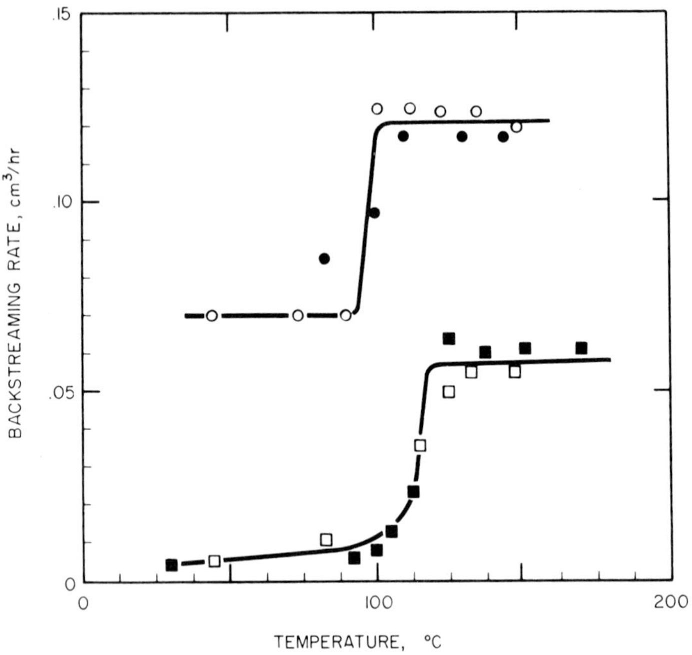

FIG. 26. Dependence of backstreaming on the temperature of the cold cap.

fluctuating density of the vapor ejected from the nozzle causing variations in the pumping speed as well as in the pressure ratio for light gases.

Vigorous boiling in vacuum causes droplets to splash upward and be blown out through the nozzle opening causing momentary blockage as well as wet running in the nozzle. A splash baffle located inside the boiler is used to eliminate this effect. Low pressure ratio for light gases can cause pressure instabilities. An increase in the heater power, as well as an increase of the number of stages, will be beneficial. Bance and Harden[24] describe tests showing random pressure instabilities due to overcooling of the top jet cap leading to discontinuities in the vapor stream.

Outside the pump, gas bubbles from the elastomer seals are the major and frequent source of pressure bursts. Gas evolved from and permeating through the elastomer seals is sometimes trapped by oil films. As the pressure builds up in the bubbles, they break the oil film, and release the gas into the system. The inlet flange of the pump is the most vulnerable location. Careful design of the O-ring groove can minimize this instability. Fluid dripping from the baffles on the hot surface of the top jet cap can evaporate producing vapor and temporarily affect the pumping

[24] U. R. Bance and E. H. Harden, *Vacuum* **16,** 619 (1966).

speeds. Pumps having pressure fluctuations for light gases of $\pm 1\%$ at the pump inlet have been reported.[25,26]

4.3.7.5. Air Cooling and Water Cooling. There are no principal differences between air cooling and water cooling. The major concern is the temperature at the pump inlet which determines the rate of re-evaporation of the condensed pumping fluid. Ideally, the inlet region of the pump or the baffle should be the coldest part of the vacuum system which has no cold trap. A system designer should pay attention to inlet duct temperature distribution, especially when the vacuum system is placed in a cabinet. Cool air should be drawn from outside for pump cooling. Additional fans may be sometimes required to cool the inlet duct near the pump, keeping in mind the tendency of condensable vapors to migrate toward the coldest areas.

Only small pumps are usually air cooled (maximum diameter 10 cm). Air cooling of larger pumps is practically inconvenient (especially in summer time) because of heat exhausted into the room.

[25] M. H. Hablanian and A. A. Landfors, *J. Vac. Sci. Technol.* **13,** 494 (1976).
[26] J. D. Buckingham and N. Dennis, *Vacuum* **25** (1975).

4.4. Other High Vacuum Pumps*

4.4.1. Molecular Pumps

In many aspects of performance, mechanical molecular pumps and turbopumps have similarities with diffusion pumps. Instead of transferring momentum to the pumped gas molecules by collision with vapor molecules, we have here collisions with fast moving solid surfaces.

There are two types of molecular pumps. The first has unbroken moving surface (either a cylinder or a disk) which "kicks" gas molecules toward the discharge side of the pump while they "bounce" between moving (rotor) and stationary (stator) surfaces. The second type has a series of bladed disks similar to axial flow compressors. Each compression stage consists of one moving and one stationary disk. One advantage of the latter design is that the same surface is not periodically exposed to high and low pressure.

In general, the basic performance of most turbomolecular pumps is similar to small- and medium-sized diffusion pumps. Many comments on pumping speed, throughput, and pressure ratio relationships hold for both. Unusual designs go beyond the common inlet pressure range of up to about 10^{-2} Torr and carry out the pumping all the way to atmosphere.[1] Such pumps are used not for repeated rapid evacuation, but for continuous pumping after initial evacuation.

The significant critical point of performance is the maximum throughput point. This alone determines the maximum amount of gas removed from a given system. Detail discussion of turbomolecular pumps will be found in Chapter 5.4.

4.4.2. Oil Ejectors

Oil ejector pumps may be viewed as a cross between booster–diffusion pumps and steam ejectors (see Section 4.2.5.). Booster–diffusion pumps externally resemble diffusion pumps. They are designed to displace the performance curve to inlet pressures roughly ten times higher that it is common for diffusion pumps. They use pumping fluids with higher vapor

[1] L. P. Maurice, *Int. Vac. Congr.* **6** (1974), Kyoto, Japan.

* Chapter 4.4 is by M. H. Hablanian.

ISBN 0-12-475914-9

pressure so that their boiler pressure and forepressure tolerance are correspondingly higher. They are used in high gas load applications (such as vacuum metallurgy) and usually have peak pumping speed near 10^{-2} Torr inlet pressure. In the past, they were also used for backing diffusion pumps. During last 15 years, the use of oil booster pumps has been reduced due to the development of more powerful diffusion pumps and more common use of mechanical booster pumps (Roots blowers, Section 4.2.3).

Oil ejectors go further in the direction of higher inlet pressures. They resemble externally the steam ejector except they have integral boilers and water cooled diffusers rather than separate condensers.

4.4.3. Mercury Diffusion Pumps

The use of mercury as a pumping fluid permits a greater latitude in boiler pressure and inlet pressure range. Small mercury pumps exist with a discharge pressure as high as 50 Torr. In principle, even atmospheric pressure can be achieved. The use of mercury pumps may be of advantage, for example, when the pumped device is to be filled with mercury vapor, or when occasionally high gas loads must be handled by the pump and hydrocarbon contamination cannot be risked.

Mercury vapor pressure at room temperature is near 1.5×10^{-3} Torr. Thus, backstreaming and trapping of mercury vapor must be given more careful consideration. Usually, mercury pumps are used with baffles and liquid nitrogen traps. Rapid accumulation on the trap surfaces compared to oil diffusion pumps cannot be prevented.

Mercury vapor does not condense on water-cooled surfaces as easily as oil vapor and previous hydrocarbon contamination (for example, from forevacuum pumps) may interfere with condensation and result in reduced pumping speed. Mercury vapor is toxic. Handling of mercury vapor must be done carefully to avoid spillage (especially on rough concrete floors). Also, the venting of mechanical pump exhausts and vacuum chambers after defrosting of traps must be made with special precautions. The detail operation and performance of mercury diffusion pumps is discussed in technical literature.[2]

4.4.4. Sorption Pumps (High Vacuum)

Physical adsorption or sorption of gases by ordinary surfaces or porous materials can be used to obtain high vacuum. Adsorption or sorption can

[2] B. D. Power, N. T. M. Dennis, and D. J. Crawley, *Trans. Am. Vac. Soc.* p. 1218 (1961). Pergamon, New York, 1962.

be used at any pressure, but due to practical limitations only certain specific practices have been evolved in vacuum technology.

Sorption pumps are often used for rough pumping ultrahigh vacuum systems to eliminate the concern of the possibility of hydrocarbon contamination from the oil sealed mechanical pump. They are often used together with dry (oil-less) mechanical pumps. By using two or three such pumps in sequential evacuation steps, the bulk of the air can be removed from the vacuum chamber. Sufficiently low pressures can be obtained to transfer the pumping to ion-getter pumps without overloading them.

Sorption can be used to obtain high vacuum either by elevating the temperature to degas the sorbing substance and then returning to room temperature or by cooling to cryogenic temperatures.

Commercially available equipment is usually designed for forevacuum pumping. However, high vacuum can also be obtained by multistage evacuation with pumps of similar basic design but having higher access conductances.

For additional comments in sorption pumping effects, see Sections 4.2.6 and 5.3.1. Hermetically sealed devices are often evacuated and sealed under elevated temperature. During cooling, the number of adsorption sites is sufficient to provide significant pumping action reducing the pressure by one or more orders of magnitude. There must be countless examples of sorption effects in high vacuum practice whenever large temperature changes are encountered. For example, a graphite heater element in a high vacuum furnace can maintain high vacuum overnight in a chamber isolated from pumps with the element still hot and power shut off. This will occur as long as the furnace is adequately leak tight and the residual gases are adequately sorbed.

5. PRODUCTION OF ULTRAHIGH VACUUM

5.1. Fundamental Concepts in the Production of Ultrahigh Vacuum*†

5.1.1. Projected Performance of an Ideal Vacuum System

From the kinetic theory of gases one learns that for a gas in a container at constant temperature

$$PV = \text{const}$$

differentiation by time, and integrating under the conditions that the vessel is being evacuated by a pump of constant volumetric speed $S = dV/dt$, then if the pressure is P_0 at $t = 0$ we obtain

$$P = P_0 e^{-(S/V)t}.$$

A system filled with an ideal gas, which has no interaction with the walls other than reflection, exhausted by a pump of constant volumetric efficiency, should be expected to exhibit a pressure versus time performance as expressed by the above equation. During the initial evacuation of a large vacuum vessel by a small mechanical fore pump, the conditions postulated are nearly met, and a semilogarithmic plot of the pressure as a function of the time is nearly a straight line.

Theory predicts that the pressure should fall exponentially with time, at a rate determined by the "time constant" of V/S. In many vacuum systems found in today's laboratories the pumping rate (liters per second) exceeds the volume (liters) by a factor of 10, resulting in a time constant of 0.1 sec.

It is most interesting to observe the predictions of this concept! The pressure would drop from 10^{-3} to 10^{-9} Torr in 1.4 sec. If the time constant were unity, it would take only 21 sec to go from 10^{-3} to 10^{-12} Torr! This obvious discrepancy from reality is not the result of faulty theory, nor of the failure of the pumps to perform with essentially constant speed

† See also Chapter 5.1 in Vol 4B of this series.

* Chapter 5.1 is by W. M. Brubaker.

ISBN 0-12-475914-9

over the assumed pressure range. The real world does not correspond to our assumed model.

5.1.2. Performance of a Real Vacuum System

In the real world, pressure versus time data bear little resemblance to that predicted for an ideal system. The pressure drops rapidly at first, although the initial rate is many times slower than the calculated time constant would predict. On continued pumping the rate of fall becomes ever slower, as though the time constant were becoming longer. If the system has not been baked since being exposed to atmospheric pressure, it may be days or weeks before a steady state situation is achieved.

5.1.3. Discussion of Gross Differences between Performances of Ideal and Real Vacuum Systems

The discrepancy can be interpreted as a time varying time constant V/S. Obviously, the volume is constant and there are many data which indicate that the speed of the pump remains relatively constant over the range of pressures under consideration.

A closer look at the time constant shows that the volume enters as an indicator of the number of molecules which have to be removed. An additional source of molecules, other than those found in the openness of the space enclosed by the surrounding vacuum walls, would account for observations.

The main reason why pressures fall so slowly in a vacuum system that has been exposed to the atmosphere is that the molecules which must be removed in order to obtain the lowest pressures spend most of their time sitting on the exposed surfaces of the system. They can be removed by vacuum pumps only when they are moving through the space between the surfaces of the system.

The discussion which follows is intended to be first order, and with no consideration of second-order effects. Its purpose is to emphasize the magnitudes of the numbers involved in relating the density of molecules in the space to the density of molecules which may reside on any or all of the surfaces exposed to the vacuum region.

5.1.4. Practical Considerations

5.1.4.1. Molecular Density versus Pressure. The kinetic theory of gases states that one gram–mole of any substance is composed of 6×10^{23} molecules. At a pressure of 760 Torr and a temperature of 0°C this gram–mole occupies 22.4 liters. At a room temperature of 20°C the volume is increased to 24 liters. The number of molecules per cubic centimeter, as a function of the pressure (in torr) is found

$$n = [6 \times 10^{23}/(2.4 \times 10^4 \times 760)]P = 3.3 \times 10^{16}P.$$

At atmospheric pressure there are 2.5×10^{19} molecules/cm^3, and when the molecular density drops to 1 molecule/cm^3 the pressure is 1×10^{-17} Torr. To more dramatically display this relationship, it appears as a log plot in Fig. 1.

5.1.4.2. Surface Conditions. In almost all instances the surfaces of the vacuum wall and all other surfaces exposed to the volume of the vacuum system are highly contaminated. That is, the base material of which the surfaces are made is covered by molecules which may be many layers deep. First, the lattice structure is thought to have regular sites onto which molecules of another substance may be bound. The density of these binding sites is of the same order as that of the loci of the molecules of the base material. Thus there can be an adsorbed layer, one molecule thick, over all of these surfaces. Additionally, molecules of other species may be condensed onto the surfaces. This is particularly true of polar molecules, like those of the ubiquitous water. Some experimenters have estimated that water may be condensed on vacuum surfaces to depths of hundreds of monolayers!

5.1.4.3. Ratio of Surface Area to System Volume. To get a crude estimate of the ratio of the surface to the volume of a vacuum system, con-

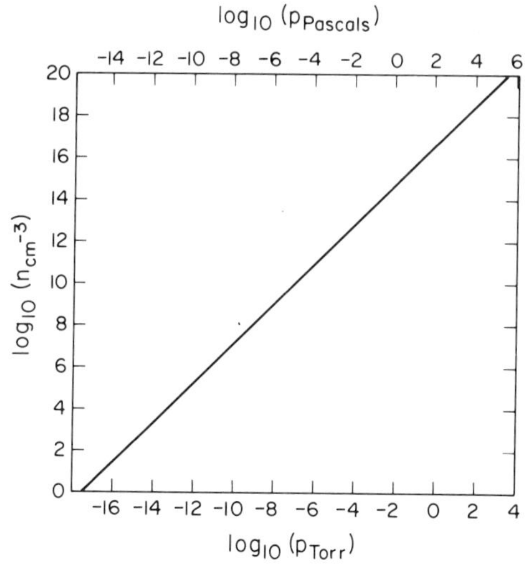

FIG. 1. Number density as a function of gas pressure at 0°C (273 K).

sider first what this ratio is when the vessel is a sphere of radius r:

$$\text{surface of sphere/volume of sphere} = 4\pi r^2/(\tfrac{4}{3}\pi r^3) = 3/r.$$

Vacuum systems come in all sizes and shapes, but it seems that for most the surface (cm^2) is equal to the volume (cm^3) within a factor of 10, in either direction. When noting the surface area, the surfaces of all internal parts, electrodes, etc., must be included.

5.1.4.4. Role of Surfaces in High Vacuum. On the assumption that for every cubic centimeter of volume in the vacuum system there is one square centimeter of surface area, what is the increment in pressure if there is a change in the thickness of the condensed–adsorbed molecules of one monolayer? The number of molecules in a monolayer is a function of the size of the molecules. An order of magnitude figure is 10^{15} cm^{-2}. From the equation relating molecular density and pressure in Section 5.1.3.1,

$$\Delta P = \Delta n/(3.3 \times 10^{16}) = 3 \times 10^{-2} \quad \text{Torr!}$$

In a system of the assumed geometry at a modestly low pressure of 10^{-8} Torr, a condensed or adsorbed layer of gases on surfaces to the extent of a micromonolayer would involve as many molecules as there are in the enclosed volume at any given instant. In other words, if there were an equivalent of one monomolecular layer of gases on the surfaces and if, on the average, each molecule spent one millionth of the time in the volume, and 99.9999% of its time on a surface, this would provide sufficient molecules to support a pressure of 10^{-8} Torr.

5.1.4.5. Relation between Vapor Pressure and Temperature. When a vacuum system has been exposed to air and then pumped down, a partial pressure analyzer will show that the highest peak is at mass 18 (H_2O) if the system is leak tight and has not been baked. This situation may exist for weeks or months. Tables of water vapor pressure as a function of temperature show that a modest temperature increase of 12°C causes the vapor pressure to double in the vicinity of room temperature. While there is no liquid water in the system after prolonged pumping, it is probable that a very similar pressure–temperature relationship exists between the adsorbed surface layers of water and the pressure of the vapor in equilibrium over the adsorbed layers.

The vapor pressures for the elements over a wide range of temperatures have been plotted on a log scale by Honig and appears in Figs. 2–4. Of particular interest for this discussion is the steepness of the curves. At room temperature a temperature increment of 20°C is typically associated with a pressure change by a factor of 10!

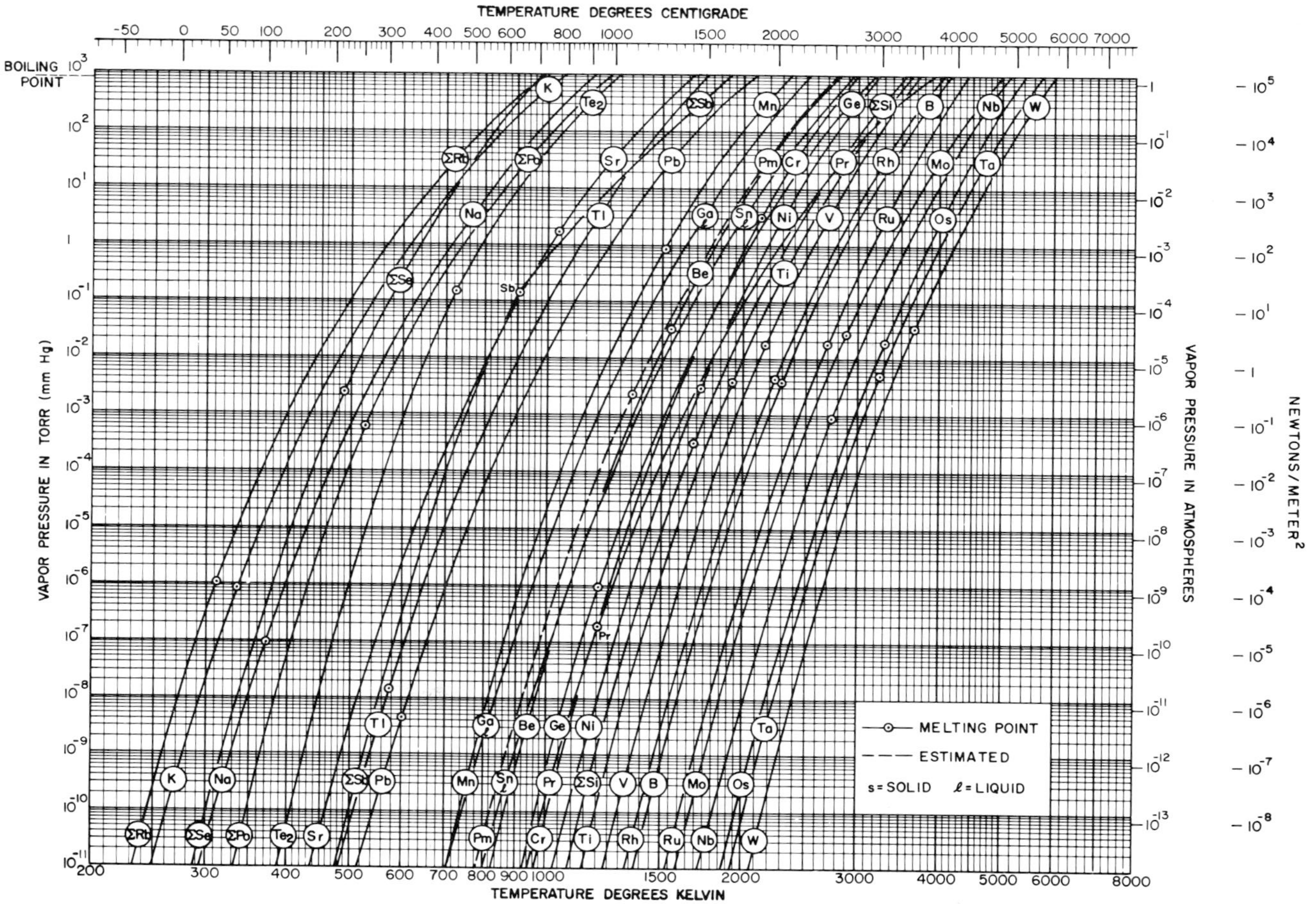

FIG. 2. Vapor pressure of elements as a function of temperature. Melting points are indicated by the small circles. These data were compiled by R. E. Honig and D. E. Kramer, *RCA Rev.* **30**, 285 (1969).

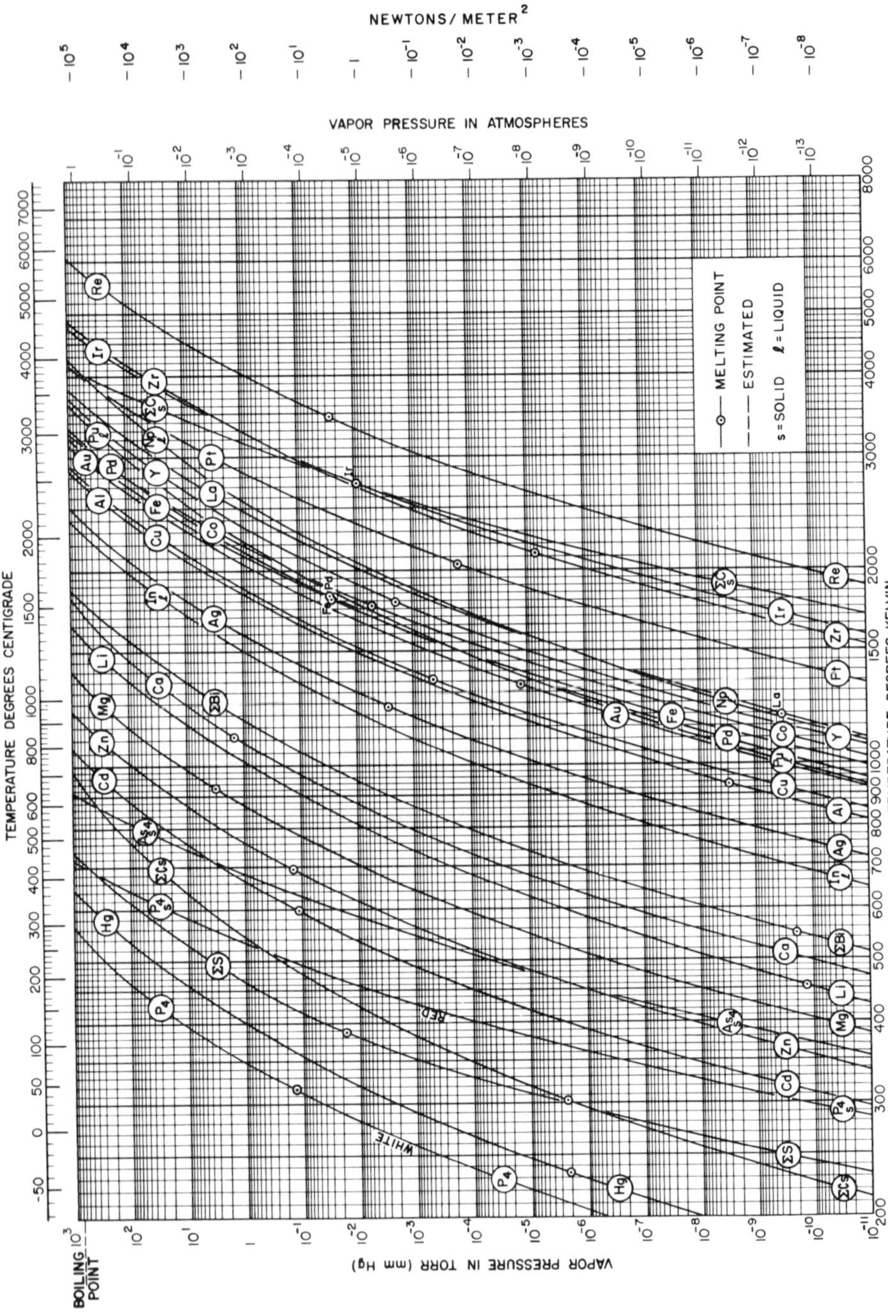

Fig. 3. Same as Fig. 2.

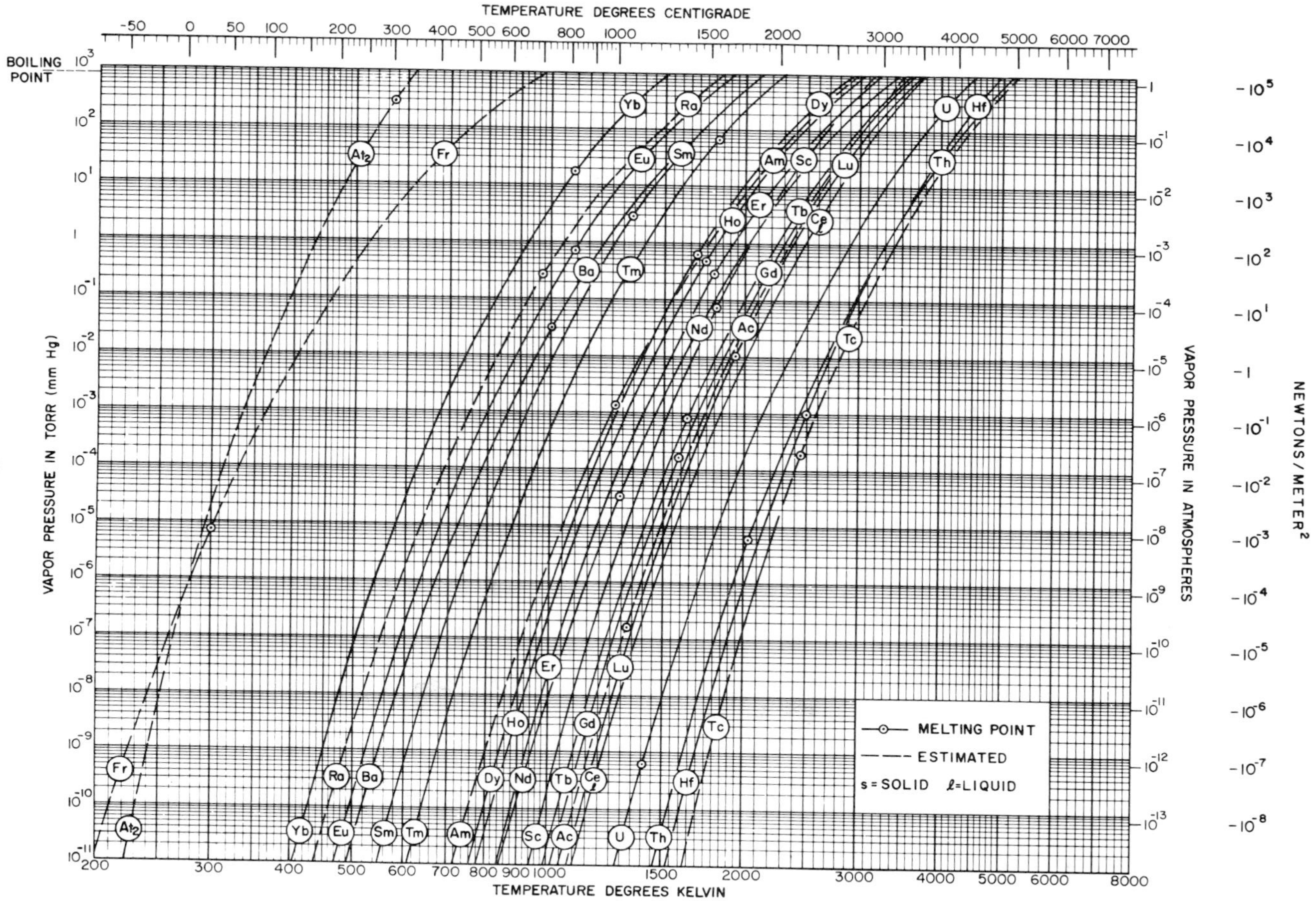

FIG. 4. Same as Fig. 2.

5.1.4.6. Time Required to Remove a Monolayer of Gas by Use of High Vacuum Pumps. It has been assumed that the frequently encountered vacuum system has at least (and probably much more than) one square centimeter of exposed surface area for each cubic centimeter of volume. In order to estimate the time required to remove a monolayer of gas through the use of the pumps, it is necessary to estimate the capacity of the pumps.

A great majority of the vacuum systems have a ratio of pumping speed per unit volume which lies between one and ten. In order to favor the pumps, assume that the pumping speed (liters/second) is ten times the volume in liters. These two assumptions combine to give an estimated pump speed of 10^{-2} liter/sec for every square centimeter of exposed surface. This 10^{-2} liter/sec represents a rate of molecules removed per second of

$$dn/dt = 3.3 \times 10^{17} P.$$

The time required to remove the 10^{15} molecules in the monolayer then becomes

$$t = 3 \times 10^{-3}/P \quad \text{sec.}$$

A plot of this relation is shown in Fig. 5. It is apparent that in the UHV region it is entirely impractical to remove a monolayer of gas by pumping on it at these low pressures. Life is not that long!

5.1.4.7. Importance of Clean Surfaces. It has just been noted that a single monolayer of a gas on the exposed surfaces of a vacuum system concerns a quantity of gas which is capable of supplying essentially an infinite source of gas to completely load the vacuum pumps. Realizing that the wavelength of visible light is in the range of 5×10^{-5} cm and that the intermolecular spacing in the monolayer is about 10^{-7} cm emphasizes the minuteness of a monolayer. Chemical cleaning of surfaces as the vacuum system is assembled must be done with liquids which leave entirely negligible amounts of residue as they evaporate to dryness. Further, it is essential that the vapor pressure of the last liquid used in cleaning be sufficiently high in order to achieve evaporation to dryness in times of modest length.

5.1.4.8. Importance of Baking the Entire Vacuum System. It has been noted that a temperature change of 20°C may be expected to increase the pressure of condensed or adsorbed molecules by a factor of 10.

Reference to the pressure versus temperature data presented in Figs. 2–4 shows that if the temperature is raised from 20 to 200°C the pressures increase by a factor of about 10^5. This is equivalent to saying that

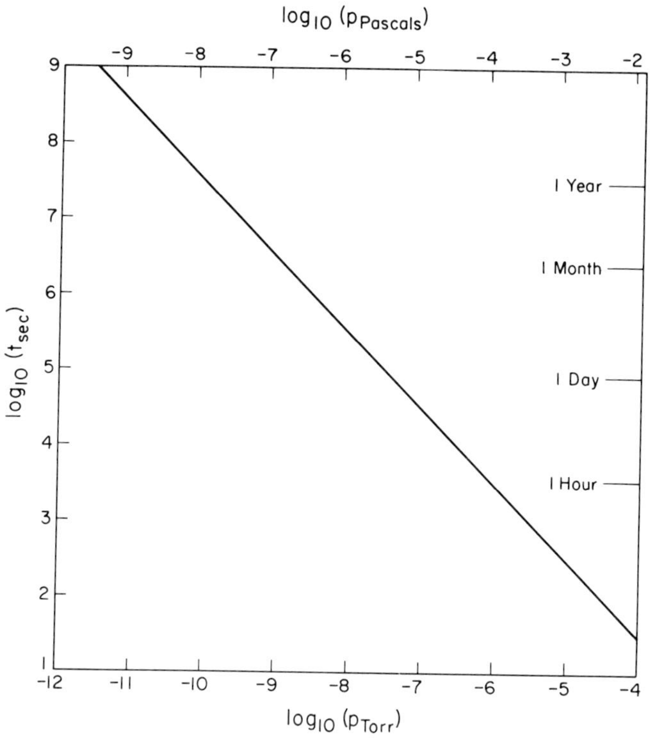

FIG. 5. Time required to remove a monolayer of molecules adsorbed on the surfaces of a vacuum vessel.

pumping for a second at the higher temperature will remove as many molecules from the system as would pumping for a whole day at the lower temperature. If the baking is done at the higher temperature of 400°C, the pressure is seen to increase over its room temperature value by a factor of 10^8. This is the ratio of a year to a second.

Of course the advantages of heating are not as dramatic as suggested by the order-of-magnitude considerations. The pumping time must be much longer than one second. However, continuous pumping for a period of hours at an elevated temperature is seen to be most effective in driving the molecules off the surfaces and into the volume of the system, where they may be removed by the pumps. These considerations show why, particularly at the starting of the heating period, the rate of rise of the temperature must be limited in order to prevent overloading the high vacuum pumps.

It is necessary to heat the entire system at one time because molecules

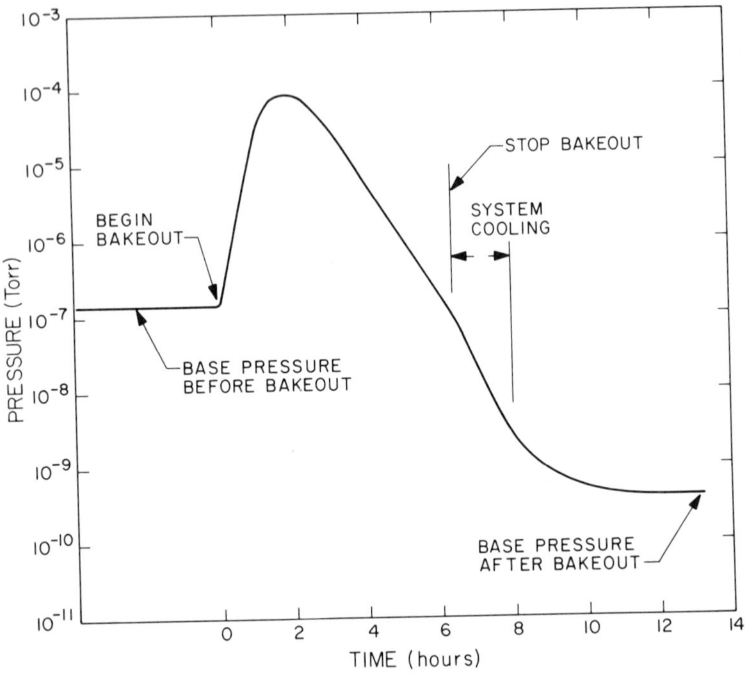

FIG. 6. An example of pressure–time relations for a baked vacuum system. In this case baking improved the ultimate pressure by $2\frac{1}{2}$ orders of magnitude.

will condense or be adsorbed on to cooler surfaces, where they can again become "infinite" sources of gas at the very low pressure levels. This is why one speaks of baking the entire system in an "oven."

The actual pumpdown data are very unique. They are functions of *all* of the *exposed* surfaces in the system. Surface conditions of polish, roughness, and the presence of substances which may remain even after bakeout are all important and will vary among systems. Figure 6 illustrates the pressure improvement that may be obtained by baking a vacuum system.

5.2. Getter Pumping*

5.2.1. Introduction

Getter pumping by titanium films was developed as a commercial vacuum technique during the early 1960s. This technique, better known as titanium-sublimation pumping, is not a stand-alone pumping method and is always used with auxiliary pumps such as sputter-ion pumps. Vacuum systems using titanium sublimation and sputter-ion pumps can be completely sealed off from the rough pumping means, after initial pumpdown from atmosphere, and consequently do not require continuously operating roughing pumps. Such systems can be made inherently free of oil vapor and are attractive for applications where an ultraclean, noncontaminating environment is essential. To reduce the pump down time in systems that must be cycled frequently to atmosphere, these high vacuum pumping mechanisms are located below an isolation valve and remain under vacuum when the experimental apparatus is vented to atmosphere. Figure 1 illustrates a typical vacuum system using titanium sublimation pumping. Means for rough pumping the bell jar from atmosphere are not shown. Figure 2 illustrates a section of the vacuum wall where a fresh titanium film has been deposited.

There is a major contrast between the traditional getter pumping of vacuum tubes and the getter pumping systems. In the former, the getter is flashed once after the tube is sealed off and the resultant getter film pumps throughout the life of the tube. In vacuum systems, a fresh titanium film may pump for only a few hours or even less, making frequent depositions necessary. When pumping high gas loads, continuous deposition may be required. Continuing development through the 1970s has advanced the methods of achieving sublimation pumping performance in such areas as pumping capacity, operating reliability, ease of routine maintenance, and improved sublimator life.

5.2.2. Principle of Operation

Chemically active species resulting from wall outgassing or vacuum processing are readily gettered by fresh titanium films. Titanium has be-

* Chapter 5.2 is by David J. Harra.

METHODS OF EXPERIMENTAL PHYSICS, VOL. 14

ISBN 0-12-475914-9

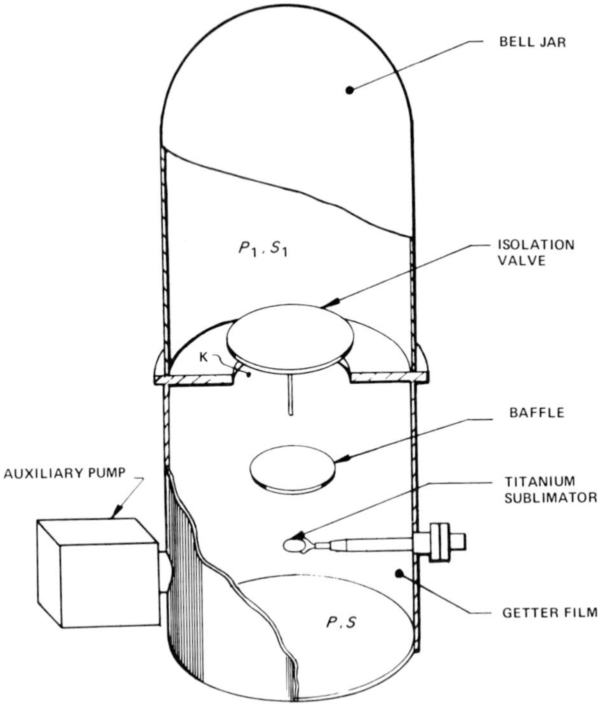

FIG. 1. Typical vacuum system using titanium sublimation pumping.

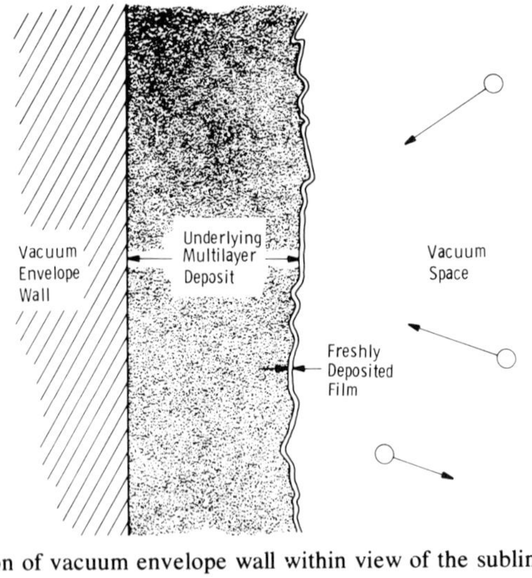

FIG. 2. Section of vacuum envelope wall within view of the sublimator.

TABLE I. Apparent Chemical Reactions during Pumping When the Supply of Titanium Is Limited

$O_2 + Ti \longrightarrow TiO_2$
$N_2 + 2Ti \longrightarrow 2TiN$
$CO + Ti \longrightarrow TiCO$
$CO_2 + Ti \longrightarrow TiCO_2$
$H_2 + Ti \rightleftarrows TiH_2$
$H_2O + 2Ti \longrightarrow TiO + H_2\uparrow + Ti \longrightarrow TiO + TiH_2$

come the preferred getter for general vacuum-pumping applications because of its relatively high vapor pressure characteristic and its broad spectrum chemical reactivity. It provides high speed pumping action for gases such as O_2, N_2, CO_2, H_2, CO, and water vapor. A titanium source is heated to sublimation temperatures around 1500°C, causing a highly active titanium film to be deposited on surrounding surfaces. Active gas molecules impinging on this film react with the titanium, forming stable solid compounds such as titanium nitrides and oxides. By thus transforming gaseous molecules into solids, a highly effective vacuum-pumping action is achieved. An auxiliary pump is needed to remove any remaining nongetterable gases such as helium, neon, argon, and methane. Table I indicates the apparent chemical reactions that take place during pumping when the amount of titanium is limited. Titanium hydride is the least stable, thermally, and its reaction is shown to be reversible. TiH_2 must be kept near or below room temperature in order to limit equilibrium dissociation pressures of hydrogen to acceptable values.[1]

5.2.3. Titanium Film-Pumping Speed Parameters

Unlike other kinds of vacuum pumps that have specified pumping speeds, titanium sublimation pumps are often custom built into the space available, and their speed characteristics vary accordingly. The speed provided by a sublimation pump is a function of several variables, and each of these must be considered to obtain optimum pumping performance. The main parameters are the sticking coefficient for the particular gas of interest, the projected surface area of the film, the conductance per unit area for the gas of interest, and the overall conductance limitations imposed by the associated vacuum envelope. These, in turn, are influenced by such things as the temperature of the film during deposition, the thickness of the freshly deposited film, and the subsequent temperature of the film during the gettering activity. The nature of the substrate upon which the film has been deposited and the configuration of the sub-

[1] C. W. Schoenfelder and J. H. Swisher, *J. Vac. Sci. Technol.* **10,** 862 (1973).

strate are also important. Other variables include the gas temperature and pressure, the sublimation rate, the pumping time, and the effects of nonisotropic gas flow patterns.

The difficulties encountered when attempting a precise analysis of any specific pumping case are formidable because of the large number of parameters involved and the uncertainties associated with many important parameters. Approximate methods for dealing with some of the variables have been developed and, although their accuracy is limited, they provide a useful guide for predicting performance and selecting optimum operating procedures (See Section 5.2.4 below.)

5.2.3.1. Sticking Coefficient: Definition, Related Factors, and Applicable Data. When a gas molecule collides with a fresh titanium film, a number of events may result: It may (a) simply rebound away from the film, (b) enter a weakly bound absorption state where it may reside for awhile before desorbing, or (c) enter a strongly bound state where it may reside more or less indefinitely. Events (b) and (c) both act in principle as pumping mechanisms, although the significance of the former depends entirely upon whether or not the residence time is long in comparison with the time scale of practical interest.

Separating the contributions of events (b) and (c) adds considerable experimental difficulty, and, to date, the literature pertaining to titanium film sticking-coefficient measurements has not provided definitive information on this question. To correspond with experimental practice therefore, the instantaneous sticking coefficient (s') will be defined as follows:

$$s' = (f_1 - f_2)/f_1, \tag{5.2.1}$$

where f_1 is the component of molecular flux density for a particular species impinging on the surface and f_2 is the component of flux density for that species leaving the surface at any given time from both events (a) and (b).

The sticking coefficient has been measured both during continuous deposition of titanium and subsequent to flash deposition. The results are usually presented as shown in Fig. 3. These two techniques are analogous to the continuous and batch titanium–sublimation pumping methods commonly used. In the former, the sticking coefficient is measured as a function of the sorption ratio, and, in the latter, as a function of the quantity sorbed. The sorption ratio is the net rate of sorption of gas molecules divided by the rate of deposition of titanium atoms.

The "initial sticking coefficient" (s) obtained at low coverage or low sorption ratios, the quantity sorbed at saturation, and the number of ti-

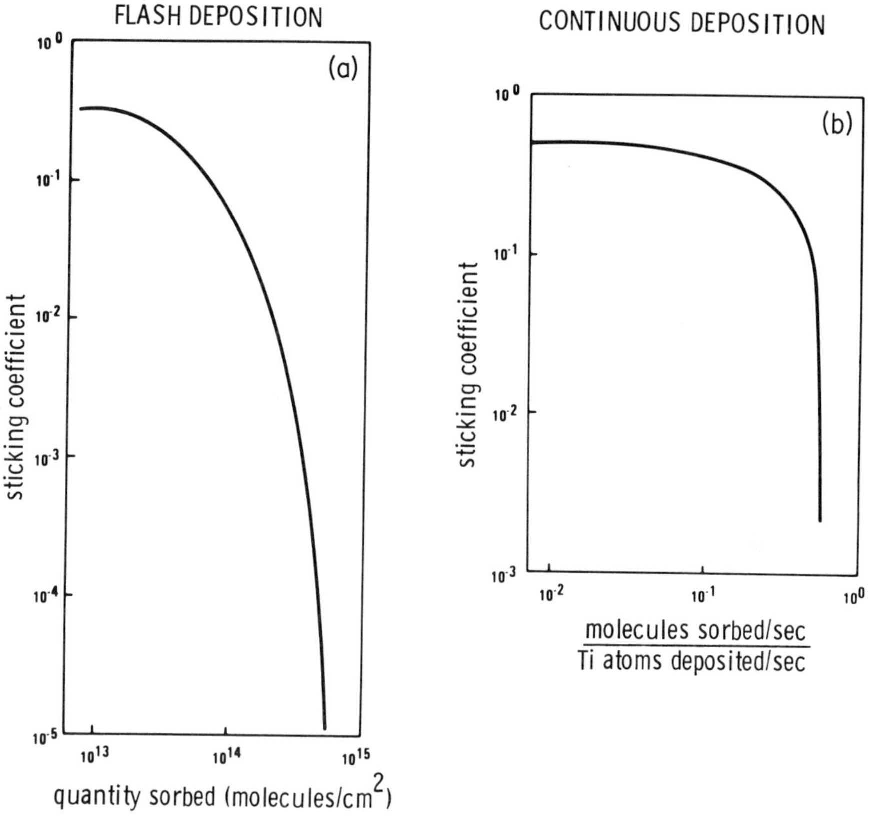

FIG. 3. Typical sticking coefficient curves from flash deposition and continuous deposition experiments.

tanium atoms required per gas molecule sorbed (which is obtained from the maximum obtainable sorption ratio) are used to characterize the film reactivity for a particular gas. A summary of the available data from the literature[2] is presented in Table II. The data in the table are believed to be most representative for the broadest range of film conditions in typical sublimation pumps. However, since experimental results from sorption measurements at different laboratories vary widely, considerable variation from the values in the table can be expected in practice. Further discussion of problems associated with sticking coefficient measurement techniques can be found elsewhere.[2,3]

[2] D. J. Harra, *J. Vac. Sci. Technol.* **13,** 471 (1976).

[3] D. O. Hayward and N. Taylor, *J. Sci. Instrum.* **44,** 327 (1967).

TABLE II. Sorption Data for Gases on Titanium Films at 300 and 78 K

	Initial sticking coefficient(s)		Range of quantity sorbed[a] (Z) at saturation ($\times 10^{15}$ molecules/cm^2)		Minimum number (B) of titanium atoms required per gas molecule
	300 K	78 K	300 K	78 K	
H_2	0.06	0.04	8–230[b]	7–70	1
D_2	0.1	0.2	6–11[b]	—	1
H_2O	0.5	—	30	—	2
CO	0.7	0.95	5–23	50–160	1
N_2	0.3	0.7	0.3–12	3–60	2
O_2	0.8	1.0	24	—	1
CO_2	0.5	—	4–24	—	1

[a] For fresh film thicknesses of $>10^{15}$ Ti atoms/cm^2.

[b] The quantity of hydrogen or deuterium sorbed at saturation may exceed the number of Ti atoms/cm^2 in the fresh film through diffusion into the underlying deposit at 300 K.

5.2.4. Methods for Estimating Pumping Speed

This section presents an approach for estimating the pumping speed of continuously deposited titanium films for H_2, D_2, H_2O, N_2, CO, and O_2. The approach is based upon an empirical formulation described elsewhere.[4] The following simplifying assumptions are required: the titanium film is deposited on a fraction of the inner surface of a large chamber, either the sticking coefficient or the area covered by the film is sufficiently small that the gas flow within the chamber is essentially random, and the gas temperature is uniform. Under these ideal conditions the steady state pumping speed produced within the chamber can be expressed in liters per second by

$$S = sAC \bigg/ \left(1 + \frac{sACBGFP}{R}\right), \tag{5.2.2}$$

where s is the sticking coefficient at low coverage, A is the area covered by the film in square centimeters, C is the conductance per unit area, B is the number of titanium atoms consumed per gas molecule sorbed, G is the number of gas molecules per Torr-liter, F is the calibration factor of the gauge used to measure the pressure P in the chamber, and R is the number of titanium atoms deposited onto the film area per second. Here, the steady state speed is defined as the speed obtained after time dependent surface coverage effects due to changes in pressure, sublimation rate or diffusion into the film, have subsided.

[4] D. J. Harra, *Jpn. J. Appl. Phys. Suppl.* **2** (1), 41 (1974).

The quantities C and G are given by

$$C = 3.64(T/M)^{1/2} \tag{5.2.3}$$

and

$$G = 9.63 \times 10^{21}(1/T), \tag{5.2.4}$$

where C is in liters/sec/cm², T is the gas temperature in degrees K, and M is the molecular weight. F may also be a function of temperature depending upon the type of pressure gauge used. To aid in sublimation rate conversion, 1 gm/hr of titanium equals 3.50×10^{18} atoms/sec. Approximate values for s and B can be found in Table II.

The speed delivered to a vessel separated from the sublimation pump by a passage having a conductance K is usually the quantity of interest. If S_1 is the speed delivered in the vessel (see Fig. 1) Eq. (5.2.1) and the reciprocal conductance addition formula yield

$$S_1 = sAC \bigg/ \left(1 + \frac{sAC}{K} + \frac{sACBGFP}{R}\right). \tag{5.2.5}$$

Assuming the gas load is getterable and emanating entirely from the vessel, the pressure P_1 in the vessel is given by

$$P_1 = PS/S_1. \tag{5.2.6}$$

The following special cases of Eq. (5.2.5) are useful. At sufficiently low pressures, Eq. (5.2.5) reduces to

$$S_1 \approx sAC \bigg/ \left(1 + \frac{sAC}{K}\right), \tag{5.2.7}$$

which gives the maximum possible pumping speed in the vessel. At sufficiently high pressures, Eq. (5.2.5) reduces to

$$S_1 \approx R/BGFP, \tag{5.2.8}$$

which shows that at high pressures, speed is proportional to sublimation rate and inversely proportional to pressure. The pressure at which S_1 falls to half the maximum value defined by Eq. (5.2.7) is

$$P = [(K + sAC)R]/KsACBGF. \tag{5.2.9}$$

The assumptions underlying Eqs. (5.2.2) and (5.2.5)–(5.2.9) are not always satisfied in practical sublimation pumps and attempts to measure the speed S and pressure P within such pumps often fail due to the highly nonisotropic gas flow patterns which exist for values of s on the order of 0.3 or higher.[3] However, since the conductance K usually limits severely the resultant speeds S_1, reasonably accurate estimates of S_1 and P_1 are often obtainable. (See the end of Section 5.2.4.1 below for an indication of the magnitude of possible errors.) In any case, caution should be used when applying the above equations, and speed measurements or Monte Carlo calculations of S_1 are recommended if greater accuracy is required.

5.2.4.1. Sample Calculations. Assume the system shown in Fig. 1 to be 46 cm in diameter and that the bottom circular area and the cylindrical walls are coated with titanium to a height of 46 cm. What steady state pumping speed is produced at 1×10^{-4} Torr for hydrogen in the vicinity of the film at a sublimation rate of 0.1 gm/hr with the film at room temperature?

Neglecting the baffle, the film area is 8200 cm².

$A = 8200 \text{ cm}^2$,
$s = 0.06$ (from Table II),
$B = 1$ (from Table II),
$C = 44 \text{ liters/sec/cm}^2$ [from Eq. (3)],
$G = 3.3 \times 10^{19}$ molecules/Torr liter (from Eq. 4),
$R = (0.1)(3.5 \times 10^{18}) = 3.5 \times 10^{17}$ Ti atoms/sec,
$F = 1$ (assuming pressure gauge calibrated for hydrogen).

From Eq. (5.2.2),

$$S = 21600/(1 + 204) = 105 \quad \text{liters/sec.}$$

This value may or may not be a significant contribution to the overall system speed, depending upon the speed provided by the auxiliary pump. Note that at this pressure, the speed is less than 0.5% of the 21,600 liters/sec that will be provided in the vicinity of the film at lower pressures.

Assuming an isolation valve conductance for hydrogen of 10,000 liters/sec, what will be the maximum speed delivered to the bell jar?

From Eq. (5.2.7),

$$S_1 = 21600 \Big/ \left(1 + \frac{21600}{10000}\right) = 6800 \quad \text{liters/sec.}$$

At what pressure in the vicinity of the film and at 0.1 gm/hr sublimation rate will the speed in the bell jar fall to 3400 liters/sec, half its maximum value?

From Eq. (5.2.9),

$$P = [(10{,}000 + 21{,}600)3.5 \times 10^{17}]/[(10{,}000)(21{,}600)(1)(3.3 \times 10^{19})(1)]$$
$$= 1.6 \times 10^{-6} \quad \text{Torr.}$$

Note that the corresponding pressure in the bell jar from Eq. (6) and (2) is 2.4×10^{-6} Torr. Also, if the sublimation rate were increased to 0.5 gm/hr, the bell jar speed would drop to half its maximum value at 8×10^{-6} Torr.

The getter film pumping speed characteristic corresponding to the sample calculation above is shown in Fig. 4. An auxiliary pumping speed of 100 liters/sec is also shown in the figure. Lines AB, CD, and EF drawn at 45° angles represent constant throughput operation at three different hydrogen flow rates. Points A, C, and E represent pressures maintained by the auxiliary pump without assistance from the getter film. Points B, D, and F represent the corresponding steady state pressures maintained with assistance from the getter film. If the sublimator is operated intermittently, pressure excursions along AB, CD, and EF will occur as the condition of the titanium film alternates between freshly deposited and fully saturated.

In the above calculations for hydrogen, where the initial sticking coefficient of 0.06 is small relative to unity, the assumptions underlying Eqs. (5.2.2) and (5.2.5)–(5.2.9) are essentially satisfied. The comparable case for nitrogen affords an opportunity to illustrate the kind of errors introduced when the limitations of these assumptions are approached.

For nitrogen,

$$s = 0.3 \quad \text{(from Table II)}$$
$$K = 10{,}000(2/28)^{1/2} = 2700 \text{ in liters/sec.}$$

Equation (5.2.2) at low pressures yields $S = 29000$ liters/sec in the vi-

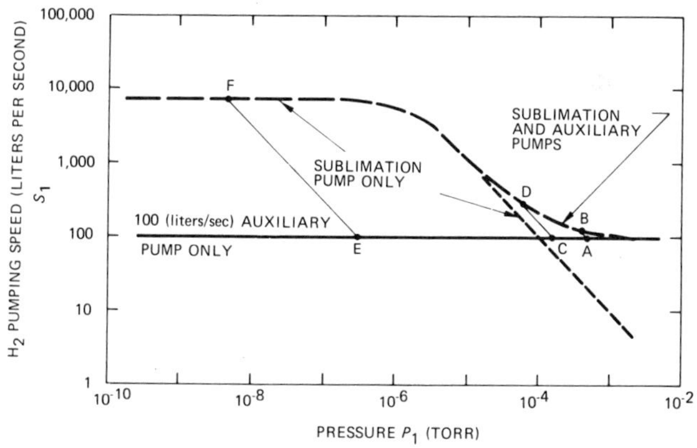

FIG. 4. S_1 vs P_1 for hydrogen for the system shown in Fig. 1 with an auxiliary pumping speed of 100 liters/sec, a sublimation rate of 0.1 g/hr, and a film substrate at room temperature.

cinity of the film and Eq. (5.2.7) yields 2470 liters/sec maximum in the bell jar. However, if the lower 46 cm of the system in Fig. 1 were replaced by a 46-cm diameter end plate having a unity sticking coefficient for nitrogen, an aperture speed of only 19,000 liters/sec would be available. This is considerably less than the 29,000 liters/sec obtained from Eq. (5.2.2) and illustrates the degree of uncertainty associated with such estimates under these conditions. Note, however, that when 19,000 liters/sec is combined with the 2700 liters/sec valve conductance using the reciprocal conductance formula, a bell jar speed of 2370 liters/sec is obtained. This is within 4% of the value obtained from Eq. (5.2.7) and illustrates the better accuracy obtained from Eqs. (5.2.5)–(5.2.8) referred to at the end of Section 5.2.4.

5.2.5. Titanium Sublimators: Desirable Characteristics and Types of Construction

The important characteristics of titanium sublimators include the total available titanium, the range of sublimation rates available, the sublimation rate controllability and the operating power required. Also of interest are pumping and outgassing of the source, the operating pressure range, the cost, and reliability.

Available Titanium. The quantity of available titanium contained in the source influences the frequency of vacuum system shutdown for source replacement. Consequently, operating time and convenience are significantly improved with increased available titanium. Barring catastrophic failure, sublimator life is proportional to the available titanium and inversely proportional to the sublimation rate selected.

Sublimation Rate. The sublimation rate determines the available pumping speeds at high pressures and the time required to deposit a fresh titanium film of given thickness on the substrate. Sublimation rates up to 0.5 gm/hr are desirable for most vacuum systems.

Sublimation Rate Controllability. In practice, one must either deposit a complete film of titanium and allow it to pump until saturated, and then replenish it, or deposit continuously at a rate sufficient to pump at the required speed and pressure. In either case, it is desirable to have simple, convenient adjustments that will cause consistent and reproducible sublimation rates to occur. It is also important that the sublimation rates remain constant or vary only slightly throughout the life of the sublimator at any given control setting. The latter has been achieved only recently with modern sublimator designs.

Power and Heating. The power put into a source is essentially all lost by radiation to surrounding surfaces. By heating adjacent surfaces it can increase coolant consumption and cause desorption of previously pumped or otherwise adsorbed gases such as hydrogen or water vapor. Consequently, power input must be minimized and allowance for its effects must be made.

Pumping and Outgassing by the Source. When the source is first heated, it releases gas from the bulk titanium and its supporting structure, and when it cools, it repumps some gas. Subsequent heat-up from room temperature always releases gas but never the large quantity released at first. Keeping the source hot but below sublimation temperatures is an alternative to turning it off completely and helps to eliminate pressure bursts upon warm up. Operation below sublimation temperature at active gas pressures greater than 1×10^{-2} Torr can form thick oxide or nitride layers at the sublimator surface and lead to reduced sublimation rates subsequently.[5] The radiantly heated sublimator[6] described below will recover from most such exposures after operating for a few minutes at full power and 1×10^{-3} Torr or less.

Operating Pressure Range. Resistively and radiantly heated source designs are capable of operating at pressures up to 5×10^{-2} Torr and momentary exposures up to 2×10^{-1} Torr can be tolerated. In contrast, electron bombardment heated sublimators are not generally operable above approximately 5×10^{-4} Torr due to the occurrence of high voltage discharges at higher pressures. Operation at pressures in the 10^{-8}–10^{-10} Torr ranges is possible with most sources provided they are thoroughly outgassed during system bakeout. At such low pressures however continuous operation of the sublimators is typically not required.

Cost. A paramount feature of getter pumping is the large volumetric speed that can be achieved with an inexpensive source of titanium depositing onto a large surface. Getter film areas must be made large, with large conductances, to achieve the potential of this method. Near and long-term maintenance costs should be considered in selecting a given source design.

Reliability. Sublimator source reliability has continued to improve over the last few years. The resistively heated titanium–molybdenum alloy wire[7] and the radiantly heated sources are generally the most reliable.[8] In

[5] M. V. Kuznetsov, A. S. Nasarov, and G. F. Ivanovsky, *J. Vac. Sci. Technol.* **6,** 34 (1969).

[6] D. J. Harra and T. W. Snouse, *J. Vac. Sci. Technol.* **9,** 552 (1972).

[7] G. M. McCracken, *Vacuum* **15,** 433 (1965).

[8] R. Steinberg and D. L. Alger, *NASA TM X-2650* (1972).

general, reliability must be evaluated in light of the specific requirements and hazards of each situation.

Types of Construction. Several devices have been described based upon different methods of construction and different techniques for heating titanium and are referenced elsewhere.[4,6,8] Some of the more widely used approaches that have performed with varying degrees of success are outlined below.

Resistively Heated Titanium Wire. This method, conceptually one of the simplest, does not achieve good utilization of the available titanium. Local regions of reduced diameter and, therefore, increased resistance, develop as evaporation proceeds. These regions run hottest and melt, causing an open circuit, before other parts of the wire have been used.

Resistively Heated Titanium and Molybdenum Wire Wrapped around a Tungsten Wire. This was the first successful design, commercially, for general vacuum pumping. The tungsten provided a mechanically and electrically stable heating element. Wetting of the molybdenum wire maintained an even distribution of titanium along the tungsten wire even when melting temperatures were reached. Although this approach achieved workable sources, the variation of rate throughout life, the limited amount of available titanium, and the tendency to catastrophic burnout stimulated a continued search for better methods.

Resistively Heated Titanium–Molybdenum Alloy Wire (85 wt.% Ti—15% Mo).[7] This material provides a source that sublimes 20–30% more titanium for a given length and operating current as those constructed of titanium and molybdenum wound around tungsten. It is less susceptible to burnout because it achieves the sublimation rates without melting. Hot spots that occur simply cause depletion of the titanium without melting of the molybdenum. At operating temperature, this material develops crystal boundaries passing all the way through the wire, causing a very fragile condition. A source of this type is shown in Fig. 5. It contains three filament wires that are operated sequentially. Each wire yields 1.2 g of titanium at rates of up to 0.2 gm/hr. Power required is 50 A at approximately 300 W.

Large Resistance Heated Sources. In general, attempts to achieve higher rates and large usable quantities of titanium by scaling up the previously described methods have not been attractive. The following two examples illustrate the nature of the problem.

(a) Longer wires. Rate and usable titanium can be increased in direct proportion to length. For example, consider a filament source of length L compared to one length $2L$:

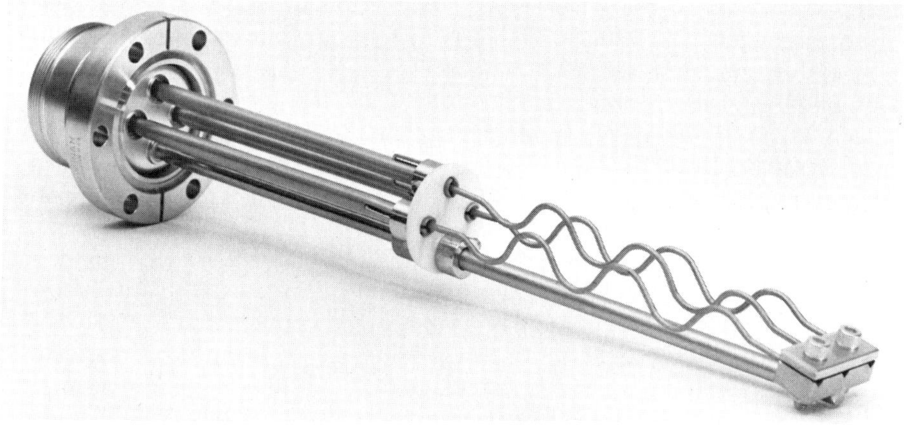

FIG. 5. Titanium sublimation source and holder using three alloy wire filaments.

Length = 15 cm	Length = 30 cm
R = 0.15 gm/hr	$2R$ = 0.3 gm/hr
Ti = 1.2 gm usable	2Ti = 2.4 gm usable
I = 50 A	I = 50 A
V = 7 V	$2V$ = 14 V
VI = 350 W	$2VI$ = 700 W

A limitation of this approach is the extra space required. Where space is available, long lengths have been used successfully, although supports for long lengths may be necessary.

(b) Larger diameter filaments. If the filament length is held at 15 cm and the rate and quantity are increased by increasing the diameter, the following approximate values emerge:

Diameter = d	Diameter = $2d$
R = 0.15 gm/hr	R = 0.3 gm/hr
Ti = 1.2 gm usable	Ti = 4.8 gm usable
I = 50 A	I = 141 A
V = 7 V	V = 4.9 V
VI = 350 W	VI = 700 W

This approach gains a factor of 4 in usable titanium and a factor of 2 in rate. The high current is a disadvantage because of the heavy electrical feedthroughs and cables required. From the above comparison of methods, it can be seen that simple resistance heated arrays with larger quantities of usable titanium inherently draw high currents, require high power, and provide only moderate sublimation rates.

Electron Bombardment Sources. Electron bombardment has been used to heat rods, wires, and billets of titanium to sublimation temperatures. Characteristically, a large amount of usable titanium is available, while only a small portion of the material is heated to sublimation temperature. Relatively high capacities and sublimation rates at relatively low input powers result because the small heated portion can be operated at temperatures very near the melting point. The quantity of usable titanium is limited only by the space and design of the feed mechanism.

Several disadvantages have been experienced with this type of sublimator in addition to the occurrence of high pressure discharges referred to above. Operating at or near the melting point requires very precise control of both the power input and feed mechanisms. In practice, the molten zone is relatively unstable and, gravity permitting, can occasionally drip off and cause electrical short circuits or mechanical interference between moving parts. This difficulty is made even more troublesome by a drop in emissivity of titanium that occurs upon melting, which has a tendency to further increase the temperature at a fixed power input.

The critical power and feed adjustments require more frequent operator decisions. Most feed mechanisms require manual initiation at periodic intervals in order to maintain a satisfactory sublimation rate. Between feed operations, the sublimator surface recedes from the hot zone, causing the sublimation rate to fall off. This falloff represents a significant uncertainty in assessing the actual sublimation rate at any given time.

Sublimators of this type tend to evaporate material onto their heating and control mechanisms producing significant deposits. These deposits frequently peel or break loose, and, if left in place, result in unreliable operation.

Radiant Heated Sublimators. A radiantly heated sublimator has been described[6] that provides large amounts of usable material, uses resistance heating, employs no moving parts, and provides reliable operation. It consists of a hollow shell of titanium with an internally mounted tungsten filament heater. The resistance heated filament radiates a controlled amount of power which heats the titanium to sublimation temperatures. This source design yields over 35 gm of titanium at sublimation rates that are known within $\pm 30\%$ throughout its useful life. Sublimation rates up to 0.5 gm/hr are provided at an operating current of 32 A and 710 W. Up to 3500 hr of operating life are obtainable from a single source, depending upon the sublimation rate required. The simplicity of this new type of source, coupled with its predictable and controllable operating characteristics, have significantly reduced the skill required to achieve efficient sublimation pumping performance. A smaller radiantly heated source has also been developed for operation at reduced power. It provides up to 0.25 gm/hr at an operating current of 50 A and 350 W and can dispense over 15 gm of titanium. These two sources are shown in Fig. 6.

Some disadvantages have been experienced with this type of sublimator. Outgassing during initial turn-on after installation may cause the pressure to rise above 5×10^{-2} Torr. This can be eliminated within 10 to 15 min by outgassing it into the roughing pump during the first pumpdown. Of course, initial outgassing is required with all types of sublimators.

These sublimators operate at up to 750 W of radiated power. The latter is about twice the power used by prior titanium molybdenum alloy filament sublimators and, consequently, the system heating effects can be considerably worse.

A comprehensive evaluation has been reported[8] of this type of subli-

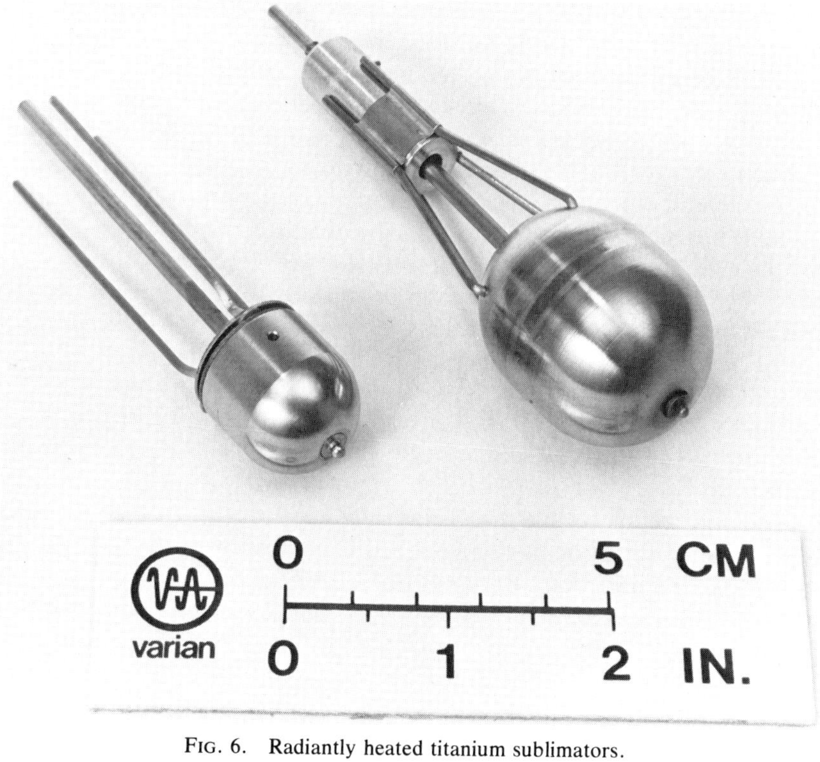

FIG. 6. Radiantly heated titanium sublimators.

mator design versus rod and wire fed electron bombardment sublimators relative to the requirements of neutron generators. The radiantly heated sublimator was found to have the highest reliability. Also, during an extensive long term pumping test in our laboratories, 30 radiantly heated sources were expended without a single failure with an average titanium yield of 39.9 gm. This suggests a source reliability of greater than 95% at the 90% confidence level.

5.2.6. Substrate Design Considerations

Once the desired pumping speed and operating pressure range have been specified for the particular gases of interest, the required film area, sublimation rate, and conductance K can be determined using Eqs. (5.2.2)–(5.2.9). The configuration of the substrate can then be selected based upon the available space, access for sublimator replacement, and other interface requirements. If the getter pump will remain at 1×10^{-7} Torr or less for long periods of time and if the distance between the sublimator and the substrate is sufficient to prevent film temperatures from exceeding 100°C when the sublimator is operating, cooling of the substrate by ambient air should be sufficient. At pressures on the order of 1×10^{-8} Torr, operation of the sublimator may only be required for a few minutes every few hours. Hence the average power radiated by the source can easily be dissipated.

At pressures in the 10^{-6} Torr range and higher, continuous operation of the sublimator is required, and water cooling of the substrate is essential. In light duty applications where no more than a few tens of grams of titanium will be dispensed over the life of a getter pump, little attention need be given to means for eventual removal of the accumulated titanium deposit. In heavy duty applications, however, where a few hundreds of grams of titanium may be dispensed per year of operation, peeling of the accumulated film becomes an important factor. When the film peels, it is no longer in good thermal contact with the substrate, and may release considerable quantities of previously pumped gas when heated by the sublimator.

A solution to this problem was described recently[9] and is based on a demountable sublimation pumping array (Fig. 7) which completely surrounds the sublimators. By collecting all the sublimed titanium on the demountable array, difficulties associated with peeling of titanium from hard-to-clean surfaces are avoided. The accumulated titanium film can be etched away when necessary by removing and then chemically cleaning the pump array. The down time can be kept within minutes by installing a spare sublimation pumping array which has previously been cleaned.

FIG. 7. Demountable sublimation pumping array with end plate removed.

A pumping module with such a demountable array is shown in Fig. 8. The pumping speeds at the baseplate produced by the sublimation pumping array for N_2, H_2, and H_2O were 3500, 8000, and 3800 liters/sec, respectively,[9] at pressures less than 1×10^{-5} Torr. The module shown in Fig. 8 also contains a liquid nitrogen cryopanel, an isolation valve to keep the pumping components under vacuum when the base plate area is at atmospheric pressure, and 240 liters/sec of triode sputter-ion pumps. The isolation valve seal plate swings clear of the valve aperture permitting at least twice the conductance and approximately twice the baseplate pumping speeds obtained with a comparable poppet-type valve. The cryopanel increases the total speed for water vapor to 6000 liters/sec at the base plate. As is well known, water vapor is the major gas present during a normal pumpdown after an air exposure. Besides removing the nongetterable gases, the triode ion pump serves to maintain vacuum when the sublimation and cryopumps are in a standby condition. Typical speed curves for nitrogen are shown in Fig. 9 for various sublimation rates. Figure 10 illustrates the character and extent of peeling of the film after 370 g of titanium had been deposited. Important methods for delaying the onset of peeling in a given structure are (a) increase the distance between the sublimator and the substrate as much as possible during the

[9] D. J. Harra, *J. Vac. Sci. Technol.* **12,** 539 (1975).

FIG. 8. Pumping module with demountable array installed. Overall height of the module is 84 cm and the baseplate diameter is 66 cm.

initial design; (b) keep the substrate as clean as possible prior to depositing the first titanium film; and (c) adopt operating methods that help conserve the amount of titanium used during operation (see Section 5.2.7). The former reduces the thickness of the deposit inversely proportional to the square of the distance for a given quantity sublimed. Peeling typically occurs when the deposit reaches thickness on the order of 0.003–0.006 in.

The use of liquid nitrogen cooled substrates has diminished over the past few years for a number of reasons. Although lower ultimate hydrogen partial pressures and higher intrinsic pumping speeds can be achieved by depositing onto liquid nitrogen surfaces, the gas released due

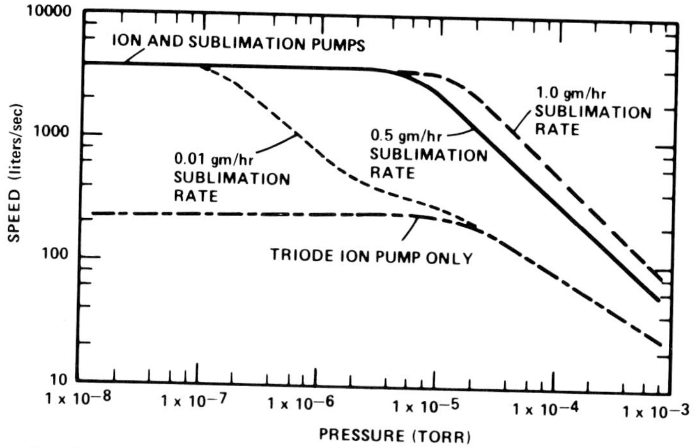

FIG. 9. Steady state nitrogen pumping speed vs pressure at the inlet flange for the pumping module shown in Fig. 8.

FIG. 10. Condition of film illustrating the need for easy access to the pumping surface in heavy duty applications. Loss of good thermal contact between the loose film and the substrate leads to pumping performance degradation. Onset of peeling occurred after subliming 240 gm of titanium onto the array shown and 370 gm had been collected when the photo was taken. Removal of the film by chemical cleaning is normally done after 280 gm have been collected.

to slight temperature increases of the cryopanel during deposition can present a problem. Also, the additional routine cleaning of the cryopanel to remove the titanium flakes represents a significant increase in operating cost. Finally, the increased film speeds due to the higher sticking coefficients usually do not increase the speeds delivered to the work area due to conductance limitations.

5.2.7. Achieving Efficient Operation

Achieving efficient utilization of the available titanium is an important goal in sublimation pump design and operation. Utilization efficiency E is defined for a particular species by

$$E = BGQ/R, \tag{5.2.10}$$

where B, G, and R, as defined in Section 5.2.3, are the minimum number of titanium atoms consumed per molecule, the number of molecules per torr liter, and the sublimation rate. Here, Q is the instantaneous throughput of gas being pumped by the film in torr liters per second, and, of course, $Q = SFP$. E, then, is a dimensionless number varying from zero to one, and although the most efficient utilization of titanium is obtained by operating at $E = 1$, the pumping speed produced by the film when $E = 1$ is zero. This can be seen as follows: substituting Q/S for FP in Eq. (5.2.2) and solving for S yields

$$S = sAC[1 - (BGQ/R)]. \tag{5.2.11}$$

Combining Eqs. (5.2.10) and (5.2.11) yields

$$S = sAC(1 - E). \tag{5.2.12}$$

Hence if $E = 1$, $S = 0$, and a linear tradeoff is apparent between speed S and efficiency.

The tradeoff is more favorable to S_1 due to the conductance limitation of K. Substituting E/S for $BGFP/R$ in Eq. (5.2.5) and eliminating S using Eq. (5.2.12) yields

$$S_1 = K \bigg/ \left[1 + \frac{K}{sAC(1 - E)}\right]. \tag{5.2.13}$$

Note for instance that if $sAC/K = 3$ in Eq. (5.2.13), the maximum value of S_1 at $E = 0$, is $\frac{3}{4}K$, and that S_1 drops to only half this maximum value at $E = 0.8$. Hence, in those cases where K severely limits the resultant speed S_1, higher titanium utilization efficiencies can be obtained before comparable percentage reductions occur in S_1 relative to S.

Equation (5.2.10) can be used to select the appropriate proportionality factor between P and R to provide operation at constant efficiency. Substituting $SFP = Q$ and S from Eq. (5.2.12) into Eq. (5.2.10) yields

$$R = sACBGF[(1/E) - 1)]P. \qquad (5.2.14)$$

Finally, the question of interest in many situations is how long a getter pump will last when used efficiently. The answer, of course, is very dependent upon the efficiency achieved and the operating pressure. If the sublimation rate is held constant, the operating time Ω in hours corresponding to a given weight W of titanium in grams is given by

$$W = R\Omega/(3.5 \times 10^{18}). \qquad (5.2.15)$$

Combining Eqs. (5.2.14) and (5.2.15) yields an expression for operating time

$$\Omega = (3.5 \times 10^{18}\ W)/[sACBGF(1/E - 1)P]. \qquad (5.2.16)$$

Figure 11 illustrates the relationship between operating time for 280 gm of titanium as a function of utilization efficiency, resultant inlet flange

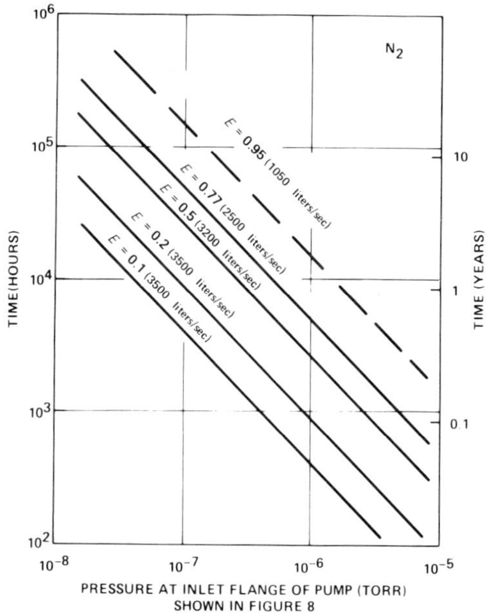

FIG. 11. Pumping time for 280 gm of titanium calculated for nitrogen using Eq. (5.2.10). Titanium utilization efficiency (E) and corresponding inlet flange pumping speed (S_1) as indicated. Substrate area (7900 cm^2) and conductance K (4100 liters/sec) correspond to the pump in Fig. 8.

speed, and inlet flange pressure for nitrogen for the pump module shown in Fig. 8. Routine cleaning of the demountable array (Fig. 10) is required after sublimation of approximately 280 gm for that device. Figure 12 illustrates a method that has been used in our laboratories to achieve higher utilization efficiencies.

5.2.7.1. Periodic Deposition. Periodic deposition at a frequency proportional to pressure is another method for improving titanium utilization efficiency. A typical plot of pressure versus time during this mode of operation is shown in Fig. 13. In the figure, the time t_1 is the *on* time and is chosen long enough to permit deposition of a fresh layer of titanium of thickness corresponding to approximately 3×10^{15} Ti atoms/cm² in accordance with the following equation

$$t_1 = 3 \times 10^{15} A/R \quad \text{sec.} \tag{5.2.17}$$

For a rate of 0.1 gm/hr and an area of 8200 cm² as in Section 5.2.4.1, t_1 = 70 sec.

Period t_2 in Fig. 13 is the *off* time and is the time required for getterable gases to nearly saturate the film. To achieve efficient titanium utilization in this mode, each deposit must become completely saturated with adsorbed species before another titanium layer is deposited. However,

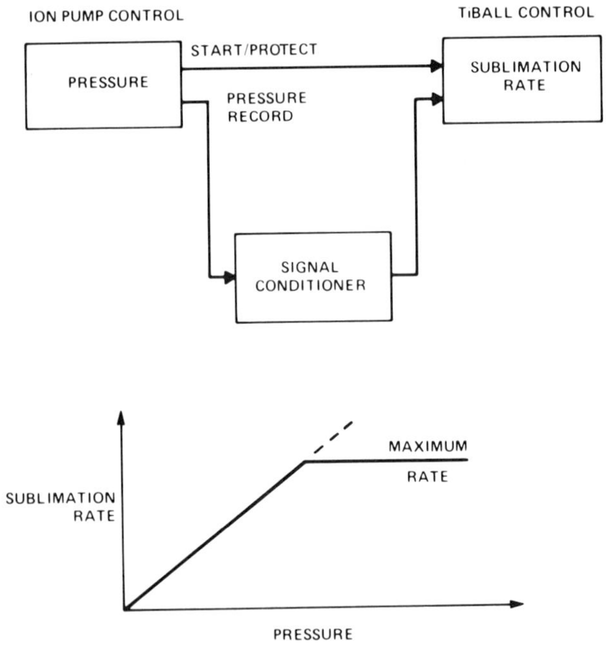

FIG. 12. A method for controlling sublimation rate proportional to pressure.

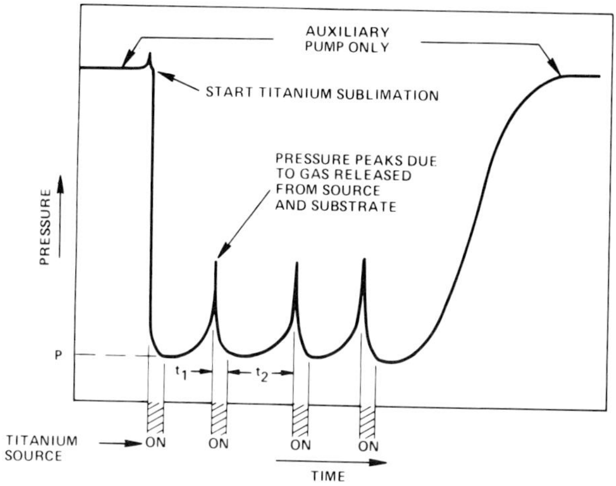

FIG. 13. Pressure vs time during the flash deposition mode of operation.

since the film pumping speed goes to zero at saturation, periodic depositions are usually made prior to complete saturation in order to maintain a minimum pumping speed. Hence, a tradeoff is again necessary between pumping speed and titanium utilization efficiency.

An estimate of t_2 can be obtained by equating one third the quantity sorbed at saturation (Z) from Table II with the number of gas molecules sorbed per square centimeter in time t_2:

$$\tfrac{1}{3}Z = sCGFPt_2, \tag{5.2.18}$$

$$t_2 = Z/3sCGFP \quad \text{sec} \tag{5.2.19}$$

Taking $Z = 3 \times 10^{15}$ molecules/cm^2 for nitrogen and assuming $P = 1 \times 10^{-8}$ Torr, Eq. (5.2.19) yields $t_2 = 15$ min.

Equations (5.2.18) and (5.2.19) assume the total pressure P to be due entirely to getterable gases and neglect the effects of increasing surface coverage on the sticking coefficient. In practice, t_2 should be determined experimentally since the total pressure over the getter film may be due in large part to nongetterable species.

5.3. Ion Pumps*

5.3.1. Introduction

The introduction of commercial ion pumps in the latter half of the 1950s revolutionized the vacuum industry. One of its major advantages is the elimination of the continuous use of the mechanical forepump. Another is the ability to pump to the uhv region without the use of cryogenic surfaces. The elimination of the mechanical pump increases the reliability of the system because there are no moving parts in modern ion pumps.

The disadvantages of ion pumps include their higher initial cost and their larger volume and mass for comparable pumping speeds. They are not best suited for pressures in the 10^{-5} Torr range or higher, nor for uses in which there is a heavy load of gases of zero or low chemical activity. These include all of the noble gases, and some of the lighter hydrocarbons such as methane.

The mechanism of operation of ion pumps is twofold: reactive gases are removed by chemical combination (gettering) on surfaces of a freshly deposited reactive metal (usually titanium); noble gases are removed by ion burial in the cathode surfaces.

It has long been known that higher power transmitting tubes tend to become "soft" during long period of storage, and that if they are operated at reduced voltages for a time they clean up and again are able to withstand their normal, high potentials. Using one half of a double triode tube as an ion gauge, and the other for normal operation, the pumping action of the normally operating triode can be observed.

Even though the phenomenon of electrical clean up has long been known, papers as late as 1950 speculated on the mechanism, because at that time it was not well understood.

5.3.1.1. Electrical Cleanup. The mechanism of the phenomenon called electrical cleanup has become better understood with the development work which has been done on ion pumps. Since there is no such thing as a perfect vacuum, passage of energetic electrons between elec-

* Chapter 5.3 is by W. M. Brubaker.

ISBN 0-12-475914-9

trodes in a "vacuum" envelope results in the formation of positive ions. The electric fields which accelerate the electrons also accelerate the positive ions in the opposite direction. Very frequently these ions strike the cathode. If the energy with which these ions strike the cathode is more than a few tens of electron volts, molecules of the cathode surface are eroded away under this bombardment. This process is known as sputtering. Additionally, an energetic ion incident upon a surface penetrates into the surface a small distance. Of course, it picks up an electron and becomes a neutral atom. However, owing to its penetration into the surface, it is captured. Thermal energies at room temperature are well below those required to remove it.

5.3.1.2. Gettering. Gettering is used today in the production and maintenance of very low pressures as a continuous process, in contrast to its historic use as a process for scavenging the gases in a vacuum tube by the flashing of a pellet. In this pratical application a continuous (or intermittent periods repeated regularly) renewal of the gettering surfaces is required.

The various types of ion pumps, magnetic or nonmagnetic, differ in the mechanism used for providing this continuous supply of gettering material. The magnetic, Penning discharge types use sputtering by the ions incident upon the cathodes. The nonmagnetic types use sublimation from sizeable chunks of the getter material. These sources of material are heated either by electron bombardment or by ohmic losses.

A number of metals are chemically very active in the elemental state. However, titanium and tantalum are most favored for use as getter materials. It is important that the vapor pressures of the chemical compounds formed by the gettering process, as well as of the getter material itself, be very low at room temperatures. The most important compounds are nitrides, oxides, and hydrides.

5.3.1.3. Noble Gas Pumping Requirements. In order for an ion pump to be practical it must be able to effectively pump all of the component gases found in the earth's atmosphere. Most of these gases are chemically active, and so can be gettered. However, most of the noble gases are present in small amounts. Argon, the dominant noble gas, is present at a concentration of 1%. Helium, krypton, and xenon occur at much lower levels. Radon is very low, because it is radioactive with a half-life of only four days. Thus, practical ion pumps must provide a pumping action for gases which are chemically completely inert as well as for those gases of high chemical activity. Differences between the practical ion pumps which are available commercially lie in the methods employed for providing each of these functions.

5.3.2. The First Commercial Ion Pump

The first commercial ion pump followed the pioneering work done by Herb and his associates at the University of Wisconsin.[1,2] The commercial version was manufactured and marketed by Consolidated Vacuum Corporation under the trade name "Evaporion Pump." The gettering action was provided by the periodic evaporation of a small (0.05-cm diameter) titanium wire. This wire was propelled in small increments from a spool onto a hot tantalum post where it first melted, then evaporated. The period between these evaporations was adjusted to provide the amount of freshly deposited titanium to give the desired gas pressure (of the reactive gases) under the impressed gas loads. Noble gases were ionized by a beam of electrons, and then driven by strong electric fields into the freshly deposited titanium, where they remained buried.

The speed of this pump for nitrogen was much higher than its speed for air, because of its very slow speed for argon, which is present at a concentration of 1%. Being a noble gas, argon must first be ionized before it can be pumped. The structure for the ionizing function resembled an enlarged version of the triode ionization gauge used at that time for pressure measurements. By today's standards, this is a weak ionizer.

5.3.3. The Penning Discharge-Type Pump

The second type of ion pump to be made available commercially employs a Penning discharge. This type has been the most popular form of the commercial ion pumps.

It has long been recognized that an electrical discharge cleans up the gases in an enclosure, and, of course, this is a pumping action. As electronic power tubes for the generation of high frequency oscillations became larger and more costly, it became important to be able to pump these tubes before they were put into service after long periods of storage. It has been reported that the concept of a small, low cost appendage pump for this purpose was one of the motivating factors that encouraged the initial development of the Penning discharge device as a practical ion pump. Gurewitsch and Westendorp[3] first described the use of a Penning discharge as an ion pump. They thought of the cathode as being an adsorber for the imbedded ions, and may not have been aware of the gettering action of the sputtered cathode material, although they mention the use of titanium as a cathode material.

[1] R. G. Herb, R. H. Davis, A. S. Divatia, and D. Saxon, *Phys. Rev.* **89,** 897 (1953).
[2] R. H. Davis and A. S. Divatia, *Rev. Sci. Inst.* **25,** 1193 (1954).
[3] A. M. Gurewitsch and W. F. Westendorp, *Rev. Sci. Inst.* **25,** 389 (1954).

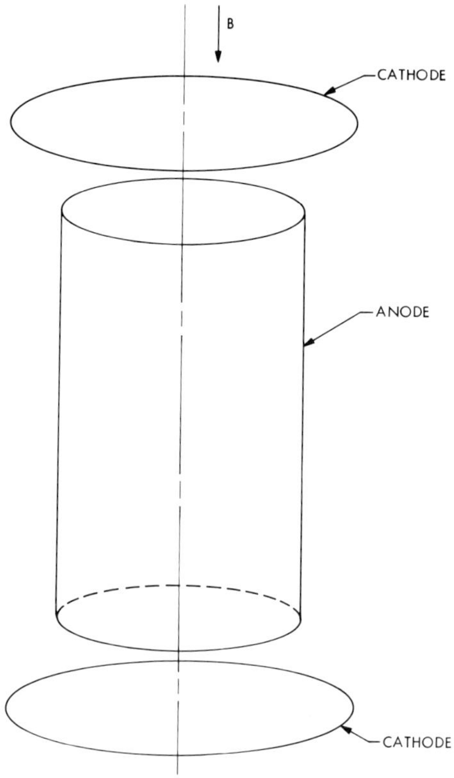

FIG. 1. Geometry of a single cell of a Penning discharge. Space charge of the trapped, circulating electrons makes the potential on the axis essentially that of the cathodes. Thus the electric field is radial. Electron density is at a maximum a short distance from the anode. Electrons can progress radially toward the anode only as they lose kinetic energy, mainly through inelastic (ionizing) collisions with molecules.

Hall and his associates at Varian developed the Penning discharge, with titanium cathodes, into a commercial ion pump.[4] The pump worked so well that it quickly became recognized as a device with potential uses much greater than merely as appendage pumps.

5.3.3.1. Introduction. The simple Penning cell illustrated in Fig. 1 consists of a cylindrical anode placed between two planar cathodes. The anode cell may have a variety of cross sectional shapes; it may be circular, square, or hexagonal. Anode structures of ion pumps consist of an array of many cells adjacent to each other. The whole assembly is immersed in an axial magnetic field, of strength from 0.1 to 0.2 T (1–2 kG).

[4] L. D. Hall, *Rev. Sci. Inst.* **29**, 367 (1958).

The function of the magnetic field is to confine the electrons to constrained trajectories which may be kilometers long. DC potentials ranging from 2 to 10 kV are impressed between the cathodes and the anode structure. The space–charge-limited discharge is self-maintaining from a cold cathode, even at the lowest pressures attainable.

5.3.3.2. The Penning Discharge. A space–charge-limited discharge takes place between the cylindrical anode and the planar cathodes in crossed electric and magnetic fields. (The electric field is radial; the magnetic field is axial.) Once the discharge is struck, space–charge of the trapped, orbiting electrons dominates the potential and field distributions. Calculations predicted[5] and experiments proved[6,7] that the potential along the axis is essentially that of the cathode. The discharge is maintained by an overabundance of electrons released from the cathode surfaces. These electrons are released by a combination of ion bombardment, photon impingment, and, to a lesser extent, by metastable molecules. Because of the avalanche effect in which an electron released at the cathode forms such copious numbers of ion pairs on its way to the anode, it is not necessary for the rate of release of electrons from the cathode to be particularly great. Only a very small portion of the electrons reaching the anode originate at the cathode; most of them come from the formation of an ion and an electron pair through the process of ionization in the gas by electron bombardment. Because more electrons are released at the cathode than the space–charge conditions can accept into the discharge, many are returned to the cathode. This is the mechanism which sets the potential of the axis very close to that of the cathodes. This region of cathode potential exists beyond the axis to a cylinder of appreciable radius. The radius of this cylinder adjusts automatically to take account of varying rates of electron emission. This varies with the cathode material, but discharges seemingly run between electrodes of all metals.

The density of the circulating electrons builds with increasing distance from the axis of the cell, reaching a maximum just before the anode surface. Most ions are formed where the electron density is high, near the anode. They are impelled by strong fields toward the cell axis, their paths being curved by the magnetic field. They may oscillate back and forth a number of times before they approach an end of the cell. Here components of axial electric field drive the ions into the cathode. Erosion patterns on the cathodes show that the rate of incidence of the ions is greatest in the vicinity of the axis. When the anodes are square in cross

[5] W. M. Brubaker, *Ann. Conf. Phys. Electron. 20th.* Massachussetts Institute of Technology, Cambridge, Mass., 1960.

[6] W. Knauer and M. A. Lutz, *J. Appl. Phys. Lett.* **2,** 109 (1963).

[7] W. G. Dow, *J. Appl. Phys.* **34,** 2395 (1963).

section, the erosion patterns have lines of heavier erosion extending from the axis toward the corners of the cells.

One of the limitations of the early diode pumps of the Penning discharge type was their inability to pump against an air leak in a continuous manner. Periodically the pressure in a system with an air leak rises to the 10^{-4} Torr level for a short time, and then returns to a level somewhat lower than that from which the runaway experience started.

Using a mass spectrometer to observe the partial pressures of the various components of the atmosphere during these pressure excursions, Brubaker[8] discovered that the pressures of nitrogen, oxygen, and water vapor changed little during the excursion; the composition of the gas at the height of the excursion was mainly argon. Thus it was established that these pressure bursts were due to releases of previously pumped argon. The effect is aggravated when pure argon is admitted to the system. Figure 2 shows a mass spectrometer record obtained by Brubaker under these conditions. The period between excursions lengthens when the pressure of argon is reduced.

5.3.3.3. Noble Gas Pumping Mechanism. Brubaker[8] described an experiment designed to demonstrate the mechanism by which noble gases are pumped. A small mass spectrometer was used to give a continuous indication of the argon pressure in a vessel with a very slow diffusion pump. Argon was leaked into the system at a constant rate during the experiments. The Penning cell geometry of Fig. 3 was used to learn that the deposition of titanium to the anode only makes a negligible change in the pumping speed for argon. The cell of Fig. 4 allows titanium to be deposited on to the cathodes, but not on the inner, active, surfaces of the anode. When the filament was heated, the argon pressure fell by a factor of 32. Within experimental error, the pumping speed of the device for argon supports the hypothesis that the discharge current of the cell is carried entirely by argon ions at the cathode. Similar measurements made with a diode pump indicate that in steady state conditions the probability of an ion of argon remaining in the cathode is of the order of 1%. These definitive experiments clearly demonstrate that the mechanism by which atoms of noble gases are removed (pumped) is by ion burial in the cathode plates. The pumping action at the anode by "plastering" may exist, but for all practical purposes the locus of the pumping for inert gases is the cathode.

5.3.3.4. Enhancement of Noble Gas Pumping. Once the mechanism for noble gas pumping has been established, it is easy to understand why

[8] W. M. Brubaker, *Trans. Nat. Symp. Vac. Technol. 6th,* p. 302, American Vacuum Society. Pergamon, Oxford, 1959.

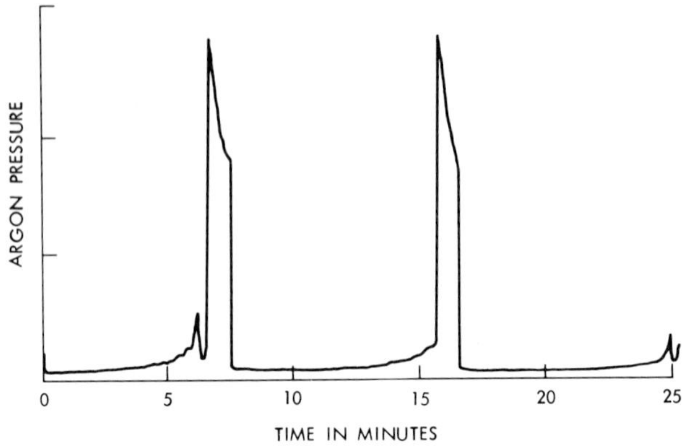

FIG. 2. Argon instability of a simple diode ion pump. The pressure bursts are due to the release of previously pumped argon.

simple diode pumps have so much difficulty in pumping them. An ion incident upon the cathode with a high velocity penetrates a small distance into the cathode surface. In so doing, it erodes away or sputters several atoms of the cathode material which may condense on any surface within line of sight. After prolonged pumping of noble gases, this sputtering uncovers and releases previously buried atoms of these gases.

If the density of ionic bombardment were completely uniform over all of the cathode surfaces, the ability of the diode Penning cell to pump inert gases on a continuous basis would approach zero. Fortunately, the ions are incident on to the cathode surfaces in a very nonuniform manner. As is so clearly shown by the cathode erosion patterns after prolonged use, the density of arrival is the greatest on the axis, and falls rapidly with increasing radius. Some of this sputtered material goes to the opposing cathode, with a deposition rate which is nearly uniform over the entire surface. If some ions are assumed to be incident over the entire cathode surface, it is possible for there to be a small region around the cell periphery in which the buildup of the cathode surface exceeds the erosion rate. Ions deposited in this region are likely to remain (as neutral atoms). This concept has been proved by pumping radioactive krypton and then laying a photosensitive surface on to the bombarded cathode. The developed pattern imprinted by the decaying krypton atoms outlined the peripheries of the cells, in the vicinity of the anodes.

When the concept of the mechanism by which noble gases are pumped became clear, there followed a number of electrode arrangements for improving the performance of the pumps in this regard. All of these schemes had a common goal: to increase the area of the cathode surfaces over which the deposition of sputter material exceeds the rate of erosion by sputtering.

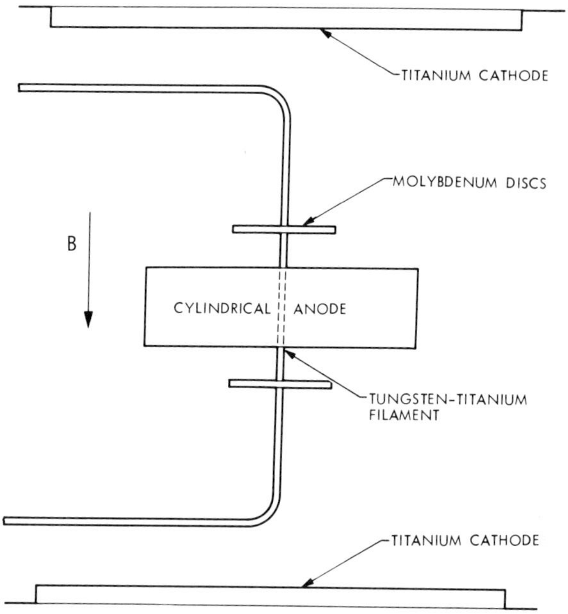

FIG. 3. Penning cell geometry used to investigate the influence of titanium deposition on the anode in rare gas pumping.

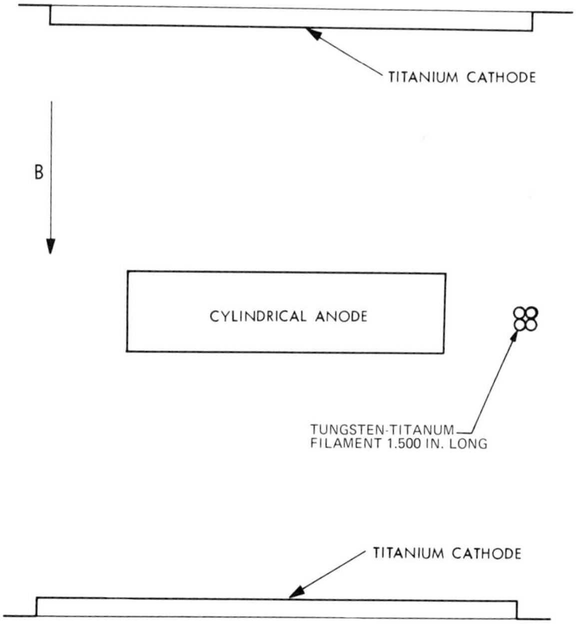

FIG. 4. Penning cell geometry used to test the effect of titanium deposition on the cathode surfaces for rare gas pumping.

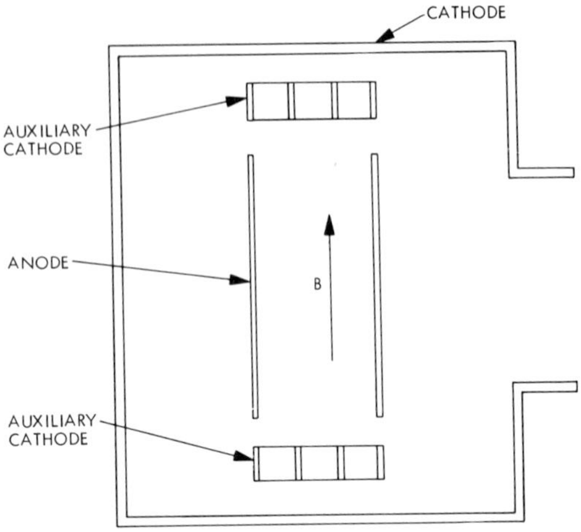

FIG. 5. Geometry of the three-electrode ion pumps for enhanced noble gas pumping. Sputtering occurs at the auxiliary cathodes and showers material to the main cathode (the collector surface), covering previously collected atoms.

5.3.3.5. Triode Structure. The first geometry suggested to enhance the pumping of noble gases was described by Brubaker.[8] It is the triode arrangement which is offered today as the "noble gas" version of the ion pump by a prominent manufacturer. Figure 5 shows the geometry of this triode pump.

In the original triode design the auxiliary cathodes were made negative relative to the triode or main cathodes. An ion incident at a grazing angle sputters several times as many molecules as one incident normally. Thus, ions which strike the auxiliary cathode at higher energy are quite effective in showering sputtered material on to the main cathode, there to cover the ions which had previously impacted and were neutralized. The sputtering rate of the ions incident at a near normal angle, with lower energy, is less so there tends to be a buildup of material on the main cathode, thus permanently trapping noble gas ions there.

Hamilton[9] observed that the triode pump works nearly as well when the auxiliary cathodes are placed at the same potential as the main cathodes, thus eliminating a separate power supply.

5.3.3.6. Slotted Cathode Diode and Differential Sputter Cathodes. Another way to cause the sputtering process to cause a net buildup of material over a portion of the cathode surface is to cut grooves into the

[9] A. R. Hamilton, *Trans. Nat. Vac. Symp. 8th*, p. 388, American Vacuum Society. Pergamon, Oxford, 1961.

cathodes.[10,11] Ions striking the vertical edges of the grooves are effective in showering cathode material into the bottoms of the grooves, thus producing a net buildup in the groove bottoms, adjacent to the vertical side walls.

The grooved cathode approach to noble gas pumping has sufficient merit that it was offered for sale for a period of time.[12]

Another scheme which increases the cathode area where there is a net buildup of sputtered material is that of using opposing cathodes of different sputtering yield rates. The high yield cathode erodes faster than the other, and some of this sputtered material traverses the cell and deposits on the opposing cathode.

Experiments show that the erosion rate is the greatest on the portion of the cathode near the cell axis, and that it is a minimum at the periphery of the cell. Thus, the increased deposition rate onto the cathode of lower sputtering yield causes an increase in the area of the region where there is a net buildup of cathode material.

The success of the differential sputtering arrangement indicates that stable pumping of noble gases is achieved when there is a net buildup over only a portion of the cathode surfaces. These pumps are available commercially.

5.3.4. Orbitron Ion Pump

The Orbitron ion pump, like its predecessor, the Evaporion pump, uses only electrostatic fields to constrain the paths of the electrons to provide long paths for more efficient ionization of the molecules in the gas phase. It also was developed by Herb at the University of Wisconsin.[4,5] The Orbitron gets its name from the nature of these electron paths. As shown in Fig. 6, electrons which leave the filament cannot reach the outer wall of the pump because the potential of the wall is negative relative to the filament. The electrons are attracted toward the central electrode, but as they leave the filament they receive a component of angular acceleration from which they gain appreciable angular momentum. Conservation of this angular momentum prevents the electrons from moving directly to the central, positive electrode. The geometry of the cell produces fields which reflect the orbiting electrons at the ends of the structure, so the electrons are effectively trapped, and they can advance toward the anode

[10] R. L. Jepson, A. B. Francis, S. L. Rutherford, and B. Kietzmann, *Trans. Nat. Symp. Vac. Technol. 7th,* p. 45, American Vacuum Society.

[11] W. M. Brubaker and C. E. Berry, U.S. Patent 3,112,683 (1963).

[12] Noble Gas Ion Pump, Varian Associates, Palo Alto, California.

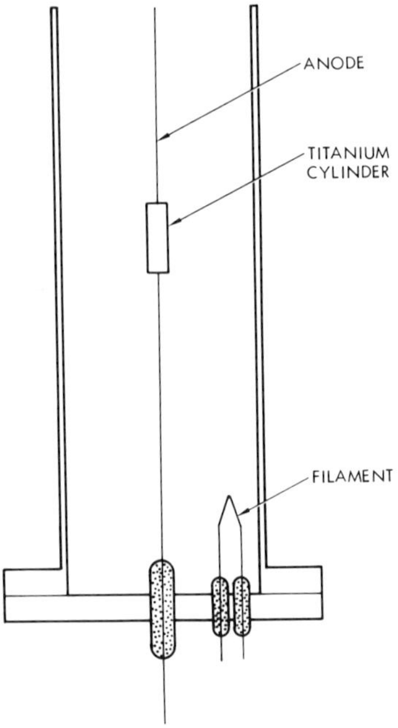

FIG. 6. Geometry of the Orbitron ion pump. The filament is biased positively with respect to the outer cylinder. The electrons travel long orbital paths before collection by the center anode. Titanium sublimation from the hot anode provides gettering action in the pump.

only as they lose kinetic energy, particularly that associated with the circumferential components of their velocities.

The electrons can lose energy by colliding with each other, with molecules, or by interacting with ac fields. Although the geometry of the Orbitron resembles the magnetron high frequency generator, it lacks the asymmetries which are required for the efficient generation of these oscillations. Even with complete azimuthal geometrical symmetry, there is yet another manner in which electrons can lose circumferential velocity through the interaction with electric fields. This is through the ohmic losses associated with the motion of the image charges in the outer wall. When there is complete azimuthal symmetry of electronic space–charge, these losses are minimal. However, nonuniformity, which is always present to a small degree, makes this form of energy loss finite. The fact that an electron current makes its way to the anode even at pressures below 10^{-10} Torr indicates that electrons lose energy by means other than by collision with molecules.

Gettering action is provided in the Orbitron by the sublimation of cylindrical chunks of titanium attached to the central electrode, the anode. Potentials approaching 10 kV are applied to this electrode. Thus, the current required to impart several hundred watts of power to heat the titanium chunks is moderate.

Based on the rate at which argon is pumped by an Orbitron ion pump, the average length of the electron paths is at least 30 m. This indicates that the orbitron geometry is quite effective in providing generous ionizing means for facilitating the pumping of inert gases. Further, with the continuous supply of fresh titanium on the surfaces which receive the ions, there is no instability in the pumping of these gases, as is observed with the magnetic types.

Some commercial versions of the Orbitron ion pump use ohmic heating for the sublimation of titanium. In the larger sizes, one sublimator may serve several Orbitron-type cells. The use of the ohmic heating greatly improves the starting characteristics, because at the higher starting pressures it is sometimes difficult to impress sufficient voltage on the center electrode to achieve adequate sublimation.

5.3.5. Practical Aspects of Ion Pumps

Ion pumps are uniquely different from diffusion pumps in their principles of operation. Diffusion pumps are properly spoken of as having "throughput." During the time of their normal operation, ion pumps have only input, no output. Ionically pumped systems are completely sealed from the atmosphere, while diffusion pumped systems require mechanical pumps to compress the gas exhausted from the very low level of the exhaust port to atmospheric pressure. These systems are vulnerable to the failure of the mechanical pump. The failure modes include power outages, bearing failures, and frequently drive belt failures. The operating ion pumped system is immune to these potentialities.

The startup of all ion pumps from atmospheric pressure requires the use of a forepump of some kind. The conventional rotary mechanical pump is satisfactory for this purpose, particularly if the pumping line is trapped to prevent the entry of oil vapors from the pump into the system. Alternately, liquid nitrogen chilled zeolite sorption pumps are very satisfactory because they are completely noncontaminating. The starting pressures for most ion pumps lie between 1 and 20 μm. The lower the pressure, the easier is the starting.

The starting of ion pumps is hindered by at least three factors. The discharge scours gases from surfaces which it touches. Fortunately, this gas source is of short duration. As the pump is warmed by the energy put into the discharge, gases are released from all of the heated surfaces. If the pump has seen much use, the walls and the anode are likely to be covered with heavy deposits of sputtered titanium which may not be adhering well. It may be peeling and so presents an unexpectedly large surface area. The outgasing of this deposit on warming of the pump on startup slows the process. As mentioned above, the operation of the pump at lower potential decreases the sputtering rate, adding to the time required to achieve a pumpdown. Cleaning the sputtered titanium from the surfaces of the anode and the other exposed portions of the pump facilitates easier starting.

During the starting period the pump surfaces are being relieved of lightly adsorbed gases which must be removed from the volume of the system either by the roughing pump (mechanical or sorption) or by gettering on freshly deposited surfaces. Low voltage operation favors the former. As the rate of release of surface gases decreases and the gases are removed, the pressure falls, the potential rises, sputtering rate increases, and the net pumping speed exceeds the rate of gas release. At this time the roughing pump may be valved off and the pressure continues to fall.

At pressures conducive to the formation of a cold cathode discharge, there is an inescapable rule that the external power supply must be current limited in order to prevent an intense discharge. Ion pumps operate at anode voltages between 4 and 10 kV when the pressure is below 10^{-6} Torr. At pressures above this, warming of the pump by the energy dissipated in the discharge becomes appreciable, and to avoid excessive heating of the pump at the higher pressures, the power supply is current limited by an impedance device so that at half voltage (maximum power output) the power is limited so that the pump does not overheat even with prolonged operation at that pressure.

At startup pressures the discharge is so copious that the potential on the pump is of the order of 1 kV. As the pressure in the system falls and the voltage on the pump rises, the percentage change in the applied potential is much greater than that of the discharge current, because the operating potential is so low. This remains true until the half voltage, maximum power point is reached. At this point the percentage change in the potential and the current are equal. At higher pressures the change in the current is the more sensitive.

Well used ion pumps can become more difficult to start, mainly because of the buildup of titanium deposits. When this situation exists, starting is facilitated by first letting the pump run until it reaches its full temperature for that pressure and then turning it off. Permit the forepump to continue. When power is restored to the cooled pump, the pressure will be lower, the discharge current lower, the potential and the sputtering rate higher. These conditions result in a greater pumping speed, combined with a lower gas load from the outgassing surfaces of the pump and the portions of the system warmed by the hot pump. Low pressures may be achieved before appreciable warming occurs.

The aging of an ion pump is directly proportional to the total amount of gas it has pumped. Since the pump speed is essentially independent of the pressure over the ranges at which pumps are used, the wearing or deterioration of a pump is proportional to the product (or integral) of the pressure and the time of usage. Hard starting, not the limiting pressure, is usually the first sign of aging. A degree of rejuvenation is achieved by a disassembly of the pumps followed by the removal of the loosely adhering deposits of titanium on the anode and all other portions which are in line-of-sight positions from the cathodes.

There are emissions from ion pumps which may be undesired in the rest of the vacuum system. First of all, the very name of the device indicates that there is a gaseous discharge of sorts going on in an ion pump. That the pressure is very low merely attenuates the intensity of the discharge, it does not eliminate it. Where there is ionization of molecules by electron bombardment, there is also excitation of molecules with the subsequent release of photons as the electrons return to their lowest energy states. These photons are unaffected by electric and magnetic fields, and even reflect from surfaces which they may strike. Ions and electrons may escape, even in the presence of the fields. They, too, seem to have a finite probability of bouncing from surfaces. Sputtered cathode material, in small amounts, may also escape from the ion pump.

Fortunately, it is very easy to greatly attenuate the progress of these particles from the ion pump. A right angle turn in the pumping line may be all that is needed. For more sensitive applications further line-of-sight baffles may be found desirable.

Pumps of the Penning discharge type all require rather intense magnetic fields. The degree to which these fields extend beyond the pump is a function of the design of the magnet, particularly the yoke. At distances from the magnet which are large relative to the magnet gap, the magnet appears as a dipole, and the field falls as the inverse cube of the distance. For more critical applications it is possible to effectively contain the field very close to the pump by using a pillbox type of yoke which completely encloses the unit. Pumping can be accomplished through slits parallel to the flux in the yoke which leak negligible magnetic field.

Since ion pumps are constructed of materials with very low vapor pressures, the need for the use of baffles or shields between the ion pump and the rest of the system is less apparent than in the case of oil diffusion pumps. There is no backstreaming from an ion pump of condensable fluids which would be harmful if permitted to enter the system. However, there is usually a low level reemission of previously pumped gases, particularly under heavy loading. Since these generally are permanent gases, they are not readily trapped.

Because the ion pump uses intense beams or currents of energetic, ionizing electrons, which produce copious quantities of photons as they collide with molecules or with the anodes of the discharges, there is a stream of photons with a wide range of energies which leave most ion pumps. Additionally, electrons and/or ions are emitted by ion pumps. Generally, these photons and charged particle streams are easily attenuated to negligible levels by line-of-sight baffles. Right angle turns in the pumping line are quite effective in this regard. So are staggered inserts in the line which preclude any line-of-sight paths through it.

5.4. Cryogenic Pumping*

5.4.1. General Considerations

To begin, we must decide upon a suitable definition of the term "pumping." For reasons which shall become apparent, it is most useful to adopt the following:

DEFINITION: A gas atom or molecule shall be said to have been "pumped" if it is removed from active circulation within the system for a "long" period of time.

The qualifier "long" is quantified solely in terms of the uses to which the vacuum system is being applied. The sense of this definition lies in the fact that pumps of all sorts function generally in one of two ways, namely, they either exhaust the gas atoms or molecules to the outside world or they *trap* and *store* them in some way. All the "fluidless" or "dry" pumps fall into this latter category, and so it is to the dual problems of trapping and storing of gas atoms and molecules that we now turn our attention.

When a gas atom or molecule sticks to a surface, we say that it has been adsorbed. A crucial point to consider is the strength of the bond between the surface and the adsorbed gas. This bond strength is measured in terms of the energy of adsorption. By convention, if this energy is greater than 10 kcal/mole, the process is called chemisorption or chemical adsorption (although an actual chemical reaction may not have taken place in a simplistic sense of the word), and physisorption or physical adsorption if the energy of adsorption is less than 10 kcal/mole. For reference, 1 kcal/mole is the equivalent of about 0.0434 eV/atom. Thus, it is clear that the line of demarcation between physisorption and chemisorption lies at about ten times the thermal energy equivalent of room temperature (kT for 300 K is about 0.03 eV). The probability that a molecule will be physisorbed upon striking a surface is called the *condensation coefficient* and the probability that it will be chemisorbed is called the *sticking coefficient*. Under suitable circumstances either of these probabilities may be interestingly close to unity and a useful amount of *trapping* may occur.

---

* Chapter 5.4 is by Lawrence T. Lamont, Jr.

METHODS OF EXPERIMENTAL PHYSICS, VOL. 14

ISBN 0-12-475914-9

A further concept we need is that of *mean residence time*. This parameter simply measures the average time a molecule remains bound to an adsorption site, and so describes the *storage* aspect of the pumping function. Functionally, it depends upon the activation energy for desorption E_d and the absolute temperature of the surface in accordance with

$$1/\tau = (1/\tau_0) \exp(-E_d/kT_s),$$

where τ is the mean residence time, τ_0 is a constant, characteristic of the system, typically 10^{-12} sec, and k is the Boltzmann constant, 8.63×10^{-5} eV/molecule K.

We shall consider here two types of cryogenic pumping devices; namely, cryopumps and cryosorption pumps. Both of these devices function by condensing the gas (physical adsorption) at temperatures far below laboratory ambient. The term "cryopump" will be reserved for devices in which the adsorption takes place upon a simple surface, and the term "cryosorption pump" will be applied to any device using a molecular sieve sorbant as the primary adsorbing medium.

As with all uhv pumps, the net pumping rate is the difference between the condensation rate and the desorption rate. Following Redhead, Hobson, and Kornelsen,[1] we may write

$$d\sigma/dt = c\nu P - (\sigma/\tau),$$

where σ is the coverage in molecules/cm^2, c the condensation coefficient, ν the specific arrival rate in molecules/cm^2 sec Torr, P the pressure in torr, and τ the mean residence time in seconds.

It is a simple matter to convert this equation into more familiar terms. Recall that the throughput to such a surface must be proportional to the rate of change of the coverage of that surface, so that we may write

$$Q = \frac{A}{G}\frac{d\sigma}{dt},$$

where A is the adsorbing area in square centimeters and G the number of molecules per Torr liter.
At room temperature, $G = 3.27 \times 10^{19}$. We may thus infer a pumping speed S from the pumping equation

$$Q = SP + V\frac{dP}{dt}.$$

[1] P. A. Redhead, J. P. Hobson, and E. V. Kornelsen, "The Physical Basis of Ultrahigh Vacuum." Chapman & Hall, London, 1968.

In the limit that τ approaches infinity and equilibrium pumping (i.e., dP/dt goes to zero), these equations may be simplified and combined to obtain

$$S = cvA/G.$$

Substituting from the efflux equation for the specific arrival rate, we finally have

$$S = 3.64[T/m]^{1/2}cA \quad \text{liters/sec.}$$

For water vapor at room temperature condensing upon a liquid nitrogen chilled cryopanel, this amounts to about 14 liters/sec for each square centimeter of pumping area.

For real cryogenic pumps, we must drop the idealization that the mean residence time is infinite, since this physically means that the *storage* aspect of the pumping function is ideal and imposes no limitations on the performance. Again following the analysis of Redhead, Hobson, and Kornelsen,[1] we note that the desorption term (i.e., σ/τ) for the surface is proportional to the equilibrium pressure over that same surface in the absence of gas flow in an isothermal system. This equilibrium pressure is either the saturation vapor pressure of the gas *at the temperature of the surface of the condensate,* or some value less than the saturation vapor pressure when the quantity of gas adsorbed is less than the amount required to form a "bulk" layer of condensate. The first case describes what we have defined as true cryopumps, and the latter is appropriate to cryosorption pumps. For a more detailed discussion of these considerations, the reader is referred to Ref. 1.

5.4.2. Cryopumps

True cryopumps are subject to the limitation that the equilibrium pressure appropriate to the reevolution term in the pumping equation is ultimately the saturation vapor pressure of the gas being pumped. Thus, these pumps naturally fall into two general categories; namely, those which function primarily as *traps* for the "condensables" in a system such as water vapor and pump oils, and *pumps* operating at far lower temperatures which provide useful pumping of all the common residual gases in vacuum systems with the exception of helium, in the case of 4.2 K cryopumps, and hydrogen, helium, and neon for the more common 20 K units.

5.4.2.1. **Liquid Nitrogen Chilled Cryopumps.** Nitrogen has a normal boiling point of 77.4 K. At this temperature, only the condensables such as water vapor and pump oils are effectively pumped, and thus the device is generally viewed more as a trap for these gases than a true pump. Of course, the distinction is purely one of function, since the physical processes and limitations are identical to those for pumps operating at far lower pressures.

The very selective nature of this pumping action makes liquid nitrogen chilled cryopumps of considerable utility for certain applications. Recall that the most common residual gas in *unbaked* vacuum systems is generally water vapor. Since a relatively small cryopanel is capable of producing a very high speed for this gas (recall that the specific pumping speed for water vapor can be as high as 14 liters/sec cm^2), their use can be an extremely economical way to improve the performance of *unbaked* systems. For a baked uhv system, however, the residual gas composition is generally quite different and the advantages of this device tend to lose their significance. Of course, oil contamination is a serious fault in a uhv system, so its consideration should be unnecessary here. Parenthetically, we note that the use of liquid nitrogen chilled cryotraps with oil diffusion pumps not only satisfy the trapping requirement (See Sections 4.3.5, 4.3.6, and 4.3.7), but also provide a significant additional pumping speed for water to the system. Since these systems are not usually baked, it is not at all unusual for the cryopumping action of the trap to dominate the pumpdown characteristic. This behavior is dramatic in the pumping of systems for sputtering where the diffusion pump must be severely conductance limited during processing. It has been shown that even when the conductance limitation is imposed *during the initial pumpdown,* the pumping speed provided by the liquid nitrogen chilled cryotrap is sufficient to ensure that the pumpdown performance is only slightly degraded.[2] A similar advantage in turbomolecular pumped systems may be attained by using a suitable cryogenic trap as a pump for condensables, even though it is asserted that hydrocarbon backstreaming from turbomolecular pumps is not significant.

In addition to the use of cryotraps designed primarily for use with diffusion pumps, one may fabricate simple cryopanels by brazing tubing to a panel of appropriate size, or simply expose a cryocoil within the system (called, historically, a Meissner coil). Despite the high thermal mass of copper, it is the material of choice for such pumps. Stainless steel is not useful due to its poor thermal conductivity, and aluminum is difficult to weld and most aluminum braze alloys are not acceptable for uhv system applications. This raises the important point that *all* materials introduced to the uhv environment must be uhv compatable. As obvious as this may

[2] V. Hoffman, "Electr. Pkg. and Prod." (1973). (Available as VR-79 from Varian Associates, 611 Hansen Way, Palo Alto, Ca.)

seem, it is often ignored on the assumption that the effect will be small. The reader is referred to Parts 8 and 9 of this text for guidelines. The experimenter has great latitude in the physical design of the cryopanel or cryocoil itself, and it usually makes the most sense to allow convenience of use to dictate the basic form of the pump. Several guidelines may be useful here. First, if the chilled portions of the pump are to be exposed to laboratory ambient each time the system is cycled, then it is generally desirable to have the thermal mass of the pump be as small as practical. This will ensure both the most economical use of the cryogen and minimize the time required to warm up the pump prior to venting the system. Second, the most efficient *pumping* will be attained when proper attention is paid to the location and nature of the sources and sinks of gas within the system. After all, the purpose of the additional pumping is to *meaningfully* improve the vacuum environment and not merely cause an ionization gauge to indicate a lower pressure! Frequently, this means closely coupling the cryopump to the region of the system in which the experiment is being performed. This technique may also be profitably applied when cryogenic trap/pumps are used in conjunction with diffusion and turbomolecular pumps, assuming that the user has some latitude in system design.

Heat shields are not generally used with these pumps unless some large heat source is present within the system. The reasons for this are economic rather than fundamental, since it is true that the dominant heat load to the cryosurface is usually radiation under uhv conditions. When heaters or their equivalent are present in the system, a simple ambient temperature (laboratory) heat shield, water cooled if necessary, will usually suffice.

5.4.2.2. True Cryopumps Operating at 20 K. These pumps usually operate with closed circuit refrigeration rather than by the gross addition of cryogen as required with the various liquid nitrogen chilled devices. The economics of this situation changes the details of operation considerably. As before, the dominant heat load to the pumping surface is radiation, but now heat shielding is a necessity due to the high cost of the refrigeration required. By properly designing such a pump, it is possible to maintain pumping speeds the order of 10^3 liters/sec with as little as 2 W of refrigeration at pressures in excess of a few tenths of a Pascal (i.e., in the low milliTorr range). It should be understood that the shielding used has a function beyond simple radiative heat load protection; namely, the removal of condensables such as water vapor from the gas population to be pumped by the 20 K cryopanel. Obviously, this means that the heat shields must themselves be cryopanels—usually at liquid nitrogen temperature—with additional ambient baffling provided, if necessary, to shield them from large radiated heat loads such as ovens, Knudsen cells, and the like.

Protection of the 20 K cryopanel from high gas loads of condensables such as water vapor is *not* simply a matter of reducing the thermal loading on the refrigerator. Rather, the problem is that the thermal conductivity of the adsorbed layer limits the capacity of the pump. The question of thermal conductivity of adsorbed layers is quite complex, since the adsorbed gas film has a very complicated—and variable—structure. One safe statement, however, is that the thermal conductivity is generally quite small compared to "bulk" values. Since the condensation coefficient is a very strong function of surface temperature, it is clear that even the thin layers of condensate in question here can be limiting. Very little quantitative information is available, but Hands[3] has reported thermal conductivities for adsorbed nitrogen at 20 K which range from 0.05 to 0.5 W/m K under practical cryopumping conditions. However, there is another consideration involved here. The limitations of large throughputs of water vapor are of significance primarily with *unbaked* systems. For the baked uhv system, water vapor is far less important *after the bakeout has been terminated,* and carbon dioxide becomes important. This gas may be a source of difficulty since its condensation coefficient may be as large as 0.62 on a liquid nitrogen chilled cryopanel/heat shield[4] and yet its vapor pressure is quite high—by uhv standards—being about 10^{-6} Pa (about 10^{-8} Torr). An optimized bakeable uhv system, then, may involve some careful techniques such as valvable cryopumps operating at differing temperatures to avoid these problems. For example, if the system is to be pumped cryogenically throughout its entire uhv cycle, it may be necessary to avoid exposing the main 20 K cryopump to the system during bake to avoid the water vapor problem, and then to avoid the use of liquid nitrogen chilled baffles surrounding the 20 K cryopanel to avoid the CO_2 problem. Of course, if the liquid nitrogen cryoshields are omitted, then additional refrigeration will have to be provided. This may simply be the price of optimized performance for these pumps. Fortunately, it does not appear to be a common problem, since there are relatively few systems where the gas load of CO_2 would be large enough to cause a problem and concomitantly have a working pressure requirement low enough that a problem would occur. Other gases may have similar effects, however, so when a gas is to be ad-

[3] B. A. Hands, *Vacuum* **26**(1) (1976).

[4] R. C. Longsworth, "Characteristics of Cryopumps Cooled by Small Closed-cycle 10 K Refrigerators." Paper presented at the joint Meeting of the New York and Philadelphia Chapters of the American Vacuum Society, 12 May 1976, Princeton, N.J. (Reprint available from Air Products and Chemicals, Inc., Allentown, Pa.)

mitted to a uhv system using a cryopump, the operator must consider such interactions.

5.4.2.3. True Cryopumps Operating at 4.2 K. At these temperatures, pumps almost invariably operate with closed circuit refrigeration due to the extreme cost of the requisite quantities of cryogen (liquid He) and various technical considerations. The same general considerations as were mentioned in Section 5.4.1 apply here, except that the problems of thermal conductivity may be greatly magnified. For example, the thermal conductivity of OFHC copper, which had risen by a factor of about 4 as the temperature had dropped from laboratory ambient to 20 K, drops again to the laboratory ambient value at 4.2 K. Other materials exhibit similar behavior, and so it is not unreasonable to anticipate the same phenomenon in the adsorbed layers on the cryosurface. Such an effect would certainly limit the capacity of the pump even further than might be expected based upon experience at higher temperatures.

Operation at 4.2 K may be desired when there is a need to pump neon. However, hydrogen is still not satisfactorily pumped since its saturated vapor pressure is about 10^{-5} Pa (about 10^{-7} Torr) and auxiliary pumping for that gas will have to be provided. The best pump for the requisite auxiliary pumping is probably a titanium sublimation pump, since between the two, all gases except for helium are pumped well. When helium pumping is a consideration, there are two realistic possibilities: (1) add a noble gas enhanced sputter-ion pump, or (2) abandon the *true* cryopump and use a 20 K *cryosorption* pump as discussed in Section 5.4.3.2.

5.4.3. Cryosorption Pumps

Recall that with true cryopumps, the equilibrium pressure appropriate to the re-evolution term in the pumping equation was the saturation vapor pressure of the gas being pumped. For *cryosorption* pumps, the quantity of gas pumped is not generally sufficient to form a monolayer due to the presence of "large effective surface area" adsorbants. A word of caution is due here. The very concept of simple surface area can be quite misleading here. Now, the sorbants used are commonly either one of the Zeolites (metal aluminosilicates) or activated charcoal. Either fortunately or unfortunately, depending upon ones point of view, it is possible to measure the quantity of gas adsorbed by these materials in terms of the area of a simple cryopanel (i.e., *true* cryopumps) that could adsorb the same quantity of gas. It is therefore occasionally stated that a certain sorbant has an effective surface area per unit weight such as is given for certain common materials in Table I. The misleading aspect of this notion is simply that the "pore size" in these materials is of atomic dimen-

TABLE I. Sorbant Properties

Material	Form	Gas	Area/unit wt
Coconut charcoal	Granule	Nitrogen	889
Linde 5A	Pellet	Nitrogen	600
Linde 13X	Powder	Nitrogen	514

sions and the gas atom or molecule is no longer interacting with a simple sorbing surface or, for that matter, even with a *single* surface. It seems most likely that one should think of cryosorption as a case where the *adsorbed gas is bound to more than one surface*. Not only does such a mechanism explain the large sorption capacity of the pumps, but their ability to pump *all* gases, including helium, at the relatively high temperature of 20 K becomes more reasonable as well.

5.4.3.1. Cryosorption Pumps Operating at Liquid Nitrogen Temperatures. In current practice, these pumps are used primarily for rough pumping uhv systems. This technique was reported by Jepsen *et al.*[5] for activated charcoal, by Bannock[6] for the various Zeolites, and by Knor[7] for activated charcoal and Linde-type 13X molecular sieve material. In all these cases, the common residual gases were pumped with high efficiency with the exceptions of helium, hydrogen, and neon. Since all of these gases are present to some extent in a normal laboratory atmosphere, they must be considered here. Sorption may be a highly selective process, even for the gases that are pumped efficiently. It is tempting to think that the equilibrium pressures could be predicted from the original gas composition and the adsorption isotherms. In practice, this is found to be only approximately true since the "problem gases" are minor constituents and are significantly effected by the sorption of large quantities of nitrogen by the sorbant. In particular, with Linde sieve 5A, it is found that neon is pumped to a lesser extent than predicted due, probably, to the filling of adsorption sites with nitrogen. On the other hand, it is also found for this material that helium pumping is slightly enhanced. The latter effect may be due to a mutual solubility effect. The composition of an atmosphere before and after a typical sorption pumping run is given in Table II.

Since the limiting partial pressure in an atmospheric pumpdown is clearly neon (and, to a lesser extent, helium) one may reasonably ask how the load of these gases might be reduced. One obvious way is to pre-

[5] R. L. Jepsen, S. L. Mercer, and M. J. Callaghan, *Rev. Sci. Inst.* **30,** 377 (1959).
[6] R. R. Bannock, *Vacuum* **12,** 101 (1962).
[7] Z. Knor, *Czech. J. Phys.* **13,** 302 (1963).

TABLE II. Gas Composition after Pumping to a Pressure of 1 Pa using a Cryosorption Roughing Pump

Gas	N.B.P. (K)	Air partial pressure (Pascals)	Partial pressures at 1 Pa
N_2	77.4	7.93×10^4	1×10^{-1}
O_2	90.1	2.11×10^4	5×10^{-2}
H_2O	373.0	2.26×10^3	1×10^{-1}
A	87.4	9.44×10^2	3×10^{-3}
CO_2	194.6	3.06×10^1	2×10^{-3}
Ne	27.2	1.86	7×10^{-1}
He	4.2	5.05×10^{-1}	7×10^{-2}
Kr	120.3	1.01×10^{-1}	Nil
H_2	20.4	5.05×10^{-2}	3×10^{-3}
Xe	165.3	9.04×10^{-3}	Nil

pump the chamber with a mechanical pump to some point (in pressure—not time!) and then continue with a second stage of cryosorption pumping, as seen in Fig. 1. For very sensitive applications, however, the use of an oil sealed rotary pump may not be acceptable, even for the prepumping function described here. Note that additional stages of cryopumping would not be expected to reduce significantly the ultimate pressure for cascaded cryosorption pumps. This is because the limit imposed by the adsorption isotherms for neon and its pressure in the system would be reached in each case. Two pumps in parallel produce a similar (true) ultimate pressure for precisely the same reason. Now consider the fact that in the higher pressure regime, the gas flow is far from free molecular. Under conditions of high flow and high pressure, we may reasonably expect that some of the neon and helium will be swept along with the major constituent gases into the pump. Only when the pressure and/or flow rate drop to allow back diffusion of these gases will they reenter the system. This is seen in Fig. 2 where the total pressure and the partial pressures of nitrogen and neon are plotted as a function of time. It is evident that to optimize the sequential pumpdown of a chamber by sorption pumps alone, we must terminate the first stage of pumping when the neon partial pressure has reached its minimum value. To maximize this effect, one must also pump the neon from the other sorption pump(s) prior to chilling. Thus, the best performance for pure cryosorption rough pumping is obtained by the following technique.

(1) Chill the first stage pump per the manufacturer's recommendations. (A 15 min prechill is usually adequate with properly designed pumps.)

(2) Rough the system, including the other *unchilled* sorption pumps

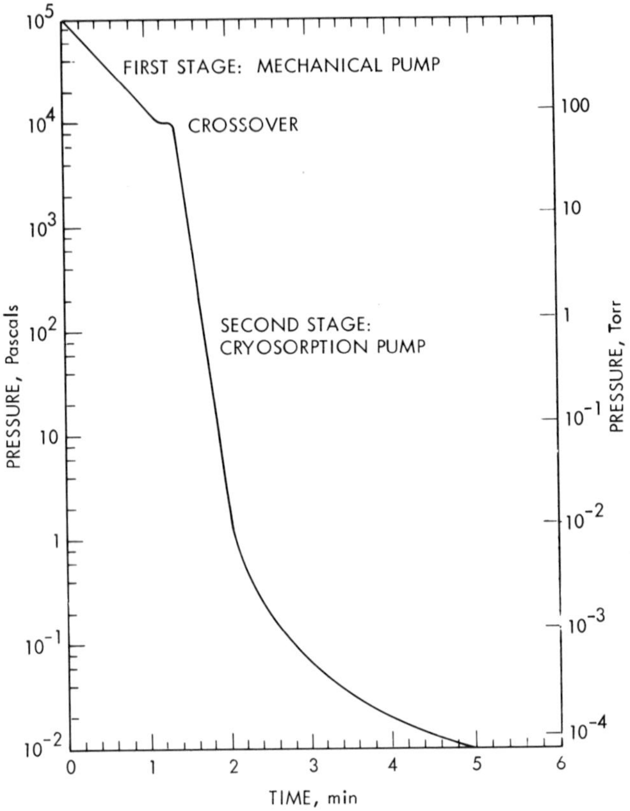

FIG. 1. Pumpdown of a 150-liter vacuum chamber in two stages: the first stage is a 350 liters/min mechanical pump (carbon vane) and the second stage is a high capacity cryosorption pump. Crossover is taken at a total pressure of 10^4 Pa.

until the neon partial pressure is minimized. This will typically occur when the total pressure is about 100 Pa (about 1 Torr). With a Varian VacSorb pump and a 100 liter volume, this takes about 90 sec (!), so be sure that you do not pump beyond this point—particularly with small systems.

(3) Valve off the first stage pump(s) as rapidly as possible and chill the second stage. This prechill will take longer since there is no gas within the pump to conduct heat away from the sieve material.

(4) Open the valve to the second stage pump(s) and pump down to base pressure.

A comparison of the above technique to the pumping of the system with the same two pumps operated in parallel is seen in Fig. 3. This technique

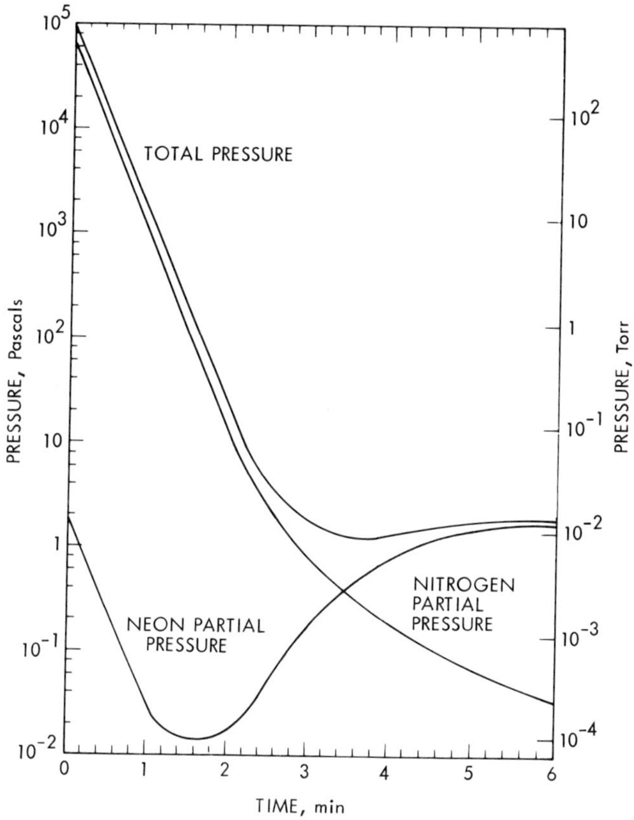

FIG. 2. Total pressure, partial pressure of nitrogen, and partial pressure of neon as a function of time. Note that helium behaves in a manner much like neon, but is not shown on this graph since it rarely constitutes a problem for roughing vacuum chambers.

was first reported by Turner,[8] but is not so widely understood as it should be since publication was outside the normal technical journals of the field.

5.4.3.2. Uhv Cryosorption Pumps Operating at 20 K and Below. Uhv cryosorption pumps operating at 20 K are currently enjoying an upsurge of popularity due to advances made recently in performance and reliability in refrigerator technology. Since cryosorbants are used, the "noncondensable" gases are effectively pumped and the characteristics of these devices approach the ideal. It seems likely that the ultimate utility of these pumps will now be determined by proper attention being paid to their special design problems and to uhv technique. The latter consider-

[8] F. T. Turner, *Environ. Q.* **10**(2) (1964). (Reprint available from Varian Associates, 611 Hansen Way, Palo Alto, Ca.)

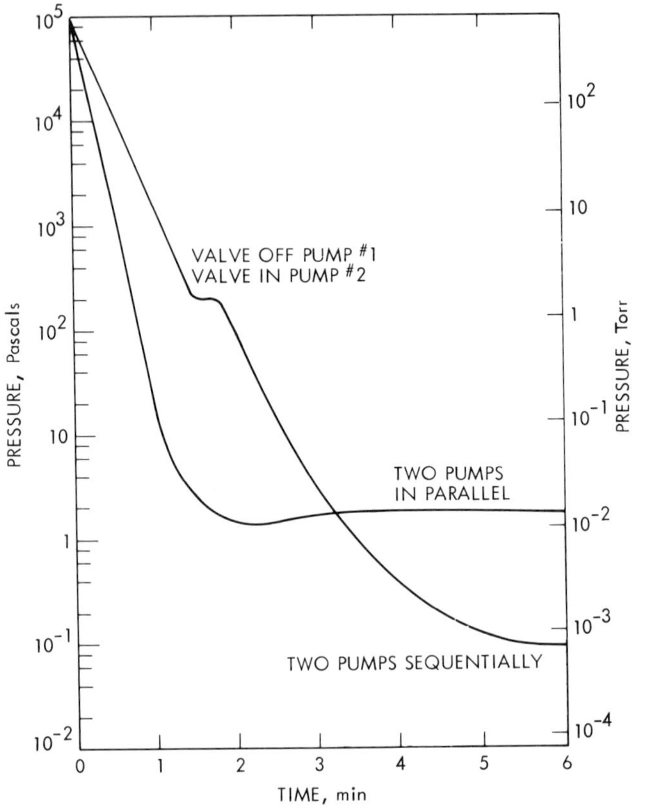

FIG. 3. Comparison of two cryosorption pumps on a chamber operated in parallel (i.e., both valved in to the chamber at the same time), and sequentially. In the sequential operation, the crossover time is selected so that the partial pressure of neon is at a minimum. This point depends upon the size of the chamber and, of course, the size of the pumps and the conductances involved, so the user must scale as appropriate to his situation. The pumps used here were standard capacity Varian VacSorb pumps and the chamber volume was 100 liters.

ation may be a particularly troublesome pitfall since much of the cryosorption technology has developed through the rough pumping application and many techniques and practices will have to be relearned.

Construction of a cryosorption pump of this sort requires melding of several disciplines and so is probably beyond the "average" user. Indeed, the lead time that the major manufacturers seem to have required to enter the marketplace with these pumps may be taken as a reasonable measure of the difficulty of the design problem. Thus, rather than attempt to specify how one might design and fabricate such a pump, we will concentrate on what sort of design flaws to avoid in purchased items.

One major consideration is the reliability and required maintenance of the refrigerator used on the pump. As previously mentioned, it was these very advances which resulted in the current upsurge of interest. The current claims are typically 9000 hr field maintainable service intervals. The prudent prospective purchaser will check with other users for recent experience in this matter.

Another matter of considerable significance is simply good vacuum practice. Much of these details will be buried within the pump and hence not accessible to user scrutiny. However, certain aspects, such as choice of materials of construction and assembly techniques, bear mention. For example, stainless steel is a highly desirable material of construction for all parts of the pump envelope where heat transfer is not a consideration, but is not suitable for the cryogenic elements themselves. Elastomers and epoxies are generally unacceptable, as they are with any uhv application. Aluminum is an excellent material in many respects, but it is difficult to weld properly and difficult to clean. Of course, it is an excellent thermal conductor with relatively low specific heat, so one may reasonably expect to see it used often despite the difficulties it poses.

Most pumps will be supplied with liquid nitrogen baffles to reduce the heat load on the refrigerator. As noted before, for 20 K true cryopumps, this may generate problems if a gas such as CO_2 must be pumped which may condense with high probability upon the liquid nitrogen chilled cryopanel and yet has an unacceptably high saturated vapor pressure at that temperature. What will happen in that case is simply that the pressure will "hang up" at a point somewhat below the saturated vapor pressure of the gas at liquid nitrogen temperature, falling only slowly until all the gas has been transferred to the lower temperature cryosorption part of the pump. This process will be complicated further by the fact that some fraction—probably in excess of 50%—of the re-evolved gas will *not* go to the 20 K pump but rather be re-emitted back into the system. From there, some fraction of it will be recondensed upon the liquid nitrogen cryopanel to make its way back through the procedure once more! The moral of this is simply to avoid the use of the liquid nitrogen cryopanel/shield when the pressure limits imposed by such a process will be a significant effect. As with the 20 K true cryopump, the only viable option for extremely sensitive applications may be the use of multiple valvable pumps. At this writing, there is insufficient experience with commercially available pumps and the classes of service which they must satisfy to adequately judge the practical limitations this may impose on the use of these units.

It is generally true that activated carbon is used as the sorbant in these pumps. This appears to be primarily due to the relative ease of regeneration of the carbon, which requires only room temperature regeneration cycles rather than the bakeout required by the zeolites. There is a slight problem here in that it seems unlikely that one will be able to avoid at least a mild bake of the carbon sorbant—to speed up the regeneration cycle if nothing else—and activated carbon is not particularly amenable to bakeout. A further problem is simply that *all* the common sorbants are very poor thermal conductors in their bulk and the granular nature of the charge makes the situation far worse. This is because the heat is normally conducted between two "contacting" pieces of material by a thin film of gas on the contacting surface that conducts and convects the heat. In the absence of the gas under high vacuum, the only heat transfer mechanisms of significance are point contact conduction and radiation. This means that the thermal design of the pump insofar as it impacts heat transfer *from the sorbant* is crucial. One may hope for some advances to be made in this area, as the present state of the art is rather primitive.

5.4.4. Some Special Considerations for Cryogenic Pumps

It is important to understand that cryogenic pumps of all sorts actively modify the vacuum environment in ways other than simple removal of gas from active circulation within the system. For example, for many gases of interest, the condensation coefficient is quite high and the distribution of gas atoms or molecules in phase space may be significantly altered in the vicinity of the pump, since most vacuum "pressure gauges" actually measure something which is proportional to the number density. In the limiting case, half of the distribution may be missing and the gauge will be in error by a factor of 2. Of course, the same problem occurs with gauge placement near any pump with a high capture coefficient. The other aspect of this is simply that the temperature of the gas re-emitted from the pump will be that of the cryosurface it just left. Therefore, if a large fraction of the interior of the chamber is cryogenic, the temperature of the gas will be altered. In some cases, it becomes difficult to unambiguously specify what pressure even *means* in the system. There may be cases, such as in space simulators, where this is a concern.

Another aspect of cryosorption and true cryopumps is that their capacity to trap and store gas is limited—in some cases severely—and they must be regularly recycled. For some pumps, the nature of the operational cycle is such that they have to be warmed up and rechilled upon each pumpdown. While this is not a serious objection for the laboratory worker in general, such a situation would probably prove intolerable in production. Thus, if one is performing developmental research, it would seem prudent to avoid any element in the process which could not be scaled to the final form of the sequence.

Cost will be a factor in many situations. At present, uhv cryosorption pumps are cost competitive with turbomolecular pumps in the smaller sizes (less than about 10^3 liters/sec) but not with trapped diffusion pumps. As the required speed increases, the cost per unit speed falls, and very large cryopumps of all sorts eventually become the most economical choice. Liquid nitrogen chilled cryosorption pumps for system roughing have no real competition when oil-free roughing is required. Even so, they are generally competitive with oil sealed rotary pumps when liquid nitrogen is easily available.

Finally, one must note that all cryopumps are to some degree subject to performance degradation by long wavelength infrared radiation. As discussed briefly in Section 5.4.1, the reemission term in the pumping equation is strongly temperature dependent. Since the layer of condensate is a very poor thermal conductor[3] and tends to have an absorptance typically the order of 0.9 for long wavelength infrared,[9–11] it may become necessary to design appropriate heat shields when heaters are used in a cryopumped or cryotrapped system. Examples of this include space simulation chambers, vacuum furnaces, and vacuum deposition systems, including heaters. Design of such shields so that they adequately protect the cryosurfaces from excessive heat load while not seriously conductance limiting the pumping speed delivered to the chamber, certainly constitutes an important and possibly quite difficult engineering problem.

5.4.5. Safety Considerations

Apart from the hazards of working with the cryogens themselves, these pumps are relatively safe. There are, however, two serious points to consider when working with a high capacity valvable cryosorption pump; namely, (1) a foolproof relief valve *absolutely must* be provided to allow the escape of previously pumped gases should the pump warm up with the valve to the system closed, and (2) such pumps must not be used to pump dangerous gas mixtures such as H_2 and O_2 in combination, or pyrophoric gases such as SiH_4, or most certainly not a combination such as SiH_4 and O_2! The question of relief valves has been addressed by all

[9] B. C. Moore, *Trans. AVS Natl. Vac. Symp. 9th*, p. 212. Macmillan, New York, 1962.

[10] R. P. Caren, A. S. Gilcrest, and C. A. Zierman, "Advances in Cryogenic Engineering," Vol. 9, p. 457. Plenum, New York, 1964.

[11] T. M. Cunningham, and R. L. Young, "Advances in Cryogenic Engineering" Vol. 8, p. 85. Plenum, New York, 1963.

manufacturers of such pumps and it remains to the user to ensure that they are functional and not to defeat them in any way.

When the design of a high capacity cryosorption pump is contemplated, the pressure relief valve question must be faced by the designer. To decide when the capacity of the pump is sufficiently high to require a relief valve, the following may be used as a general guideline

$$(Qt)_{\text{pump}}/V_{\text{pump}} > 10^5,$$

where $(Qt)_{\text{pump}}$ is the pump capacity in Pascal liters of gas and V_{pump} is the volume of the pump in liters.

While it is certainly possible to design a 20 K uhv cryosorption pump which satisfies this criterion and therefore requires a relief valve, it is tempting to observe that a uhv pump is operated at such low pressures that the maximum quantity of gas adsorbed may be far less than the maximum possible for the pump due to the detailed shape of the adsorption isotherm. It seems prudent, however, to design to the "worst case" considering the explosion hazard in question, so the use of a relief valve is strongly recommended.

5.5. Turbomolecular Vacuum Pumps*

5.5.1. Introduction

This chapter is intended to aid the academic and industrial physicist in selecting and applying commercially available turbomolecular vacuum pumps.

5.5.2. History

Turbomolecular pumps first became commercially available in 1957.[1] Before that time, as early as 1912,[2] "molecular drag" pumps were available from time to time covering roughly the same pressure range as the turbomolecular pump but, because of insufficient flow capacity and a tendency to mechanical seizure, were not commercially successful. Since the introduction of the turbomolecular pump in 1957, six major manufacturers of vacuum equipment have brought their own turbomolecular vacuum pumps to the market, thus establishing turbomolecular pumps as a new "workhorse" for vacuum technology.

5.5.3. Distinguishing Features

Turbomolecular pumps are characterized by "bladed" rotor and stator construction (Fig. 1) with running clearances in the millimeter range. In contrast, "molecular drag" and "viscous drag" pumps are characterized by continuous cylindrical or spiral working surfaces with running clearances in the 0.01 mm range. One commercially available pump is a hybrid of turbomolecular, molecular drag and viscous drag pumps.[3]

5.5.4. Application Area

For physicists with previous vacuum pumping experience, who have never worked with turbomolecular pumps, the application area can best be described as essentially the same as the venerable diffusion pump.

[1] W. Becker, *Vak. Tech.* **7,** 149 (1958). Also in *Proc. Int. Congr. Vac. Technol, 1st, Namur, Belgium, 1958.*

[2] W. Gaede, *Ann. Phys.* **41,** 337 (1913).

[3] L. Maurice, *Proc. Int. Vac. Congr., 6th, Japan; J. Appl. Phys.* **2** (1) (1974).

* Chapter 5.5 is by G. Osterstrom.

ISBN 0-12-475914-9

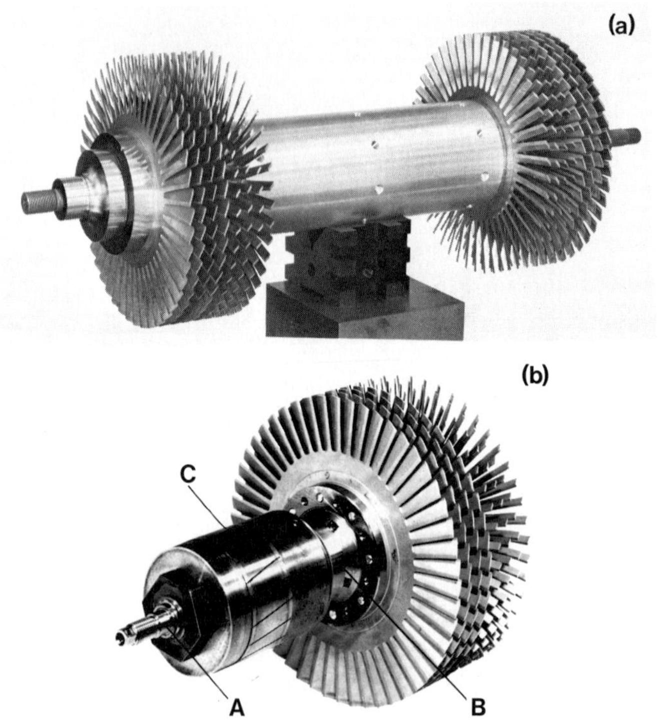

FIG. 1. Typical turbomolecular pump rotors. RPM-50,000. Tip speed 374 m/sec. (a) Double-ended form. Usually horizontal. Bearings at both ends. Each of the two eight-stage pump rotor disk groups pumps half the flow endwise from the center. Note the open high conductance passages between the inlet blades and the closed low conductance passages between the outlet blades. (b) Single-ended form. Usually vertical. Bearings at one end (A) and at center (B). The eight-stage pump section at the right pumps from right to left. The section at the left incorporates a normal induction motor (C) which is driven by a stator powered by a high frequency source.

Both pumps can discharge indefinitely to the atmosphere through some sort of forepump. Both have, as compared with capture pumps, a pumping action nonselective with respect to molecular species, from 1×10^{-2} Torr to lower than 1×10^{-9} Torr inlet pressure. Both are available in the same capacity ranges: from about 70 liters/sec to 10,000 liters/sec. Diffusion pumps to 60,000 liters/sec are available. However, no general purpose turbomolecular pumps larger than 10,000 liters/sec have been built to date, although there is no technical reason why this could not be done.

Wherever very small throughputs at extremely high pumping speeds are required, as in large ultravacuum systems, a parallel combination of turbomolecular and capture pumps, (such as ion pumps, sublimation pumps, or cryopumps), may be more economical than the turbo alone. In such applications, the turbomolecular pump can remove helium and hydrogen, both of which can present difficulties, especially for sublimation pumps and cryopumps. The turbomolecular pump system could also be used alone to "rough pump" the system to, say, 1×10^{-7} Torr to minimize the captured gas quantity.

5.5.5. Theory of the Turbomolecular Vacuum Pump

The most frequently cited theory of the turbomolecular pump is based on the work of Shapiro and his students at the Massachusetts Institute of Technology.[4]

Using the classical kinetic theory of gases and no empirical data, they investigated the special case of a bladed turbine wheel treated as a two-dimensional cascade operating in the free "molecular" pressure regime wherein intermolecular collisions do not occur and all velocity changes are due to collisions with the blade surfaces. The theory was checked experimentally only for a single moving blade row and found to be in good agreement. The theory was extended in approximate form to multirow machines having an assembly of rotors and stators. Major factors considered in the two-dimensional theory were blade angle, blade spacing, and ratio of molecular velocity to blade speed. Shapiro has continued to develop the theory to include three-dimensional effects associated with variations along the radius, leakage effects, the effects of changes between successive blade rows, the optimal matching of successive rows, and the special geometrical considerations imposed by the manufacturing problems. What follows is a description of the original theoretical model. However an alternative pictorial representation is offered in Fig. 2 that may be more meaningful for those not mathematically inclined.

Figure 3 is an idealized representation of a two-dimensional cascade of flat plates. Such factors as casing wall surfaces and geometrical and kinetic variations that occur as a consequence of variable radii are not considered. The pumping capability of such a row, when it is moving, is caused by molecules incident on the rotor from the two sides having unequal probability of reaching the opposite side. The principle of operation is most easily illustrated by looking at the limiting case where the blade speed V_b is large compared to the molecular speeds. Consider the blade at rest and the molecules in relative motion. Then, as shown in Fig. 3

[4] C. H. Kruger and A. H. Shapiro, "Vacuum Pumping with a Bladed Axial-Flow Turbomachine." *Trans. Natl. Symp. Vac. Technol. 7th*, pp. 6–12 (1960). Pergamon, New York, 1961.

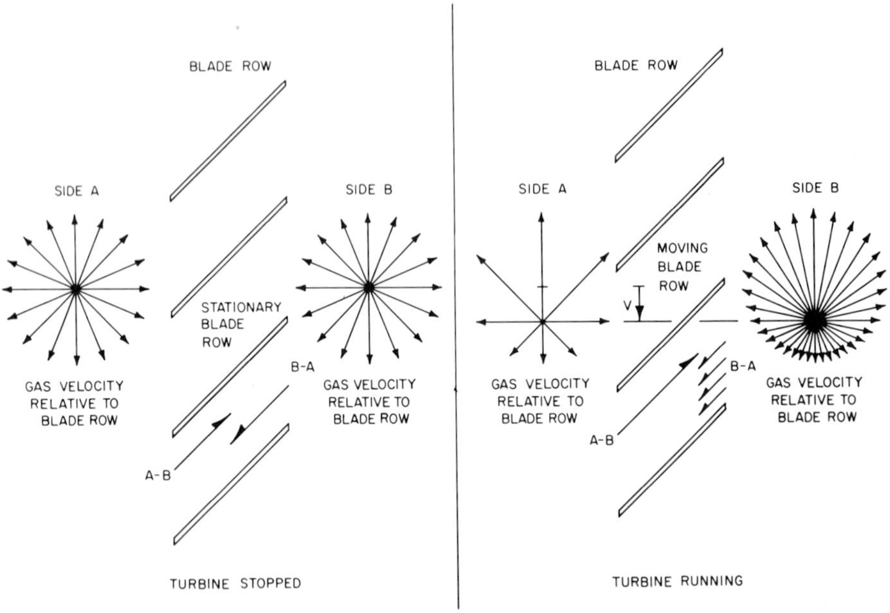

FIG. 2. Nonmathematical representation of how pressure gradient is developed in a turbomolecular pump. The vector circles represent molecular velocity distribution. The numbers of vectors represent the density and/or pressure of the gas. The pumping action is caused by nonsymmetrical molecular velocity orientation relative to the angle of the passages between the blades. The asymmetry results in greater probability of passage from side A to side B than vice versa, hence the pressure gradient. At shut off (no net flow), the pressure gradient is maximum with equal numbers of molecules constantly changing sides. The densities and velocities of the counterstreams are then in exact balance to give equal numbers of molecules.

almost all molecules impinging on side one can be considered incident on an inclined surface near point C. Assuming diffuse reflection (nonmirror reflection) those molecules reemitted in the angle c_1 will return to side 1, those in the angle c_3 will pass into side 2, and those in the angle c_2 will escape to both sides of the blade row. Correspondingly, in Fig. 4, all molecules striking the blade row from side 2 will arrive near point D. Those emitted in the angle d_1 will return to side 2, those in the angle d_3 will pass to side 1, and those in angle d_2 will escape to both sides. By comparing the relative sizes of the various angles, it is seen that molecules from side 1 have a much greater probability of transmission to side 2 than the converse.

Let Σ_{12} be the probability that a molecule from side 1 striking the blade row will be ultimately transmitted to side 2, and let Σ_{21} be the probability that a molecule incident on side 2 will be transmitted to side 1. Let N_1

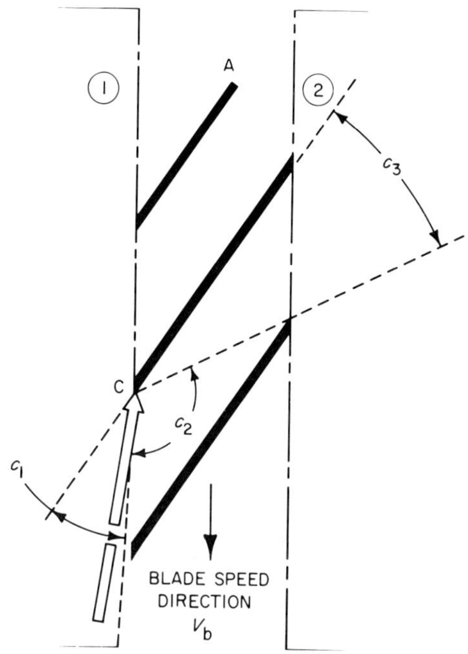

FIG. 3. Blade row transmission angles in forward direction.

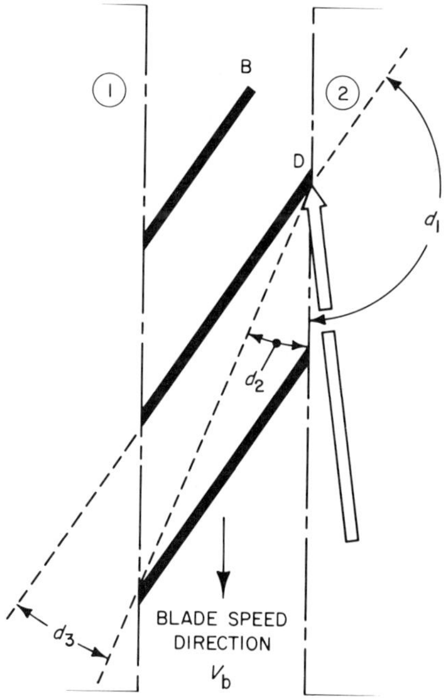

FIG. 4. Blade row transmission angles in backward direction.

represent the number molecular flux incident on the blade row from side 1, and N_2 represent the number molecular flux incident on the blade row from side 2. Let W be the ratio of the net molecular flux passing from side 1 to side 2, to the incident flux N_1. (W is elsewhere in vacuum technology known as the "Ho coefficient.") Observing conservation of the number of molecules the following steady state equation may be written:

$$WN_1 = N_1\Sigma_{12} - N_2\Sigma_{21}$$

or

$$N_2/N_1 = \Sigma_{12}/\Sigma_{21} - W/\Sigma_{21}.$$

The ratio N_2/N_1 is the number density ratio n_2/n_1, which is equal to P_2/P_1. Therefore,

$$n_2/n_1 = P_2/P_1 = \Sigma_{12}/\Sigma_{21} - W/\Sigma_{21}.$$

This latter equation shows us that the pressure ratio decreases linearly as W increases. Or, for zero flow ($W = 0$) the pressure ratio is

$$P_2/P_1 = \Sigma_{12}/\Sigma_{21}.$$

For no pressure difference ($P_2 = P_1$) the flow is

$$W = \Sigma_{12} - \Sigma_{21}.$$

For a high pressure ratio, it is desirable that Σ_{12} be large compared with Σ_{21}. Referring back to Fig. 4, we see that, for high pressure ratio, d_1/d_3 should be maximized and that this condition is promoted by minimizing either or both blade angle α and the channel width ratio s/b. Referring back to Fig. 3, we see that for high transmission probability, c_3/c_1 should be maximized and that this condition is promoted contrariwise by maximizing either or both blade angle α and the channel width ratio s/b. The pump must have both flow capability and pressure ratio capability to function. Since blade geometry is contradictory for these two aspects of the pumps, the practical blade geometry compromise must be made with the pressure ratio requirements for the application as a basis for optimization.

Molecules passing in either direction through the moving blade row contact the blade surfaces at least once, according to our model, and, therefore, have a component of velocity the same as the row when they emerge and approach the adjacent stator row. Thus, molecules striking the stator rows have the same relative velocity to the stators as our rotor row model. If rotor rows and stator rows were geometrically identical, the transmission coefficients and pressure ratio would be identical for both rotors and stators.

In a pump of eight rotor rows and eight stator rows, the overall pressure ratio from inlet to outlet for gases such as air and water is at least of the order 1×10^6. Because of the enormous decrease in volume, the outlet stages obviously do not need as much pumping speed as the inlet stages. Therefore, pumps are made with inlet stages having high conductance and low pressure ratio, and with outlet stages having low conductance and high pressure ratio. The optimization for a particular end use of a multistage pump, especially where rotor and stator construction differ because of the strength requirements, can become quite complicated because of the large number of variables involved. With a computer and a workable theory, however, thousands of possible pump combinations can be explored to select the best choice without building models. The accuracy of the calculations is only satisfactory if all stages of the pump are in the free molecular flow condition. The calculations permit optimization for only one gas at a time. The higher the blade speed relative to the gas mean thermal speed, the more effective the pumping principle operates. Hydrogen, with the highest thermal molecular speed is therefore the most difficult gas to pump at high pressure ratios. One practical approach is to calculate pumps for two widely different molecular weights such as hydrogen and nitrogen and pick a design which has satisfactory pressure ratio for hydrogen and adequate pumping speed for nitrogen. This gives a general purpose pump.

5.5.5.1. Plateau Pumping Speed. The pump functions because gas molecules entering the passages between the blades are struck by the blades. From this we can see that the pumping speed is determined by the simultaneous effects of the blade velocity and gas conductance up to and into the passages between the blades. If the pressure conditions at the pump inlet are in the molecular flow regime, pumping speed does not vary with inlet pressure because molecular flow conductance does not vary with pressure, and because blade velocity is a constant. This explains the "plateau region" (Fig. 5) of the pumping speed curve extending from a high pressure region where intermolecular collisions are beginning to occur in the inlet stages of the pumpdown, to the low pressure "roll off" region where the pressure ratio imposed by the turbopump design approaches the zero flow pressure ratio capability of the pump.

5.5.5.2. "Blankoff" Pressures. The low pressure "roll off" (Fig. 5) occurs first for hydrogen and is the determining factor in the pump's "blankoff" total (ion gauge) pressure. In estimating the hydrogen pressure at which this will occur, one should keep in mind that it is the partial pressure of hydrogen at the outlet of the turbopump that is divided by pump's pressure ratio capability to get the inlet pressure. The partial pressure of hydrogen in the outlet of the turbopump depends primarily on the tendency of the forepump to generate hydrogen more so than on any-

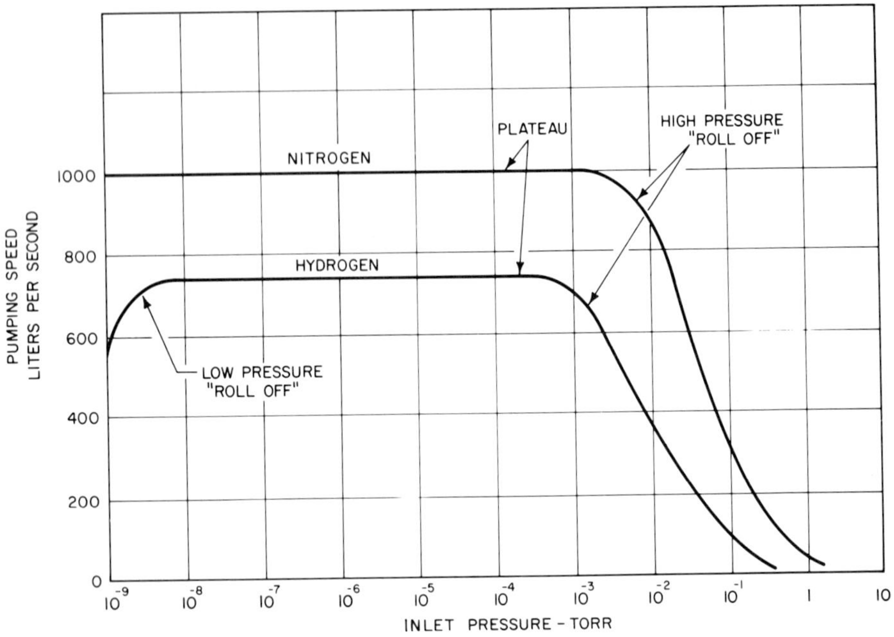

FIG. 5. Typical turbomolecular pumping speed curves. A mechanical oil-sealed forepump is in tandem with the turbomolecular pump and discharges to the atmosphere. The speed for the hydrogen plateau is often dependent on the pumping speed of the mechanical forepump.

thing else. The hydrogen probably comes from decomposition of the forepump oil under the localized high temperature conditions of sliding vanes.[5] One would expect that low rubbing speeds and stable oils would give lower hydrogen partial pressures than the contrary. From the fact that turbopumps having a hydrogen pressure ratio of about 100 have a blankoff total pressure (mostly hydrogen) of less than 1×10^{-9}, it has been deduced that the partial pressure of hydrogen in the forepump of a low speed mechanical forepump must be in the vicinity of 1×10^{-7} Torr. The pumping speed roll off pressure for gases other than hydrogen is so low that measurements become difficult and are seldom experimentally determined. Because the molecules entering the pump are independent of one another, at the ultimate pressure of the pump, the pumping action for gases other than hydrogen continues. In principle, this could be shown if a partial pressure gauge instead of a total pressure gauge were used.

[5] L. Laurenson, L. Holland, and M. A. Baker, "Degradation of Lubricating Fluids in Vacuum, Lubrication in Hostile Environments." Inst. Mech. Eng., London, 1969.

5.5.5.3. Effect of Molecular Weight on Pumping Speed. Differences in plateau pumping speed for different gases are caused by differences in conductance for the various molecular weight gases up to the inlet blade row and by the transmission probability of the early blade rows. Conveniently, the pumping speed for all gases except hydrogen is roughly the same because conductance is inversely proportional to the square root of the molecular weight, while the transmission probability is, roughly, directly proportional to the square root of the molecular weight. These two factors tend to counter one another so that, to the degree of accuracy required for most practical vacuum systems, the pumping speed of all gases except hydrogen is about the same. In some turbopumps, the hydrogen pumping speed can be substantially improved by using larger forepumps.

One important consequence of the nearly uniform pumping speed for all molecular weight gases is that the molecular composition of the gas in the vacuum state is not drastically altered. When operating in the plateau region this is an important consideration if the pump is a component in a gas analyzer.

5.5.5.4. Effect of Rotor Speed. In an optimally designed turbopump, pumping speed is roughly directly proportional to rotating speed. Maximum pumping speed can be made by running turbopumps at the highest possible blade tip speeds providing the design is optimized for that speed (Fig. 6). The number of stages required also decreases as the blade speed is increased. In the past 20 years available turbopump blade tip speeds have increased from 143 to 374 m/sec without resort to exotic materials.[6] This, plus optimization, has permitted an eight-fold increase of Ho coefficient for nitrogen from 0.05 to 0.4. (The Ho coefficient is the fraction of molecules entering the pump flange opening which are captured by the pump.) Based on a rotor hub diameter half the pump inlet diameter, the maximum obtainable Ho coefficient would be 0.75 because the area blocked by the hub structure cannot pump.

5.5.5.5. Effect of Forepump Capacity. At the high pressure end of the turbopump performance curve, the transition range from turbopump "plateau" speed to forepump speed covers three decades of pressure from about 1 to 10^{-3} Torr inlet pressure. Over this transition range, blade rows are functioning imperfectly because of too many undesirable intermolecular collisions. The pumping action begins to deteriorate when the molecular mean free path is less than 10 times the interblade spacing of the outlet blade disks. Because the inlet stages are always being pumped on by the outlet stages, the gas density at the inlet blade rows is lower, the

[6] G. E. Osterstrom and A. H. Shapiro, *J. Vac. Sci. Technol.* **9** (1), 405–408 (1971).

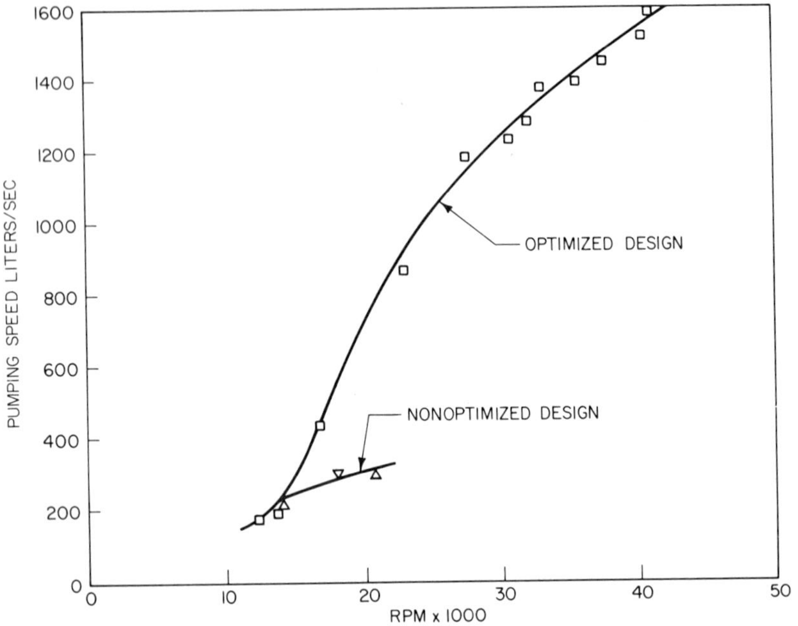

FIG. 6. Pumping speed versus rotating speed in two types of turbomolecular pumps. The two pumps had rotors of about the same size. The nonoptimized design is showing marked effects of conductance limited flow somewhere in the rotor–stator combination or in the inlet passages as indicated by the low rate of pumping speed increase with rotating speed increase. There is relatively little to be gained by attempts to run such a design faster. On the other hand, the optimized pump shows worthwhile gains to the highest speeds tried. Eventually even the optimized pumps will also be conductance limited and the curve will flatten out. Forepump size must be kept large enough so that outlet stages are always in the free molecular flow range for both types.

number of intermolecular collisions is fewer, and the stages are functioning better than at the outlet stages where pressure is primarily controlled by forepump capacity. Therefore, over the transition region, the larger the forepump used, the greater the pumping speed of the turbo because the inlet stages are brought closer to the molecular flow condition. At about 1 Torr turbopump inlet pressure, pressure in all the stages is so high that no pressure ratio is developed; the effective speed at the turbo inlet is the forepump speed, reduced slightly by the impedance of the passages through the turbopump. Motor power requirements may be affected by the shape of the high pressure end of the pumping speed curve. In the transition range, the frictional drag of the substantial mass of gas in the blade rows causes frictional torque on the motor which will probably cause the pump to operate at less than full speed because of the speed–torque characteristics of the motor. The more torque the motor has, the faster the turbopump can push gas through itself to evacuate the system and get out of the transitional range. However, while this temporary overload capacity of the motor can shorten the evacuation period, it can overheat the pump rotor if continued indefinitely.

5.5.6. Residual Gas Composition

The "cleanliness" of a turbomolecular pump is usually judged on the basis of a residual gas analysis taken at the pump entrance. A mass spectrometer is used for this purpose. The mass spectrum shows the outgassing of the surfaces of the pump together with the mass spectrometer and its attachments. Outgassing is time, temperature, and "history" dependent. A clean residual gas spectrum of the turbomolecular pump is shown in Fig. 7. The peak at $m/e = 2$ is hydrogen and is generally considered to come from thermal decomposition of lubricating oil in the backing pump. If water is present, some hydrogen could come from dissociation of the water molecule by the mass spectrometer. Water, if

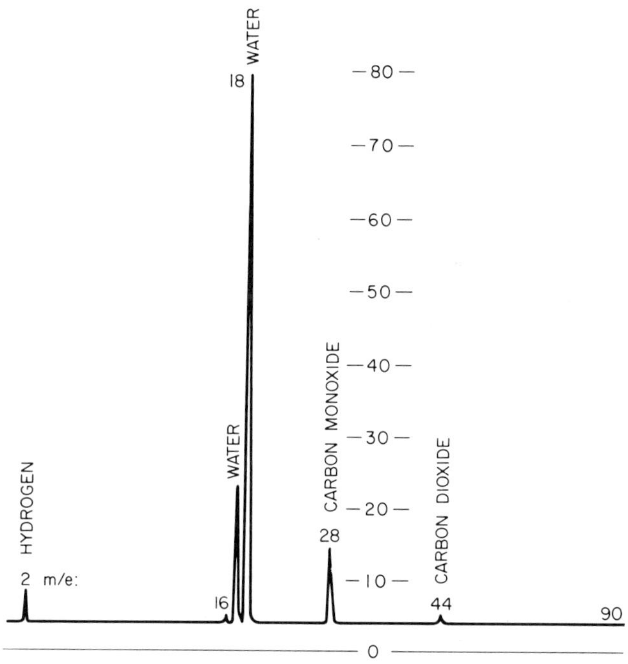

FIG. 7. Mass scan of a clean turbomolecular pump. Total pressure (BA gauge) is 1×10^{-9} Torr. The pump is not yet fully dehydrated and will eventually reach lower pressures as water is desorbed. The carbon monoxide and carbon dioxide are attributed to sources within the mass spectrometer with extremely long pumping hydrogen becomes the dominant peak. The mass spectrum shown is what one usually sees on systems.

present, gives a peak at 18 and a smaller companion peak at 17 from the dissociation fragment OH. Peaks at 28 and 44 are carbon monoxide and carbon dioxide introduced by the hot filament in the mass spectrometer itself. Paraffinic forepump oil contamination, if present in detectable quantities, gives a "repeating" series of peaks quite different from those just described (Fig. 8). The mass spectrum of a turbopump which has little running time since assembly typically shows water and hydrocarbon peaks from the solvent. If there are no liquid hydrocarbons present and all the gas is coming from desorbtion of the metal surfaces, the peaks diminish at a rate inversely proportional to the elapsed time. This gives a straight line on a log–log chart and one can extrapolate the plot to predict when the peaks will reach a satisfactory level. The desorbtion can be speeded up by heating the pump.

When a pump which shows a "clean" mass spectrum is vented to dry nitrogen, it will usually be observed that on the subsequent pumpdown, the water peak reappears as the dominant peak. It would seem that the backfill gas has "stirred up" subsurface water and brought it closer to the surface more rapidly than it would have come through normal thermal diffusion. This may be a useful way to speed the degassing of the pump surfaces as well as the vacuum chamber wall without application of heat.

5.5.7. A Comparison with Diffusion Pumps

Perhaps the single greatest advantage of the turbomolecular pump over the competing diffusion pump is the absence of a working fluid. Diffusion pump users must be constantly on guard against contamination of the working chamber by misdirected pump fluids diverted by such means as pump jet break up at high air pressures during air inrush, or "creep," along duct surfaces. The turbomolecular pump requires much less caution. Because of their relatively high volatility, turbo lubricating oils are far easier to remove from system walls in the event of contamination than are the low vapor pressure fluids used in diffusion pumps. As a further improvement, turbomolecular pumps with grease, gas film, and magnetic levitation type bearings are now coming onto the market; these promise nearly total freedom from organic contamination.

The principal relative disadvantages of the turbomolecular pump are noise, vibration, and the possibility of damage by rotor–stator collision. The price, capacity, and size of commercially available turbomolecular pumps has improved over the last 20 years to the point where they can be competitive with diffusion pumps equipped with the trap, baffle, and valves required for clean pumping. Turbomolecular pumps are available which are smaller physically than equivalent diffusion pump arrangements.

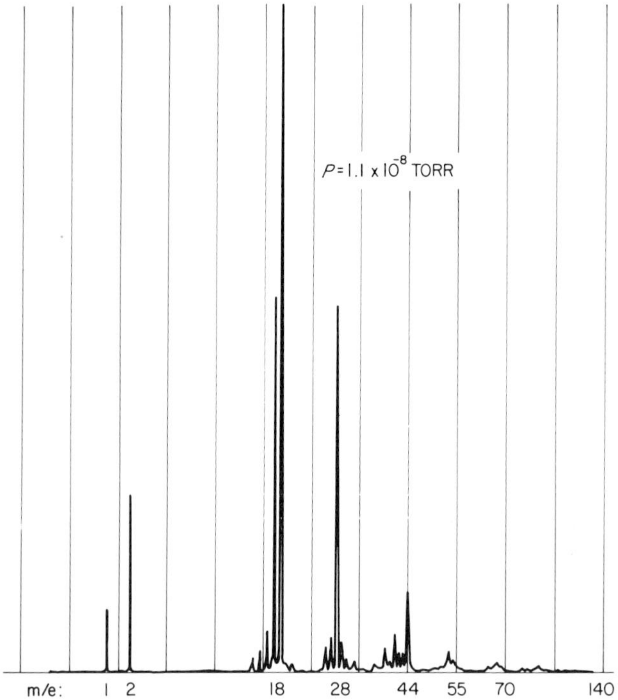

FIG. 8. Mass scan of a turbomolecular pump deliberately contaminated with forepump oil. Note the six groups of mass peaks. The ionizing process in the mass spectrometer has shattered the molecules into fragments smaller than about $m/e = 100$. Different types of oils give different patterns. Mixtures of different types of forepump and turbopump oils would be impossible to apportion at a glance.

5.5.8. Pumping Hydrogen

Because turbomolecular pumps typically have a pressure ratio several orders of magnitude lower for hydrogen than for higher molecular weight gases, one might conclude that turbos cannot pump hydrogen effectively. Catalogue hydrogen pumping speed curves for turbomolecular pumps show that this is not true. Hydrogen pumping speeds typically range from roughly 50% to slightly greater than 100% of the pumping speed for nitrogen. Over its normal constant pumping speed range ("plateau" region), turbomolecular hydrogen pumping speed is limited by the dimensions of the passages between the blades, not by the hydrogen pressure ratio at no-flow conditions as is sometimes thought. In selecting among the various turbomolecular pumps available, the no-flow hydrogen pressure ratio has no direct significance to the user except in the special case where the turbomolecular pump is being used in an analyzer mode to separate hydrogen from other gases.

In optimizing turbomolecular pump design (all major dimensional and speed parameters being equal), pumping speed can only be increased by giving up pressure ratio, and vice versa. Pumping speed is the costly aspect of a pump needed in a practical system. Therefore, only enough hydrogen pressure ratio need be incorporated to establish overall conductance limited (free molecular) flow conditions with the hydrogen inlet partial pressure required for the task to be performed by the pump. The prospective user will be able to see this in the price–capacity–dimensional relationship of competing pumps.

5.5.9. Choice Between "Horizontal" and "Vertical" Types

The terms "horizontal" and "vertical" are only important in the historical context. Turbomolecular pumps with inlet flow parallel to the axis of rotation include the vertical type (Fig. 9). These pumps have one rotor–stator cascade. Turbos with the flow entering perpendicular to the axis of rotation at the center of the rotor include the horizontal type (Fig. 10). In this form the entering flow divides symmetrically as it makes opposing right angle turns, half passing into each of two identical cascades. At the exit of the cascades, the flows are rejoined in a common outlet. The two types will be referred to as "single-ended" and "double-ended" pumps. Both types are commercially available either with horizontal or vertical axes of rotor rotation.

The main advantage of the single-ended turbomolecular pump is its ability to replace a diffusion pump in an existing system without extensive alterations to the system to fit it in. All other things being equal, a single-ended type is cheaper than a double-ended type because it has only half as many blade rows, and because the simple tubular body shape is less costly to make than is the "T" form of the double-ended pump. The main disadvantage of the single-ended pump when operated vertically with entrance upward, as it would be in the case of a diffusion pump replacement, is that objects falling into the pump may cause a wreck. A protective screen can seriously choke a high performance turbomolecular pump. It is difficult to decide when the openings in the screen are small enough to give complete protection. Screens in use to date can give only partial protection. Another problem with the single-ended pumps is the need to cope with large axial forces when the pump interior is suddenly raised to atmospheric pressure. In a 6 in. single-ended pump, axial rotor thrust forces of hundreds of pounds then occur. Bearings and bearing supports capable of opposing this thrust must be incorporated into the design. A third disadvantage of the single-ended pump is instability: a tend-

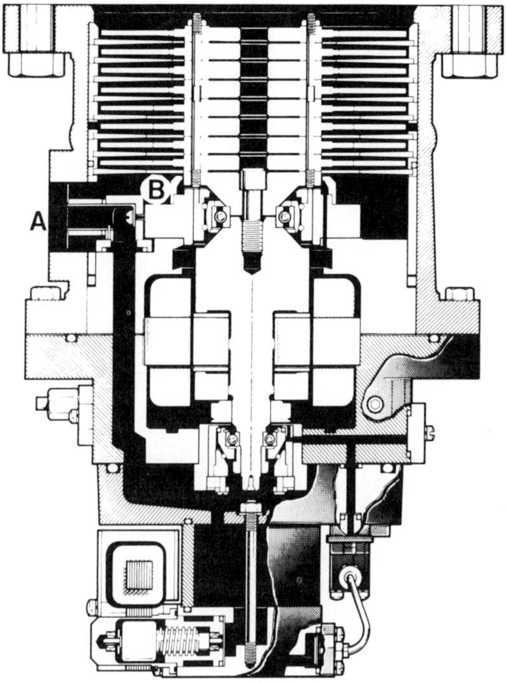

FIG. 9. Typical single-ended or "vertical" pump. Pump rotor and motor are bolted together at the center bearing to form the complete rotor. Gas passes in through the pump rotor–stator cascade and thence out a side port to the forepump. An oil pump is at the bottom to catch oil drain and recirculate oil to the bearings. A special feature of this pump is an outlet throttle in the outlet port which establishes zone "A" (forepump side of throttle) at a lower pressure than zone "B." The oily sections of the pump are connected to zone A by a flow passage such that, during evacuation, the foam from the deaerating oil passes to the forepump and does not rise up through the bearings into the turbobody cavity. This reduces the possibility of contamination by turbopump oil. The single-ended type of pump places the pumping members directly adjacent to the port. There is almost no unused space.

ency to develop vibrations with relatively slight changes in balance, bearings, or bearing support conditions. In current single-ended pumps, the mass center of the rotating system is not centrally located between two bearings but rather concentrated at one bearing or beyond the bearing system altogether in a manner analogous to a rotating cantilever beam. Greater care is required in making and maintaining stable rotating machines of this type than in machines wherein the center of gravity of the rotor is centered between two relatively widely spaced bearings, as in a double-ended turbomolecular pump. A fourth disadvantage to the single-ended pump is the problem of replacing the central "trapped"

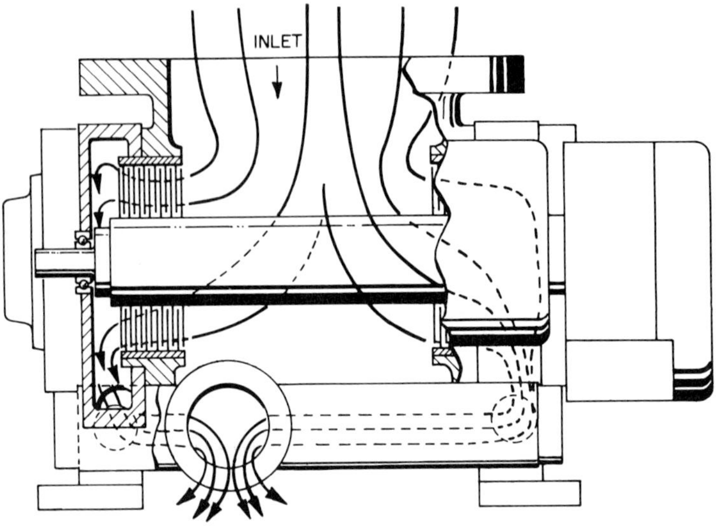

FIG. 10. Typical double-ended or "horizontal" pump. Because the pumping members are not immediately adjacent to the inlet flange, a large low conductance passage must be established between flange and pumping members. This is "lost space" as compared with the single-ended pump and the inlet flange size is also larger for the same capacity than the single-ended pump. However, the space can be utilized for water cryopumping and as a trap for loose objects that might otherwise fall into the pumping members.

bearing. In all current designs, the rotating system must be taken apart, usually between the motor rotor and the pump rotor, in order to get to this bearing. Because the natural spin axis is slightly different with each bearing replacement reassembly, the rotor system can never be in perfect balance on its bearings. The bearings must be allowed to "float" in free vibration. A sufficiently compliant bearing mount is therefore required for bearing replacement by the user without rebalancing facilities. Nevertheless, manufacturers of turbomolecular pumps have solved most of the foregoing problems sufficiently well to make the single-ended turbopump a very useful machine.

Double-ended pumps can have top, bottom, or side inlets available one at a time or simultaneously. If the axis of rotation is horizontal, as it normally is, objects entering the center section of the pump can fall harmlessly to the bottom. If the bottom or side inlet is used, there is even less likelihood that objects will find their way into the blade cascade. All other parameters being equal, a double-ended pump is shorter, i.e., lower, in the inlet direction (or "lower") than the single-ended form, but longer perpendicular to the entry axis. Other advantages and disadvantages of the double-ended pump are implied in the preceding discussion of the single-ended pump.

5.5.10. Bearings

5.5.10.1. Oil Lubricated Bearings. At the present time most turbomolecular pump rotors are spun on oil lubricated ball bearings. Since these bearings are at the discharge side of the rotor disk cascade, oil vapors do not penetrate upstream to the turbo inlet. The oil is recirculated to the bearings by some sort of pump such as a conical bore up the rotor axis, an external oil pump, or a wick arrangement. The oil circulating means should provide a controlled amount of oil—not too much and not too little—during acceleration, deceleration, full speed running, reduced speed running, and power failure. Assuming that the high speed operation will probably cause some sludging or gelling of the oil, the oil circulating system must be able to function without getting plugged. When a separate oil pump is used, or when the oil flow control system contains small orifices which might be blocked, or if oil is liable to be lost through foaming on frequent and sudden pumpdown, the oil circulating system should contain an oil flow or oil level interlock to warn the user and possibly shut the system down if lubrication is in doubt.

As discussed elsewhere in this chapter, rotor and bearing temperatures in excess of 150°F may occur. Ordinary oil starts to decompose above this temperature. The higher the temperature, the faster sludge or gels form. The prospective user should inquire about the operating condition limitations of the pump to avoid these undesirable decomposition products. Straightforward operation under high vacuum conditions will usually cause no problems. Under normal operating conditions, oil lubricated pumps can run for years without any attention to the oil or bearings.

At the upper end of the turbomolecular pump pressure range, the flow of oil through the bearings may serve to carry away the gas friction heat. This is the reversal of the more common condition wherein the rotor operates at such low pressures that the principal contributor to temperature rise in the rotor is the oil friction itself.

5.5.10.2. Grease Lubricated Bearings. Recently developed extreme low vapor pressure bearing greases of low effective viscosity have made it practical for turbomolecular pumps to have the advantages of conventional ball bearings, yet avoid the disadvantage in vacuum work of highly mobile liquid oils. Oil lubricated pumps of both horizontal single- and double-ended forms require special precautions during pumpdown and venting, especially with the rotor operating at high speed, to avoid gross movements of oil into the inlet section. Grease bearings avoid the need for such precautions. Thus, the only mechanism left for contamination is by gaseous phase diffusion through a stopped pump. In practice, only a very small proportion of users have had trouble with contamination from this source. It has been the liquid oil that has been troublesome. Fortunately, it is possible to convert many oil lubricated pumps over to grease lubrication.

The main disadvantage of grease lubrication is the eventual loss, probably by evaporation, of the liquid components of the grease. The user must be concerned with periodic regreasing intervals and methods. By using a grease filled hypodermic syringe to pierce a rubber plug, one can replenish the grease without stopping the pump or breaking the vacuum. Under straightforward application situations, the manufacturer can recommend a practical regreasing schedule of, for example, once a year of nonstop operation or once every 10,000 hr total of intermittent operation. If the bearings are unusually hot, as well they might be under prolonged relatively high pressure applications (as in sputtering, for example), the regreasing interval is less predictable. Here, various symptoms such as increased electric motor current, rotor speed change, or peculiar noises must be relied upon. For most users, these uncertainties have not proven a serious problem and grease lubrication has proven a major step toward realizing the inherent reliability and "foolproof" nature of the turbos.

5.5.10.3. Gas Film Bearings. Turbomolecular pumps with air film bearings are now commercially available.[3] This type of bearing uses only metal in the vacuum system and offers unlimited bearing life. However, the prospective user should make a thorough investigation of the peripheral air system requirements both for normal steady state operation and under conditions of unscheduled power interruption because air bearings require a continuous supply of high pressure air, free from particulate matter and from water. Failure of this air supply will result in high speed metallic rubbing contacts with consequent instant destruction of the bearing.

The air film strength must be sufficient to prevent contact under abnormal conditions such as jarring the pump or, in the case of single-ended pumps, rotor thrust under air inrush conditions. Air pressure sufficient to resist these occasional thrusts may substantially increase the air supply pressure that must be constantly maintained.

Air bearing turbomolecular pumps available to date use a helically relieved portion of the rotor shaft as a viscous gas pump mechanism to pump bearing air away from the vacuum space. When the shaft stops, the air to vacuum sealing capability of this pump ceases, and the turbo will be vented. An isolation valve may therefore be needed between the pump and the vacuum system if venting is not permissible.

5.5.10.4. Magnetic Levitation Bearings. Magnetic bearing turbos have been offered. The magnetic bearing system offers unlimited life because of the absence of physical contact, as does the air bearing. However, the magnetic bearing does not require the air pressure and shaft seal of the air bearing. A turbomolecular pump equipped with magnetic bearings can operate lubricant free and completely isolated from the atmosphere (in closed circulating systems). However, if organic insulation materials are used on the electromagnetic support coils, a source of possible hydrocarbon contamination exists, especially if the coils get overheated. Magnetic suspensions are more compliant than are ball bearings or air bearings. This added compliance gives a high degree of vibration isolation between the rotor and the body of the pump. Thus, pumps with magnetic bearings exhibit an extremely low vibration and noise level. The magnet control circuits require so little power that battery operation during spin down in the event of power failure is quite practical.

Magnetic bearing systems require an electromechanical servosystem to maintain the position of the rotor. Such a system is necessarily electronically complex and thus subject to individual component failure. To protect against this, a magnetic bearing system must have a restraining mechanical bearing system to bring the rotor safely to a stop in the event of electronic component failure. Upon failure of the magnetic bearings, air inrush, or jarring, the whirling rotor moves into contact with this bearing which must thereupon accelerate to full rotor speed in an extremely short time, probably in milliseconds. Oil-free ball bearings capable of surviving such an acceleration have a very short and uncertain life and replacement might well be considered after each contact event. The high compliance characteristic of magnetic bearings could make contact events difficult to avoid with ordinary handling.

5.5.11. Drives

5.5.11.1. External Motor. Until 1967, almost all of the thousands of turbomolecular pumps in use were powered by ordinary line frequency motors with speed-increasing belt drives transmitting torque into the vacuum space through a shaft seal; the seal usually incorporating an oil bath on the atmospheric side. This type of seal requires periodic refilling of the oil bath reservoir on a highly variable (from pump to pump) basis and so is always an item of concern to the user. Nevertheless, this type of drive, because of its low cost, its tremendous overload capacity and its maintainability by generally available mechanical skills, could still be the best choice in some applications.

5.5.11.2. Internal Motor. Many currently available turbomolecular pumps have direct high frequency electric motors operating in the region of forepump pressure. No shaft seal is required. In most cases, the motor is of the induction type wherein induced electric currents flow in the rotor. These rotor currents cause ohmic heating and consequent temperature rise of the motor rotor, bearings, and the pump rotor itself because the pump rotor is joined to the motor rotor. Because of the vacuum insulation effect, even a few tens of watts can cause a slow but impressive temperature increase until radiation heat transfer increases sufficiently to establish equilibrium. Heat input to the rotor comes from bearing friction and pumped gas friction (above a few milliTorr for most gases), and this heat is further increased by the motor rotor ohmic heating; this is itself a function of the friction load. Excessive temperature in the rotor system may deteriorate the lubricant, damage ball bearings, aggravate differential expansion problems, and weaken highly stressed rotor metal. The prospective turbomolecular pump user should inquire about time limitations he must place on the various rotor temperature increasing factors including gas pressure, gas type, rotor speed, bakeout heating, acceleration and deceleration cycles, coolant temperature, room temperature, and ambient magnetic fields.

The foregoing situation is somewhat eased in several designs. One uses a ceramic permanent magnet motor rotor which has no electric currents flowing within it and consequently no ohmic heating. Another has the motor rotor isolated from the turbo rotor by a nonmetallic gear. These two arrangements offer cooler pump rotors, all other things being equal.

5.5.11.3. Solid State High Frequency Power Supply. Most of the currently available turbomolecular pumps utilize high frequency polyphase solid state power supplies. With the maturing of solid state electronics, these power supplies have become very attractive; most justifiably so because of the inherent capability of variable speed. However, because of the very special electronic skills required for repairs, solid state power supplies must be extremely reliable. The high required quality level to achieve this makes these power supplies more costly than one might think.

5.5.11.4. Electromechanical High Frequency Power Supply. Electromechanical high frequency alternators should be considered instead of solid state power supplies whenever radiation levels are high enough to damage solid state electronics or whenever solid state electronics repair skills are unavailable. Because of extremely high temporary overload capability and voltage transient resistance, a single high frequency alternator can serve as a common power source for a number of sequentially started turbomolecular pumps. Each pump must have its own control station including overcurrent protection.

The most readily available alternators are for a single frequency. Such alternators might be unsatisfactory for prolonged turbo operation in the milliTorr range because of excessive temperature rise due to gas friction at full speed operation.

Where damaging radiation is present, the alternator should not incorporate solid state diodes in the rotating field circuitry. A permanent magnet type alternator is one available alternative.

5.5.12. Fast Opening Inlet Valves and Implosions

Although some turbopumps can withstand more violent implosions than others, implosions are enough of a risk to justify protective interlock switches and operator caution.

If an air operated slide valve is used at the inlet of a turbopump, a pressure switch on the vacuum chamber side of the valve should be so connected that the valve cannot be opened unless the pressure is low enough so that the turbopump can stand the resultant inrush when the valve is opened. Air operated slide valves, because of the effective elasticity of the compressed air operating them, tend to "jump" open, thus discharging the air into the turbopump almost explosively. The consequent force of the implosion can cause rotors and stators to strike together and wreck the pump. To reduce the pressure on the vacuum chamber side of the turbopump inlet valve to a safe level, a small valve by-passing the main turbo inlet valves can be used.

The maximum safe rate of pressure rise (for example, atmospheres per second) should be obtained from the manufacturer. Pressure rise tests can be made by opening the fast valve into the system with the turbopump at a standstill to check for safe rise rate limits.

5.5.13. Vibration and Noise Reduction

Turbopumps are successfully used on such instruments as mass spectrometers and electron microscopes wherein vibration free mechanical alignment of parts is crucial. There are, nevertheless, potential problems if such parts have the same frequency of mechanical resonance as the turbopump unbalance rotating frequency. To make a superficial check for possible problems, pluck the components in the manner of a string instrument. The tone heard is usually the natural resonance frequency of the component. Compare the tone with the calibrated tone such as an adjustable acoustic oscillator or a musical instrument whose tone frequency is known, such as a harmonica or tuning fork. Resonant amplification, or "tuned," coupling between the turbopump unbalance vibration and the component vibration will be most likely if the observed component frequency is between 75 and 125% of the pump rotating frequency. If "detuning" seems necessary, the component natural frequency can be lowered by decreasing the support stiffness or by adding or subtracting mass to raise the natural frequency, until the component is "detuned."

A turbomolecular pump, although quiet and almost vibration free by itself, can nevertheless excite undesirable noise generating vibrations in connected light wall pipes and vessels. The amplitude of these vibrations is often greater than that of the pump itself. The nodes and antinodes of vibration can be determined through fingertip sensations and the zones of flexure can be located. Lightweight sheet metal ribs to stiffen the flexing zones usually eliminate the disturbing vibrations.

Another method of noise reduction is the application of sheet acoustic materials. One effective commercially available material consists of an outer layer of lead weighted vinyl sheet separated from the vibrating metal surface by a compliant sponge layer. The material is available with contact adhesive for easy application.

A third method of noise reduction, which also serves as a means to reduce vibrations of sensitive system components, is to use a compliant vibration isolation joint between the turbopump and the structure. At the turbopump inlet, an elastomer inlet gasket, such as an O ring, can serve this purpose. The higher the rotating speed of the turbo, the more effective the isolation. The vibration isolator must not be bypassed with metallic components such as tight flange bolts. In an evacuated flanged system, atmospheric pressure is usually sufficient to compress the gasket and to hold the joint together even if the flange bolts are loose. The flange bolts must be so adjusted that when the system is not evacuated they are effectively holding the flanges together and sealed. If the O ring is "captured" in supporting metal rings or grooves, the faces of the support rings should not be in contact with the flange faces when the system is evacuated.

A metal isolation bellows, as a type of compliant joint, has its own set of problems and should be used at the turbopump inlet only if the elastomer arrangement described above is not practical because of outgassing considerations. The only readily compliant direction of a metal bellows is axial and if all vibration were axial there would be no problem. However, lateral vibrations are usually present and adequate lateral compliance is necessary, requiring considerable bellows length to accommodate consequent assymetrical flexing of the convolutions. Further, torsional compliance is inherently low, and in a straight single bellows section there is nothing that can be done about it. With these limitations in mind, the following qualities for a bellows are nevertheless suggested. The bellows must not be too stiff. When evacuated a "free" bellows should shorten about 1/10 its length. Adjacent convolutions must not touch. The bellows length should be not less than half the diameter. The compliance of the bellows must not be bypassed by rigid support rods.

Metal bellows convolutions are usually thinner than the light walled vessel to be isolated and can themselves be excited into noisy vibrations. One method of subduing such bellows vibrations is to wrap the bellows lightly with paper tape so that the convolutions slide on the tape when flexing occurs. The resultant damping effect keeps the amplitude of vibration down.

The isolating effect of a compliant joint between the turbopump and the system must not be bypassed through ground by rigid supports to ground on both the turbopump and system. One support must be compliant. The best arrangement is to suspend the pump by the complaint joint from the system. The forepump connection must also be compliant. The turbopump feet must not touch ground. The system from which the turbopump is hung can be rigidly supported to ground.

If the turbopump is transmitting vibrations to a responsive structure or ground through its feet, compliant rubber mounts can be inserted under the feet. Rubber stoppers as found in a chemical laboratory are convenient for this purpose. The stoppers should deflect at least 3 mm under the weight of the turbopump.

5.5.14. Starting and Stopping Turbopumps

If a turbomolecular pump is replacing a diffusion pump in an automatic system, the safest arrangement is to retain the same valves and valve program as that used in the diffusion pump. The acceleration and deceleration time of the turbopump corresponds to the heating and cooling time of the diffusion pump.

In oil lubricated turbopumps, experience has shown that the most common cause for hydrocarbon contamination is airborne spray of oil which can occur if the pump is operated at high speed with air pressure inside the pump over 10 Torr. A turbopump then behaves like a positive displacement compressor with its outlet closed. The "trapped" air recirculates in an unstable eddying manner between the inlet and the outlet of the turbo. Typically, the flow is toward the outlet on one side, reverses at the end, and comes to the inlet on the other side. Secondary vortices develop extending into the bearing cavities which carry oil outward into the blade rows where it is distributed by the gross recirculation within the turbopump. This can happen in both horizontal and vertical turbopumps. This problem can be dealt with by startup and shutdown procedures which will be described in succeeding paragraphs. It can also be avoided by a special construction as shown in Fig. 9. Here, a restriction is placed in the forepump connection in such a way that the forepump side is connected to the bearing system where the liquid oil is normally contained. When undesirable air is in the turbopump body, the forepump, which is to be kept running as long as the turbopump rotor is spinning, draws a positive stream of air from the pump cavity through the bearing system, and thence into the forepump. By correctly sizing the restriction, the diverted air flow through the bearing system blocks the effect of previously described minor vortexes which would otherwise carry bearing oil out into the turbopump cavity.

The following starting procedure is recommended in lieu of contrary instructions by any particular turbopump manufacturer. With the turbopump off and not rotating, the forepump is started. When the pressure of the turbopump outlet connection reaches 2 Torr, the turbo is started. When the turbopump is up to normal running speed, the turbopump inlet valve, if used, is opened provided the pressure on the vacuum chamber side of the valve is below about 2 Torr.

The streaming velocity of the air entering the forepump during the rough pumping process is greater than the backward diffusion rate of oil contamination until the forepump has reduced the pressure to about 1 Torr. When the turbopump is started at about 2 Torr, it has sufficient speed by the time the forepump has reduced the pressure to 1 Torr that the turbopump now will block the backstreaming of contamination from the forepump as it progresses to the turbo outlet after the rough pumping air stream density is sufficiently low to permit back diffusion. The maker of the turbopump may have specific instructions how to stop the pump and to vent air. However, in the absence of such instructions, the safest method from the point of view of noncontamination of the turbopump is to reverse the above pumpdown procedure. Leave the forepump running. Turn off the turbopump. Immediately open a vent at the turbopump inlet and raise the pressure at the turbo inlet to about 5 Torr. When the turbopump has stopped, stop the forepump and let the system come up to atmospheric pressure. This procedure slows the turbo down with low density air so that violent eddy current previously described are not present. It also maintains a flow of purged air through the connection to the forepump so that warm oil vapors do not pass back up to the turbopump as the turbopump approaches a stop. It is essential that the vent valve be placed at the turbopump inlet. The final rise to atmospheric pressure with the forepump stopped occurs with air always flowing into the forepump from the inlet of the turbopump. If this procedure makes an objectionable oil fog at the forepump outlet, use an outlet filter on the backing pump.

An alternative way to stop the turbopump unvented is simpler and sufficiently clean for most pump applications. Cut off the turbopump motor power and the forepump power at the same time. The turbopump will coast long enough to permit the oil vapors from the forepump to subside as the forepump cools. When the turbopump has stopped, vent the system through the turbo inlet side.

For most turbopump applications, venting with normal humidity atmospheric air is satisfactory. The amount of water brought to the system and pump surfaces with a single venting is small compared to the amount of water that will quickly sorb onto the surfaces when the system is opened. However, if dried vent gas is to be used, boiloff nitrogen from liquid nitrogen storage vessels is very satisfactory. If pressurized nitrogen cylinder gas is used, it should be specified "water pumped" to avoid hydrocarbon oils from the gas compressor.

5.5.15. Protecting against Turbopump Stoppage

The second most common cause of contamination in turbopumps is stopping of the turbo while the forepump continues to run. It should be noted that this is a far less serious source of contamination than the liquid oil contamination previously described. Gaseous phase warm forepump oil vapor and forepump oil decomposition products obviously transport contamination thousands of times slower than oil spray. The simplest counter measure is to arrange the electrical circuitry so that when the turbo motor power is cut off for any reason such as power failure or interlock trip, the forepump will also stop. The forepump should not subsequently automatically restart unless the turbopump can also automatically start. With this arrangement, the turbopump will coast for a half hour or so and effectively block the contamination from the stopped but cooling forepump. If the forepump is totally self-sealing from the atmosphere when stopped, a small contamination flow will occur due to thermal diffusion while the pumping system is still warmer than the vacuum chamber. If the vacuum chamber is kept deliberately slightly warmer than the forepump becomes by the time the turbopump stops, the thermal diffusion can be blocked.

If an automatic vent at the turbopump inlet is provided to establish a pressure of about 10 Torr after the turbopump has stopped, the thermal diffusion can be greatly retarded because the contamination gases must diffuse through the backfill gas to reach the clean surfaces. A vent arrangement of this type should have a self-contained power source because power cutoff is one of the primary sources of the trouble in the first place. One very simple, almost natural alternative, is a forepump that loses its oil seal at about the same time it cools to room temperature. This will take care of the problem without special devices and with only a small contamination from venting backwards. Of course, older forepump designs which foam back up the inlet connection if stopped under vacuum cannot be used.

For applications where the most extreme precautions are justified, an isolation valve between the turbopump and the vacuum chamber should automatically close when the turbopump power or the forepump power is cut off or for any other reason either pump stops functioning. Rotating speed sensors, or pressure sensors, may be necessary to cover every possible situation. A valve isolating the turbopump from the forepump should also automatically close. The aforementioned turbo inlet valves and turbo outlet valves should be capable of closing when all power is cut off. Upon power restoration or correction of other faults, the system must be automatically or manually restarted so that the forepump inlet pressure is normal before the forepump is opened to the turbopump and the turbopump should be up to speed with normal turbo inlet pressure before the isolation valve between the turbopump and the system is opened either automatically or manually. The foregoing extreme measures are usually not justified.

A sorbent trap between the turbopump and the forepump is often suggested. Such a trap, if effective, would be useful in the event of the turbopump stopping with the forepump still running. It might also give added security against inadvertent delay in starting the turbopump after the forepump has evacuated the system to the point where oil vapor might move through a standing turbopump. Because the conditioning of such traps is of critical importance in their effectiveness, and there is no clear evidence of their state, one should not depend on traps to avoid contamination.

5.5.16. "Baking" Turbomolecular Pumps

Ultravacuum work systems are usually made entirely of metal and can be baked to 400°C. At this temperature degassing of surfaces occurs very rapidly and hydrocarbon contaminants are decomposed into readily pumped gasses. Commercially available turbomolecular pumps are not suited for such service because they have rotors of aluminum alloy which begins to lose needed strength at about 148°C. Lubricants can present an even more severe limitation on turbomolecular pump rotor temperatures. Oil can be decomposed at sustained temperatures much over 76°C. With these lower temperature limitations, degassing is orders of magnitude slower than at 400°C. A new turbomolecular pump can take several weeks to initially arrive at roughly the same pressure and contamination level achieved by a 400°C baked system overnight. However, if a thus conditioned turbomolecular pump is subsequently vented to dry atmosphere for only short periods, subsequent pumpdowns to ultravacuum level can take place within a day in most cases.

Most commercially available turbomolecular pumps are typically specified as heatable to 100°C. However, even though the housing, inlet flange, and stators may be at 100°C, the rotor in an oil or air lubricated turbomolecular pump will be substantially cooler. The rotor is feebly heated by radiation from the stators and strongly cooled by contact with the flowing lubricant. Thus, the net effect of heating the outside of the pump might only be a rise of a few degrees over the normal temperature rise. Special rotor heating methods can be employed such as shutting off the coolant, magnetic eddy current heating, passing purge gas through the pump, or applying strong radiant heat to the rotor. A radiation thermopile device can be used to monitor rotor temperature.

6. METAL VACUUM SYSTEMS AND COMPONENTS*

In each specific application, what are the most important vacuum concepts that will further the achievement of practical ends? In outline, a set of decisions are often made as follows. A manufacturing process or scientific experiment must be carried out under vacuum. What sort of system is needed and what measurements must be taken to assure success and to minimize the chance of failure? In each specific instance, how can the practitioner predict and then achieve a successful solution? As in many other arts and sciences, successful solutions need not be arrived at only by logical processes, intuition plays an important role. Failure can result from overdesign, from rigidly following recipes to the letter, and from using standard practice where standard practice must be discarded in favor of an approach specific to a situation.

When coming to grips with each specific vacuum problem, we offer this common-sense advice: List the general vacuum requirements as you see them now; do not strive for details, but watch out for so-called trivial effects that can halt or compromise achievement. How can one spot what is important and what is not? The question of developing judgment can be approached from many angles and on many levels.

The vacuum business may be looked at in terms of a tool box of possible approaches. From this one can select a particular mode with the knowledge that the limitations of that particular mode, its strengths or weaknesses, must always be kept in mind. One shifts from one mode to another when that shift is fruitful, and the right way (often more than one) may be defined as the way that works. Many so-called standard practices and attitudes in the vacuum business should be carefully weighed for acceptance or rejection. Healthy skepticism must be maintained at all times. For example, it is often convenient to treat a vacuum as a fluid which can be pumped, and one uses analogies to ohmic electrical circuits, extracts useful formulas from the kinetic theory of gases, and makes pressure the single parameter that defines a vacuum. On the other hand, it

* Part 6 is by N. Milleron and R. C. Wolgast.

METHODS OF EXPERIMENTAL PHYSICS, VOL. 14

ISBN 0-12-475914-9

should be understood that the molecular concentration of a species in the gas phase has no simple direct relationship to the kind and amount of material that may be desorbed from the walls, for example. More specifically, one must be alert to the misuse of the "sum-of-the-reciprocals" approach in computing free molecular conductance to assess the likelihood of system speed (see Chapter 1.3), to the dependence of conductance on wall temperature, the misuse of the term equilibrium vapor pressure, the incorrect assumption that gases in vacuum systems are generally Maxwellian in character, the manufacturer's often over optimistic statements about the pumping speeds of their equipment, the loss in pumping speeds due to connecting plumbing to a chamber, the origin and removal of contaminants in vacuum systems, the role and value of hot filament ionization gauges in the prediction process, and to other specifics.

Since vacuum effects are dominated by boundary conditions, one must be concerned with how many molecules, where, and of what kind are present in the condensed, as well as the gaseous, phases. For example, the molecular composition of the gas phase is unlikely to be the same as the molecular composition of either the adsorbed phase on surfaces or the absorbed phase in bulk materials. What kinds of molecules are likely to be associated with the condensed phase? With the gas phase? How do these molecular concentrations change as a function of parameters such as temperature and heat of desorption? Beginning with such considerations, an intuitive appreciation can be built up showing why and to what degree the condensed phase may dominate the gas phase and determine the achievable vacuum.

When we consider metal vacuum systems and components, we need to bear in mind how such vacuum systems respond to temperature changes, etc. Thus, if we heat or cool a vacuum chamber we find that the molecular composition and total pressure of the gas phase will change markedly. Suppose, for example, that we raise the temperature from 295 K (room temperature) to 373 K (boiling point of water), a factor of 1.26. The ideal gas law predicts that the pressure P will increase by 1.26. But if our system pumping speed is small compared to the theoretical maximum speed associated with our vacuum system, then P will rise initially by about a factor of 10 and then assume a quasi equilibrium value (at 373 K) of approximately two times greater than P at 295 K.

All too often, textbooks advocate erroneously that a vacuum can be specified completely in terms of a clean, dry, and empty chamber, and the reader is then left with the impression that the base pressure predicts sufficiently well how a system will respond when dynamical processes are initiated. Clearly, the situation is much more complex.

6.1. A Diffusion Pumped Vacuum System and Its Schematic Representation

A vacuum brazing furnace with its high vacuum system is shown as an example and is represented approximately to scale in Fig. 1. In routine operation, this "bell jar" furnace maintains a very low level of contamination even when rapidly cycled. The design allows direct access along its vertical axis and through its bottom area. It is provided with three molybdenum and one stainless steel heat shields; the furnace is being used for vacuum brazing and degassing at hot zone temperatures up to 1200°C. The bell jar dome opens at flanged joint 7, and may then be fully raised thus allowing complete assembly and positioning of pieces to be brazed or treated on the stationary hearth grid.

This "offset" design permits routine operation and cycling of the fur-

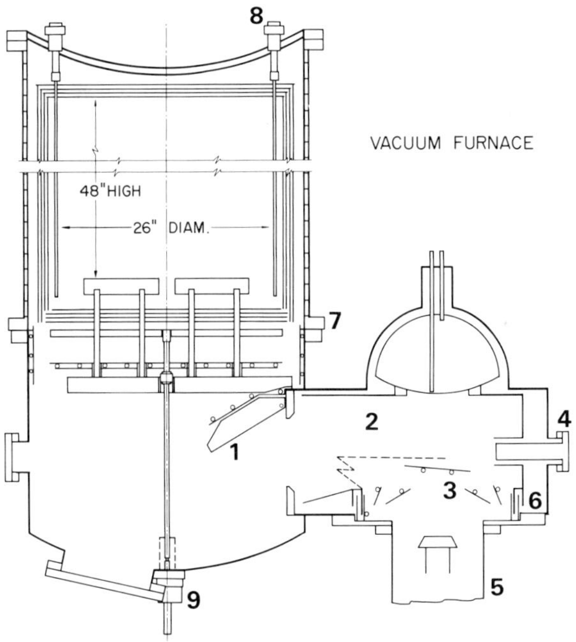

FIG. 1. A vacuum brazing furnace designed for low contamination, for brazing large metal-to-ceramic joints. (1) Valve; (2) liquid nitrogen cooled trap; (3) water cooled baffle; (4) valve actuator feedthrough, bellows sealed; (5) diffusion pump, oil, 25-cm diam; (6) creep barrier; (7) flange separation for opening the furnace; (8) electrical feedthroughs for lowering the bottom heat shields (48 in. = 122 cm and 26 in. = 66 cm).

nace without sacrificing control of contamination, access, speed for condensibles or speed for noncondensibles. An optimized relationship is nearly realized between the bell jar, 60° swing leaf valve, LN trap, baffle for D.P. oil and the plane of action for a diffusion pump jet (Note: a full size D.P. was not available, hence the D.P. actually used is 25 cm nom, much smaller than optimum).

The figures of merit for this system are as follows: valve open area equals 0.38 of the cross sectional area of the inside diameter of the furnace (66 cm); the volumetric speed for water vapor is $\sim 0.9 \times 18.8 \times 10^3$ liters/sec.

No "gas pips" occur during the filling of the LN trap and the temperature of the trap surfaces (2) does not vary more than ~1/100°C with the liquid nitrogen level change in the reservoir. The high conductance water cooled baffle (3) provides a minimum restriction of flow to the diffusion pump (5) and returns backstreamed pump oil along the walls of the pump. A thin stainless steel anticreep barrier (6) cooled by radiation to the liquid nitrogen chilled annulus, prevents surface migration of pump oil to the furnace. The bottom heat shields in the furnace may be lowered by rod (9) during the warming part of the furnace cycle to provide high conductance from the hot zone to the trap inlet.

A schematic of this furnace vacuum system, including the roughing and fore pumps, is shown in Fig. 2. The symbols for vacuum components are defined in the American Vacuum Society Standard 7.1.

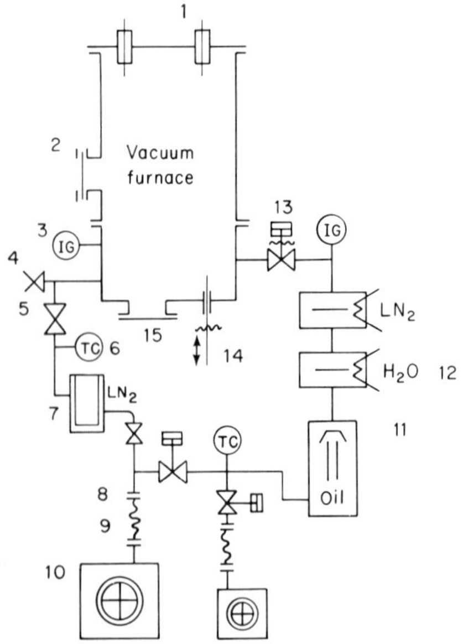

FIG. 2. Schematic representation of the vacuum furnace shown in Fig. 1. The graphic symbols used are among those contained in the American Vacuum Society Standard 7.1. (1) Electrical feedthrough; (2) viewport; (3) vacuum gage, ionization, hot filament; (4) air admittance valve; (5) valve; (6) vacuum gauge, thermocouple; (7) thimble trap; (8) demountable coupling; (9) flexible line; (10) mechanical pump, liquid sealed, two stage; (11) diffusion pump; (12) baffle, water cooled; (13) valve, pneumatic, bellows sealed; (14) linear motion feedthrough, bellows sealed; (15) blind flange port.

6.2. Vacuum Flanges

Separable joints are used when needed for accessibility, to subdivide large systems into practical sized units, or to introduce replaceable components. The convenience provided by flanged joints that are mechanically sealed by bolting or clamping must be balanced against significant drawbacks. Elastomer sealed joints may be relatively inexpensive and reliable but contribute outgassing and permeation that may be unacceptable or require large pumping capacity. In many systems they are the largest source of unwanted gases and contamination.

All-metal joints are necessary in systems which require very high or low temperatures or operate in a high radiation field. They are also important in achieving the desired performance of unbaked systems. All-metal joints that are bolted or clamped are expensive and must be designed, fabricated, and installed carefully, especially in the larger sizes or when subjected to extreme temperatures. Thermal stresses due to nonuniform temperature changes and high temperature creep may cause troublesome leaks, which usually occur during the cooling part of the cycle. Whether elastomer or metal gasketed, the flanged joints may represent a significant fraction of the cost and problems of vacuum systems.

In cases where high or low service temperature or very high temperature bakeout is required, or access is infrequent or needed only for possible repairs, one can consider cutting the chamber and rewelding it[1] instead of providing a flanged joint sealed by mechanical force. If the chamber is not too large, the cutting can be done on a lathe or milling machine. Cutting is possible *in situ* with tubing cutters or high speed carbide tools, either hand held or fixtured. Some specific cut and weld designs are discussed in Section 9.3.2. Rewelding the chambers can be inexpensive and reliable for stainless steel and aluminum. The cut–weld technique is finding increasing application.

Flange dimensions are only partially standardized. Three "Tentative Standards" of the American Vacuum Society concern flange dimensions (but not the gasket or seal design). These AVS standards were written in 1965 and 1969 and did not consider an international metric standard. Generally, diffusion pumps, traps, and gate valves in the United States use bolt circle diameters and number of bolts which follow the dimensions of the American Standards Association (ASA) "150 pound" (150 psi working pressure) (10^5 kg/m^2) series. Since these flanges are stronger and heavier than is usually necessary for vacuum applications, one or

[1] N. Milleron, "Some Recent Developments in Vacuum Techniques for Accelerators and Storage Rings," *IEEE Trans. Nuc. Sci NS-14,* No. 3, 794 (1974).

more of the other flange dimensions, such as the inside diameter, thickness, or bolt diameter, often differ from the ASA standard. The bolt drilling for elastomer sealed gate valves may usually be custom specified at little additional expense. Most metal-gasketed flanges used in the United States follow the nominal dimensions of the Varian ConFlat design or the Batzer foil–seal design. A recent article about flange dimensions[2] states:

> The International Standards Organization (ISO) has specified a series of flange standards based upon metric considerations. ISO flanges are now used exclusively for new construction in Europe, and are in increasing use in the United States.

Flanges of all kinds are available from many suppliers. The flanges should be made from vacuum quality material, free from porosity or undesirable material. Filamentary voids occur frequently in stainless steel plate or other shapes 5 mm or more thick. The voids are eliminated from flange blanks by ring forging, cross forging, or by using vacuum remelted ingots. Many companies will supply custom flanges made from vacuum quality blanks. Type 303 stainless steel contains a high percentage of sulfur to improve machinability, and must be avoided in any elevated temperature application to prevent sulfur contamination of the vacuum. Type 303 cannot be welded satisfactorily for vacuum service.

6.2.1. Characteristics of the Ideal Flanged Joint

A flanged vacuum joint must meet two basic requirements: (1) the sealing surfaces must conform sufficiently to be leak tight; and (2) the sealing force must be maintained during the subsequent history of the joint. Additional qualities desirable in an ideal flange are: the sealing surfaces are recessed or otherwise protected to avoid damage in handling; the flanges are sexless; the joint can transmit mechanical forces without jeopardizing the seal; the joint can withstand high temperature and cooldown cycles; the flanges are self-aligning; the joint requires little assembly space; and requires few bolts. The joint should be able to withstand corrosive conditions in its working environment. An example is the presence of atmospheric gases and high temperature during bakeout.

6.2.2. Elastomer Gasketed Flanges

In the elastomer gasketed flanges described below, the gasket provides properties to meet the two basic sealing requirements: it conforms to the

[2] N. Peacock, "Dimensional Tables of Some Common Vacuum Flanges in International Use." HPS Corp., Boulder, Colorado 1976.

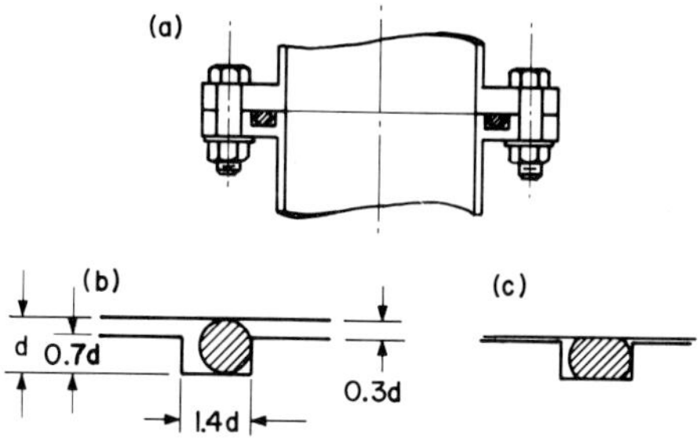

FIG. 3. Common flange seal using elastomer O ring. (a) Assembled flange; (b) seal detail before closure showing the most commonly specified depth and width of the seal gland or groove; (c) seal detail after closure.

seal surface because it is soft; and it maintains a springback sealing force because it is elastic.

The most common elastomer flange seal uses an O-ring gasket, a thin torus of synthetic rubber, as shown in Fig. 3. The design provides a compression of the O ring to about 70% of the initial section diameter, to effect the seal without overstressing the O ring. Enough gland volume is provided so the O ring is not forced out of the groove and between the flat mating surfaces. In normal use the O ring may be reused indefinitely. Because the flanges are drawn tight metal to metal, vibration and mechanical forces on the joint need not affect the seal. The deflection of the O ring at assembly (the "squeeze") is large so the tolerance on the flatness of the flange can be large, approximately 30% of the squeeze. This consideration affects the selection of the O-ring section diameter (d in Fig. 3) to be used in a given joint. Usually the smallest section diameter is chosen to minimize outgassing from the seal consistent with the flatness and machining tolerances of the flanges. When the outgassing is not a dominant consideration, larger section diameters may be chosen to relax fabrication tolerances or for reasons of standardization or ease of handling. O rings are available in a wide range of stock sizes and materials. Dimensions, tolerances, material specifications, and gland dimensions are available from the manufacturers.

The specifications of the O-ring groove should allow a slope on the groove sides up to 5° and a radius on the inner corners equal to about two-tenths of the O-ring section diameter to ease machining. These specifications are shown in Fig. 4. The surface finish of the seal areas should be at least as good as 1.6 μm (63 μin.) and preferably 0.8 μm (32 μin.). Outer edges should be broken to about 0.13 mm (0.005 in.) and smoothed to avoid scratching the O ring during assembly.

Commercial seals are available for flat-faced flanges.* An example is

* CVC Corp. and Parker Seal Co.

shown in Fig. 5. These seals eliminate the machining of the O-ring groove and the flanges are sexless.

O rings are adaptable to various seal configurations. The groove may be dovetailed on one or both sides as shown in Fig. 6 to help retain the O ring during assembly. The double dovetail is useful on valve plates where gas flow during operation may otherwise blow the O ring out of the groove.

The grooves may be made narrower than the standard width if necessary, using the dimensions given in Fig. 7. In these examples, the O-ring compression is less than standard but the sealing force is maintained due to the lateral support.

The planform of the groove may be any desired shape if the corners are not too sharp. A typical planform of a rectangular corner is shown in Fig. 8(a). This design provides some retention of the O ring and is less expensive to machine than the planform shown in Fig. 8(b), unless the groove is to be cut on a computer-controlled milling machine. O rings of any seal length may be made by splicing a length of round O-ring stock. A vulcanizing process is available or a straight butt joint may be made using an adhesive such as Eastman 910 (with skill). A commercial splicing kit is available.*

A compression seal to a tube is shown in Fig. 9. For a static seal the gland volume should be about 10% greater than the initial O-ring volume. When the tube diameter is much greater than the O-ring section diameter the gland width, dimension "A" in Fig. 9, should be 1.31 times the O-ring section diameter. For relatively smaller tube sizes the actual O-ring and gland volumes must be used to determine gland width and tolerances.

The most commonly used O-ring materials for vacuum applications are Buna-N and Viton which have a Shore A Durometer hardness of 70. The sealing force required is about 2.4×10^6 Pa[N/m^2] (or 25 kgf/cm^2 or 350 lb/in.^{2}) based on initial O-ring projected area. Viton may be used at 125°C in the compressed condition for long time periods, and at 250°C for shorter periods in the uncompressed condition (the seal of an open valve, for example). Some hardening of the Viton may occur at the higher temperature. This temperature tolerance permits a significant amount of system degassing by baking. The outgassing from Viton contains lower quantities of hydrocarbons than that from Buna-N. Silicone based elastomers are usable at higher temperatures but have higher permeation rates. (See other sections of this book for a discussion of material properties.)

* Loctite Co.

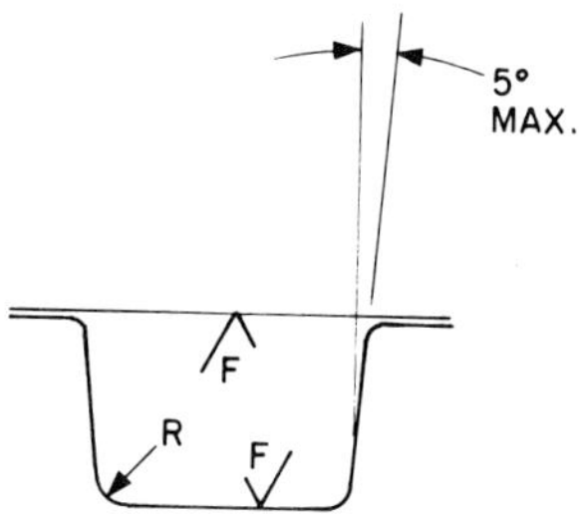

FIG. 4.

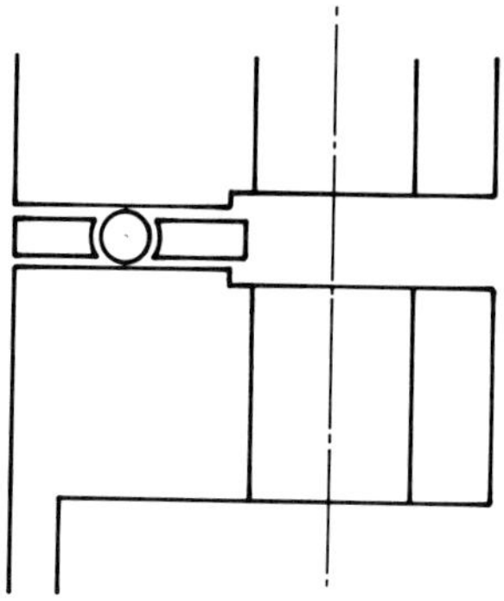

FIG. 5.

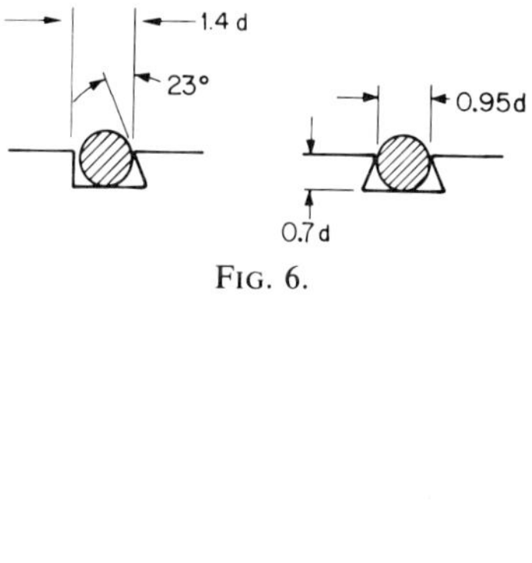

FIG. 6.

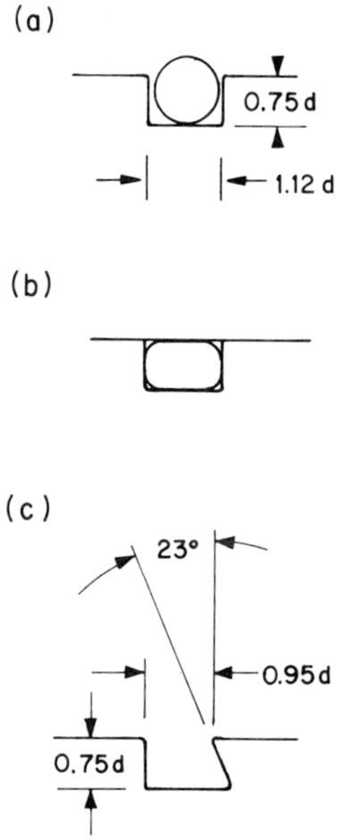

FIG. 7.

FIG. 4. Tolerances on O-ring groove shape and finish. Radius R should be less than two-tenths of the O-ring section diameter. Finish F should be 1.6 μm rms and preferably 0.8 μm. Edges should be broken to about 0.13 mm and smoothed.

FIG. 5. Example of a commercial seal that eliminates the O-ring groove.

FIG. 6. Dovetail grooves that retain the O ring.

FIG. 7. Dimensions for narrow grooves. (a) Straight-sided narrow groove; (b) after closure; (c) narrow dovetail.

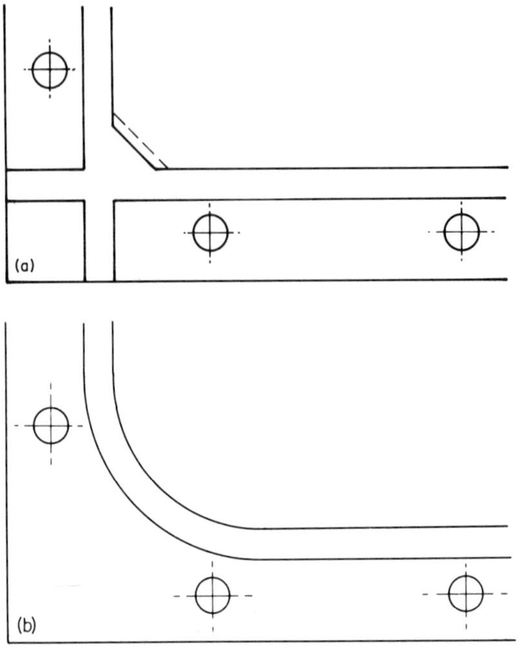

FIG. 8. Typical planforms for rectangular corners. (a) Dovetailed corner retains O ring; (b) preferred planform when machined on computor controlled milling machine.

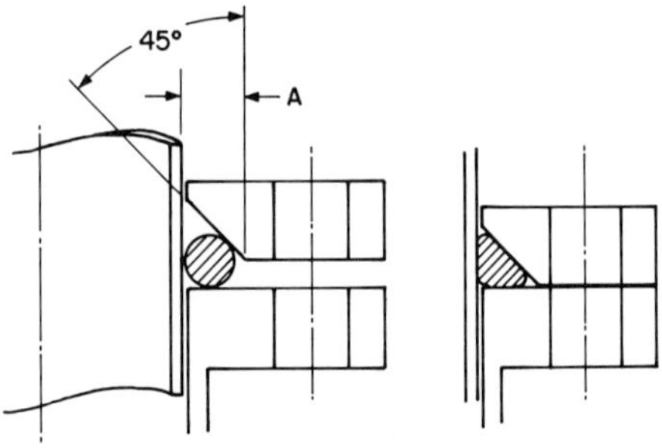

FIG. 9. Compression seal to a tube before and after closure. The gland volume should be 10% greater than the O-ring volume for a static seal.

Any grease applied to the O ring to assist in making the seal will generate an increased outgassing load from the grease, and some perhaps from the skin oils picked up from the technician's hands. The grease is not necessary in a static seal if surface finish and cleanliness are good.

A commercially available tubing joint that does not require machined and bolted flanges is shown in Fig. 10.* They may be significantly less

* Aeroquip Co., Marmon Division.

Joints have been made with square and rectangular gaskets as shown in Fig. 11.[3] The only advantage in using square gaskets over the O ring is that the gasket is retained and the cost of machining the groove for large rectangular seals may be less than that of the dovetailed groove required to retain an O ring. Rectangular gaskets can provide a large gasket deflection to accommodate out-of-flatness of large seals. The selection of readily available gasket materials for high vacuum use may be less extensive than for O rings. The gasket is usually made by splicing square stock.

The manufacture of the O ring usually entails a molding flash (a thin extrusion) beyond the inner and outer diameters. Methods used to remove the molding flash may leave some irregularities. It may be necessary to use vacuum grease when O rings were used to seal in the radial direction as in Fig. 12. In making radial seals attention must be given to the assembly procedure in order that the O ring is not cut by sharp corners or by passing over burred screw threads. If the assembly requires sliding the seal surface over the O ring under pressure some lubrication is usually necessary such as molybdenum disulfide or vacuum grease. Ethyl alcohol may serve as a temporary lubricant to aid assembly.

In some cases of O-ring seals it may be useful to have no bolts or clamps. One example is the metal bell jar where the atmospheric load is more than adequate to provide a reliable seal force. Even small flanges are sealed effectively if there are no disturbing forces. A radial seal may be used to avoid this limitation in many cases. A useful example is the thimble trap where bolts may be avoided to encourage prompt and frequent cleaning of the trap.

In some systems it may be possible to avoid using elastomer seals except for the bell jar seal or valve seal for example. In these cases special attention to minimize the O-ring gas loads may be worthwhile. These attentions may include preinstallation bakeout of the O ring, small vent holes to the bottom of the O-ring groove to eliminate virtual leaks from the volume of air that may be trapped in the lower corner of the groove, double seal with intermediate pumpout to decrease permeation, or baking the seal and system during pumpdown and then cooling the seal during operation. Polyimide can be baked to quite high temperatures and is used for valve seals, but it is relatively hard, and the surface finish and seal forces required begin to approach those of metal seals.

Elastomer seals are restricted in application by their limited resistance expensive than bolted flanges for large diameter joints, and can be assembled quickly.

[3] A. Guthrie, "Vacuum Technology." Wiley, 1963.

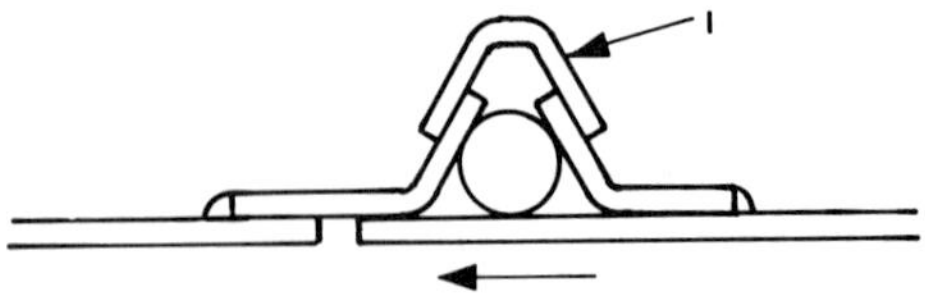

FIG. 10.

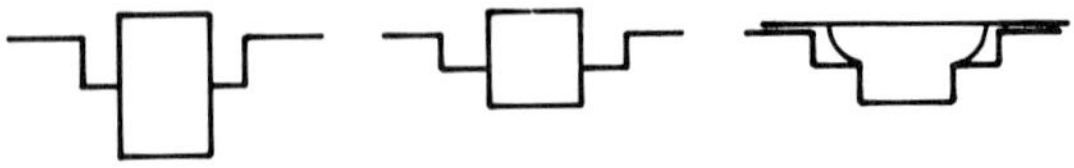

FIG. 11.

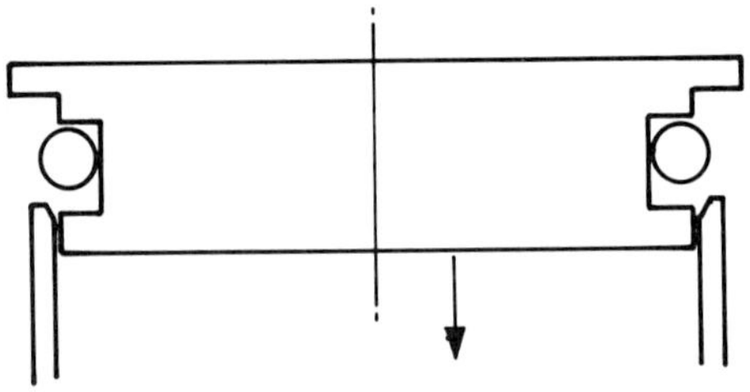

FIG. 12.

FIG. 10. Commercial tubing joint (1) is a split vee band clamp that draws the tubes together and compresses the O ring. (This joint is a patented product of Marman Div., Aeroquip Corp., Los Angeles, Calif.)

FIG. 11. Cross sections of rectangular and square gaskets.

FIG. 12. Radial seal. Smooth champfered edges aid assembly.

to radiation and by the effects of temperature. Buna-N and Viton O rings harden and take a permanent set at high temperature, and harden and shrink at low temperature. However, a seal for use at liquid helium temperature has been made using polyester sheet (polyimide would do as well). Such a seal could be designed as shown in Fig. 13. The conical deflection of the flange supplies the elasticity to compensate for the limited shrinkage of the plastic.

FIG. 13. Low temperature seal. (1) Gasket made of polyester or polyimide sheet 0.08–0.25 mm thick. The seal width W and bolt torque should be chosen to provide plastic deformation of the gasket. The seal is suitable for flanges of aluminum and other alloys.

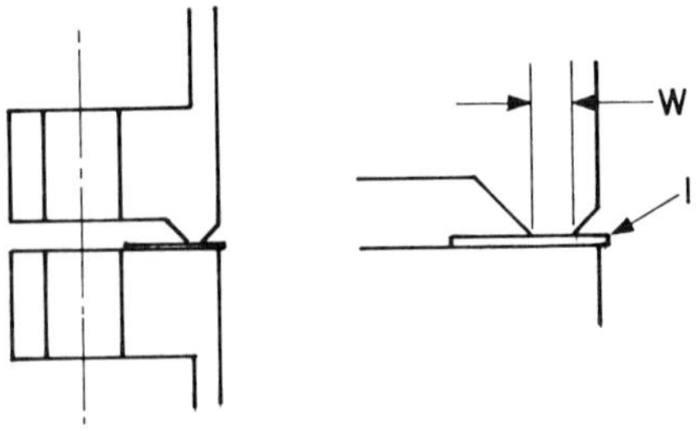

6.2.3. All-Metal Flanges

In the design of all-metal flanged joints, the first basic sealing requirement—that of providing a conforming leak-tight interface—is usually met successfully by the use of a soft metal gasket. The second basic requirement—that of maintaining adequate sealing force during subsequent operation or bakeout—is not as easily satisfied.

Most joints consist of stainless steel flanges and gaskets of a soft metal such as nickel, copper, aluminum, or indium. The stress relief and creep temperatures of the materials affect the design of the joint. In the region of 300–400°C some stress relief takes place in the 300 series stainless steels; and significant creep occurs in copper. Aluminum creeps at lower temperatures and indium creeps at room temperature. In joints subject to creep, longer bakeout times at a given temperature are more likely to cause leaks, which usually occur on the cooldown part of the cycle. Other material properties are also important. Indium becomes corrosive to stainless steel at sufficiently high temperatures, for example. Clean surfaces in vacuum at high temperature and high contact pressure may diffusion weld to each other to a troublesome extent.

If the joint is subjected to elevated temperatures it is important to design for enough elastic springback to maintain adequate sealing force in spite of the stress relief and creep accompanying a high temperature cycle or by plastic deformation caused by stresses due to nonuniform temperature. The latter may be caused by low as well as high temperature cycles. This is illustrated by the diagram in Fig. 14(a). If the force-deflection characteristic of the joint is too steep, i.e., if there is too little elastic springback or "stored energy," the corresponding diagram is given in Fig. 14(b).

A joint which has been designed to provide sufficient elastic deflection as well as high seal force is the Batzer flange, shown in Fig. 15.[4] When the joint is bolted up, the flanges deflect elastically to a slightly conical shape, providing a deflection at the seal of about 0.15 mm for each flange. Elastic stretch of the through bolts is about 0.05 mm, giving a total deflection of 0.35 mm for a flange pair. The potential creep of the gasket, about 0.02 mm, is limited to the thickness of the gasket under the seal surfaces after the joint is bolted. The potential creep of the gasket is much less than the elastic deflection so most of the initial sealing force remains after bakeout. The angle A is about 1.5°, about 0.8° greater than the conical deflection of the flange. This ensures that the seal is made in a definite region at the inside diameter. The actual radial width of the compressed

[4] T. H. Batzer and J. F. Ryan, "Some New Techniques in Ultra-High Vacuum," *Trans. Natl. Vac. Symp*, **10**, p. 166 (1963).

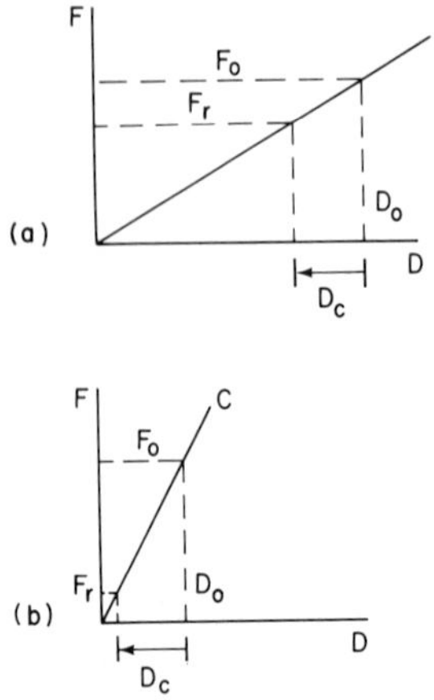

FIG. 14.

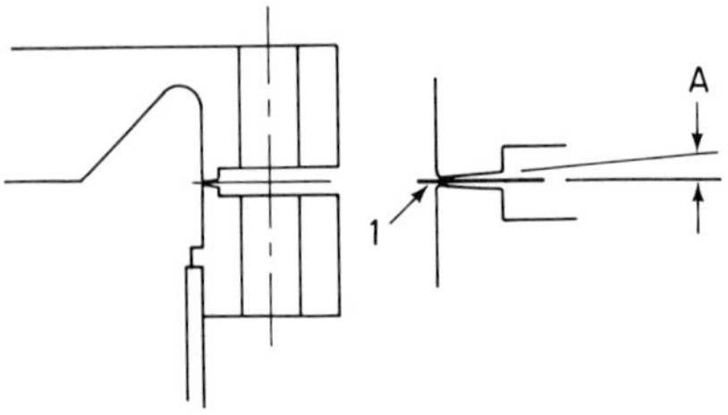

FIG. 15.

FIG. 14. Force–deflection diagrams for flanged joints illustrate the reduction of sealing force due to plastic flow of gasket or structure during operation or bakeout. F is the force on the seal, D the deflection at the seal, C the elastic characteristic of the joint, F_0 the initial seal force, D_0 the initial deflection, D_c the unloading deflection due to plastic flow, and F_r the remaining seal force. (a) Joint with adequate stored energy or springback; (b) joint with little stored energy and small remaining seal force.

FIG. 15. Batzer or foil-seal flange. (1) Aluminum foil 0.02–0.08 mm thick, usually household foil or alloy type 1145-0. Angle A, about 1.5°, is shown exaggerated for clarity.

area of the gasket is about 1 mm. The sealing force is about 3.5×10^5 N/lineal m (360 kgf/lineal cm or 2000 lb/in.) which is obtained from using 9.5 mm diam with 6.3 threads/cm (0.375 in. UNC 16) type 304 stainless steel bolts, spaced about 5 cm (2 in.) apart, and bolting torque of 4×10^5 N m (400 kgf cm or 350 lb in.).

To obtain the proper sealing force and to prevent the bolts from seizing to the nut at high temperature, the bolts should be lubricated with a high temperature molybdenum disulfide based lubricant.* This lubricant is definitely preferred to "Silver Goop" and is less expensive than silverplating the bolts. The flange design is based on the fully annealed condition of the steel which provides the best nonmagnetic and anticorrosive properties. The flanges are commercially available. The Batzer design may be applied to indefinitely large seal diameters, and is also used on rectangular flanges.

The aluminum foil should be cut from a single piece if possible. In the larger sizes it is easier to use somewhat thicker, 0.8 mm (0.003 in.), indus-

* Such as "Fel-pro" from the Felt Products Co.

trial foil type 1145-0 in the soft condition. The foil is pressed gently onto the flange and trimmed with a scalpel, the foil extending over the bolt holes to the outer diameter of the flange. If two or more pieces of foil are used with lapped edges the initial seal will be satisfactory, but the flange edge is indented slightly under the overlap and the flange may have to be remachined for the next seal. Foil of copper or nickel may also be used.

Another approach to joint design is to provide initial plastic deformation of the soft metal gasket to effect the leak-tight interface, with the gasket captured in a rigid structure. With long accumulated bakeout times and high temperatures the internal stresses in the gasket can become quite completely relaxed and the force on the seal quite low. The joint can remain leak tight as long as differential thermal expansion is limited.

A frequently used flange of this type is the "ConFlat," shown in Fig. 16. The sloped edge presses into the copper gasket (2 mm thick) to make the seal. Outward radial flow of the gasket should be limited by the outside diameter of the gasket groove. Vent grooves are provided in each flange for leak checking. On the 15 cm (6 in.) tube size, bolts are spaced 2.84 cm (1.12 in.) apart. The assembly procedure calls for the flanges to be bolted metal to metal. There is little elastic springback in the design and a potential for creep of the thick copper gasket at elevated temperatures. Reliability of the larger sizes under 300–350°C bakeout is improved by using the copper alloy which has a few percent silver to increase the gasket creep strength, using bolts with a high creep strength, and maintaining a uniform temperature distribution while heating and cooling.

In both the Batzer and ConFlat designs the seal is in the area just outside of the flange edge and does not depend upon the sharpness of the edge. Reliability of the seal is not lessened by chemical or electropolishing the flange, which removes about 0.013 mm (0.0005 in.). It is important that there are no sharp scratches crossing the seal area either in the flange or the gasket.

Square Viton gaskets are available for the ConFlat flanges. They are

FIG. 16. ConFlat flange.® (1) Copper gasket about 2 mm thick; (2) vent grooves to eliminate trapped volume and to aid leak checking. (ConFlat® flanges are a proprietary design of Varian Assoc., Palo Alto, Calif.)

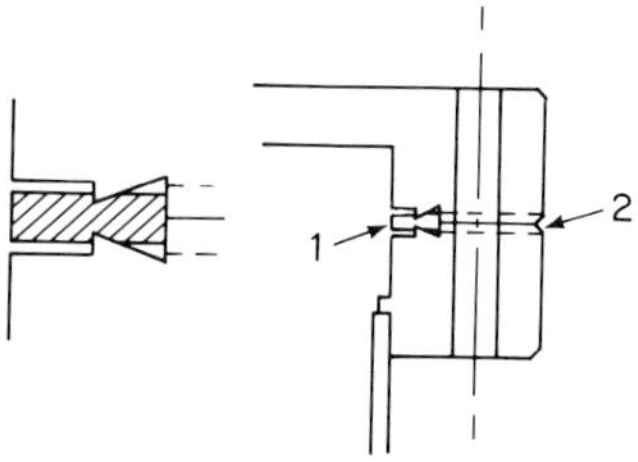

useful for temporary assembly for leak checking or for joining to aluminum sections of a vacuum system.

Wheeler flanges, shown in Fig. 17, are available for sizes of 27 cm diam (10.75 in.) and larger. The copper wire gasket is captured on three sides.

Another capture geometry for a thick copper gasket is shown in Fig. 18.

The corner gold wire seal, shown in Fig. 19, is reliable and may be encountered on many existing instruments. A new gasket must usually be made for each closure. It can be seen that this joint provides elastic springback by the same flange deflection characteristic of the Batzer flange.

Another seal that is useful for circular and noncircular planforms is Milleron's ridged or coined copper seal,[5] shown in Fig. 20, requiring only flat faced flanges. It has been tested as an RF joint and through 600°C bakeout.

A very extensive and useful compendium of vacuum seals has been made by Roth.[6] "New" seal designs are reinvented periodically in the literature. Reliability data are hard to acquire and we hope the reader will be able to make his own judgment in selecting a seal consistent with quantitative technical analysis.

There are a considerable number of commercial metal seal designs that appear to stem from the need for a pressure seal at high temperatures, such as the tubular metal O ring, "K" seals, etc. Generally they do not provide both the high seal force and the amount of springback required for reliable bakeable vacuum seals. A design that does have some good characteristics and may have potential as a bakeable vacuum seal is shown in Fig. 21. It has the advantages of lightweight flanges, and low closure force while providing a high seal force. Only a small amount of springback is provided by the radial compression of the gasket, but the gasket can be stainless steel with adequate creep strength. Its thermal expansion matches that of the stainless steel flanges, which minimizes relative thermal distortion. If the gasket edges were rounded and the gasket silver plated, the reliability of the larger sizes as a vacuum seal might be enhanced. The "vee" band must be tapped around the circumference during tightening to insure a uniform distribution of the seal force.

None of the all-metal flanges described above have all the characteristics of the ideal flange. Another step toward the ideal could be the design shown in Fig. 22.

For joints opened very frequently or remotely, accurately lapped

[5] N. Milleron, "Some Component Designs Permitting Ultra-High Vacuum with Large Oil Diffusion Pumps," *Trans. AVS Natl. Vac. Symp.*, p. 143 (1958). Pergamon, Oxford, 1959.

[6] A. Roth, "Vacuum Sealing Techniques." Pergamon, New York, 1966.

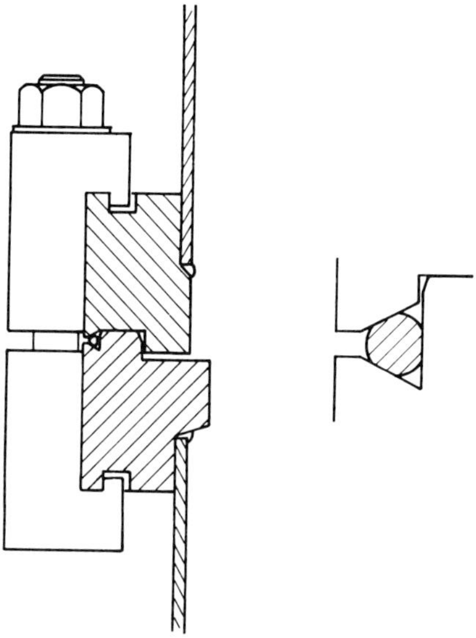

FIG. 17.

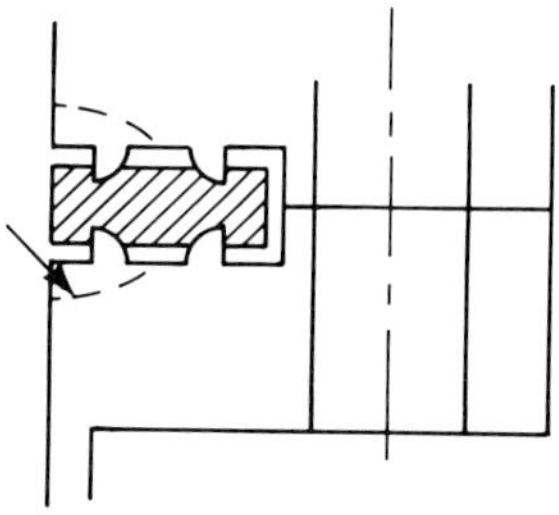

FIG. 18.

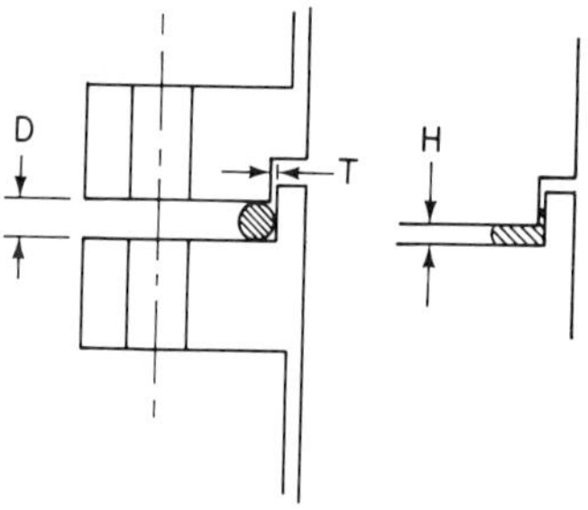

FIG. 19.

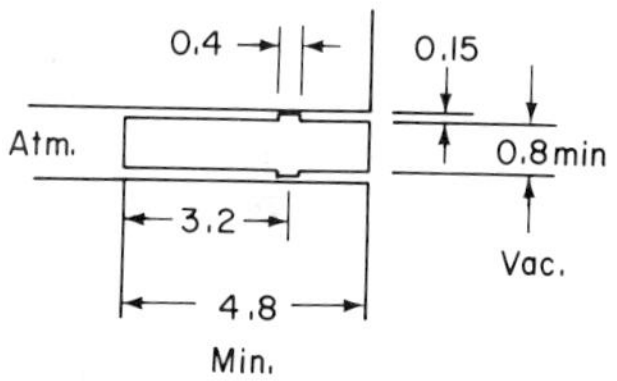

FIG. 20.

FIG. 17. Wheeler flange with copper wire gasket (the Wheeler™ flange is a proprietary design of Varian Associates Inc.).

FIG. 18. Another capture geometry for a thick copper gasket. Vent grooves are indicated by the arrow.

FIG. 19. Corner gold wire seal. The wire diameter D is 0.5–1.0 mm. The radial clearance, T, is small, with T/D = 0.05–0.2. After closure H/D is approximately 0.5.

FIG. 20. Ridged or coined copper seal. Dimensions are in millimeters.

mating surfaces with labyrinth seals[7] and intermediate pump out channels have been used, as shown in Fig. 23. Appropriate pumping capability must be provided for each channel.

An all-metal access port is shown in Fig. 24. Because of the small cone angle of the seal plate, a relatively low axial load generates high seal forces. Deflection of the thin seal plate provides elastic springback.

Seals that employ an inflatable bellows to supply a well maintained seal force have been used for large glass windows with an indium gasket for bubble chambers and in an all-metal bakeable gate valve. The cost of such a seal mechanism may be justified for a large joint that must be opened frequently.

Seals are subject to damage from corrosion on the atmospheric side. Copper oxidizes readily in air at normal bakeout temperatures, and copper gaskets have been silver plated to minimize this problem. Dripping or condensed water can cause corrosion and subsequent leakage of an aluminum seal. Seal designs that provide a definite width of seal of about a millimeter are less susceptible to corrosion failure than seals that rely on a narrow knife edge.

A bakeable metal seal using the Batzer principle has been used to replace an O-ring seal. It is shown in Fig. 25. The number of bolt holes was increased to provide 9.5-mm-diameter (0.375 in.) bolts every 50 mm (2 in.) or less of bolt circle circumference. If the bolt spacing is less than 50 mm the Batzer design bolt torque should be correspondingly reduced to provide the design seal loading of 3.5×10^5 N/lineal m (360 kgf/lineal cm or 2000 lb/lineal in.). Shoulder washers may be used to center the 9.5-mm bolts in oversize bolt holes that may exist in the O-ring flange.

"C" seals have also been used to replace elastomer O rings, shown in Fig. 26. They may be made of Inconel, which has a higher yield point than stainless steel, plated with silver or indium. At the first closure the "C" is collapsed plastically but a small elastic springback (about 0.1 mm) remains. The seal force is quite small. They are relatively expensive but may be reused in the same groove, or grooves of identical uniform depth, by replating when necessary. The techniques of cleanliness and surface flatness and finish are more critical than for elastomer seals. The "C" seal is not usually recommended for the design of new equipment.

The seals discussed in this section should be adaptable to flanges made of wrought aluminum or other alloys provided they are designed and used within the temperature, thermal expansion, and stress–strain characteristics of these materials. Soft aluminum or plated gaskets would be suitable for use with aluminum alloy flanges.

[7] E. Donath, "Differential Pumping of a Narrow Slot," *Trans. Natl. Vac. Symp.*, **10,** p. 271 (1963).

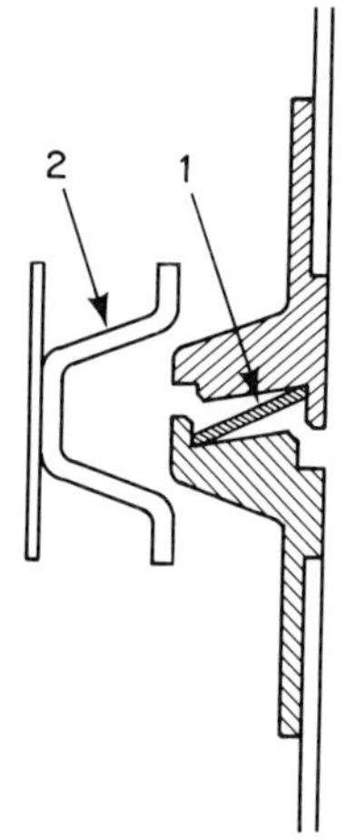

FIG. 21.

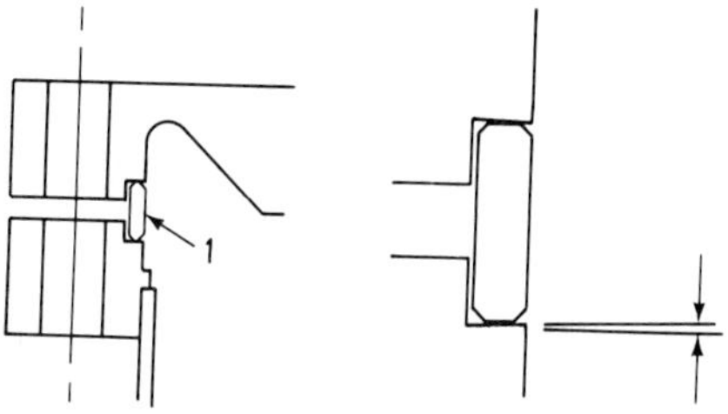

FIG. 22.

FIG. 21. Conoseal flange®. (1) Conically preformed stainless steel gasket; (2) "vee" Band clamp ring, segmented. Conoseal flanges are a proprietary design of the Aeroquip Corp., Los Angeles, Calif.

FIG. 22. Possible modification of the Batzer flange. (1) Stainless steel gasket ring plated with aluminum or other ductile metal to a thickness of about 0.05 mm.

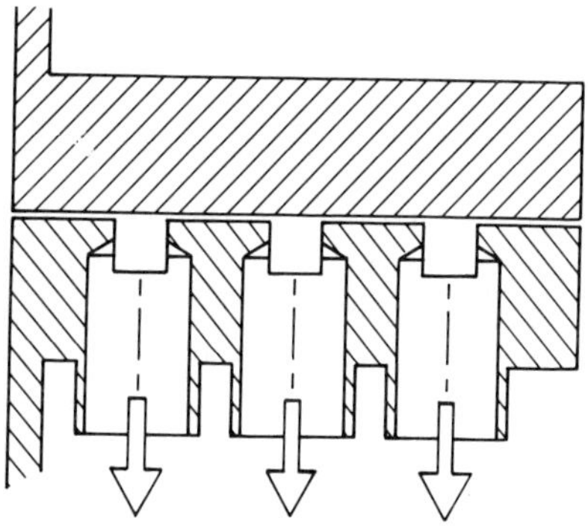

FIG. 23.

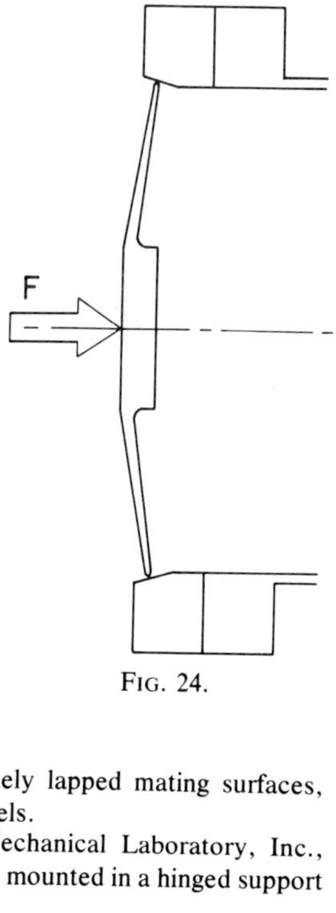

FIG. 24.

FIG. 23. Labyrinth seal. The seal is formed by accurately lapped mating surfaces, with intermediate pumping at successively lower pressure levels.

FIG. 24. Access port (patent pending by Huntington Mechanical Laboratory, Inc., Mountain View, Calif.) Closure force F is supplied by a screw mounted in a hinged support frame.

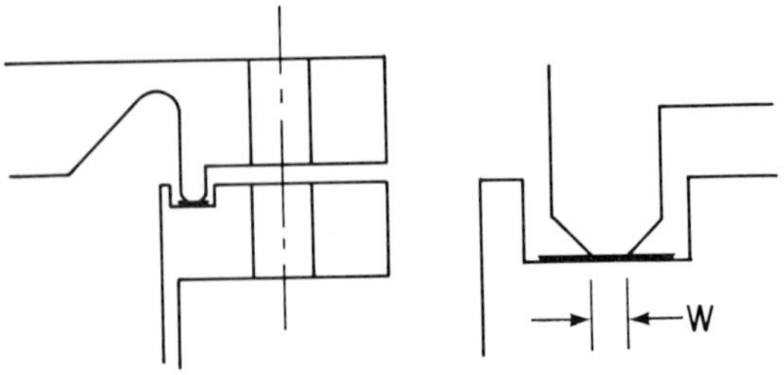

FIG. 25. Bakeable Batzer-type seal used to replace an elastomer O ring. Soft aluminum foil, about 0.08-mm thick is placed in the O-ring groove. The width *W* of the seal is 1.2 mm.

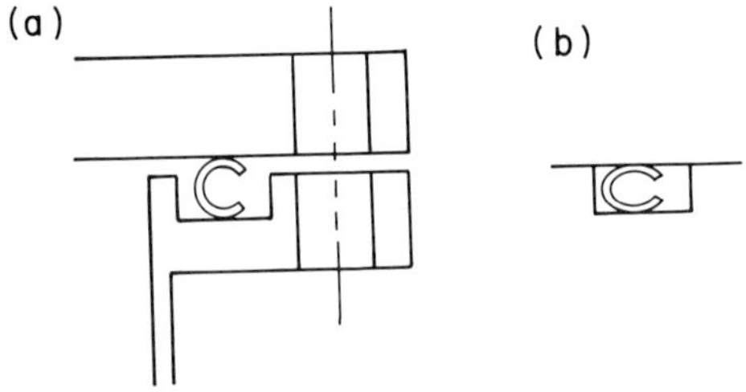

FIG. 26. Metal "C" seal used to replace an elastomer O ring. The C ring is usually Inconel plated with indium.

6.3. Vacuum, Water, and Gas Connectors: Function and Choices

Connectors permitting joints to be readily connected and disconnected must often be used both inside and outside vacuum systems. Thus leak-tight connectors for hoses, flexible, and rigid lines carrying gases and liquids at high and low pressure form an essential part of the arsenal of vacuum designers and users. In this discussion, the term connector is used for demountable joints less than about 2 in. in diameter. The term connector is a generic one, there being screw, flange, and ball and socket varieties. Appropriate to each style connector, seals may be designed for grease, elastomer gaskets, soft, or hard metals. Connector designs employ male and female and sexless mating parts and are available for joining metal to metal and metal to nonmetal components. Specific applications may require especially designed connectors. For example, a connector may have to be assembled in a confined situation. Some connectors require more engagement length than can be allowed for, have a gasket that is difficult to position, or be too large in a given dimension. Other questions, such as reliability of service, initial cost, lifetime, and availability, may have to be considered.

Commercially available connectors come in two general types with variations and modifications possible.

6.3.1. Types of Available Connectors

In the past, large organizations have often designed and stocked their own in-house types of connectors. For example, 40 years ago the Lawrence Radiation Laboratory implemented a series of brass fittings and flat rubber washer connectors called Rad Lab fittings. Having these fittings stocked and readily on hand proved to be of inestimable value allowing these connectors to be used as is or in modified form. Today, connectors closely related to the Rad Lab design, using O-ring elastomer gaskets are commercially available. Figure 27 provides some idea of the Rad Lab series as an introduction to current commercial catalogs. It is worth noting that Rad Lab style fittings, with brass, hard aluminum alloy, or stainless steel materials, have been used successfully with aluminum washer gaskets made of kitchen foil. The Rad Lab type fitting, modified to receive an elastomer O ring in a chamfered seat, also works well in joining metal to glass tubing (see below).

Connectors joining stainless steel tubing for either high or low temperature service, where mechanical vibration and/or torques must be transmitted through the joint, are especially demanding. There are a number of types of stainless connector fittings on the market. One commercial design* that has withstood a variety of tests and has performed well in service is singled out here. This connector can be used either with aluminum, copper, or nickel washer gaskets. Miniature flange connectors are

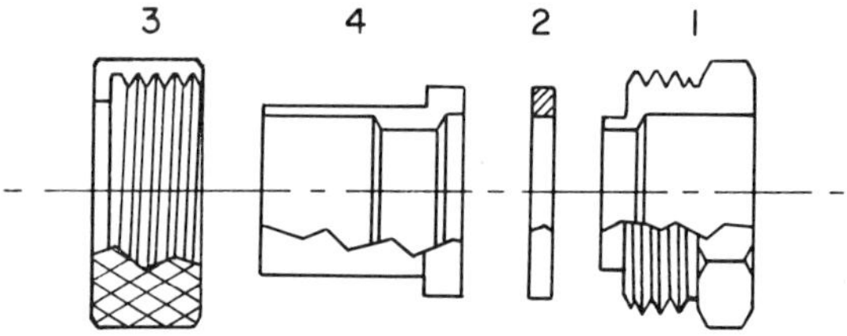

FIG. 27. Rad lab union coupling fittings.* (1) Threadpiece—solder tube socket connection is shown; threadpiece could also be a solid plug, male or female pipe thread, hose barb, or double ended to form a union; (2) square gasket or O ring; (3) nut, knurled; (4) tailpiece—solder tube socket shown; could also be solid cap, male or female pipe thread, or hose barb. (These fittings are from the University of California Radiation Laboratory, now the Lawrence Berkeley Laboratory and the Lawrence Livermore Laboratory. [For tubing sizes with outside diameters of: 6.35 mm ($\frac{1}{4}$ in.), 9.525 mm ($\frac{3}{8}$ in.), 12.7 mm ($\frac{1}{2}$ in.), 19.05 mm ($\frac{3}{4}$ in.), 25.4 mm (1 in.).])

* Trade name Cajon.

also commercially available. At date of this writing, the miniature flanges are in general more expensive for joining conduit of approximately 1.9 cm ($\frac{3}{4}$ in.) diameter and smaller. However, the flange style of connector does not require engagement and for reasons of space may be better suited to a particular application. As a word of caution, in general it is a poor bargain to try to save money on connectors for service required inside vacuum systems. Such connectors are often used to carry water or high pressure gases. Much time and money can be lost if such a connector proves to be faulty.

6.3.2. Indium Wire Seals, Metal–Glass, Metal–Metal

In addition to the elastomer O-ring seals in Rad Lab style screw connectors mentioned above, glass tubing can be readily connected to metal tubing using an indium wire seal. To effect the seal, one replaces the elastomer O ring in a champfered seat (Fig. 45) with a loop of indium wire. Since indium readily can be pressure welded to itself, the indium gasket can be formed simply by wrapping the indium wire around the glass tubing, being sure that the ends of the indium wire lap past one another. The indium will readily pressure weld to itself when the connector is properly assembled and tightened together.

6.4. Vacuum Valves

Of all the components in a vacuum system the vacuum valves have traditionally offered the most trouble. Trouble from valves is also much more likely as the surface temperature or cycling temperature of the valve increases or decreases from room temperature. From a sealing and operating standpoint, valves for the lowest temperature service are comparatively trouble free. In low temperature service the principle trouble with the valve is the thermal heat leak introduced into the system.

Troubles in valves are likely to be associated with certain aspects of the valves' operational cycle. Briefly, a valve is a component wherein a hole or through passage can be sealed or unsealed by means of the translation or rotation of some form of a solid or liquid component. Our purpose here is not to produce a catalog of the various designs that have been built and tested, but rather to give some guidance about the selection and operation of the relatively few designs that are known to be reliable. Valves have two basic geometries, the straight through or gate type and the angle type. In some specific instances, however, the function of a valve should be incorporated into another component, for example, into a cold trap and baffle assembly as in Fig. 1 (see also Chapter 6.5).

Some of the simple facts of life about valves are as follows. The interior action and body of a valve can harbor substantial amounts of contaminants. Actuating a valve can introduce contamination by exposing interior parts, leakage from the air, or from the frictional scraping by one surface sliding or rolling on another. Valves leak from a variety of causes including foreign matter on the valve seat.

6.4.1. Selecting a Valve for Routine Vacuum Service

Conventionally, routine vacuum service is identified by one parameter, the operating pressure within the vacuum system. As we have discussed at the beginning of Part 6, this single parameter method of characterizing a vacuum may be profitably replaced by a multiparameter designation. However, when we say a system is routine, we generally mean we are not fussy about it. Valves suitable for routine use employ elastomer seals in the valve body construction, in the motion actuator, and for the main valve seat. The body seals are invariably static and therefore offer little or no trouble, provided the valve operates near room temperature and is not located in a destructive chemical or radiation environment. The elastomer seal in the valve actuator often gives rise to the most serious trouble. Preferably this seal should allow for rotary action rather than sliding action. The main elastomer seal in the valve seat should last for the mechanical lifetime of the valve provided that injury to this main seal does not occur.

6.4.2. Leak Valves

Two categories of leak valves may be identified, coarse and fine. Coarse valves, for example, those suitable for letting a routine vacuum system up to air, are straightforward and need not concern us.

However, although many kinds of leak valves are commercially available, getting a sufficiently fine and constant control at a price one can afford could be a problem, which may be compounded if many leak valves are required, for example, so that several gases may be controlled in parallel. The criteria for leak valve selection should be based upon whether the valve can be cleaned by heating, the closed conductance of the valve as a function of the number of open–close cycles, how constant a given flow rate can be maintained, the degree of backlash or hysteresis that a valve exhibits in moving from one setting to another and back again, and the initial cost and lifetime of the valve. Several manufacturers make all-metal, stainless to stainless closure valves that are suitable for adaptation as leak valves. These valves are usually not supplied with a high degree of mechanical amplification for the fine control of the valve. Such an actuator can be contrived by the user of the valve if many valves are required. As an alternative to commercial leak valves it is possible to build a homemade version based on the simple pinching and unpinching of a small diameter heavy wall copper tube. Using a pair of hardened steel jaws that are removable, the action depends upon squeezing the tube shut between flat surfaces and squeezing the tube open afterwards by another set of jaws operating at right angles to the first set. The copper tubing without the jaws can be readily heated with a torch and this type of leak valve can be produced on a minimum budget.

6.4.3. Bakeable Valves

To qualify as bakeable a valve must be able to withstand heating in air to a temperature of at least 350°C. In vacuum services we are not aware of any valves requiring bakeout above 500°C.

These bakeout temperatures refer to baking in the open position. While at times desirable, baking of valves in the closed position with oxygen or air sealed from vacuum by the valve is a difficult requirement to satisfy. This requirement has only been satisfied to date by valves of aperture less than about 1 cm. A great deal of government and private money has been spent designing, constructing, and testing bakeable valves of apertures up to 30 or 40 cm. Cycling a valve through repeated bakeout cycles presents a high risk of failure even for valves of the best design and where the most careful procedures are followed. There are other possible methods for cleaning all metal valves that have been contaminated chemically. Perhaps the most effective of these is to bakeout at a temperature not to exceed 200°C, while flowing either a chemically inert gas such as argon or, in some cases, a chemically active gas such as oxygen through the valve. Even much lower temperature bakeout may prove to be sufficient. For example, indium and alloys with other metals have been successfully used as valve seat materials, some of which are renewable by heating and others by mechanical reformation in place.

As with nonbakeable valves, the sealing of the valve actuator is a crucial component. Motions are introduced through the valve actuator by means of all metal bellows. If the valve is designed for a long translational stroke, a long bellows is needed which requires a great many convolutions, usually formed by a welding technique and thus presenting a large interior area associated with the narrow spaces in each convolution. Such a bellows is very costly. Less expensive bellows of more open corrugation may be used to allow a tipping action or pseudorotation to be introduced to translate or rotate the valve disk. As a general point of reference, several all metal valves of apertures up to 30 cm are now commercially available and will function satisfactorily in vacuum systems provided that high temperature bakeout is avoided.

6.5. Traps, Baffles, and Valves in Combination

Typically, a baffle condenses a vapor flow to a liquid so that the liquid can drain off, e.g., return to the working cycle of a diffusion pump. A well designed and useful trap, on the other hand, catches and retains the higher vapor pressure fraction of the working fluid of a pump. At first sight, the terms baffle and trap would seem to concern the same type of function, namely, to slow down but not prevent the rate at which some species of molecules pass through. In ordinary practice, as well as under very demanding requirements, however, traps can completely stop some molecular species.

Molecules arrive at the surfaces of traps and baffles by two main mechanisms, volume flow and surface creep. Trapping molecules in vacuum systems can be done in a combination of either of two ways: (1) by binding molecules with energies much greater than kT of the surface, and (2) by lowering the temperature of the surface so that kT is less than the heat of physisorption of a molecular species on a surface (see Chapter 1.4), where k is Boltzmann's constant and T the absolute temperature.

If we put a trap directly over a diffusion pump without any baffling action being present, all of the working fluid of the diffusion pump will be caught by the trap in a time predicted by the rate at which the jets of the diffusion pump feed the working fluid to the trap. This rate may be many orders of magnitude greater than the rate predicted from the equilibrium vapor pressure of the working fluid of the pump at the wall temperature of the pump.

As mentioned in Chapter 6.4, it is possible to combine a valve, trap, and baffle in one design such that an elastomer can be used in the valve and still realize excellent control of contamination from the valve actuator mechanisms and the elastomer gasket of the valve plate seal.

The question of the role, design, and maintenance of creep proof barriers in traps, especially those above oil diffusion pumps, remains to be fully explored. As a rule of thumb, uncracked oil from a diffusion pump is completely inhibited from creeping by a surface $\leqslant -50°C$. On the other hand, a cold trap, to perform effectively in an ordinary vacuum system, must be $\leqslant -100°C$ due to the vapor pressure of water, and $\leqslant 80$ K due to the vapor pressure of CO_2. For ultrahigh vacuum, liquid nitrogen (LN) temperature or lower is required for CO_2 (amounts of CO_2 must be less than monolayer coverage at the boiling point of LN).

It cannot be overstressed that the principle reason that cold traps are frequently ineffective in preventing the passage of oil or mercury is due to warming of the trap and its internal filling lines when LN is added. LN trap designs exist (see below) that eliminate this problem.

6.5.1. Liquid Nitrogen Traps

A liquid nitrogen trap need not have an active chemical surface, but in some cases it is advantageous to cool a chemically active surface to liquid nitrogen temperature in order to get effective trapping for methane and other light-weight molecular species. Here we will consider plain metal surfaces cooled to liquid nitrogen temperature. Such a trap has two primary functions. One is to act as a cryopump for water vapor and the other is to prevent contamination, typically from an oil charged diffusion pump, from contaminating a given vacuum environment. It should always be borne in mind that a well designed liquid nitrogen trap can provide pumping speed of at least 10 liters/sec/cm^2 of system entrance area for room temperature water vapor under free molecular conditions. Tests have shown that a well-designed properly used liquid nitrogen trap can reduce oil contamination from diffusion pumps to negligible levels. Speaking more quantitatively, measurements have shown that oil molecules and cracked products with molecular weights >60 pass through a particular design, properly used, at rates $<10^5$ molecules/cm^2/sec.[1] These rates correspond to a vapor pressure at room temperature of less than about 10^{-15} Torr.

A simple test can indicate the effectiveness of an LN trap by looking for "pressure" pips on an ionization gauge when LN is replenished in its LN reservoir.

6.6. Flexible Connectors

Flexible connectors are used where some relative movement is required to accommodate thermal expansion or positioning adjustment, to isolate vibration, or to protect weak components from mechanical stress. If the vacuum requirements are not stringent, rubber or plastic hose may be used. Vacuum hose is typically made of butyl rubber or poly(vinyl chloride) with especially thick walls or reinforcement to resist collapse. Such hose is often used on roughing or foreline systems to isolate mechanical pump vibrations. If the system includes a trap, it is better to use the elastomer hose only on the forepump side of the trap.

Metal bellows are usually desirable for high vacuum applications. Brass, bronze, stainless steel, and inconel are typical materials. They are made by roll-forming or hydroforming (sylphon) as in Fig. 28, or by welding (diaphragm) as in Fig. 29. Both types are available in a wide range of diameters and lengths and with various end shapes. The welded bellows provide more stroke per unit length, but are more expensive and can only be cleaned by baking.

Flexible couplings with bellows lengths prewelded to commonly used flanges are commercially available in elastomer and metal gasketed versions, as shown in Fig. 30. Bellows terminated by metal-to-glass seals to glass tube ends are available for use in glass systems, shown in Fig. 31. Glass bellows have been made for systems that need to be all glass.

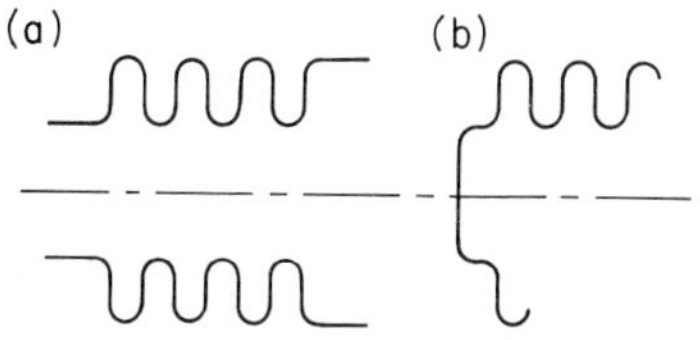

FIG. 28.

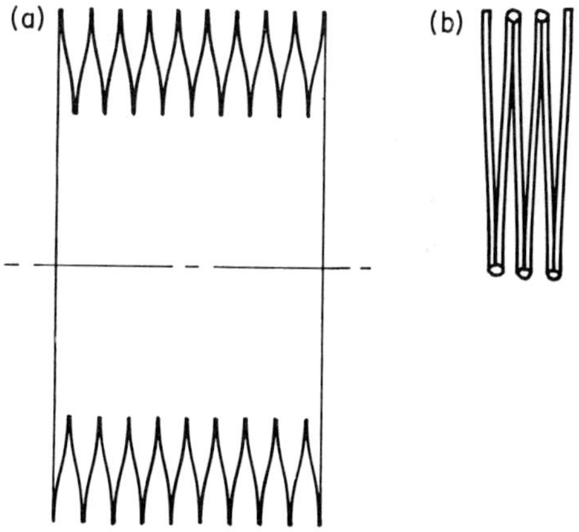

FIG. 29.

FIG. 28. (a) Roll-formed or hydroformed bellows (sylphon type); (b) roll-formed bellows with closed end.

FIG. 29. (a) Welded bellows (diaphragm type), stretched configuration of originally flat washers; (b) detail of fusion welds.

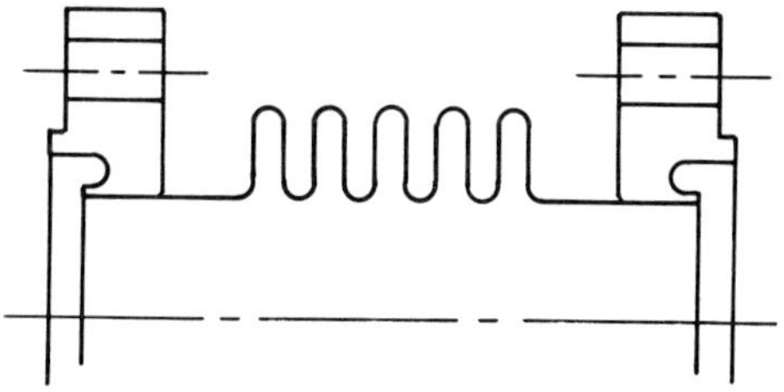

FIG. 30. Sylphon bellows welded to flanges.

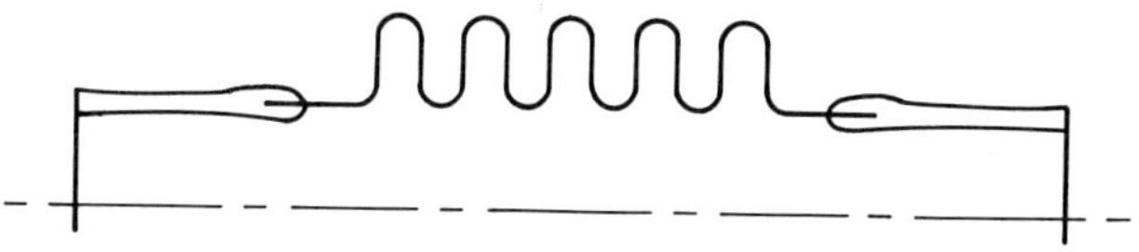

FIG. 31. Metal bellows sealed to glass tube ends.

Rugged sylphon metal hose is available in long lengths in copper. It may be cut to desired length and hard soldered to fittings or flanges to make a flexible connector, often useful for mechanical pump connections. The hose may be stretched somewhat initially to widen the convolutions for solvent or acid cleaning.

When put under vacuum, atmospheric pressure tends to shorten the bellows, requiring that the mounting of the two parts of the system provide sufficient mechanical restraint. Where this is impractical, postfitted wooden blocks may be strapped across the bellows if the movement in compression is not needed after installation. Otherwise a preloaded spring assembly may be used to resist the "vacuum" forces and to limit movement.

Bellows can be a weak spot in a system. If not designed for operating stresses below the endurance limit of the material they have a finite lifetime. Thin-wall bellows cannot be cleaned by strong chemical polish and etch methods, and they are vulnerable to corrosion from the atmospheric side. Joints to thin-wall brass and bronze bellows are best made by soft rather than hard solder, unless such joints can be made in a vacuum or inert gas furnace.

6.7. Mechanical Motion Feedthroughs

6.7.1. Rotary Motion Feedthroughs

A simple reliable rotary feedthrough can be made with an O-ring seal as shown in Fig. 32. The seal area of the shaft must be polished and the O ring lubricated with vacuum grease. A well-made seal is leak tight during rotation, but the speed of rotation is limited by frictional heat and wear. The degree of compression of the O ring is less than that for static seals. The radial dimension of the gland space should be about 80% of the O-ring diameter. The O-ring manufacturer's design data provide dynamic seal gland dimensions. A modification of this seal is shown in Fig. 33. Experience indicates that the gland volume should be about 18% larger than the O-ring volume.

A rotary motion feedthrough can be bellows sealed. It is shown in Fig. 34 in its simplest form. It is useful in many applications, as drawn, where the outgassing of the lubricated O ring must be avoided. Commercial feedthroughs using this sealing principle, "wobble motion," and provided with bearing mounted input and output shafts are available. A considerable surface area, and output shaft bearings, and lubricating medium are exposed to the vacuum. These feedthroughs provide a generous range of rotational speed and torque.

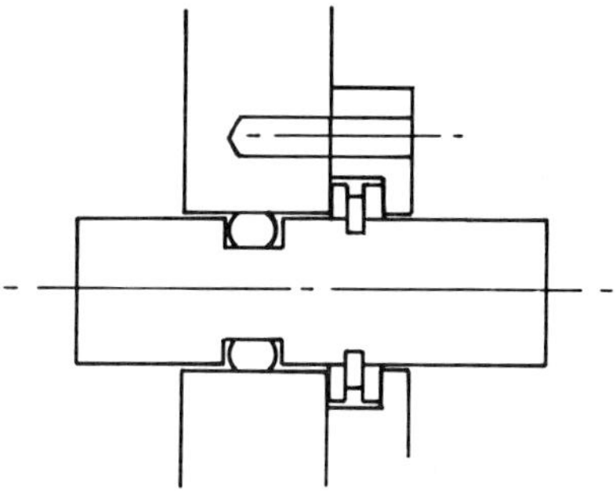

FIG. 32.

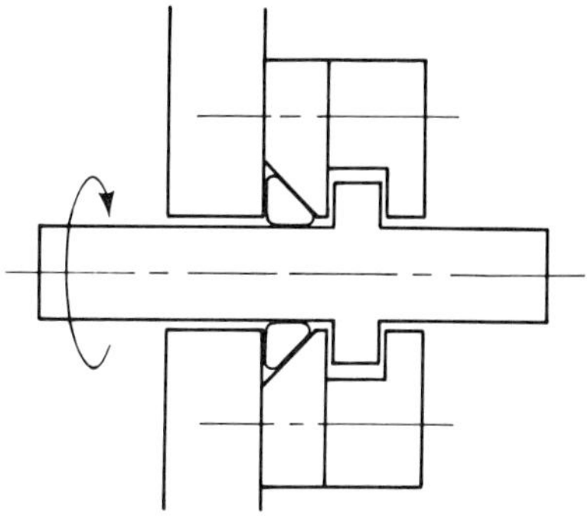

FIG. 33.

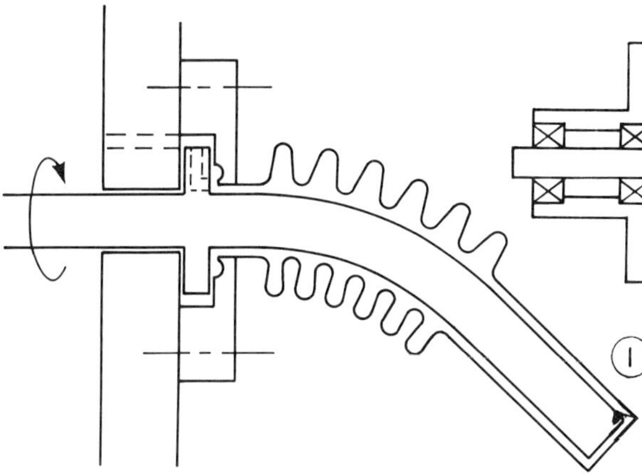

FIG. 34.

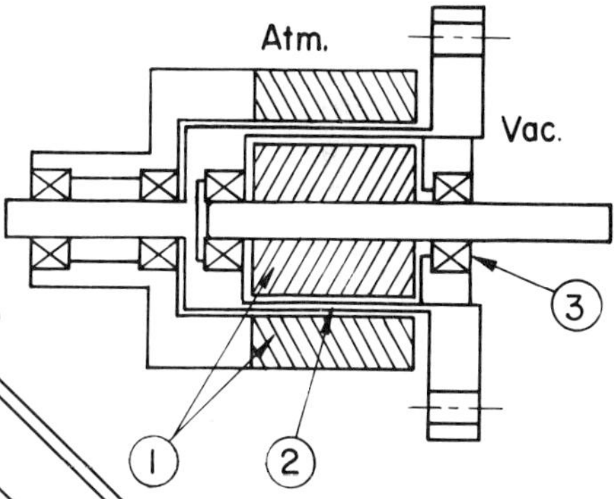

FIG. 35.

FIG. 32. A simple rotary motion feedthrough using an O-ring seal.
FIG. 33. Rotary motion feedthrough using a triangular gland for the O-ring seal.
FIG. 34. Bellows-sealed, rotary motion feedthrough.
FIG. 35. Principle of the rotary motion magnetic feedthrough. (1) Permanent magnet material; (2) thin, nonmagnetic vacuum wall; (3) bearings.

Rotary feedthroughs are also made using permanent magnets, coupled magnetically through a thin nonmagnetic vacuum wall as shown schematically in Fig. 35. The torque transmitted is limited by the magnetic field strengths. When overtorqued the mating pole sets will shift one or more lock-in intervals, therefore there is not a one-to-one correspondence between the input and output shaft positions.

A rotational feedthrough using a magnetic liquid suspension as a seal is available.* The suspension is composed of magnetic particles in a liquid

* Ferrofluidic Corp., Burlington, Mass.

vehicle, such as diffusion pump fluid, and exhibits a corresponding vapor pressure. Permanent magnets in the housing and a steel shaft provide the magnetic field. The sealing principle is shown in Fig. 36. Many suspension retaining ridges are required to withstand full atmospheric pressure. During pumpdown and during initial rotation some leakage occurs as the pressure gradient is equalized across the ridges. The feedthrough is capable of high torques and rotational speeds.

A rotary swing motion over part of a revolution may be bellows sealed as shown in Fig. 37. Freedom of movement throughout a large solid angle is possible if the pivot is mounted on an external gimbal.

6.7.2. Linear Motion Feedthroughs

The feedthrough of a round shaft can be elastomer sealed using the chevron geometry as shown in Fig. 38. The shaft must be polished and vacuum greased. If stoked frequently or at high speed, oil lubrication is desirable. During movement the direct leakage of atmospheric air can be kept small with a well made seal, but adsorbed gas on the shaft surface and in the lubricant is wiped into the vacuum. In a given system this gas may be tolerable or there may be time to pump the gas before the work process in the vacuum is carried out. The gas load can be reduced by using double seals and intermediate pumping, as shown in Fig. 39. If space is available the intermediate pump region can be lengthened to encompass the full stroke; and the length of shaft entering the fine vacuum is not exposed to atmosphere. Then the gas load from the feedthrough is limited to the outgassing of the elastomer and lubricant, depending on the pumping speed and quality of the intermediate vacuum.

The use of O-ring seals in the geometry shown for the rotational feedthrough is not recommended for linear motion because of the tendency of the O ring to roll and twist in the gland.

An alternative to using rubberlike elastomers with grease or oil lubrication is to make use of the self-lubricating quality of Teflon, which is one of the better plastics with regard to outgassing properties. Teflon creeps continuously under load at room temperature, it "cold-flows," and it must be fully captured or spring loaded to maintain the sealing force. An attractive seal for feedthrough applications (not tested by the authors of this section) is shown in Fig. 40. It should not require lubricant under moderate speeds and duty factors. Teflon used in captured gaskets, valve stem seals, and ball valve seals has been tested and found satisfactory.

Linear motion in vacuum can be obtained by driving a nut with a screw operated by a rotational feedthrough as shown in Fig. 41. A screw–nut

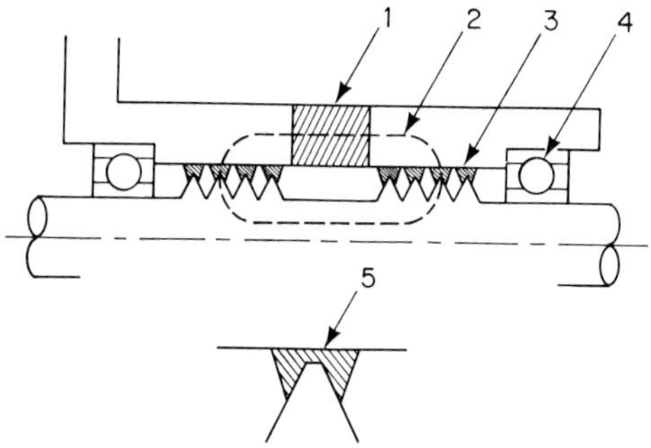

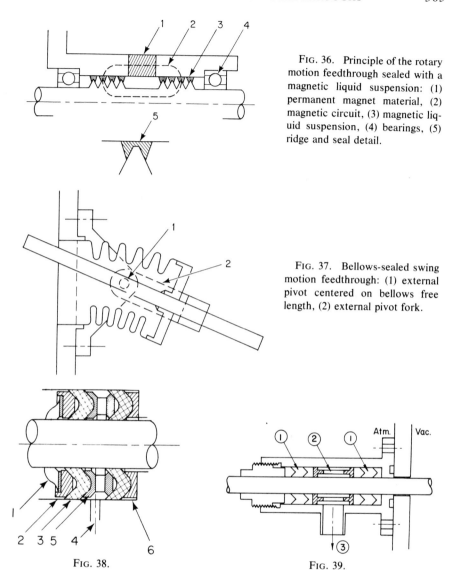

FIG. 36. Principle of the rotary motion feedthrough sealed with a magnetic liquid suspension: (1) permanent magnet material, (2) magnetic circuit, (3) magnetic liquid suspension, (4) bearings, (5) ridge and seal detail.

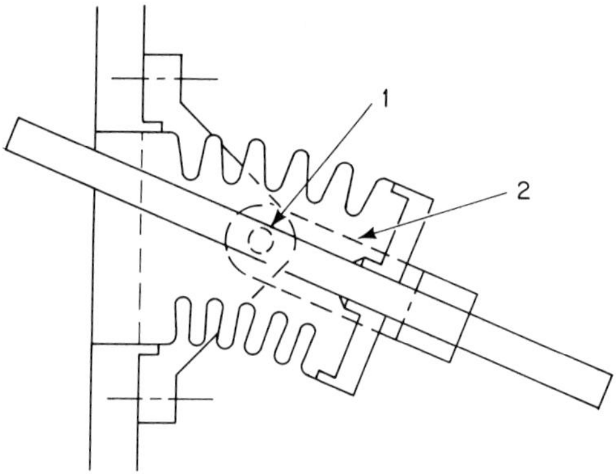

FIG. 37. Bellows-sealed swing motion feedthrough: (1) external pivot centered on bellows free length, (2) external pivot fork.

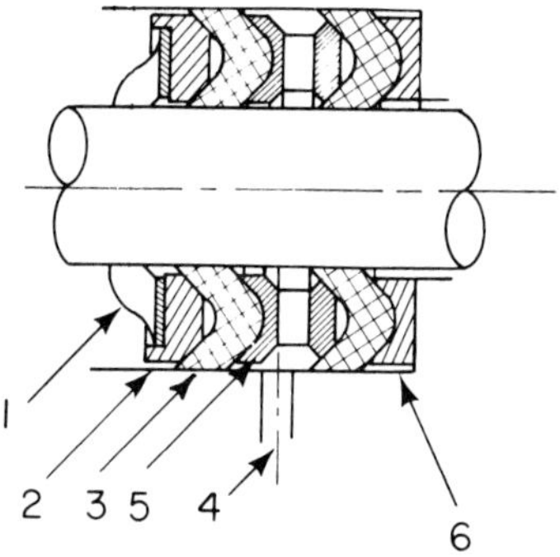

FIG. 38.

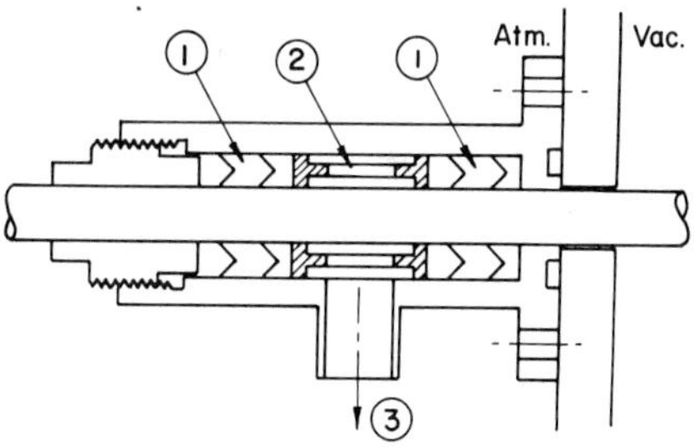

FIG. 39.

FIG. 38. Chevron feedthrough for rotary or linear motion: (1) spring washer to maintain proper compressive force, (2) pressure ring, (3) chevron gasket, (4) oil hole, (5) lantern ring, and (6) seat ring.

FIG. 39. Double-sealed linear or rotary motion with intermediate pump out: (1) chevron seal assemblies, (2) well-ventilated spacer ring, and (3) pumpout port.

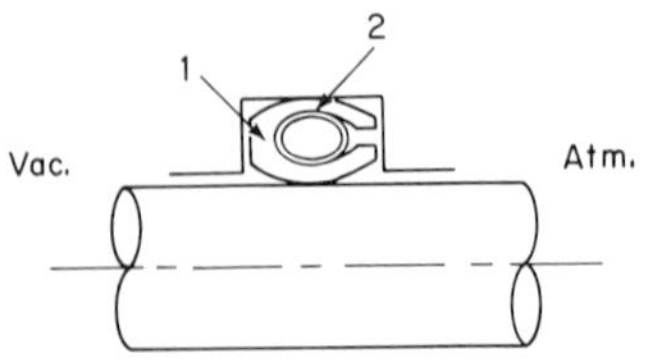

FIG. 40.

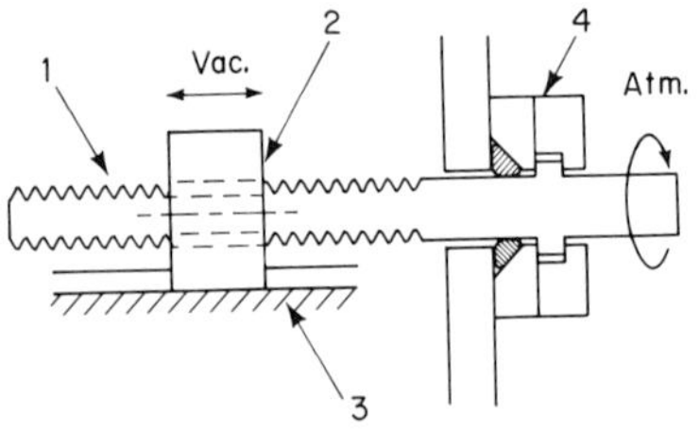

FIG. 41.

FIG. 40. "Omniseal" for linear or rotary feedthrough: (1) Teflon seal, (2) inner spring of helical strip stainless steel. (Omniseal ® is a proprietary design of the Aeroquip Corp., Los Angeles, Calif.

FIG. 41. Linear motion in vacuum using a rotational feedthrough: (1) screw, (2) nut, (3) guide, (4) feedthrough.

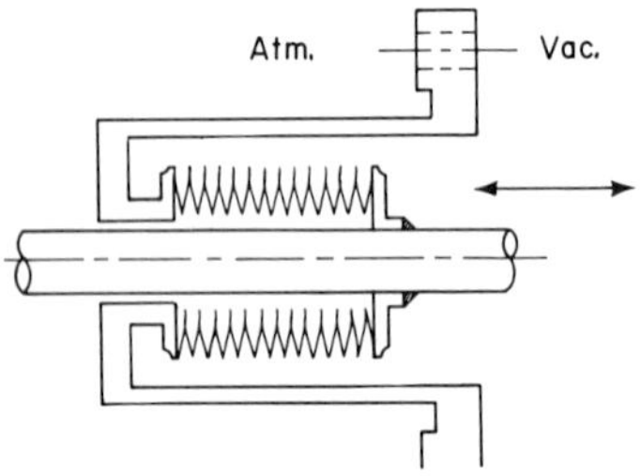

FIG. 42. Bellows-sealed linear motion feedthrough.

combination using semicircular helical grooves and recirculating balls may be used with dry lubricant.

Although relatively expensive, probably the only linear motion feedthrough satisfactory for ultrahigh vacuum use is bellows sealed as shown in Fig. 42. Welded bellows are generally used for longer strokes.

A great variety of motion feedthroughs, both elastomer and bellows sealed, providing combined as well as single motions, and some with high positional accuracy, are available commercially.

6.8. Electrical Feedthroughs

When the outgassing from plastics and elastomers are tolerable in the vacuum application, an endless variety of feedthroughs may be obtained commercially or made easily in the laboratory. A simple design for low voltages adaptable to large currents is shown in Fig. 43. Good materials for the in-vacuum insulators are high density polyethylene, acrylic plastic, epoxy, or Teflon (when the Teflon is not stressed enough to flow plastically). When greater mechanical strength is needed the insulator may be machined from fiberglass reinforced epoxy, NEMA G-10 (Natl.

Electrical Mfrs. Assoc. Grade 10). There may be filamentary leaks along the glass fibers, which can be sealed by vacuum epoxy potting the finished piece and wiping off excess epoxy back to original dimensions before curing. In high voltage applications electrical breakdown along the glass fibers has been experienced; newer production of this material may not have this problem. Insulators not exposed to the vacuum may be made from nylon, bakelite, and other materials. Another easily fabricated feedthrough that can operate at higher voltages and is mechanically strong is shown in Fig. 44. When many small instrument leads are required, the design shown in Fig. 45 may be useful; it shows many small leads cast into an epoxy insulating plug. Many of the ordinary multipin connecting plugs are available in vacuum-sealed versions.

If the thermocouple wire is used as leads the design of the figure above provides a feedthrough for thermocouples. Some commercial feedthroughs for thermocouples are made with small tubing so the thermocouple wires can be continuous as shown in Fig. 46. The vacuum seal is made by brazing the through wire to the tube end, on the vacuum side if possible. Feedthroughs are also available with a number of pairs of iron and constantan pins (and other metal pairs) to which matching thermocouple wires may be connected, often by spot welding.

In some cases it may be desirable to mount a combination of electrical feedthroughs on one flange, several medium current electrodes with a generous number of instrumentation pins, for example. For some applications the need for a particular combination may recur frequently. Some multipurpose feedthroughs are available commercially. Combination feedthroughs supplied on factory built systems may be available as replacement parts.

Small flanges with multiple connecting pins sealed by fused glass frit are available, which may be soft soldered to the system or to flanges. Many stock commercial ceramic insulators are provided with brazed metal flanges to which O-ring seals can be adapted, or which may be welded to the system or to flanges.

When plastics must be avoided the choice of insulating materials is largely limited to glass and ceramics. A nonporous machinable glass–ceramic is available* which is a better choice than boron nitride which outgasses strongly.

Most feedthroughs for ultrahigh vacuum use ceramic to metal brazed assemblies, which are usually bakeable to about 300°C. A low-current feedthrough using a commonly available assembly, may be made in the

---

* MACOR™ glass–ceramic (code 9658), MACOR Glass–Ceramic Dept., Corning Glass Works, Corning, N.Y. 14830; also Cotronics Corp., Brooklyn, N.Y. 11235.

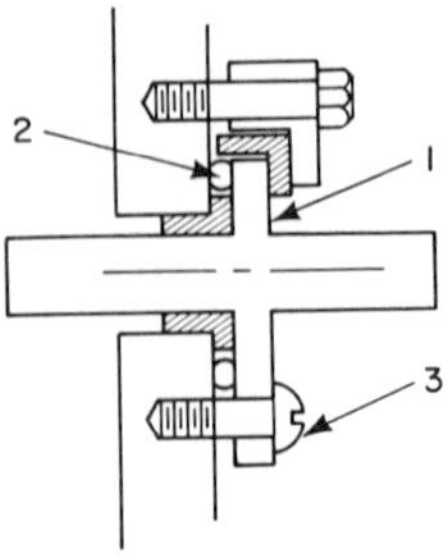

FIG. 43.

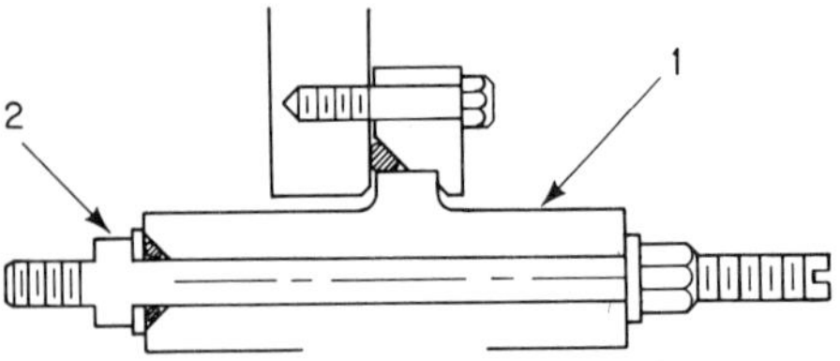

FIG. 44.

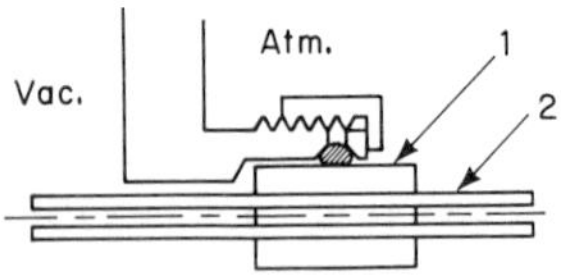

FIG. 45.

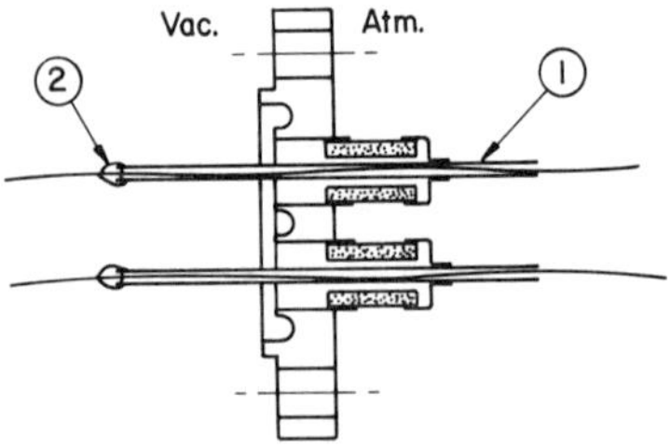

FIG. 46.

FIG. 43. Low voltage, high current feedthrough easily made in the laboratory. (1) Collar hard soldered to the electrode; (2) O ring. Lower half of figure is alternate design using: (3) nylon screw when high strength is not required.

FIG. 44. Easily fabricated feedthrough, adaptable to high currents and medium voltages, mechanically strong. (1) Insulator of high density polyethylene, acrylic or epoxy fiberglass; (2) shoulder hard soldered to the electrode, provided with a wrench flat in larger sizes.

FIG. 45. Multilead low current feedthrough in a standard compression fitting. (1) Epoxy plug cast around the leads; (2) clean, bare, or Formvar insulated leads.

FIG. 46. Feedthrough for thermocouples which allows the thermocouple leads to be continuous. (1) Small tubes factory brazed to insulator and flange assembly; (2) hard soldered vacuum seal.

laboratory as shown in Fig. 47. For uhv applications there is an extensive selection of high quality ceramic feedthroughs brazed to commonly used metal sealed flanges.

Feedthroughs are also available for coaxial and radio frequency inputs.

The current and voltage limits of electrical feedthroughs can vary with each installation. The current limit depends on the dissipation of the joule heating in the conductor, but as an example, a 1-cm diameter copper conductor will usually carry a current of 200 A. In the high voltage case a 1.3-cm length of glazed ceramic will have an average flashover voltage in one atmosphere of 10 kV. If corona discharge and current leakage must be avoided the voltage must be kept considerably lower, roughly one half of the flashover voltage. Voltage holding in a vacuum of 10^{-5} Torr or lower is generally better than in 1 atm. However, at pressures in the range of about 10^{-2} Torr or higher glow discharges will occur at 500 V or less.

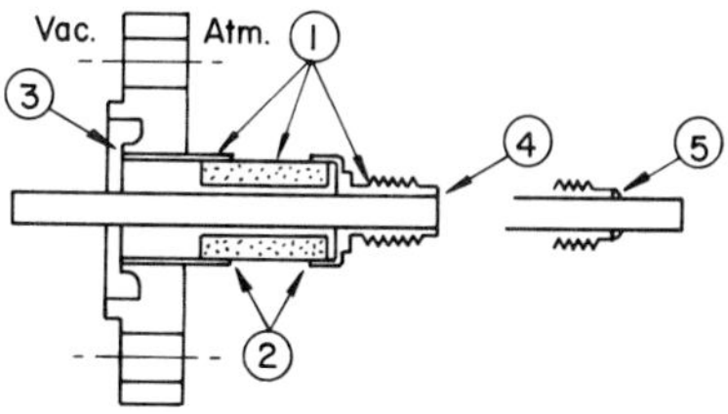

FIG. 47. Uhv feedthrough made from: (1) commonly available ceramic and metal brazed assembly; (2) factory ceramic to metal brazes; (3) fusion weld to uhv flange; (4) fusion weld to stainless steel rod; or (5) tungsten arc—inert gas braze to copper rod.

6.8.1. High Current, Low Voltage, Water Cooled Feedthrough

High currents may require water cooling because no convective gas cooling is available in vacuum, and the conductor would then be cooled by radiation, effective only at high temperatures, or by heat conduction along the length of the conductor, usually very limited. Water cooling may also be necessary when the conductor connects to a high temperature component in the system, such as a furnace heating element. The cooling water is usually supplied by a coaxial tube in the hollow electrode as shown in Fig. 48. The water flow provides also a current leakage path, therefore suitable lengths of insulating water line are required as well as low electrical conductivity of the water. Deionized cooling water circuits are available at many laboratories. Some electrolytic corrosion at metallic ends of the insulating hose may be caused by such current leakage.

6.8.2. High Voltage, Electrically Leak Proof

High quality insulators and clean surfaces are required on feedthroughs that are used to transmit extremely low currents, such as the signal currents from a mass spectrometer. High density, high alumina content ceramic (greater than 96%) is a good insulating material for this purpose. Leakage by electrical corona discharge on the atmospheric side can be limited by keeping the voltage gradient (electric field) low. This is accomplished by providing long air path lengths and by smoothly shaped

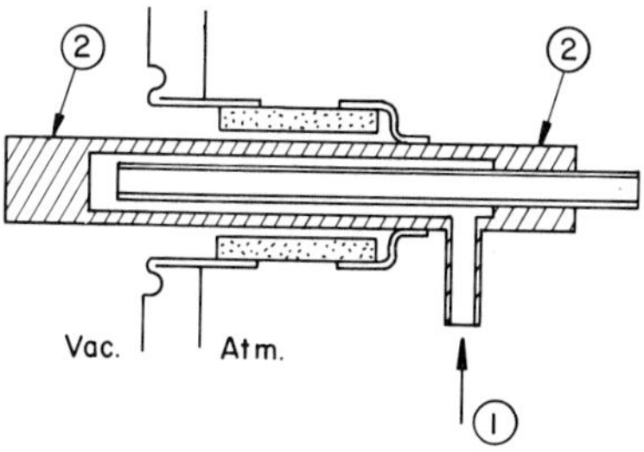

FIG. 48. Water cooled feedthrough for high currents. (1) Water inlet through insulating hose; (2) clamping areas for electrical cable.

metallic terminals. The insertion of insulating dielectric material between terminals may raise the breakdown voltage but does not decrease corona discharge and current leakage because the voltage gradient in the smaller remaining air space is increased.

6.8.3. Sputter-proof Feedthrough

In the sputtering process (see Part 16) a glow discharge is formed in a gaseous atmosphere at a pressure of a few millitorr. The discharge is generated by applying a high negative voltage to a target electrode, which is composed of the material to be sputtered. Ionized gas atoms from the plasma bombard the target electrode and knock out atoms of the target material, which then form a coating on the substrate. The gas ions would also attack and erode the conductor of the high voltage feedthrough and introduce contaminants into the system. This can be avoided if a completely closed grounded shield surrounds the conductor with a spacing less than the cathode dark space, usually about 5 or 6 mm. With this close spacing, no glow discharge occurs between the shield and the conductor. Since sputtering systems are often operated with reverse polarity or alternating rf voltages, other feedthroughs in addition to the target electrode may also need ground shielding.

Large numbers of electrons are part of and emerge from a discharge plasma. Combinations of geometry and electric and magnetic fields which trap electrons can lead to high density local plasmas and high current discharges in the arc mode. A thin wire at a high positive potential is an example of such an electron trap. Another famous example of electron trapping occurs when a magnetic field is combined with an electric field as in a Penning discharge configuration (described in another section of this book).

The sputtered material forms deposits throughout the chamber as well as on the substrate, and the feedthrough insulator needs to be well shielded to prevent metallic deposits from short circuiting the insulator, and a direct line of sight from any sputtering source in the chamber must be avoided in order to provide a long coat-over time. When electric and magnetic fields are present, metallic ions are not restricted to straight line paths. Some trial and error may be necessary to obtain adequate shielding. Since the mean free path at sputtering pressures is short, close shielding may be required, rather than the relatively open shielding characteristic of evaporation systems in molecular flow.

A limited choice of commercial feedthroughs for dc sputtering are commercially available. Ground shields, when not supplied on sputtering systems at the factory, are usually custom fabricated.

6.9. Viewports

A viewport may be provided most simply by using acrylic or polycarbonate plastic to make a blank-off flange for an O-ring sealed flanged port. A glass disk may be sealed to a flanged port with a flat elastomer gasket whose width is selected to provide enough unit loading to effect a good vacuum seal. Butyl rubber would be a good material for the gasket. Additional elastomer gasketing outside the vacuum is needed to prevent high stress glass-to-metal contacts. In larger sizes the additional reliability of an O-ring seal can be provided as shown in Fig. 49.

An indium wire seal, which requires no vacuum grease and thus keeps the window clean, could be used with careful retorquing of the bolts since the indium "creeps."

For uhv applications the optical element is furnace braised to thin metal sleeves of kovar or invar which in turn are braised or welded to uhv flanges as shown in Fig. 50. Such flanged viewports are available from a number of suppliers with glass, quartz, or synthetic sapphire windows.

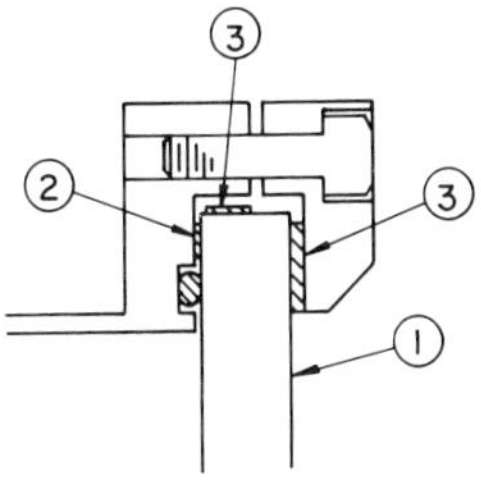

FIG. 49.

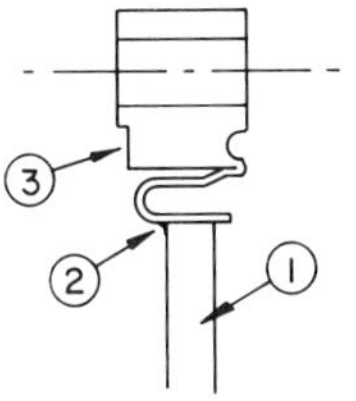

FIG. 50.

FIG. 49. Viewport with O-ring sealed glass window. (1) Glass disk; (2) protective gasket—thickness and groove depth selected to provide proper compression of the O ring; (3) protective gasket—provide gaps through gaskets 2 and 3 for leak checking. Avoid high pressure on the glass edges.

FIG. 50. Uhv viewport. (1) Optical element; (2) furnace brazes; (3) flange seal location.

7. GLASS VACUUM SYSTEMS*

7.1 Use of Glass as the Vacuum Envelope

Traditionally, since 1643 when Toricelli produced the first vacuum by inverting a column of mercury in a closed glass tube, glass has been the most common material used in the construction of small laboratory vacuum systems.

7.1.1. Advantages of Glass

There are numerous advantages to using glass as the vacuum container such as low cost, ease of fabrication, transparency, chemical inertness, insulation qualities, and low permeability to gases.

Cost and ease of fabrication go hand in hand. Not only are most common glasses more inexpensive than most other materials acceptable for high vacuum usage, but ease of fabrication also eliminates the need for costly welding, machining, or flanging which is necessary for most metal systems. Most small glass systems can be made from readily available tubing, rods, and flasks with a minimum of experience in the art of glass-blowing.

Good electrical insulation permits the use of glass for electrical high voltage leadthroughs. The transparency of glass to most radiation allows one to observe experiments within the vacuum system as well as to expose evacuated materials to various types of radiation. An additional advantage because of its transparency and insulating characteristics is the ease with which a Tesla coil can be used for leak detection of a glass system.

7.1.2. Disadvantage of Glass

The primary disadvantage of glass as a vacuum envelope is its low impact strength. Although under a state of compression glass is among the strongest materials known, its mechanical strength is greatly reduced when strains cause a tensile situation to exist. For this reason, glass systems are generally limited to several liters or less in volume.

---

* Part 7 is by V. O. Altemose.

ISBN 0-12-475914-9

7.2. Systems and Components

7.2.1. Fabrication of Glass Systems

7.2.1.1. General. The word glass applies to almost any compound which cools to a solid form without crystallizing. MacKenzie[1] states "any isotropic material, whether inorganic or organic, in which the three-dimensional atomic periodicity (long-range order greater than 20 Å) is absent and the viscosity of which is greater than about 10^{14} poises, may be described as a glass." We shall be concerned here with only those inorganic glasses with low vapor pressures at elevated temperatures.

The most commonly used glasses are those whose network forming oxides are SiO_2 and B_2O_3. A list of the more common commercial glasses along with their approximate compositions, relevant viscosity data, and coefficient of thermal expansion is shown in Table I.*

The two temperature related physical properties of great importance in fabricating and using glass vacuum systems are viscosity and coefficient of thermal expansion. Glass has no specific melting point but the viscosity or rigidity of the glass decreases gradually with increase in temperature. Four viscosities (η) are generally used to define the strain point, annealing point, softening point, and working point of a glass.[2] The strain point ($\eta \sim 10^{14.5}$ P) is the temperature at which internal stresses are relieved in a matter of several hours and is considered a safe limit to which annealed glass can be used. At the annealing point ($\eta \sim 10^{13}$ P), internal stresses are relieved in about 15 min. The softening point ($\eta \sim 10^{7.6}$ P) is the temperature where glass will deform under its own weight. The working point ($\eta \sim 10^4$ P) is the temperature at which lampworking is readily possible. A knowledge of these points allows one to determine temperatures needed to fabricate a glass vacuum system, anneal it to relieve stresses, and to ascertain the temperature to which the system can safely be thermally outgassed without collapsing under vacuum.

The thermal coefficient of expansion of glasses in general, is practically a constant between 0 and 300°C. These are the expansion values quoted in Table I. Above the strain point this coefficient increases considerably.[2] Not only is it necessary to have a reasonable expansion match when sealing two kinds of glass or glasses and metals together, but for the

[1] J. D. MacKenzie, "Modern Aspects of the Vitreous State." Butterworth, London, 1961.

[2] E. B. Shand, "Glass Engineering Handbook." McGraw-Hill, New York, 1958.

* For a detailed list of the physical properties of commercial glasses of other manufacturers throughout the world, see Volf (Ref. 10, p. 8).

TABLE I. Glass Composition and Viscosity Data for Common Commercial Glasses

Corning code number	Type glass	Composition[a] (mole %)				Coefficient of expansion 0—300°C 10^{-7}/°C	Viscosity data[b]			
		Network $SiO_2 + B_2O_3$	Modifiers $Na_2O—Li_2O—K_2O$	PbO	Other Oxides		Strain point (°C)	Annealing point (°C)	Softening point (°C)	Working point (°C)
7900	96% silica	99	—	—	1	8	820	910	1500	—
7230	Borosilicate	96	—	—	4	14	—	—	—	—
Fused silica	Silica glass	100	—	—	—	5.5	990	1050	1580	—
7070	Borosilicate	98	2	—	—	32	455	495	715	1110
7160	Borosilicate	98	1	—	1	25	497	544	830	—
7740	Borosilicate	94	5	—	1	32	519	563	820	1220
7720	Borosilicate	93	4	2	1	36	484	523	755	1110
7052	Borosilicate	86	6	—	5	46	436	480	712	1115
7040	Borosilicate (Kovar seal)	92	7	—	2	47.5	450	489	702	1080
7050	Borosilicate (*w*-seal)	92	7	—	1	46.0	461	501	703	1055
7056	Borosilicate (Kovar seal)	90	8	—	1	51.5	472	512	720	—
0010	Potash–soda–lead	77	14	8	1	91	393	430	626	970
9010	Barium–alkali–silicate	75	14	—	8	88.5	406	444	646	1015
8160	Potash–soda–lead	74	12	8	5	91.0	397	433	632	975
0120	Potash–soda–lead	76	12	11	1	89.5	395	435	630	1000
0080	Soda–lime	73	16	—	11	92.0	472	512	696	1000
1715	Alumino–silicate	72	—	—	28	35.0	834	866	1060	—
1720	Alumino–silicate	66	1	—	33	42.0	668	715	915	1200
1723	Alumino–silicate	65	—	—	35	46.0	665	712	910	1175

[a] From V. O. Altemose, *J. Appl. Phys.* **32,** 1309 (1961).
[b] From E. B. Shand (2) "Properties of Glasses and Glass–Ceramics." Corning Glass Works, Corning, N.Y., 1973.

higher expansion glasses in particular, sharp thermal gradients can cause cracks in the system.

For a detailed description of needed tools, working characteristics of glasses, and how to make various seals and glass components, the reader is referred to the excellent treatment by Wheeler.[3]

7.2.1.2. Glass-to-Glass Graded Seals. Sometimes it becomes necessary to seal glasses of widely different compositions together such as a soda–lime glass to a borosilicate glass or a borosilicate to fused silica. Due to the great differences in coefficients of thermal expansions, these glasses cannot be sealed directly together, but require intermediate glasses so that the coefficient of expansion (α) of each glass differs from that of its neighbor by a small increment.

The smaller one keeps the expansion mismatch between adjacent glasses, the less the strain that will exist in the finished seal. A reasonable rule of thumb for successful glass to glass seals is to keep this mismatch such that $\Delta\alpha \leqq 10^{-6}/°C$. Wheeler[3] suggests that in going from code 0080 ($\alpha = 92 \times 10^{-7}/°C$) to code 7740 glass ($\alpha = 32 \times 10^{-7}/°C$) a good graded seal should have six intermediate glasses or steps. For the novice glassblower it usually is not an easy matter to make this type seal and most will find it expedient and economical to purchase the desired graded seals already made.

7.2.1.3. Glass-to-Metal Seals. Glass-to-metal seals are treated extensively by Partridge.[4] Such seals generally fit into one of three categories: matched seals in which the metal is sealed directly to the glass and the resulting stresses are kept to a minimum by matching thermal expansion coefficients; unmatched seals, where the expansion of metal and glass differs considerably but where high stresses are avoided by using small diameter wires, ductile metals, or graded intermediate seals; or soldered seals.

To assure a satisfactory seal, the glass must wet the surface of the metal.[3] What is called glass wetting the metal is actually the glass dissolving in the oxide layer which covers the metal. Hall and Berger[5] have shown that when a glass bead is placed on a metal plate that is free of oxide, and both are fired in a reducing or inert atmosphere, no seal is obtained. When the metal was oxidized before being placed in the furnace in an inert atmosphere, a strong bond is formed between the glass and metal, thus showing the importance of metal oxidation before attempting the glass seal. Degree of oxidation is important. Too thin an oxide layer

[3] E. L. Wheeler, "Scientific Glassblowing." Wiley (Interscience), New York, 1958.
[4] J. H. Partridge, "Glass to Metal Seals." *Soc. Glass Technol., Sheffield, England, 1949.*
[5] A. W. Hall and E. E. Berger, *Gen. Electr. Rev.*, **37,** 96 (1934).

prevents the glass from forming a strong bond with the metal and results in a weak seal. If the oxide layer is too thick, the glass bonds strongly to the oxide, but there is danger of the heavy layer of oxide breaking away from the metal.

Matched seals, such as Kovar to code 7052 glass, can be made in a multitude of shapes and sizes and are readily available from numerous vacuum products suppliers. Details in cleaning, oxidizing, and sealing are given by Wheeler[3] as well as Partridge.[4] Table II lists the various sealing metals along with their matching glasses and may serve as a guide in selecting suitable materials for glass to metal seals.

The Housekeeper seal is an example of unmatched seals where almost any type glass can be sealed to copper if the copper is prepared properly.[6] This seal depends on the ductility of copper to overcome the mismatch.

In soldered seals, the metal part is actually soldered to a layer of metal previously applied to the glass surface by one of several possible methods.[4]

7.2.2. Typical Glass Systems and Components

Glass vacuum systems are generally fabricated from a low expansion borosilicate glass, such as Corning's code 7740 glass or its equivalent[7–9] for several reasons. First, it is probably the easiest type of glass to lampwork for the laboratory technician who is usually not an expert glassblower. Most of the system will likely be made from glass tubing which in standard wall thickness is available in 3–100 mm diameters. Standard low expansion borosilicate laboratory ware such as flasks, stopcocks, test tubes, ground joints, and even bell jars are also available to form almost any shape desirable.

Other manufacturers from various countries also produce hard borosilicate glasses similar to Cornings's code 7740 (Pyrex® brand glass). Some of these glasses are the Duran 50 and Razotherm glasses in Germany, Hysil and Phoenix glasses in England, Nife glass in Switzerland, Thermosil in Poland, Terex in Japan, Pireks in the Soviet Union, Simax in Czechoslavakia, and Kimble's KG-33 in the United States. These glasses all have the same nominal composition which in weight percent is 80% SiO_2, 12% B_2O_3, 2% Al_2O_3, and 4.5% Na_2O, with some containing quantities of less than 1% each of CaO, MgO, and K_2O. For a more com-

[6] Ref. 2, p. 123; Ref. 3, p. 162.

[7] "Laboratory Glassware Catalogue." Corning Glass Works, Corning, N.Y.

[8] "Kimble Catalogue of Laboratory Glassware D." Kimble Glass Division, Owens-Illinois Glass Co., Toledo, Ohio.

[9] "Inter-Joint Glassware Catalogue." Scientific Glass Apparatus Co., Bloomfield, N.J.

TABLE II. Glass-Metal Combinations for Seals[a]

Metal or alloy	Trade name or type	Thermal conduction (cal/sec × cm × °C)	Elec. res at 20°C (ohm-cm × 10^6)	Matching glass: Corning	Matching glass: Other	Remarks
Cold-rolled steel, SAE 1010, AISI C1010	Other grades of soft steel or iron also used	0.108	18	1990 1991	5643[b]	Tends to oxidize excessively. Plating with Cu, Cr, or Ag often used to prevent this. External rings of iron are frequently sealed to Pt-sealing glasses
17% iron, AISI 430A	Allegheny Telemet			0129 9019	3720[b]	Used in metal TV picture tubes as a ring seal outside a glass plate of lower expansion, putting glass in compression
28% chrome iron	Allegheny Sealment-1 Ascaloy 446 Carpenter 27	0.06	72	9012	K-51[c]	Used in TV tubes. Pt-sealing glasses can also be used under certain conditions and types of seals. Alloy should be preoxidized in wet H_2 gas
Platinum		0.165	10.6	0010 0080	R-5[c] R-6[c]	Now used mainly for scientific apparatus. Preoxidizing unnecessary
Composite material Core, 42% Ni, 58% Fe Sheath, Cu	Dumet	0.04	4–6	0120 7570	KG-12	Copper sheath bonded to core. Surface usually coated with borax to reduce oxidation. Expansion: Radial 90 × 10^{-7} per °C. Axial, 63 × 10^{-7} per °C. Wire size usually limited to 0.020 in diam

Nickel–chrome iron, 42% Ni, 6% Cr, 52% Fe	Sylvania 4 Alleg. Sealment HC-4 Carpenter 426	0.032	34	8160 9010			Matches Pt-sealing glasses. Relatively large seals can be made between this alloy and suitable glasses. Pretreatment in H_2 atmosphere furnace essential
Nickel–cobalt iron 28% Ni, 18% Co, 53% Fe	Kovar Fernico	0.046	47	7040 7050	7055	K-650[c] K-704[c]	Low-expansion sealing alloy. Should be annealed after cold working and pretreated in H_2 atmosphere furnace
	Radar			7052	8830	K-705[c]	
Molybdenum		0.35	5.7	7040 7052	1720		Metal rods usually ground and sometimes polished. Surface should be cleaned in fused nitrite and not overoxidized
Tungsten		0.38	5.5	3320 7720	7050 8830	K-772[c]	Same as for molybdenum
Copper	OFHC grade (oxygen-free high conductivity)	0.92	1.75	Most glasses			Copper–glass seals are made with Housekeeper technique with thin metal sections. Care must be taken to prevent overoxidation of copper

[a] From Shand, see Ref. 2, p. 119.
[b] Pittsburgh Plate Glass Co.
[c] Kimble Glass Division, Owens-Illinois Glass.

plete comparison and thorough treatment of technical glasses the reader is referred to Volf.[10]

The low thermal expansion of the above glasses also allow for the rapid heating and cooling which saves time when baking out a system to 500°C or chilling a cold trap to −196°C with liquid nitrogen. Glass diffusion pumps and gages are usually made of glass that will readily seal to code 7740 glass. A comprehensive review of glass diffusion pumps and gages is presented by Dushman[11] and Reiman.[12]

Since the design criteria for glass systems, such as pumping speed, conductance, ultimate vacuum, etc., are the same as for other systems, and they are adequately discussed throughout this book, glass system components will only be considered briefly in terms of their limitations. The all-glass system of 40 years ago which utilized glass diffusion pumps, mercury cutoffs, stopcocks, and greased ground joints is quite adequate at ordinary vacuums, but becomes somewhat limited in the degree of vacuum attainable.

Mercury cutoffs require the presence of mercury in the system and therefore limit the total system pressure to that of the vapor pressure of mercury. At room temperature this is about 10^{-3} Torr.

Glass diffusion pumps are generally adequate in terms of pumping speed and attainable vacuum for small systems. Where larger sized pumps are required, the thickness of the glass increases, and then careful heating and cooling become necessary to avoid cracking.

Standard taper ground joints and ground ball and socket joints can be used for demountable portions of vacuum systems. Used with low vapor pressure grease, such as Apiezon[13] or Dow–Corning silicone grease,[14] ground joints are usable to about 10^{-5} Torr depending on their size. These same ground joints if sealed with Apiezon W,[13] a hard wax, can be used at pressures lower than 10^{-7} Torr without difficulty.

Stopcocks are convenient and economical to use as high vacuum valves. The best high vacuum stopcocks, with hollow stoppers that can be evacuated, lubricated with Apiezon N grease, are good to pressures down to about 10^{-6} Torr. Where pressures below 10^{-7} Torr are desired, greased joints and stopcocks are not practical because they cannot be baked out.

[10] M. B. Volf, "Technical Glasses." Pitman, London, 1961.

[11] S. Dushman, "Scientific Foundations of Vacuum Technique." Wiley, New York, 1962.

[12] A. L. Reiman, "Vacuum Technique." Chapman & Hall, London, 1952.

[13] Apiezon oils, greases, and waxes, of Metropolitan–Vickers Electrical Co., London.

[14] Dow–Corning, Midland, Mich.

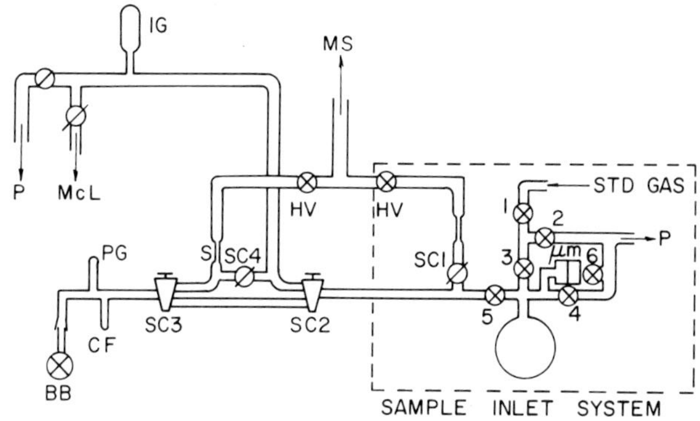

FIG. 1. Microblister system. P is the vacuum diffusion pumps, McL the McLeod gage, IG the ion gage, PG the pirani gage, CF the cold finger, SC the glass high vacuum stopcocks, S the capillary leak, HV = uhv valves (type C, Granville–Phillips Co., Boulder, Colo.), μm the micromanometer, BB the blister breaker, and MS the mass spectrometer (Altemose and Bierwiler[15]).

Due to modern vacuum technology and the ease with which high vacuum is attainable, today's typical glass vacuum system is no longer a system made entirely of glass. The use of glass to metal seals has allowed today's scientist to take advantage of the best features of both glass and metals.

A typical vacuum system used for the analysis of microblisters in glass is shown in Fig. 1.[15] That portion of the system on the blister–breaker side of the capillary leak S utilizes stopcocks and bakeout is unnecessary because pressures of 10^{-6} Torr are tolerable. On the mass spectrometer side of S, where analytical backgrounds require pressures of 10^{-8} routinely, it is customary to use all metal bakeable valves as described by Alpert.[16]

In the ultrahigh vacuum range (10^{-10} Torr), Alpert and Buritz[17] have shown that the ultimate limit in the degree of vacuum attainable in code 7740 glass systems is due to the diffusion of helium through the walls. At these pressures, aluminosilicate glass (code 1720) would provide a better system because its permeability is 10^5 times lower at 25°C. (See Chapter 7.4.) Examples of typical ultrahigh vacuum glass system are shown by Roberts and Vandeslice.[18]

[15] V. O. Altemose and T. W. Bierwiler, *Am. Ceram. Soc. Bull.* **54,** 6 (1975).

[16] D. Alpert, *Rev. Sci. Instr.* **22,** 536 (1951).

[17] D. Alpert and R. S. Buritz, *J. Appl. Phys.* **25,** 202 (1954).

[18] R. W. Roberts and T. A. Vanderslice, "Ultrahigh Vacuum and its Applications." Prentice–Hall, New York, 1963.

7.3. Outgassing of Glass

7.3.1. General

In all high vacuum and particularly ultrahigh vacuum applications the evolution of gases from the materials which make up the vacuum system is very important. Energy of any type, when absorbed by glass will cause some degree of outgassing. The major gases evolved depend upon the form of energy supplied: heat produces primarily water; electron bombardment, oxygen; ultraviolet and gamma radiation, hydrogen and water; and neutron bombardment of boron-containing glasses, helium.

For the most part, the research on the outgassing of glass by various forms of energy reported in this chapter was performed at Corning Glass Works over the last 20 years or so. Most of the work was initiated as a result of practical problems encountered in the technical usage of glasses. The need to thermally outgas glass to be used in a vacuum environment was known for many years but the mechanism was not understood. Poisoned cathodes in cathode ray tubes, particularly in color television tubes as greater electron accelerating voltages became necessary, prompted the work on outgassing caused by electron bombardment. The problem of "hard starting" in fluorescent and mercury arc lamps led to the work with ultraviolet radiation. The increased use of glassy materials of all kinds in radioactive environments and around nuclear reactors led to a study of outgassing by gamma radiation and neutron bombardment.

7.3.2. Thermal Outgassing[19]

To eliminate outgassing from glass used in high vacuum at ordinary temperature, it is necessary to vacuum bake the glass at a somewhat higher temperature. The work of Sherwood[20] and of Harris and Schumacher[21] established that when glass is heated in vacuum a rapid evolution of gas, which is primarily water vapor, occurs upon first heating, followed by an evolution which becomes increasingly persistent as the bakeout temperature approaches the softening point of the glass. The initial evolution had been attributed to surface gases and the more persistent evolution at higher temperatures attributed to diffusion from the interior.

[19] A comprehensive review of the early work on the subject is given by G. W. Morey, "The Properties of Glass." Van Nostrand-Reinhold, Princeton, N.J., 1938, and more recently by S. Dushman, "Scientific Foundations of Vacuum Techniques," 2nd ed. Wiley, 1962.

[20] R. G. Sherwood, *Phys. Rev.* **12,** 448 (1918).

[21] J. E. Harris and E. E. Schumacher, *Bell System Tech. J.* **2,** 122 (1933).

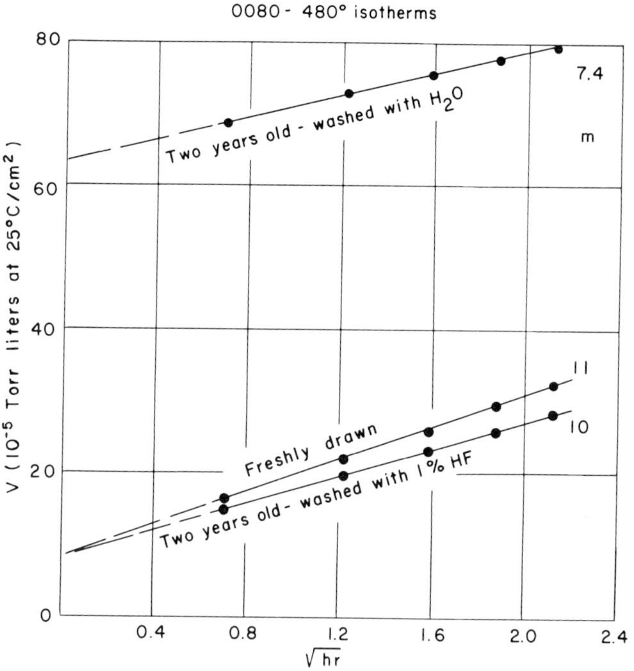

FIG. 2. Comparison of old and new code 0080 glass and old code 0080 glass washed with 1% HF at a bakeout temperature of 480°C (V vs square root of bakeout time) (Todd[22]).

The work of Todd[22] confirmed these earlier conclusions and differentiated between the water evolved from the surface and water from the glass interior. Figure 2 shows the total water evolved per unit surface area of glass as a function of the square root of time for code 0080 glass at 480°C. In Fig. 2 the 480°C isotherm of a water washed sample is compared with the same 2-year-old glass washed with 1% HF for 3 min and a similar sample with a fresh surface ½ hr after the glass tubing was drawn.

Todd proposed that the adsorbed gas, at least at temperatures above 300°C, was the portion given off rapidly, while the linear portion of the curve described water being diffused out of the bulk of the glass. The above isotherms can be expressed by the equation

$$V = mt^{1/2} + s, \tag{7.3.1}$$

where V is in Torr liters at 25°C per square centimeter area, s represents the adsorbed gas, and m is the slope of the curve and is proportional to the diffusion constant of the glass. Of great significance is the large reduction

[22] B. J. Todd, *J. Appl. Phys.* **26,** 1238 (1955).

of s by simply removing the outer surface layer of glass from the two year old sample. Todd also reports that washing with other reagents such as HCl, H_2SO_4, NH_4OH, NaOH, and the usual chromic acid cleaning solution had no effect on the value of s.

The effect of temperature on the rate of water evolution m is shown in Fig. 7.3.2. Since all samples in Fig. 3 have the same HF rinse, the value of s is essentially constant, m was shown to obey a typical Arrhenius equation so that

$$\log m = -A/T + B, \tag{7.3.2}$$

where T is absolute temperature and A and B are constants. Table III shows the values of the constants A and B, and the activation energy for diffusion obtained by Todd for the eight glasses he studied.

Todd has shown that if one assumes the adsorbed hydrate layer is completely removed from the glass during initial bakeout and m is known from Eq. (7.3.2), then it is possible to calculate the amount of water that will diffuse out of the glass with time at any temperature following some arbitrary bakeout. If the system is evacuated and baked at temperature T_1

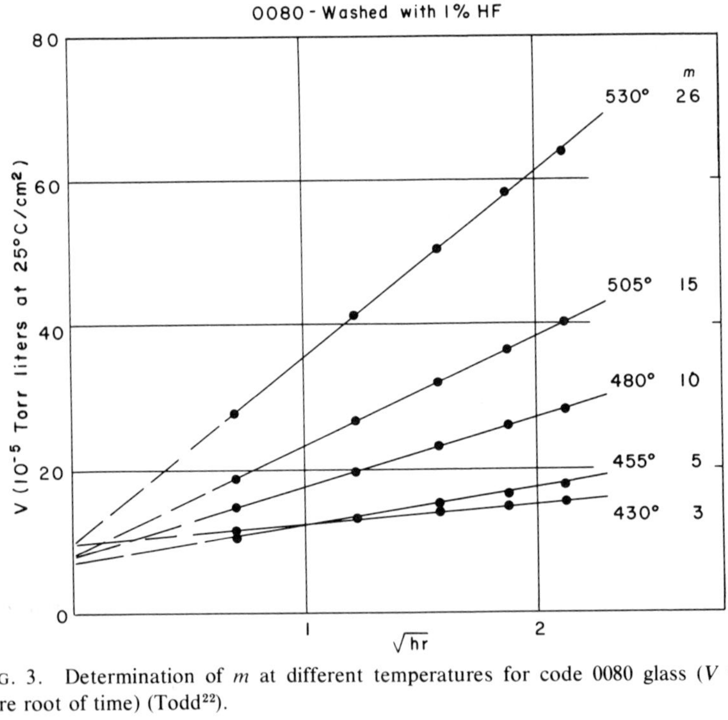

FIG. 3. Determination of m at different temperatures for code 0080 glass (V versus square root of time) (Todd[22]).

TABLE III. Constants for the Calculation of m (10^{-5} Torr liters at 25°C/cm^2/hr$^{1/2}$ at an Arbitrary Temperature from the Equation, $\log m = (-A/T) + B$

Glass code no.	Glass type	A (K)	B	ΔH_a (kcal/mole)
7911	Vycor brand 96% silica	6230	5.397	57
7910	Vycor brand 96% silica	8240	9.772	75
1720	Lime–aluminum	7000	7.952	64
7740	Borosilicate	4510	6.310	41
7720	Lead–borosilicate	4150	5.983	38
0080	Soda–lime	5420	8.153	50
0120	Potash–soda–lead	3910	6.208	36
9014	Potash–soda–varium	3840	7.799	44

(for which m_1 is known) for a time t_1, then the quantity of water evolved per unit of surface of glass, at temperature T_2, in a time interval t_2, would be

$$V_2 = (m_1^2 t_1 + m_2^2 t_2)^{1/2} - m_1 t_1^{1/2}. \tag{7.3.3}$$

This calculation is valid because even at elevated temperatures the water evolution due to diffusion will persist for very long periods of time. Todd has also shown that the water evolution can be treated as diffusion from a semi-infinite body. Bakeouts of 10 hr at 500°C affect the water concentration in only the first 10–20 μm from the surface.

Table IV shows some values of m for code 7740 glass at various temperatures calculated from Eq. (7.3.2). It is of interest to note that the outgassing rate at 500°C is six orders of magnitude faster than at 100°C and 9 orders greater than at room temperature. This indicates that for room temperature use, glass bakeouts that completely remove the surface gases are generally adequate.

So far we have considered only the outgassing from solid glass well below the softening point of the glass. There exists also the situation where glass "tipoffs" are necessary to close off static systems and where

TABLE IV. Outgassing Rates for Code 7740 Glass[a]

T (°C)	m (Torr liters at 25°C/cm^2 hr$^{1/2}$)
500	3.0
400	4.1×10^{-1}
300	2.8×10^{-2}
200	6.0×10^{-4}
100	1.7×10^{-6}
25	1.5×10^{-9}

[a] Calculated from data in Table III.

we wish to maintain reasonable vacuums for long periods of time without further pumping. Tipoffs require the actual melting of a small portion of the glass in the system to make the vacuum seal.

Dalton[23] measured as much as 200–300 Torr cm^3/gm of water from a soda–lime glass (code 0080) and as much as 500 Torr cm^3/gm from borosilicate glasses by completely melting them under vacuum. More recently[24] we have shown quantities of from 50 to 300 Torr cm^3/gm can be extracted from soda–lime glasses depending upon the water content of the atmosphere in which it was melted. Therefore, if proper care isn't taken in the way tipoffs are made, one could completely defeat the purpose of bakeout by reintroducing into the evacuated system more water than was originally removed.

Several suggestions can be made to reduce the amount of gas that might be sealed off in the system when making a tipoff seal.

(a) If the tipoff tube has not been vacuum baked, this should be done to eliminate the very large amount of water that can exist in the surface hydrate layer.

(b) Small-bore and thin-walled tubing in the tipoff region will require much less heat and less glass to be melted for shorter times to accomplish the seal. The minimum size of the tubing used will be dictated by the pumping speed required to adequately exhaust the system. (See Chapters 1.2 and 4.1.)

(c) If thinner walls (or small bore) are not practical for strength or for other reasons, the partial constriction of the tubing first, followed by another few minutes of pumping, then a final fast tipoff should reduce the residual gases due to tipoff that may be left in the system.

(d) Cold traps, if not already in use on the pump side and very near to the tipoff location, will aid greatly in increasing the pumping speed for water, the gas of primary concern.

(e) A pinch-off technique (assisting the closure by pinching with a tweezers could help in that the seal could be made at higher glass viscosity and it would require somewhat lower temperature and perhaps shorter heating times than would normally be necessary for the glass to flow together on its own.

Beyond the scope of this chapter, but worth mentioning, is the use of flashed getters in closed-off vacuum systems.[25] Since it is virtually im-

[23] R. H. Dalton, *J. Am. Chem. Soc.* **57,** 2150 (1935).

[24] V. O. Altemose, unpublished data.

[25] For a detailed discussion on getters, see Chapter 9, S. Dushman," Scientific Foundations of Vacuum Technique," 2nd ed. Wiley, 1962.

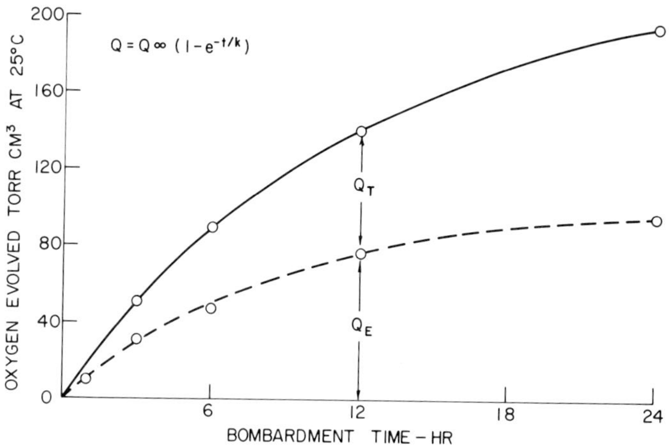

FIG. 4. Outgassing of code 8603 glass (Lineweaver[27]).

possible to thermally tipoff a glass system without some gas contamination, where the particular application will allow their use, a getter will continue chemical pumping of evolved gases for long periods of time.

7.3.3. Outgassing by Electron Bombardment[26–29]

Lineweaver *et al.* observed the outgassing of glass while bombarding it with 10–27 keV electrons. Ninety-five percent of the observed evolved gases was oxygen. The remainder was carbon monoxide and hydrogen, part of which was attributed to background gases coming from the hot portion of the electron gun used. (It should be noted that during all outgassing measurements, two cold traps at −78°C were used so that evolution of water vapor would not have been observed, if present.)

To maintain a constant temperature during sample bombardment all samples were heated to 200°C. When electron bombardment was ceased, the flow of oxygen was observed to continue if the sample temperature was maintained. By increasing the temperature to 350°C, the oxygen depletion could be accelerated indicating a diffusion controlled mechanism to remove the oxygen from the glass after it was formed.

Figure 4 is a plot of the oxygen outgassing data from code 8603 glass bombarded with 150 μA of 20 keV electrons using a 7.6 × 1.8 cm raster.

[26] B. J. Todd, J. L. Lineweaver, and J. T. Kerr, *J. Appl. Phys.* **31,** 51 (1960).

[27] J. L. Lineweaver, *J. Appl. Phys.* **34,** 1786 (1963).

[28] J. L. Lineweaver, *Proc. Symp. Art of Glassblowing* **7,** Am. Sci. Glassblowers Soc., Wilmington, Del. (1963).

[29] J. L. Lineweaver, *Suppl. Nouvo Cimento* **1,** 530 (1963).

These data were obtained using a new area of glass for each of the Q_E points. Following each bombardment, the sample was baked out until the oxygen had been depleted.

The total oxygen evolved (Q) thus consists of the sum of two components: an amount Q_E released during bombardment and an amount Q_T released thermally after bombardment. With no bombardment, Q_T is equal to zero. In most of the experiments Q_T was equal to or greater than Q_E.

For the majority of the glasses studied the electron bombardment data can be expressed by the equation

$$Q = Q_\infty[1 - \exp(-t/K)]. \qquad (7.3.4)$$

In Eq. (7.3.4) t is bombardment time in hours, K is a measure of the time dependence of the phenomenon, and Q_∞ is a maximum value of Q. Values of Q_∞ and K for eight glasses are given in Table V. The low value of Q_∞ for vitreous silica (code 7940) probably is the blank for the system.

Depth of electron penetration, thus the volume of glass depleted of oxygen varies as the square of electron energy. For code 7740 glass, this depth is estimated to be 2.7 μm for 20 keV electrons.

In terms of parameters in Eq. (7.3.4), Q_∞ was shown to vary as the depth of penetration (energy squared) and the time constant K was inversely proportional to the electron beam current per unit mass of glass affected.

After additional studies on several simple experimental glasses, Lineweaver concluded,[28] "Studies of two to four component glasses clearly indicate that of those studied, the alkali rather than the alkaline earth oxides in glass are responsible for the oxygen evolved as a result of electron bombardment."

TABLE V. Q_∞ and K for Glasses Bombarded with 150 μA of 20 keV Electrons using a $3 \times \frac{3}{4}$ in. Raster (Lineweaver[27])

Corning glass code	Torr cm³ at 25°C	K (hr)
0081	247	23.4
8603	227	12.4
9019	178	23.7
9010	170	20.6
0120	72	7.7
7740	60	5.5
7070	51	12.5
7940	<1	—

7.3.4. Outgassing by Ultraviolet Radiation

Using high pressure quartz mercury arc lamps as an ultraviolet source, Kenty[30] has shown that considerable quantities of hydrogen, as well as some carbon monoxide, carbon dioxide, and water are evolved from Pyrex® brand (presumably code 7740 glass) lamp jackets under the influence of uv radiation. He attributed these gases to the photodissociation of H_2O and CO_2 in the glass. After several thousand hours of operation, Kenty showed that the partial pressure of hydrogen within the outer part of the mercury arc lamp built up to between 2 and 3 Torr.

Using a code 7720 glass filter to cut off the shorter-wavelength radiation, Kenty also suggested that the threshold for producing hydrogen is in the neighborhood of 3000–3300 Å. Only uv wavelengths shorter than this threshold would have enough energy to form these gases.

Altemose[31] measured the gases evolved from several types of glass during irradiation with short-wavelength uv light using a mass spectrometer to measure the gas flow rate directly. He also found hydrogen to be the major gas, with some water, carbon monoxide, and carbon dioxide evolved from all glasses.

Typical gas flow curves for hydrogen are shown in Fig. 5 for a sample of code 7740 glass tubing. This sample was irradiated with about 20 W of uv radiation of wavelengths between 2000 and 2800 Å. The four curves in Fig. 5 illustrate the effect of hydrogen concentration buildup within the glass during irradiation as well as the effect of gas diffusion in allowing the gas to escape from the glass. During irradiation the uv transparent sample container was held at 100°C while the glass sample itself got to about 200°C.

For the example shown in Fig. 5 the sample was exposed to uv for 400 min, held overnight at room temperature, then exposed to uv again for 300 min. It was then thermally outgassed at 350°C for 1500 min. Initially, during irradiation, the flow rate drops off with time as shown in curve 1. Due to the great temperature dependence of hydrogen permeation in glasses, the gas flow stops immediately when the lamps are turned off because the temperature also drops sharply. While sitting overnight at room temperature the hydrogen concentration gradient within the sample is slightly relaxed. Then, when the lamps are turned on again the hydrogen production causes curve 2 to rise above where curve 1 left off until the concentration gradient returns to where it would have been if curve 1 had not been interrupted. During the 350°C bakeout the hydrogen con-

[30] C. Kenty, Report on *Ann. Conf. Phys. Electron.* **13** MIT, March 26, 1953.
[31] V. O. Altemose, umpublished manuscript, 1976.

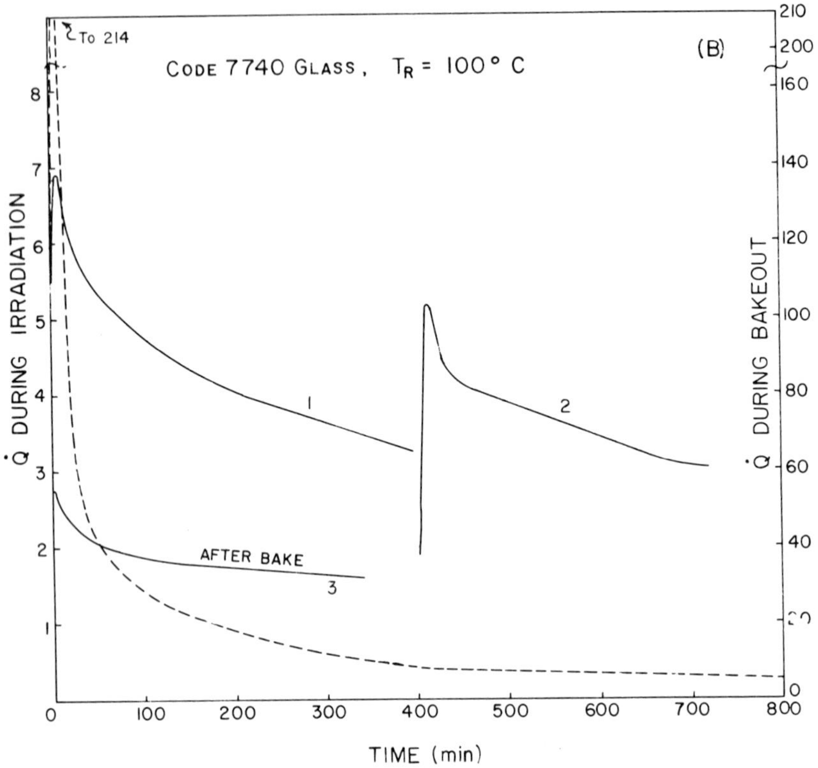

FIG. 5. Hydrogen evolution rate at 100°C irradiation temperature. Solid lines—during uv irradiation, dashed line—during postirradiation bakeout at 350°C. Units of $Q = 10^{-4}$ Torr cm³/sec (Altemose[31]).

centration is reduced very rapidly due to the increased permeability until essentially all the gas is diffused out after 1500 min. Curve 3, due to irradiation after the bakeout, shows that the actual hydrogen production within the glass has been reduced considerably with time.

Table VI shows the results of irradiating four code 7740 glass specimens at three different temperatures and a code 7916 glass (96% silica) for comparison. Obviously, the hydrogen evolved during irradiation increases with increasing temperature. In the case of the code 7916 glass, where the permeation rate is much higher at irradiation temperature than for code 7740 glass, almost no concentration of hydrogen is built up within the glass. (See Chapter 7.4. Permeability of code 7916 glass is the same as for code 7900.)

If the same amount of hydrogen is produced at all temperatures one would expect the quantities evolved during bakeout to make up for the

TABLE VI. Evolved Hydrogen—700 min uv Exposure plus 350°C Bake

Glass sample	Surface area (cm^2)	Container temperature *TR* (°C)	Hydrogen evolved (Torr cm^3) During irradiation 1st Exp. 400 min	2nd Exp. 300 min	Thermal 350°C 1500 min	Total
7740-A	147	50	6.9	3.0	77.5	87.4
7740-B	146	100	10.6	6.6	90.2	106.0
7740-C	146	150	44.7	25.9	88.4	159.0
7740-D	148	150	42.1	20.1	98.8	160.0
7916	156	100	3.5	2.6	0.25	6.4

difference between the quantities evolved during irradiation such that the totals in each case would be the same. The fact that the totals are not the same indicates that either less hydrogen is produced at the lower irradiation temperatures, or due to the larger concentration buildup, a reversible process is taking place. Nevertheless, the total amount of hydrogen is less at the lower temperatures. Hydrogen evolution from code 7740 and 7720 glasses for continuous exposure times of 1000, 1500, and 3000 min is shown in Fig. 6. During exposure the quantity of hydrogen evolved for code 7740 glass is about twice as great as for code 7720 glass. Surprisingly, during post irradiation bakeout, there is practically no hydrogen evolved from the code 7720 glass.

The only major difference in the composition of these two glasses is the presence of 2 mole % of PbO in code 7720 glass. It is believed that the lack of hydrogen evolution during bakeout and at least part of the difference during irradiation is due to a reduction of the PbO to Pb in the lead-containing glass.

The flow rate curves during exposure for all glasses tested look like those in Fig. 5 and can be expressed by the empirical equation

$$dQ_E/dt = At^{-1/4}, \tag{7.3.5}$$

where A is a constant. Integration of this equation yields the total quantity of gas

$$Q_E = \tfrac{4}{3}At^{3/4}. \tag{7.3.6}$$

The data in Fig. 6 are shown to obey this $t^{3/4}$ relationship. The upper two curves show the thermally outgassed and total ($Q_E + Q_B$) hydrogen for code 7740 glass. The amount of hydrogen evolved during bakeout does not vary appreciably for the various exposure times for code 7740 glass.

Table VII is a tabulation of the evolved hydrogen during the 1500 min

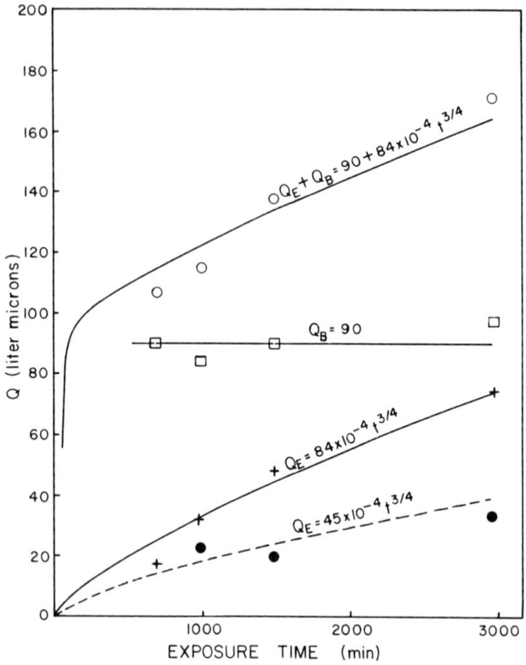

FIG. 6. Hydrogen evolved from code 7740 glass (solid line) and code 7720 glass (dashed line). Q_E—during irradiation, Q_B—during thermal bakeout (Altemose[31]).

uv exposure of various other glasses at container temperature of 100°C and during post exposure bakes at 350°C.

By using code 7740 glass filters with a uv cutoff at 2750 Å, Altemose[31] concludes that the threshold wavelength for the production of hydrogen should be 2900 Å. Coulson[32] quotes values of 119.5 kcal/mole to break the first OH bond in water and 99.4 kcal/mole to break the second OH bond. This latter, smaller quantity is equivalent to 4.32 eV/molecule and is equal to the energy associated with a photon of 2895 Å radiation. Thus, it seems reasonable that only wavelengths shorter than 2900 Å would have sufficient energy to dissociate the OH radical present in the glass structure. Where mercury arc lamps are used in or around vacuum systems, one of the primary wavelengths is at 2537 Å and outgassing as described could be a problem.

It was also concluded that the amount of hydrogen observed during irradiation was proportional to the amount of radiation incident on the sample. For uv opaque and low permeation rate glasses the hydrogen is evolved primarily from one side of the glass—the one exposed to the radi-

[32] C. A. Coulson, "Valence." Clarendon, Oxford, 1952.

TABLE VII. Hydrogen Evolved[a] from Glass Samples for 1500 min Exposure Times Plus 350°C Bakeout

Sample	q_E[b]	q_B[c]	Total q[d]	Total q[e] (normalized)
7740-J	48.1	89.8	138.0	148.0
7720-A	20.0	0.4	20.4	19.4
1720-A	13.2	0.3	13.5	13.4
1720-B	19.9	0.02	19.9	19.6
9730-A	18.9	0.6	19.5	19.4
1715-wet	22.8	0.5	23.3	23.8
1715-dry	11.4	1.5	12.9	13.7
0010-A	20.2	0.8	21.0	15.4
0081-B	14.3	0.4	14.7	15.5
0081-C	17.8	3.0	20.8	20.8
0088-A	20.1	~2.0	22.0	22.0
7740-M	61.5	90.0	152.0	152.0
7720-D	13.8	2.4	16.2	16.8
7910-A	2.2	0.4	22.6	23.9

[a] All units are Torr cm^3 at 25°C.
[b] Quantity evolved during uv exposure at 100°C.
[c] Quantity evolved during 350°C bake following exposure.
[d] Total of $a + b$.
[e] These figures are normalized for a sample surface area of 150 cm^2.

ation. For high transmission, high permeation rate glasses, the gas is evolved from both sides.

7.3.5. Outgassing by Nuclear Radiation

The increasing need for electronic equipment to remain in operation near nuclear radiation sources and the use of glass in this equipment makes it important to know something about the effects of radiation on it. Generally, the effects of glass subjected to high radiation doses are discoloration, gas evolution, and fracture.

Gases evolved from glass exposed to gamma radiation would be expected to be similar to those evolved from exposure to short wavelength ultraviolet light.[33] Both types of radiation are electromagnetic and they differ only in photon energy. The effects of α or β particles are of little concern because of their very limited range and the ease with which glasses can be shielded from them.

Usually the elements in glass have very low cross sections for fast neutrons. Therefore, the effects of these high energy neutrons should also be negligible. For some glasses, particularly those containing boron, the ab-

[33] See Section 7.3.4.

sorption of thermal neutrons will be quite high, since the capture cross section of the ^{10}B isotope is about 4000 b.

Spencer and Yarmovski[34] reported their findings pertaining to Corning code 7720, 7052, 0080, 0120, and 1723 glasses. They found no cracking of any of the glasses when exposed to a total thermal neutron flux of 10^{16} nvt. They report no ionized gas visible when sparking with a Tesla coil although discoloration was prevalent, particularly in the code 7052 and 7720 glasses. In code 7720 glass cracking was observed at or above 10^{17} nvt.

When they enclosed ionization gauges in envelopes of code 0080, 7720, and 1723 glasses, after exposure to a thermal neutron flux of 4×10^{16} nvt, they found no pressure rise in the code 0080 glass, and the pressure rose from 10^{-6} Torr to 5×10^{-3} and 3.7×10^{-3} Torr, respectively, for code 7720 and 1723 glasses.

Altemose[35] exposed small evacuated glass sample bulbs to known exposures of gamma radiation and thermal neutrons plus γ radiation. Each specimen was carefully baked at 10^{-5} Torr or better at 350°C for 3 hr with blank samples of each group kept with no radiation exposure for comparison. After a cooling-off period of several weeks (to allow decay of radioactivity), each sample was opened under vacuum and the contents analyzed with a mass spectrometer using a glass system similar to the one employed by Todd.[36]

Details of discoloration are listed in Table VIII. Although the exact γ dose is unknown during neutron exposure, it should be proportional to the neutron dose. Discoloration is believed to be primarily due to the γ radiation. All glasses except those containing lead (code 7720 and 0120 glasses) were completely bleached by baking at 350°C. The permanent darkening is caused by a reduction of lead oxide by hydrogen formed in the glass by the γ radiation. The same effect has been observed while diffusing hydrogen through code 7720 glass at elevated temperature.

Each specimen contained only a small amount of gas when it was broken open. Gas pressure in the code 7720 glass sample was about 4×10^{-3} Torr, and most of the gas was hydrogen.

Qualitatively, the results are similar to those obtained when glass is irradiated with short-wavelength ultraviolet radiation. The hydrogen is formed by the dissociation of the hydrogen bonds within the glass (presumably O–H bonds). The carbon oxides are probably produced on or very near the glass surface. Since no really special cleaning procedure

[34] R. Spencer and M. Yarmovski, *Proc. Symp. Art of Glass-Blowing* **6**, (Am. Sci. Glassblowers Soc., Wilmington, Del. (1961).

[35] V. O. Altemose, *J. Am. Ceram. Soc.* **49**(8), 446–450 (1966).

[36] B. J. Todd, *J. Soc. Glass Technol.* **40**, 192 (1956).

TABLE VIII. Discoloration of Samples and Bleaching Effect of 315°C Bakeout (Altemose[35])

Glass type	Dosage	Appearance	Degree of bleaching
	Gamma radiation only		
0080	2–5 × 10^7 rads	Quite brown	Completely bleached
0120	2–5 × 10^7 rads	Brownish-black	Completely bleached
7720	2–5 × 10^7 rads	Quite black	Partial, to light brown
	Neutrons plus gamma radiation		
	Neutron dose		
0080	7.4 × 10^{18} nvt	Dark brown to black	Completely bleached
0120	7.4 × 10^{18} nvt	Very black	No visible effect
7720	7.8 × 10^{17} nvt	Very black	No visible effect
1720	3.9 × 10^{17} nvt	Amber color	Completely bleached
7070	3.0 × 10^{17} nvt	Light yellow	Completely bleached
7720	3.2 × 10^{17} nvt	Dark amber to black	To light amber color
1720	1.4 × 10^{18} nvt	Deep amber	Completely bleached
7052		Deep amber	Completely bleached
7740	1.7 × 10^{18} nvt	Very dark amber	Completely bleached

was used in this experiment (other than bakeout), the variations in CO and CO_2 quantities could be due to varying surface conditions.

Generally, helium gas made up 94–99% of the total gas evolved from the glasses irradiated with thermal neutrons. The gases normally found in the γ-irradiated specimens, (namely, CO, CO_2, H_2, and H_2O) were also found in very small quantities. Since these gases resulted from the γ radiation in the neutron pile, they were not included in the results for neutron-irradiated glasses.

Table IX lists data for the specimens exposed to thermal neutrons. The quantity of helium expected is computed by assuming that all the neutrons were absorbed by ^{10}B atoms in the glass. It was also assumed that one helium atom was formed for each neutron absorbed and that equal quantities of helium diffuse to the inside and outside of the bulb jacket (i.e., half the helium is lost to the atmosphere).

Considering that the amount of helium formed is proportional to the neutron dosage and that a reasonable estimate of the uncertainty of the dosage is about 20%, the agreement between the expected and the observed amounts of helium is quite good. Thus, the formation of helium from alpha particles is adequate to account for all the gas formed in the glass bulbs.

Specimens 1, 2, and 3 (Table IX) were loaded with small glass tubing of the same composition to provide greater neutron absorption. This ac-

TABLE IX. Calculated and Observed Helium from Neutron-Irradiated Specimens (Altemose[35])

Specimen No.	Specimen type	Neutron dosage (nvt)	Bulb pressure (Torr)	Total expected He[a]	Total observed He[a]	Initial He (% of total He)
1	7720	7.8×10^{17}	17.0	424.0	421.0	99.9
2	0080	7.4×10^{18}		0.0	8.9	99.9
3	0080	7.4×10^{18}		0.0	8.8	99.9
4	7720	3.2×10^{17}	3.2	89.0	83.5	99.9
5	7070	3.0×10^{17}	3.2	83.5	52.5	99.8
6	7070	3.0×10^{17}	3.2	83.5	50.7	99.8
7	1720	3.9×10^{17}	0.07	55.0	19.1[b]	
8	1720	3.9×10^{17}	0.07	55.0	15.1[b]	
9	1720	1.7×10^{18}	0.31	299.0	361.0	1.9
10	1720	1.4×10^{18}	0.21	246.0	251.0	2.3
11	7740	1.7×10^{18}	12.4	460.0	397.0	98.3
12	7740	1.7×10^{18}	11.2	460.0	347.0	98.3
13	7052	1.4×10^{18}	10.7	413.0	393.0	85.3

[a] Units are cubic centimeters at 1 Torr pressure.
[b] Only part of the helium was released from glass.

counts for the much greater amount of helium expected in specimen 1 than in specimen 4.

Except for the code 1720 glass, almost all of the helium formed had already diffused out of the glass when the analysis was made (see last column, Table IX). A bakeout at 315°C for 3 hr was adequate to get the remaining gas out of the glass. Specimens 7 and 8 were baked at 315°C, while 9 and 10 were baked at 500°C. The total observed helium in each case demonstrates the effect of the helium diffusion constant. Code 1720 glass has a much tighter structure or lower permeation rate than the other glasses (see Chapter 7.4).

For most of the boron-containing glasses, very small oxygen peaks were observed during gas analysis. The presence of trace quantities of oxygen is not too surprising since for each boron atom converted to lithium, two oxygen bonds would be broken. A sufficient number of these conversions could free some oxygen in the glass. Only trace quantities would be observed, however, due to the extremely low permeability for oxygen.

Glass breakage is also of concern where glass is used in vacuum systems. Of the glasses exposed to thermal neutrons, 6 specimens were broken. Five of these (2 code 7740, 2 code 7052, and 1 code 7720) were broken during the second exposure where the dosage was 3 to 4 $\times 10^{17}$ nvt. It is inconclusive whether these glasses broke due to neutron bombardment because two previous code 7720 glass bulbs survived the 1.7×10^{18} nvt exposures with no signs of cracking. In the last exposure, a code 7052 glass specimen exposed to 1.4×10^{18} nvt showed no signs of cracking, whereas one of the same type broke (dosage = 1.7×10^{18} nvt).

7.4. Gas Permeation in Glass[37–39]

7.4.1. Vacuum Limitations Due to Gas Diffusion

In high vacuum devices, the degree of vacuum which can be maintained depends on the gases permeating through, as well as coming from, the walls of the device. In order to properly design a vacuum system, we must then know something of the process of gas permeation through the system walls.

Permeation as used here is the overall steady state flow process from the gas phase on one side of the membrane or wall to the gas phase on the other side. The term diffusion in a solid applies to the internal process by which an atom or molecule travels from one lattice position to another. Permeation of gas through a glass would then involve all the steps such as impact and absorption on the high pressure side, solution of the gas within the glass, diffusion of the atoms or molecules to the outgoing surface, and finally the desorption from the wall on the low pressure side.[40] The rate and mechanism of gas diffusion becomes important, for instance, in understanding and explaining various glass outgassing processes. (See Chapter 7.3.)

With the attainment and measurement of ultrahigh vacuum in glass systems, permeation of gas through the walls from the surrounding atmosphere was recognized as a source of gas influx, limiting the pressure. Alpert and Buritz[41] reported that, in a sealed-off Pyrex® brand (code 7740) glass system with a pressure of about 10^{-10} Torr, they observed a pressure rise of 10^{-11} Torr/min. The ultimate vacuum attainable was attributed to the diffusion of helium into the system from the atmosphere. Atmospheric helium content is only about 4–5 ppm.

Helium permeation in glass becomes troublesome in such special circumstances as the use of vacuum Dewar flasks for storing liquid helium

[37] A complete summary of work up to 1940 is given by R. M. Barrer, "Diffusion in and Through Solids." Macmillan, New York, 1941.

[38] See review of F. J. Norton, *J. Appl. Phys.* **28,** 1, (1957).

[39] Also by E. L. Williams, "The Glass Industry, Vol. 43 (in 4 parts), 1962.

[40] F. J. Norton, *J. Am. Ceram. Soc.* **36,** 3 (1953).

[41] D. Alpert and R. S. Buritz, *J. Appl. Phys.* **25** (2) 202 (1954).

where the vacuum is short lived. Norton[38] points out that for a mixture of 0.5% helium in nitrogen stored in glass, the mixture will be greatly depleted of helium within a day at room temperature due to helium take-up within the glass.

7.4.2. Helium Permeation

The amount of gas permeating through a system of surface area A_s, thickness d, in time t, with a pressure difference Δp across it, is given by

$$q = KA_s\,\Delta pt/d. \tag{7.4.1}$$

The constant K is defined as the permeation rate constant. This equation is derived for plane surfaces only, but is applicable to cylinders and spheres as well if the thickness is small compared to the specimen diameter.

K varies with temperature according to the equation

$$K = A\exp(-Q/RT), \tag{7.4.2}$$

where A is a constant, R is the universal gas constant, T is absolute temperature, and Q is the activity energy in calories per mole of gas.

In general, the permeation constant K is related to the diffusion constant D, by

$$K = SD, \tag{7.4.3}$$

where S is the solubility for a particular gas. The units of K, the permeation constant, used throughout this chapter are cm^3 of gas at STP/sec/cm^2 area/mm thickness/cm Hg pressure difference.

Figure 7(a) and (b) shows the helium permeation rate for seventeen different commercial glasses as a function of reciprocal temperature.[42] Codes 7230 and 7240 glass have values coinciding with code 7900 glass. The curve for code 7160 glass also coincides with that of code 7070.

In the above figures we see that as the glasses get tighter or in the order of decreasing permeation rates, the activation energy or slope of the curves, in general, increases. In other words, the tighter the glass is, its permeation rate becomes more temperature dependent. At low temperatures, the ratio of permeation rates from glass to glass is much greater than at high temperatures.

Table X lists the permeation rates for ten of the more common glasses at room temperature. From code 7900 to code 1723 glass, the ratio of permeation rates is over 2×10^6. One should not be misled from the fact

[42] V. O. Altemose, *J. Appl. Phys.* **32,** 7 (1961).

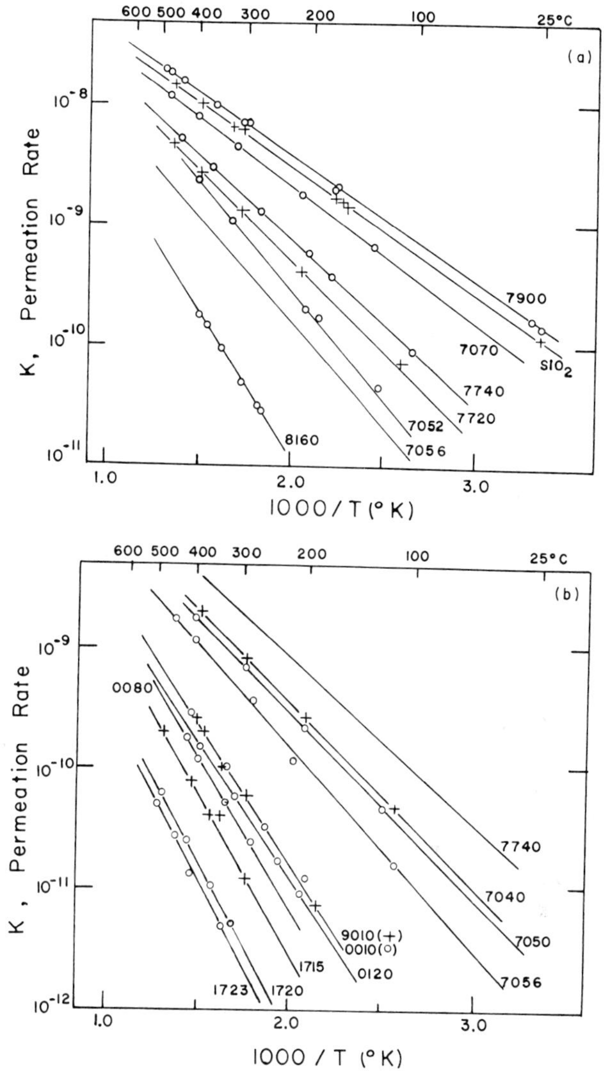

FIG. 7. (a) and (b) Permeation rate K of helium through various glasses. Log K is plotted versus reciprocal temperature, $1000/T$ (°K). Units of K are cm^3(stp)/sec/cm^2 area/mm thickness/cm Hg pressure difference (Altemose[42]).

that for ordinary high vacuum usage, the more permeable code 7740 glass, with its desirable working characteristics, will provide a satisfactory system. However, if one wants to maintain a vacuum of 10^{-7} Torr for a period of time without pumping or if one attempts to work in the aforementioned ultrahigh vacuum range (10^{-10} Torr or less) then it would be

TABLE X. Comparison of Helium Permeation Rates[a] (Altemose[45])

Glass no.	K	Glass no.	K
7900	1.4×10^{-10}	7056	7.8×10^{-13}
SiO_2	1.1×10^{-10}	0010	1.4×10^{-14}
7740	1.2×10^{-11}	0080	4.2×10^{-15}
7720	4.7×10^{-12}	1720	8.4×10^{-17}
7050	1.5×10^{-12}	1723	5.4×10^{-17}

[a] K – cm^3 (stp) – mm/sec-cm^2-cm Hg at 25°C.

imperative to use one of the less permeable glasses such as code 1720 glass.

Extrapolation of the curves in Fig. 7 is valid in the direction of lower temperatures. Determinations on code 7900 glass, for instance, show the curve to be linear to at least −78°C. In the direction of higher temperatures as one approaches the softening point of the glass, extrapolation of the curves is not valid.

Diffusion constants and solubilities for several of these glasses are shown in Table XI.

In general, the permeation rate decreases as the percentage of glass network formers (SiO_2 + B_2O_3) decreases and correlation between K and weight percent of glass formers was demonstrated by Norton.[40] Altemose pointed out that such a correlation is better satisfied by using mole percent of glass former on the grounds that it was the packing density or size of the modifier ions rather than their mass which was important.[42] Figure 8 shows this relationship for the glasses studied.

7.4.3. Permeation of Other Gases

Most of the reliable data available on the diffusion on gases other than helium in glass is for fused silica. Frank, Lee, and Swets[43] have made a very thorough study of the diffusion of helium, hydrogen, and neon through fused silica. Lee[44] has extended this work using hydrogen and discusses the hydrogen–hydroxyl reactions in silica.

Figures 9 and 10 show the permeation curves for neon and hydrogen gases in some of the more permeable glasses.[45] In Fig. 9 the neon permeation rates are lower and more temperature dependent than is the case

[43] D. E. Swets, R. C. Frank, and R. W. Lee, *J. Chem. Phys.* **34** (1), 17 (1961); *J. Chem. Phys.* **35,** 4 (1962); *J. Chem. Phys.* **36,** 1062, (1962).

[44] R. W. Lee, *J. Chem. Phys.* **38,** 448 (1963); *Phys. Chem. Glasses* **5** (2), 35 (1964).

[45] V. O. Altemose, *Symp. Art of Glassblowing* **7,** Am. Sci. Glassblowers Soc., Wilmington, Del. (1962).

TABLE XI. Diffusion Constants and Solubilities (Altemose[42])

Glass code no.	T (°C)	D (cm²/sec)	S (cm³/cm³ atm)
7740	199	5.4×10^{-7}	0.0084
7740	310	3.8×10^{-6}	0.0038
7740	492	1.4×10^{-5}	0.0046
7720	109	6.8×10^{-8}	0.0086
7720	392	2.2×10^{-6}	0.010
7052	388	6.2×10^{-7}	0.027
7040	393	2.7×10^{-6}	0.005
7050	400	2.7×10^{-6}	0.005
7056	115	3.1×10^{-8}	0.0039
7056	402	1.7×10^{-6}	0.0059
9010	393	7.2×10^{-7}	0.0022
8160	369	5.6×10^{-7}	0.0022
8160	342	4.2×10^{-7}	0.0017
0080	394	7.6×10^{-7}	0.0012
1723	414	5.7×10^{-8}	0.0016

with helium gas. The curves are similar to those of helium in that the activation energy increases as the permeation decreases.

In Fig. 10 for hydrogen permeation the permeation rates lie between those for helium and neon, but there is no general tendency for the activation energy to increase with decreasing permeation. The peculiarities of

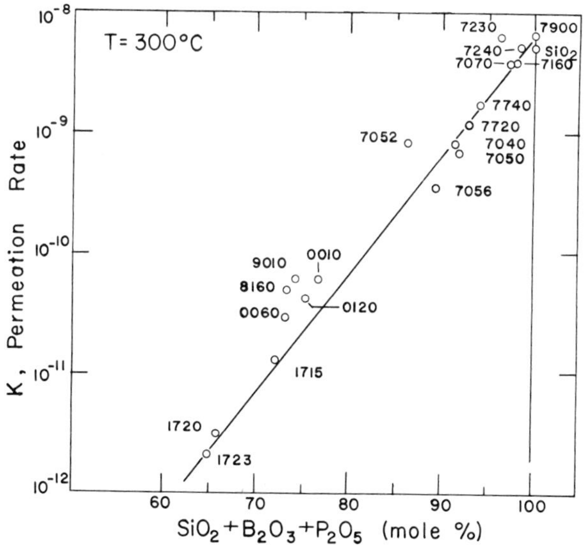

FIG. 8. Log plot of permeation rate K versus mole percent of network formers for helium at 300°C (Altemose[42]).

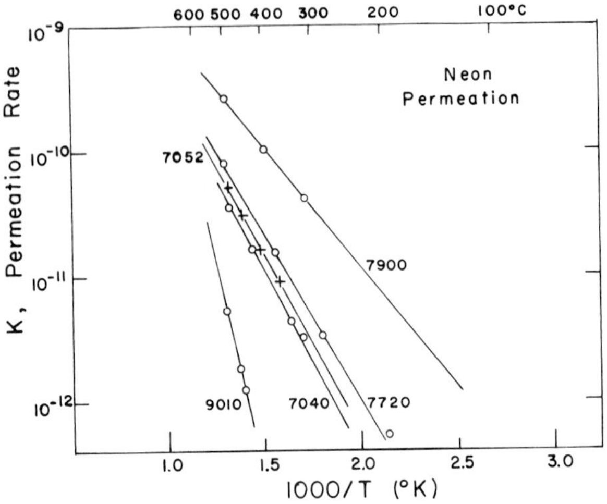

FIG. 9. Permeation rate of neon vs reciprocal temperature (Altemose[43]).

the hydrogen curves are believed due to the fact that hydrogen is a chemically active gas, whereas the other two gases are inert. The diffusion of hydrogen gas through glass is more complicated than for the inert gases. It is not difficult to visualize chemical forces that have to be overcome as a hydrogen molecule travels from one position to another in the glass network.

Figure 11 compares data of Altemose[45] for the permeation of helium,

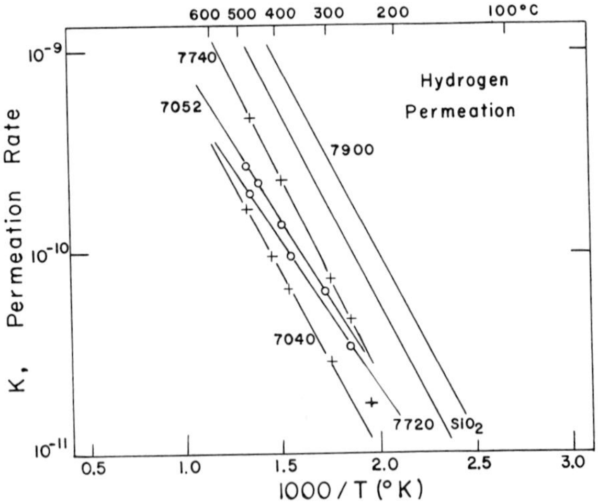

FIG. 10. Permeation rate of hydrogen vs reciprocal temperature (Altemose[43]).

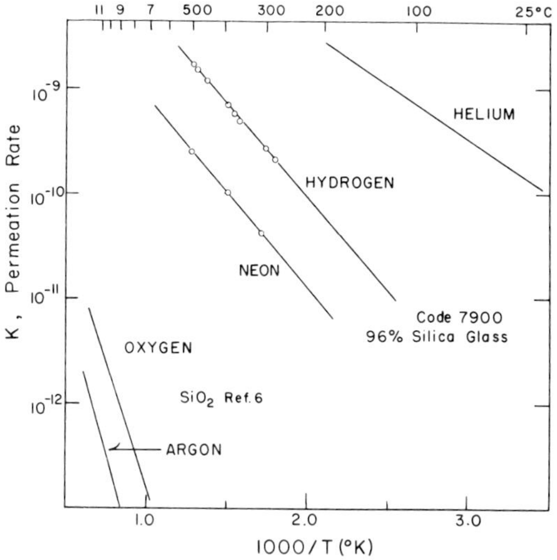

FIG. 11. Permeation of helium, hydrogen, neon through code 7900 glass (Altemose[45]) and oxygen and argon through fused silica (Norton[46]).

hydrogen, and neon through code 7900 glass along with data of Norton[46] for oxygen and argon through fused silica. This comparison gives some feeling for the effect of molecular size of the permeating species. Not only is the permeation rate smaller for the larger molecules, but the activation energy increases as the size of the molecule increases. It is worth noting that the hydrogen molecule and the neon atom have about the same effective size as does the oxygen and argon molecules. For this same open-structured glass, the activation energies for hydrogen and neon, and those for oxygen and argon, are nearly the same, thus correlating molecular size to the activation energy.

[46] F. J. Norton, *Trans. Natl. Vac. Symp.* **8,** 8 (1961).

8. PROPERTIES OF MATERIALS USED IN VACUUM TECHNOLOGY

8.1. Introduction*

8.1.1. Preface

If one is concerned with the design, construction, and use of a vacuum system, one must very early consider the choice of materials to be used. The first consideration is that a vacuum system implies existence of walls which will be subject to atmospheric pressure on one side. The total forces involved in any reasonably sized structure are considerable, and therefore, the materials must have adequate mechanical strength. The second consideration is that to maintain a vacuum environment the walls must be sufficiently vacuum tight and the seals between components must also maintain the integrity of the system. The third concern is that all materials whose surfaces face the vacuum must not overly degrade the vacuum, due to their own vapor pressure or outgassing characteristics. Once these three basic problems are solved over the temperature range of operation, one is then concerned with using the vacuum system with its internal components for the designed research or industrial use. At this stage such things as electrical, optical, magnetic, and thermal properties, must be considered.

In practice, quite a number of materials are available which satisfy the above mentioned constraints and which can be used in vacuum system construction. The specific choices one must make are quite often determined not only by the properties of the materials but also by the sometimes equally important concerns of cost, availability, shop facilities, etc. In the remainder of this section we will consider some of the specific material properties in more detail. In the remainder of this chapter, we will consider the materials themselves and attempt to provide information on their properties which will allow the individual to make reasonable choices depending on the specific detailed problems that have to be solved.

---

* Part 8 is by Y. Shapira and D. Lichtman.

ISBN 0-12-475914-9

8.1.2. General Considerations in Material Selection

8.1.2.1. Vapor Pressure. The main consideration in choosing materials for use in vacuum systems is that they do not hamper the desired degree of vacuum in those systems. This ability is primarily governed by the vapor pressure of the materials to be used. Every material, be it rubber or tungsten, has a vapor pressure, namely, the partial pressure of the atoms or molecules of the substance itself being thermally released from its surface. Although, at room temperature, the vapor pressure of some materials can be extremely low, and sometimes not even perceptible, with increased temperature it eventually rises to measurable values. For some refractory metals, the temperature has to reach more than 1500°C in order to obtain a detectable vapor pressure. An example of a bad choice for high vacuum material is cadmium which has a vapor pressure of 10^{-1} Torr at 300°C. Other materials such as some plastics or rubbers should not be heated above room temperature or not be used at all. The vapor pressure of different materials should be distinguished from the gas pressure related to desorption, outgassing, or permeation of gases associated with them. The outgassing or thermal desorption can sometimes be much higher than the inherent vapor pressure of the material but can be greatly reduced by baking under vacuum, while the vapor pressure of the clean material is intrinsic and persistent.

Materials which show high vapor pressure in the range of the operating temperatures of the system should obviously not be used. This is true also when those materials, such as Cd, Zn, or Mg, are present in an alloy to be used in bakeable high vacuum systems. According to Raoult's law, which applies to dilute solutions, the vapor pressure of the solution is lower than that of the pure solvent by an amount which is proportional to the concentration of the solute. Yet alloys such as brass containing zinc should not be used where ultrahigh vacuum and bakeability are desired because the decreased vapor pressure of the solution is still higher than the acceptable level.

The rest of this subsection includes charts (Figs. 1–3) and a table (Table I) of the vapor pressure of the elements as a function of temperature taken from "Vapor Pressure Data for the Solid and Liquid Elements" by R. E. Honig.[1]

Most of the individual works (about 75) compiled in those charts present experimental results as a function of temperature according to the general equation

$$\log_{10} p = AT^{-1} + B \log_{10} T + CT + DT^2 + E,$$

[1] R. E. Honig, *RCA Rev.* **23,** 567 (1962); R. E. Honig and D. A. Kramer, *RCA Rev.* **30,** 285 (1969).

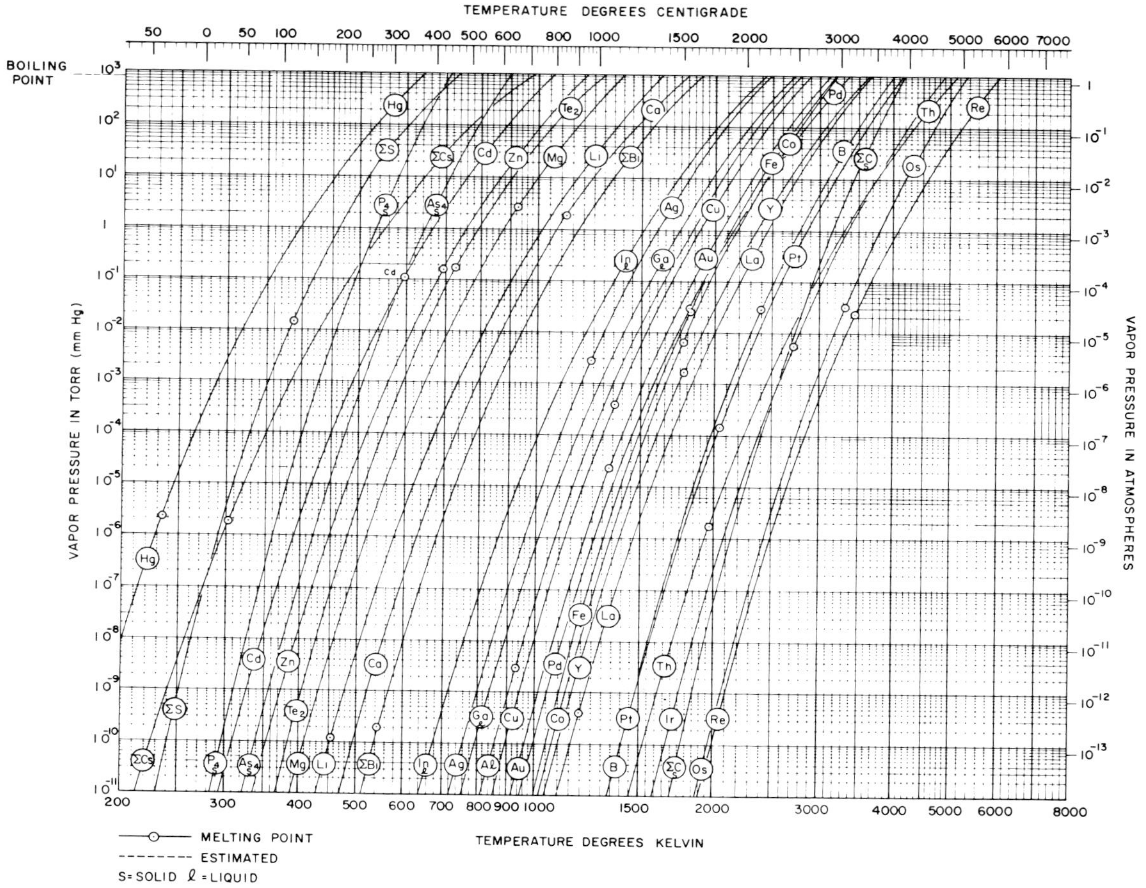

FIG. 1. Vapor pressure curves of the elements (Honig[1]).

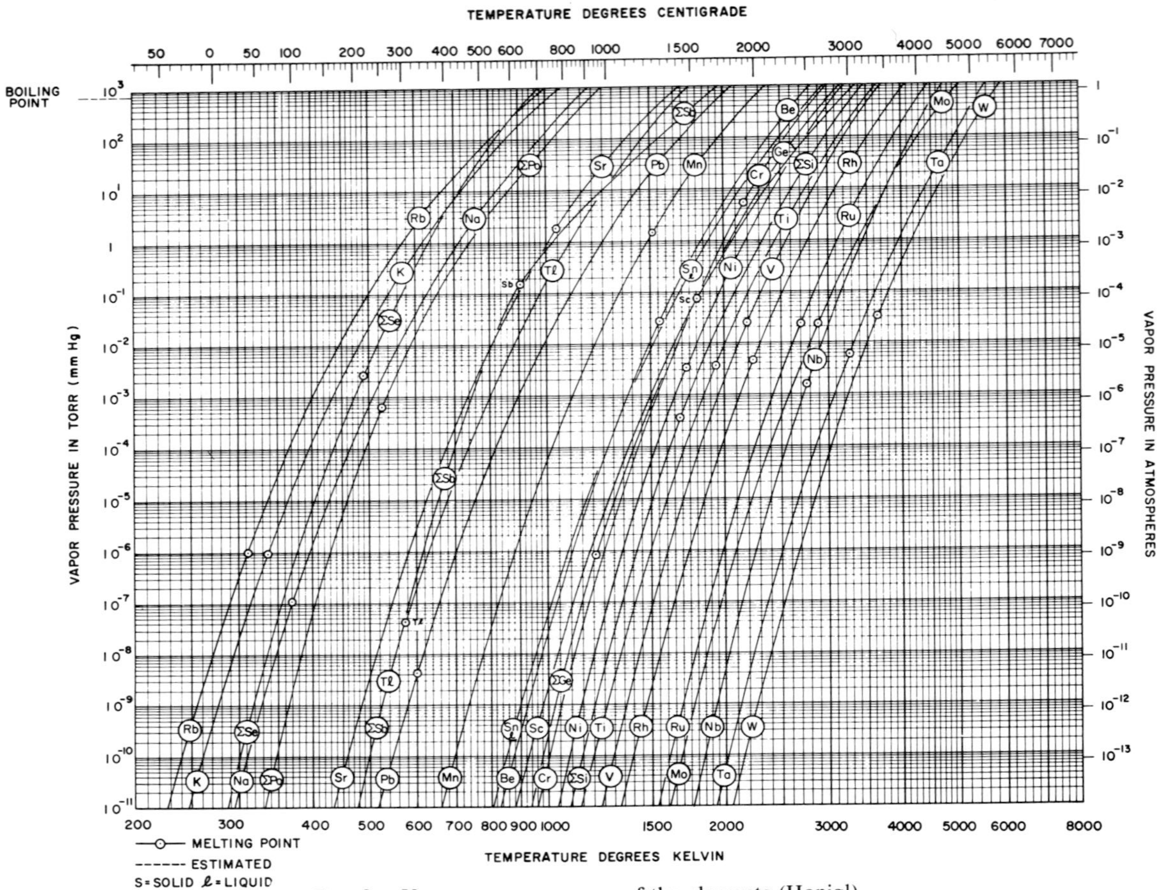

FIG. 2. Vapor pressure curves of the elements (Honig[1]).

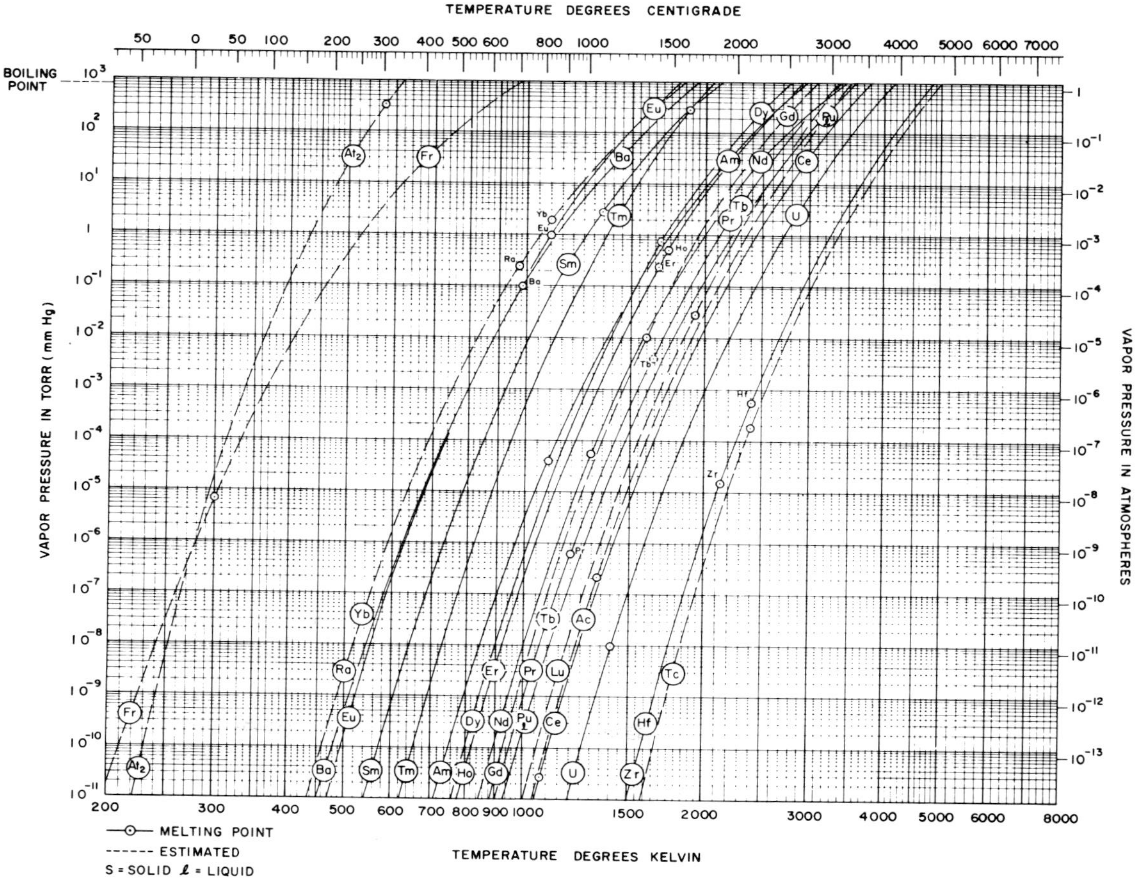

FIG. 3. Vapor pressure curves of the elements (Honig[1]).

TABLE I. Vapor Pressures for the Solid and Liquid Elements (Honig[1])

Symbol[a]	mp (K)	bp (K)	Curve in figure	Data Temp. range (K)	Temperatures (K) for vapor pressures (Torr)														
					10^{-11}	10^{-10}	10^{-9}	10^{-8}	10^{-7}	10^{-6}	10^{-5}	10^{-4}	10^{-3}	10^{-2}	10^{-1}	1	10^{1}	10^{2}	10^{3}
Ac	1320	3470	3	1873, Est.	1045	1100	1160	1230	1305	1390	1490	1605	1740	1905	2100	2350	2660	3030	3510
Ag	1234	2435	1	958–2200	721	759	800	847	899	958	1025	1105	1195	1300	1435	1605	1815	2100	2490
Al	932	2736	1	1220–1468	815	860	906	958	1015	1085	1160	1245	1355	1490	1640	1830	2050	2370	2800
Am	1200	2790	3	1103–1453	712	752	797	848	905	971	1050	1140	1245	1375	1540	1745	2020	2400	2970
As(s)	1090	886	1		323	340	358	377	400	423	447	477	510	550	590	645	712	795	900
At	575	610	3	Est.	221	231	241	252	265	280	296	316	338	364	398	434	480	540	620
Au	1336	3081	1	1073–1847	915	964	1020	1080	1150	1220	1305	1405	1525	1670	1840	2040	2320	2680	3130
B	2360	3850	1	1781–2413	1335	1405	1480	1555	1640	1740	1855	1980	2140	2300	2520	2780	3100	3500	4000
Ba	983	1895	3	1333–1419	450	480	510	545	583	627	675	735	800	883	984	1125	1310	1570	1930
Be	1556	2757	2	1103–1552	832	878	925	980	1035	1105	1180	1270	1370	1500	1650	1830	2080	2390	2810
Bi	544.5	1852	1		510	540	568	602	640	682	732	790	860	945	1050	1170	1350	1570	1900
C(s)	—	4130	1	1820–2700	1695	1765	1845	1930	2030	2140	2260	2410	2560	2730	2930	3170	3450	3780	4190
Ca	1123	1756	1	730 1546	470	495	524	555	590	630	678	732	795	870	962	1075	1250	1475	1800
Cd	594	1040	1	411–1040	293	310	328	347	368	392	419	450	490	538	593	665	762	885	1060
Ce	1077	3740	3	1611–2038	1050	1110	1175	1245	1325	1420	1525	1650	1795	1970	2180	2440	2780	3220	3830
Co	1768	3174	1	1363–1522	1020	1070	1130	1195	1265	1340	1430	1530	1655	1790	1960	2180	2440	2790	3220
Cr	2176	2938	2	1273–1557	960	1010	1055	1110	1175	1250	1335	1430	1540	1670	1825	2010	2240	2550	3000
Cs	301.8	955	1	300–955	213	226	241	257	274	297	322	351	387	428	482	553	643	775	980
Cu	1357	2846	1	1143–1897	855	895	945	995	1060	1125	1210	1300	1405	1530	1690	1890	2140	2460	2920
Dy	1680	2710	3	1258–1773	760	801	847	898	955	1020	1090	1170	1270	1390	1535	1710	1965	2300	2780
Er	1770	2850	3	1773, Est.	779	822	869	922	981	1050	1125	1220	1325	1450	1605	1800	2060	2420	2920
Eu	1099	1764	3	696–900	469	495	523	556	592	634	682	739	805	884	981	1100	1260	1500	1800
Fr	300	950	3	Est.	198	210	225	242	260	280	306	334	368	410	462	528	620	760	980
Fe	1809	3148	1	1356–1889	1000	1050	1105	1165	1230	1305	1400	1500	1615	1750	1920	2130	2390	2740	3200
Ga(l)	302.9	2676	1	1179–1383	755	796	841	892	950	1015	1090	1180	1280	1405	1555	1745	1980	2300	2730
Gd	1585	3000	3	Est.	880	930	980	1035	1100	1170	1250	1350	1465	1600	1760	1955	2220	2580	3100
Ge	1210	3100	2	1510–1885	940	980	1030	1085	1150	1220	1310	1410	1530	1670	1830	2050	2320	2680	3180
Hf	2400	4745	3	2035–2277	1505	1580	1665	1760	1865	1980	2120	2270	2450	2670	2930	3240	3630	4130	4780
Hg	234.29	629.73	1	193–575	170	180	190	201	214	229	246	266	289	319	353	398	458	535	642
Ho	1734	2842	3	923–2023	779	822	869	922	981	1050	1125	1220	1325	1450	1605	1800	2060	2410	2910
In(l)	429.3	2364	1	646–1348	641	677	716	761	912	870	937	1015	1110	1220	1355	1520	1740	2030	2430
Ir	2727	4810	1	1986–2600	1585	1665	1755	1850	1960	2080	2220	2380	2560	2770	3040	3360	3750	4250	4900
K	336.4	1031	2	373–1031	247	260	276	294	315	338	364	396	434	481	540	618	720	858	1070
La	1193	3610	1	1655–2167	1100	1155	1220	1295	1375	1465	1570	1695	1835	2000	2200	2450	2760	3150	3680
Li	453.69	1597	1	735–1353	430	452	480	508	541	579	623	677	740	810	900	1020	1170	1370	1620
Lu	1925	3300	3	Est.	1000	1060	1120	1185	1260	1345	1440	1550	1685	1845	2030	2270	2550	2910	3370
Mg	923	1376	1	626–1376	388	410	432	458	487	519	555	600	650	712	782	878	1000	1170	1400

Mn	1517	2309	2	1523–1823	660	695	734	778	827	884	948	1020	1110	1210	1335	1490	1695	1970	2370
Mo	2890	4924	2	2070–2504	1610	1690	1770	1865	1975	2095	2230	2390	2580	2800	3060	3390	3790	4300	5020
Na	370.98	1156.2	2	496–1156	294	310	328	347	370	396	428	466	508	562	630	714	825	978	1175
Nb[b]	2770	4640	2	2304–2596	1765	1845	1935	2035	2140	2260	2400	2550	2720	2930	3170	3450	3790	4200	4710
Nd	1297	3335	3	1240–1600	846	895	945	1000	1070	1135	1220	1320	1440	1575	1770	2000	2300	2740	3430
Ni	1725	3159	2	1307–1895	1040	1090	1145	1200	1270	1345	1430	1535	1655	1800	1970	2180	2430	2770	3230
Os	3318	5260	1	2300–2800	1875	1965	2060	2170	2290	2430	2580	2760	2960	3190	3460	3800	4200	4710	5340
P(s)	870	704	1		283	297	312	327	342	361	381	402	430	458	493	534	582	642	715
Pb	600.6	2016	2	1200–2028	516	546	580	615	656	702	758	820	898	988	1105	1250	1435	1700	2070
Pd	1823	3310	1	1294–1640	945	995	1050	1115	1185	1265	1355	1465	1590	1735	1920	2150	2450	2840	3380
Po	527	1220	2	711–1286	332	348	365	384	408	432	460	494	537	588	655	743	862	1040	1250
Pr	1208	3295	3	1423–1693	900	950	1005	1070	1140	1220	1315	1420	1550	1700	1890	2120	2420	2820	3370
Pt	2043	4097	1	1697–2042	1335	1405	1480	1565	1655	1765	1885	2020	2180	2370	2590	2860	3190	3610	4170
Pu(l)	913	3508	3	1392–1793	931	983	1040	1105	1180	1265	1365	1480	1615	1780	1975	2230	2550	2980	3590
Ra	973	1800	3	Est.	436	460	488	520	552	590	638	690	755	830	920	1060	1225	1490	1840
Rb	312	974	2		227	240	254	271	289	312	336	367	402	446	500	568	665	802	1000
Re	3453	5960	2	2494–2999	1900	1995	2100	2220	2350	2490	2660	2860	3080	3340	3680	4080	4600	5220	6050
Rh	2239	4000	2	1709–2205	1330	1395	1470	1550	1640	1745	1855	1980	2130	2310	2520	2780	3110	3520	4070
Ru	2700	4392	2	2000–2500	1540	1610	1695	1780	1880	1990	2120	2260	2420	2620	2860	3130	3480	3900	4450
S	388.36	717.75	1		230	240	252	263	276	290	310	328	353	382	420	462	519	606	739
Sb	903	1908	2	693–1110	477	498	526	552	582	618	656	698	748	806	885	1030	1250	1560	1960
Sc	1811	3280	2	1301–1780	881	929	983	1045	1110	1190	1280	1380	1505	1650	1835	2070	2370	2780	3360
Se	490	952	2	550–950	286	301	317	336	356	380	406	437	472	516	570	636	719	826	972
Si	1685	3418	2	1640–2054	1090	1145	1200	1265	1340	1420	1510	1610	1745	1905	2090	2330	2620	2990	3490
Sm	1345	2076	3	789–833	542	573	608	644	688	738	790	853	926	1015	1120	1260	1450	1715	2120
Sn(l)	505	2891	2	1424–1753	805	852	900	955	1020	1080	1170	1270	1380	1520	1685	1885	2140	2500	2960
Sr	1043	1640	2		433	458	483	514	546	582	626	677	738	810	900	1005	1160	1370	1680
Ta	3270	5510	2	2624–2948	1930	2020	2120	2230	2370	2510	2680	2860	3080	3330	3630	3980	4400	4930	5580
Tb	1638	3295	3	Est.	900	950	1005	1070	1140	1220	1315	1420	1550	1700	1890	2120	2420	2820	3370
Tc	2400	4900	3	Est.	1580	1665	1750	1840	1950	2060	2200	2350	2530	2760	3030	3370	3790	4300	5000
Te	723	1267	1	481–1128	366	385	405	428	454	482	515	553	596	647	706	791	905	1065	1300
Th	1968	5020	1	1757–1956	1450	1525	1610	1705	1815	1935	2080	2250	2440	2680	2960	3310	3750	4340	5130
Ti	1940	3575	2	1510–1822	1140	1200	1265	1335	1410	1500	1600	1715	1850	2010	2210	2450	2760	3130	3640
Tl	577	1710	2	519–924	473	499	527	556	592	632	680	736	803	882	979	1100	1255	1460	1750
Tm	1873	2005	3	809–1219	624	655	691	731	776	825	882	953	1030	1120	1235	1370	1540	1760	2060
U	1405.5	4090	3	1630–2071	1190	1255	1325	1405	1495	1600	1720	1855	2010	2200	2430	2720	3080	3540	4180
V	2190	3652	2	1666–1882	1235	1295	1365	1435	1510	1605	1705	1820	1960	2120	2320	2560	2850	3220	3720
W	3650	5800	2	2518–3300	2050	2150	2270	2390	2520	2680	2840	3030	3250	3500	3810	4180	4630	5200	5900
Y	1773	3570	1	1774–2103	1045	1100	1160	1230	1305	1390	1490	1605	1740	1905	2105	2355	2670	3085	3650
Yb	1097	1800	3	Est.	436	460	488	520	552	590	638	690	755	830	920	1060	1225	1490	1840
Zn	692.7	1184	1	422–1089	336	354	374	396	421	450	482	520	565	617	681	760	870	1010	1210
Zr	2128	4747	3	1949–2054	1500	1580	1665	1755	1855	1975	2110	2260	2450	2670	2930	3250	3650	4170	4830

[a] s–solid; l–liquid.
[b] Colombium.

where p is the vapor pressure (Torr), T is the absolute temperature, and A, B, C, D, and E are constants of the element, out of which only A and E are usually different from zero. The corresponding table, including the same data (Table I), indicates also the location of each element in the charts. The melting point is indicated as a small circle on the curve for each element in Figs. 1–3. In the cases where the melting point fell beyond the scope of the chart, the letters l (liquid) or s (solid) were added next to the element symbol.

8.1.2.2. Physical Properties. As has been pointed out, the walls of the system must withstand atmospheric pressure. Therefore, minimum mechanical strength requirements must be met. In this regard, not only the intrinsic mechanical strength of the material but the shape of the structure will have a considerable effect. Thus, cylindrical or spherical shapes are much stronger than flat surface structures. Since many vacuum systems are subject to temperature variations, either heating or cooling or both, the thermal properties of the materials used must be well known. One must be concerned not only with melting temperatures, but also variation of strength with temperature. For example, long before it reaches its melting point, the mechanical strength of copper will decrease such that atmospheric pressure can cause deformation of structural walls made of copper. In addition to the reaction to slowly varying temperatures, materials of the vacuum system may be subject to sudden temperature variations. Thus, the heat–shock resistance characteristic of the materials is also important.

Most vacuum systems are constructed to enable one to do certain experiments or processes. In these experiments or processes one invariably uses electrical circuits. Therefore, many components in the system must have certain desired electrical properties which are, at the same time, consistent with the requirements of the vacuum system. One must remember, for example, that most electrical components are power rated on the assumption that there will be convection cooling. The same component run at the same power level in vacuum and having available, generally, only radiation cooling will reach much higher operating temperatures. This effect can often cause catastrophic conditions for the component.

In many systems one is interested in utilizing beams of charge particles. These beams can often be readily disturbed by unwanted magnetic fields. Thus, for systems which will involve, for example, electron or ion beams, the magnetic properties of the components and walls of the system must be carefully considered. In some cases even very small magnetic fields can be a serious problem and materials considered to be generally nonmagnetic, such as some stainless steels, may still have a small but potentially detrimental magnetic property. When working with various kinds of charged particle beams one must also be concerned with the possibility of charging insulator surfaces.

In addition to the use of charged particle beams, one is often interested in the use of electromagnetic radiation. Sometimes one uses only the visible portion for the purpose of seeing inside the system. Many experiments will also require the use of not only visible but sometimes infrared or ultraviolet radiation. In this case, the optical properties of windows and other components must be known.

This discussion has indicated some of the more typical physical properties of concern when considering materials for a vacuum system. A variety of other properties such as hardness, corrosion resistance, thermal conductivity, and expansion often play an important role. Very specialized experiments may well require an understanding of additional properties.

8.1.2.3. Vacuum Related Properties. In choosing materials which have the desirable physical properties, one must simultaneously consider whether they have suitable vacuum properties. The most important parameter in this area is vapor pressure and that has been discussed in Section 8.1.2.1. However, many materials which have acceptable vapor pressure values are unsuitable because of other characteristics. For example, the material may have a very porous structure which will provide a very large surface area for adsorbed gases. It may also trap considerable quantities of cleaning fluids and/or water. These kinds of materials will cause very lengthy outgassing problems. Flaky or loosely laminated materials also fall into this category.

Materials which have smooth surfaces may exhibit unacceptably high gas permeability. Thus, many glasses permit relatively rapid diffusion of helium, which in some cases may be a problem. Other materials which are satisfactory in all other respects can have high specific gas solubility. Thus many metals including the stainless steels contain considerable dissolved hydrogen. This gas will slowly diffuse into the vacuum system, acting as a relatively constant source of outgassing. A comparative dependence of H_2 solubility of different metals as a function of temperature is given in Fig. 4. The ability to clean a material is an important consideration in choosing components for a system. As pointed out in the chapters on ultrahigh vacuum, it takes very little contamination to very seriously affect the ultimate pressure in a vacuum system.

Materials which appear to have excellent vacuum related properties are often not used because of other considerations. One does not find extensive use of, for example, platinum or platinum alloys, primarily because of cost. Use of 304 stainless steel and high alumina ceramics would be

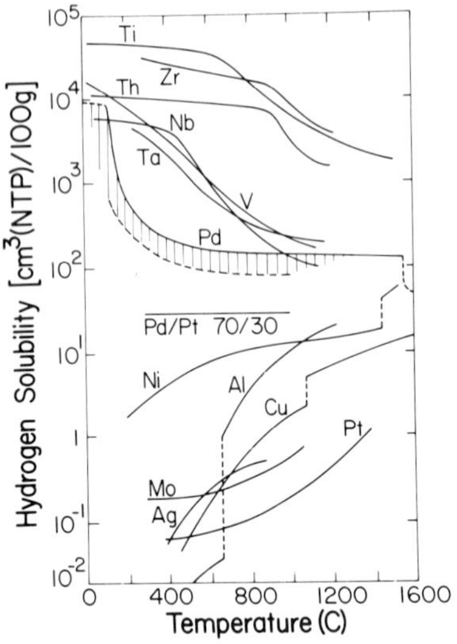

FIG. 4. Hydrogen solubility in various metals as a function of temperature (isobars for 1 atm H_2 pressure outside metal).

even more extensive if they were as machinable as brass. Use of aluminum would be very much greater if it were easier to make vacuum tight bonds to itself and other materials. Most metals oxidize on exposure to air. Some metal oxides are more stable than the base metal (e.g., titanium and aluminum) and these surface oxides are, therefore, quite acceptable in many cases. Other metal oxides develop thick unstable layers and can cause problems for the vacuum system (e.g., copper and iron). Thus one can see that the relatively very large original list of materials is rapidly reduced to a quite finite number when all of the essential properties are considered. The remainder of this part will concentrate on the specific properties of most of the materials which have been found to be acceptable for use in some or all types of vacuum systems.

With time, new materials are continually being produced and one should try to remain aware of the latest developments.

8.2. Metals and Metal Alloys

8.2.1. Steels, Stainless and Others

Stainless Steels. Stainless steels (austenitic CrNiFe alloys) and especially the nonmagnetic type (containing Cr : Ni = 18 : 8) are nowadays the prime constructional material for metal uhv systems and pumps. These alloys are corrosion resistant and strong. The characteristics of the most commonly used stainless steel (type 304) are given in Table II. The other types of stainless steels have very similar properties.

304 stainless steel stands out among the others being very strong and highly corrosion resistant, though harder to machine. It cannot be hardened by heat treatment but is hardened by cold working. It can then be annealed, preferably in a dry H_2 atmosphere, which will produce a shiny surface. The corrosion–resistance of stainless steels is due to an invisible thin chromium oxide film on the surface which is readily formed on a newly machined part. This passivation process can be speeded by immersion of the part in nitric acid solution at 50°C for 30 min followed by a thorough rinse. The existence of this protective film is also the reason why stainless steels should not be cleaned with ordinary steel wool; entrapment of Fe particles in the stainless steel sample can cause corrosion pits and defects.

The austenitic stainless steels can be brazed or welded by heliarc processing. 304 Stainless steel has good machinability and weldability characteristics though it is inferior to 303 stainless steel in that respect. The latter, however, contains sulfur, which is an undesirable component in bakeable high vacuum systems due to its vapor pressure. The temperature dependence of the thermal conductivity of 304 stainless steel as compared with other metals is given in Fig. 5. 304 Stainless steel shows superior corrosion resistance, very low vapor pressure and thermal conductivity, and it is nonmagnetic. These properties make it the most com-

TABLE II. Stainless Steel 304

Property	Temperature (°C) or form	Value	Units
Density		7.9	gm cm^{-3}
Melting point		1427	°C
Specific heat		0.12	cal gm^{-1} deg^{-1}
Thermal conductivity	100	0.039	cal cm^{-1} sec^{-1} deg^{-1}
	500	0.051	
Coefficient of linear thermal expansion	0–93	1.59×10^{-4}	deg^{-1}
	0–316	1.720×10^{-4}	
	0–538	1.800×10^{-4}	
Tensile strength	Annealed	65	kg mm^{-2}
	Cold rolled	240	
Young's modulus		19,000	kg mm^{-2}
Elastic limit	Annealed	26.5	kg mm^{-2}
Electrical resistivity	25	72.0	$\mu\Omega$ cm
	700	110	

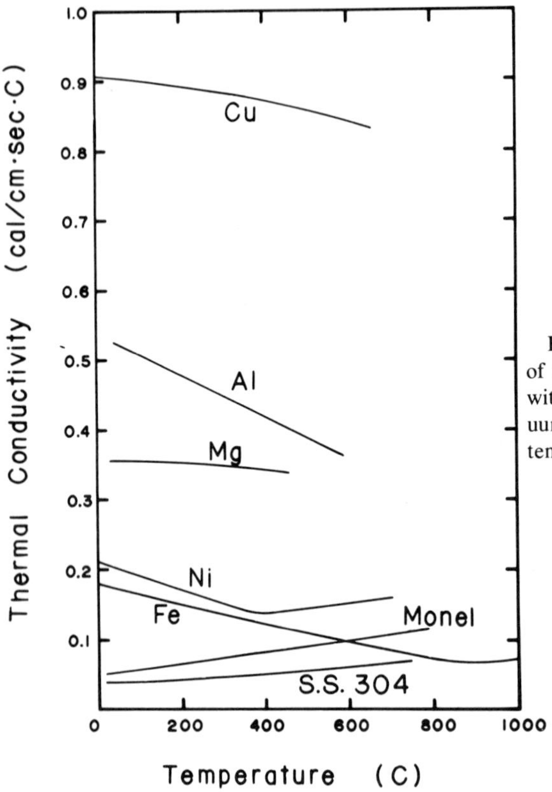

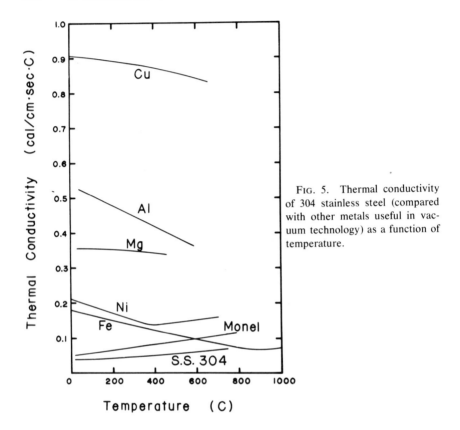

FIG. 5. Thermal conductivity of 304 stainless steel (compared with other metals useful in vacuum technology) as a function of temperature.

mon material for constructing demountable uhv chambers, flanges, ion pumps, nuts, and bolts. Stainless steel nuts and bolts should be kept separately so they are not confused with Cd-plated steel parts.

It should be noted that stainless steel is not absolutely nonmagnetic but does have a very little magnetic permeability which, as a matter of fact, can be increased by cold working.

Steels. The regular carbon or iron carbide steels have two main disadvantages for vacuum work. They are not corrosion resistant and therefore have to be kept in dry clean areas after cleaning. This fact makes their use in demountable systems highly inconvenient. They are also ferromagnetic and therefore cannot be used where magnetic effects would be a problem such as in the construction of ion pumps, magnetic mass spectrometers, or any system containing magnetic analyzers.

Iron and steels are not attacked by mercury. Cold-rolled or low-carbon steels have good to excellent machinability. They cannot be hardened by heat treatment or very little by cold working but can be surface hardened by carbonizing (case hardening). Tool steels (high-C) or high speed steels, containing higher amounts of carbon, can be heat treated to a variety of hardnesses and temper but they are harder to machine and their magnetic properties, mainly their residual magnetism, are more pronounced. They are no longer used, however, for permanent magnets due to the introduction of the better suited alnico alloys and ceramic magnets. Their main use in this area is for tools.

Other types of stainless steels, such as martensitic and ferritic, are similar to the austenitic in regard to basic properties and handling although some difference should be noted, e.g., magnetic properties, heat treatment, hardening (some of the martensitic types), Nb and Ta content that form hydrides in H_2 atmosphere and cause embrittlement (type 347, 348), etc.

8.2.2. Common Metals

Base metals such as Ni, Cu, and Al have relatively low melting points compared to the refractory metals and therefore are usually melted and formed in air. This process leads to a large degree of impurity gases trapped in the metal and to the need for thorough outgassing. They also have to be used with careful consideration of their other physical and chemical properties but their relatively low cost makes them attractive for mass production.

Nickel. Nickel is a widely used metal in vacuum technology. It is found in many vacuum applications such as base material for coated cathodes, grids, anodes, getter and heat shields, and many other mechanical structures. It is used either by itself as the base material or electroplated on other materials or as a constituent in one of the many Ni alloys. Nickel has a combination of practical properties which make it very useful. It has a high melting point relative to other common metals, it has low vapor pressure, it can be easily formed, outgassed and spot welded, and has relatively low cost.

From the application point of view one should first note the excellent resistance of Ni to a variety of corrosives. This characteristic makes it very attractive in many structural applications in corrosive atmospheres. In other cases, thin film deposition of Ni or Ni plating and coating give other materials the desirable surface resistance to corrosion. Another very notable property of Ni is its ferromagnetism. The magnetization of Ni is fairly high (see Fig. 6) but its Curie temperature is only around 350°C. The Curie temperature of Ni can be increased by adding cobalt. Although the mechanical properties of Ni are highly desirable, its magnetism restricts its use when magnetic effects should be avoided.

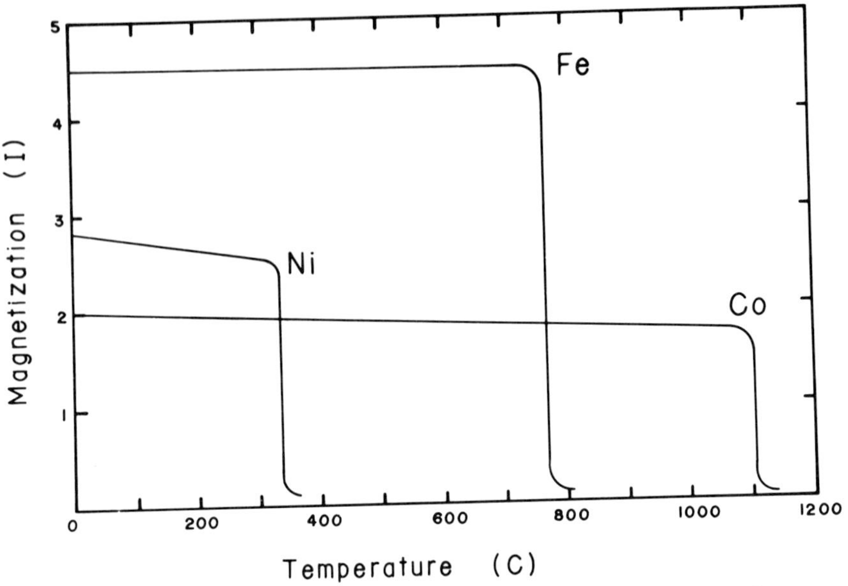

FIG. 6. Relative intensity of magnetization *I* of Fe, Ni, and Co as a function of temperature.

The commercially pure Ni (nickel 200, which contains 99.5% Ni with less than 0.25% Mn and 0.15% Fe) shows the properties mentioned above and some of its physical properties are summarized in Table III. Nickel 200 is very ductile after annealing at 600°C or above in H_2. Its tensile strength is very high. Young's modulus of Ni 200 and 98% Ni alloy as compared with other metals are shown in Fig. 7 as a function of temperature. It shows magnetic and magnetostrictive properties. It has relatively high temperature coefficient of resistivity making it a useful resistance thermometer at moderate temperatures.

Nickel is resistant to corrosion not only by atmosphere but by water, salt water, alkalies, and most organic acids as well. It is however attacked rapidly by hypochlorites, nitric acid, wet gases such as chlorine (at $T > 580°C$), bromine, SO_2, and a mixture of $N_2 + H_2 + NH_3$.

Ni shows high permeability and solubility of hydrogen which can form a solid solution in Ni. Hydrogen, oxygen, CO, and CO_2 can be made to diffuse through Ni, but it is impervious to the noble gases. By heating to 400–500°C, most of the hydrogen can be expelled without an appreciable change in hardness, but at $T > 600°C$ considerable amounts of CO evolve, resulting in embrittlement of the nickel.

Iron. Iron by itself can be used in vacuum systems only under strict limitations due to its poor corrosion resistance. It can be used where the

TABLE III. Nickel 200

Property	Temperature (°C) or form	Value	Units
Atomic number		28	
Atomic weight		58.69	
Density	25	8.9	gm cm^{-3}
Melting point		1453	°C
Specific heat	100	0.112	cal gm^{-1} deg^{-1}
	300	0.137	
	600	0.133	
Thermal conductivity	100	0.198	cal cm^{-1} sec^{-1} deg^{-1}
	200	0.175	
	300	0.152	
Coefficient of linear thermal expansion	25–100	1.33×10^{-5}	deg^{-1}
	25–300	1.44×10^{-5}	
	25–600	1.55×10^{-5}	
Brinell hardness	20		
	Hard rolled (70%)	220	kg mm^{-2}
	(25%)	180	
	(12%)	150	
Tensile strength	Hard drawn	70–100	kg mm^{-2}
	Annealed	40–55	
Young's modulus	20	18,000–22,700	kg mm^{-2}
Rigidity modulus	27	7300	kg mm^{-2}
Electrical resistivity	20°C hard rolled	9.5	$\mu\Omega$ cm
	20°C annealed	8.7	
Temperature coefficient of electrical resistivity	Hard rolled	4.6×10^{-3}	deg^{-1}
	Annealed	4.7×10^{-3}	
Electron work function		4.5–5.24	eV
Curie temperature		350–360	°C
Annealing temperature	$\frac{1}{2}$ hr	750–900	°C
H_2 solubility	200	3	cm^3[NTP]/100 gm
	600	5.5	

degree of vacuum or gas purity is not very important. Otherwise, a system containing Fe parts has to be continuously pumped to get rid of gaseous impurities. In other applications it has to be extremely pure or in special alloys.

Some of the physical properties of iron are given in Table IV. Iron is often used where its magnetic properties together with its hardness can be utilized. Its magnetic properties can, however, deteriorate considerably due to working. A moderate temperature (850°C) annealing is necessary to restore optimum permeability at all inductions.

Iron is sometimes used also because it does not react appreciably with Hg and costs less than nickel, for example. However another serious problem in its use is its H_2 permeability at $T > 200$°C. Atomic H can dif-

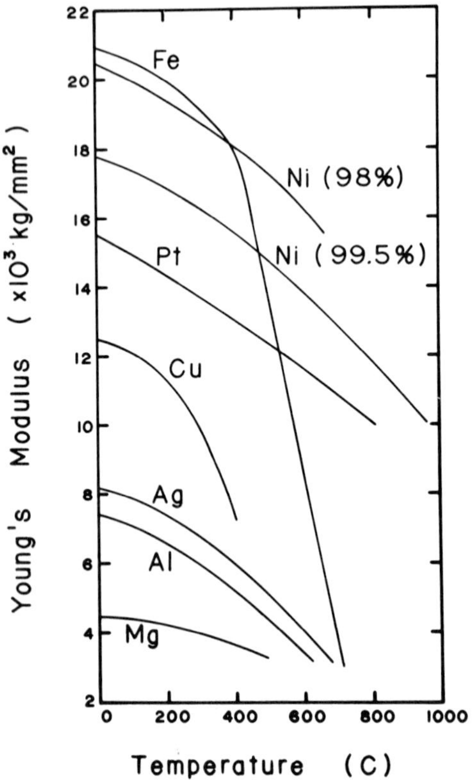

FIG. 7. Young's modulus of two kinds of Ni of different purities compared with other metals, as a function of temperature.

fuse through it even at room temperature. In general, iron is noted for strength, ferromagnetism, and inertness to mercury. However, its sensitivity to corrosion limits its use to a material for internal vacuum parts in special applications or in alloys.

Copper. The most common form of copper used in high vacuum technology is certified OFHC (oxygen-free high conductivity) copper. Table V lists some of its physical properties. OFHC copper contains at least 99.98 Cu (+ Ag) and no oxygen. It offers ductility, nonporosity, and extremely high electrical and thermal conductivities. The temperature dependence of the resistivity of Cu as compared to other metals is given in Fig. 8. Due to its high plasticity, it is widely used for gasket seals in demountable metal systems. The use of OFHC copper in high vacuum applications is necessary due to the difficulty in outgassing regular copper. The oxygen dissolved and occluded in normal copper cannot be released below the softening point of copper.

The plasticity of copper, which has essentially no elastic limit, makes it

TABLE IV. Iron

Property	Temperature (°C) or form	Value	Units
Atomic number		26	
Atomic weight		55.85	
Density	20	7.86	gm cm^{-3}
Melting point		1532	°C
Specific heat	20	0.107	cal gm^{-1} deg^{-1}
Thermal conductivity		0.175	cal cm^{-1} sec^{-1} deg^{-1}
Coefficient of linear thermal expansion	20–400	13.7×10^{-6}	deg^{-1}
Brinell hardness	Annealed	45–90	kg mm^{-2}
	Cold rolled	120	
Tensile strength	Annealed	18–28	kg mm^{-2}
	Cold rolled	40–62	
Young's modulus	Annealed	21,770	kg mm^{-2}
	Cold rolled	20,900	
Rigidity modulus	Annealed	8400	kg mm^{-2}
Electrical resistivity	20	0.096	Ω mm^2 m^{-1}
	800	106	
Temperature coefficient of electrical resistivity	0–100	5.6×10^{-3}	deg^{-1}
Electron work function		4.04–4.76	eV
Magnetic saturation		21,550	G
Annealing temperature		600–900	°C
H_2 solubility	530	1	cm^3 (NTP)/100 gm

useful for pumping stems and tubing for manifolds and coolers as well as for gaskets. For those applications, and for a variety of heat and current conduction applications in vacuum, only OFHC copper should be used. Even a higher grade phosphor-free OFHC copper is useful for sealing to glass. OFHC copper offers vacuum tightness, low gas solubility, and insensitivity to H_2 and water vapor. OFHC copper is also impervious to hydrogen and helium at room temperature. It is attacked by oxygen and tarnish builds up at temperatures above 200°C. Copper is attacked by acids only in the presence of O_2 as in strongly oxidizing acids. Mercury and mercury vapor also have a strong effect on copper, which therefore should generally not be used in Hg tubes or with Hg diffusion pumps.

Annealed copper is quite strong and it is machinable by conventional methods but due to its softness, it is not easy to machine to close tolerances. Its tensile strength falls sharply above 200°C, putting a limitation on high temperature structural applications. Generally, the use of copper is not recommended above 500°C because it has a relatively high vapor pressure, roughly an order of magnitude above that of Ni.

TABLE V. Copper

Property	Temperature (°C) or form	Value	Units
Atomic number		29	
Atomic weight		63.54	
Density		8.3–8.96	gm cm^{-3}
Melting point		1083 ± 0.1	°C
Specific heat	20	0.092	cal gm^{-1} deg^{-1}
Thermal conductivity	20	0.941	cal m^{-1} sec^{-1} deg^{-1}
	100	0.90	
	700	0.84	
Coefficient of linear thermal expansion	0–100	165 × 10^{-7}	deg^{-1}
Brinell hardness	Cast	36	kg mm^{-2}
	Annealed	45–50	
Tensile strength	Cast	16–20	kg mm^{-2}
	Annealed	20–25	
Young's modulus	Annealed	11,700–12,600	kg mm^{-2}
Rigidity modulus	Annealed	3900–4800	kg mm^{-2}
Electrical resistivity	20	0.017–0.018	Ω mm^2 m^{-1}
Temperature coefficient of electrical resistivity	20	6.8 × 10^{-3}	deg^{-1}
Electron work function		4.46	eV
Annealing temperature		450 × 600	°C
H_2 solubility	450	6 × 10^{-2}	cm^3 (NTP)/100 gm
	100	2.5	

Copper can be soldered, brazed (with Ag–Cu eutectic alloy, Au–Cu alloy, Au–Ni eutectic, and others), welded by inert gas-shielded heliarc, or joined by silver plating the contact surfaces. It should be noted that gravity and capillary action are to be considered in making joints where the eutectic melts and flows. Copper is very difficult to join by arc or resistance welding. OFHC copper, which is the type to use in vacuum application, is still a relatively cheap soft metal with very high heat and electric conductances and therefore very useful for the applications described above.

Aluminum. Aluminum is a light, fairly ductile metal, especially in its pure form, with good electrical conductivity. Some of its physical characteristics are given in Table VI. Aluminum is hardly work-hardenable. Due to its low tensile strength it can be easily rolled and bent. Its strength falls rapidly around 200°C restricting its use in bakeable systems. A much harder form of Al, whose strength is not so dependent on temperature, is sintered aluminum powder (SAP).

Aluminum corrodes very little in oxygen or even in moist air due to a

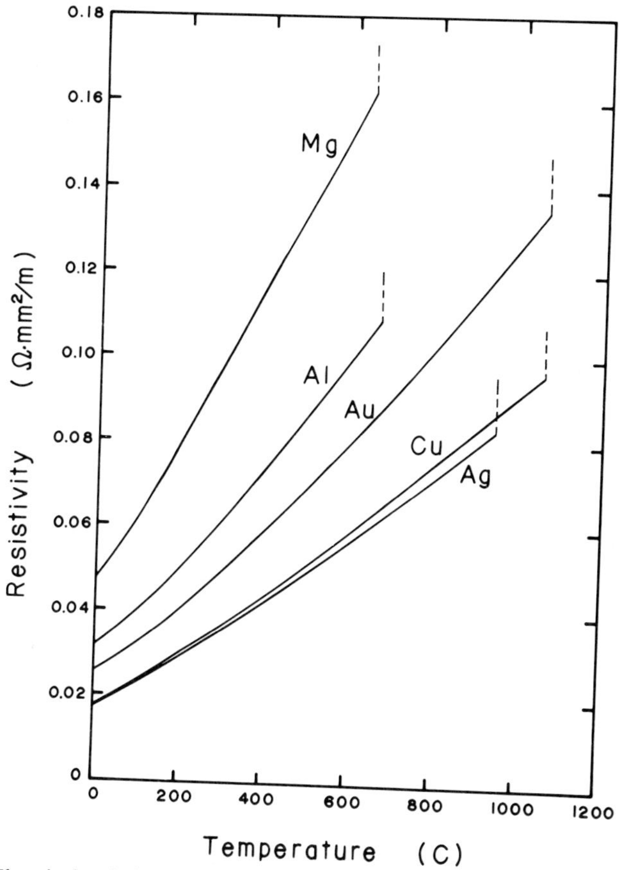

FIG. 8. Electrical resistivity of Cu compared with other good conductors as a function of temperature.

protective thin oxide layer on its surface. It is fairly resistant to HCl and HNO_3 but dissolves in HF, concentrated H_2SO_4, and alkali bases. Aluminum is also attacked by CO_2.

Due to the softness of aluminum it is not easy to machine to close tolerances. On the other hand, it can, therefore, be used as gasket material. Sometimes, Al or Al alloys as No. 1100 are used also for structural purposes, but in spite of its advantages, Al is not widely used in vacuum technology for several reasons. It is a soft material with a low melting point (660°C). Even though it can be used for gaskets, its mechanical strength decreases by about an order of magnitude at temperatures around 200°C. Aluminum also has a relatively high vapor pressure which limits its use for bakeable systems to temperatures under 300°C. However, in that

TABLE VI. Aluminum

Property	Temperature (°C) or form	Value	Units
Atomic number		13	
Atomic weight		26.97	
Density		2.70	gm cm^{-3}
Melting point		646–657	°C
Specific heat	20	0.214	cal gm^{-1} deg^{-1}
	100	0.225	
Thermal conductivity	20	0.52	cal cm^{-1} sec^{-1} deg^{-1}
	200	0.475	
Coefficient of linear thermal expansion	20–100	24.0×10^{-6}	deg^{-1}
	20–300	26.7×10^{-6}	
Brinell hardness	Soft	15–25	kg mm^{-2}
	Hard rolled	35–70	
Tensile strength		7–11	kg mm^{-2}
Young's modulus		5800–7000	kg mm^{-2}
Rigidity modulus	Soft	2760	kg mm^{-2}
	Hard rolled	2750	
Electrical resistivity	20	2.8	$\mu\Omega$ cm
	300	6.0	
Temperature coefficient of resistivity	20	4.08×10^{-3}	deg^{-1}
	250	4.25×10^{-3}	
Electron work function		4.08	eV
Magnetic susceptibility (paramagnetic)		0.65×10^{-6}	cgs
Annealing temperature		200–450	°C
H_2 solubility	580	2×10^{-2}	cm^3 (NTP)/100 gm

temperature range, it has a very low H_2 solubility. The use of aluminum is often limited to gasket material because it is also very difficult to weld or braze. Welding of Al requires special conditions. Also most soft solders, as most Al alloys besides No. 1100, contain volatile materials as lead and zinc and should not be used where bakeout is necessary.

Besides gasket material, aluminum is also used for X-ray windows, anticorrosion coatings, etc. It also exhibits an extremely high resistance to cathodic sputtering by positive ions. Although its use as a constructional material is not popular in bakeable uhv systems due to its relatively high vapor pressure and low strength, it is however advantageous to use it for large systems which are not to be baked because it is much lighter, cheaper, and more machinable than 304 stainless steel. It also shows good corrosion resistance, high thermal conductivity, and low hydrogen solubility.

Rare Metals. ZIRCONIUM. Pure clean Zr is a highly reactive metal which is used for gettering, especially of H_2, N_2, and O_2. Table VII con-

TABLE VII. Zirconium

Property	Temperature (°C) or form	Value	Units
Atomic number		40	
Atomic weight		91.22	
Density	20	6.52	gm cm^{-3}
Melting point		1857	°C
Specific heat	25	0.07	cal gm^{-1} deg^{-1}
Thermal conductivity	125	0.035 ± 5%	cal cm^{-1} sec^{-1} deg^{-1}
Coefficient of linear	20–200	5.4×10^{-6}	deg^{-1}
thermal expansion	20–400	6.9×10^{-6}	
Brinell hardness	Annealed	67	kg mm^{-2}
	Hard	150	
Tensile strength	Annealed	65	kg mm^{-2}
Young's modulus	Annealed	8000	kg mm^{-2}
Electrical resistivity	20	40	$\mu\Omega$ cm
Temperature coefficient of resistivity	0–100	0.0044	deg^{-1}
Electron work function		4.1–4.2	eV
Annealing temperatures	30 min	525	°C
H_2 solubility	375	2.4×10^{-2}	cm^3 (NTP)/100 gm

tains some of the physical properties of Zr. It has few other advantageous properties for vacuum work. It does resist corrosion very well due to a thin oxide layer on its surface. At moderate temperatures it will absorb gases as mentioned, which will not be released even at very high temperatures. Zirconium has a low neutron cross section so it can be used as a window for neutrons. It has a low secondary electron emission yield and can be used as coating on other substrates when this property would be useful.

Zirconium can be machined like brass and spot welded to Mo or W. It cannot be brazed with silver. Its strength can be enhanced by alloying with small amounts of Mo; otherwise, when hot it will not support itself. Zirconium is resistant to HCl, HNO_3, diluted H_2SO_4 and H_3PO_4, and alkali bases. It is attacked by hot concentrated H_2SO_4 and aqua regia. Although Zr exhibits a variety of useful properties for vacuum work, it is used only in small quantities, e.g., for gettering, because of its scarcity and high cost.

Zirconium hydride, ZrH_2, is also used for gettering. It has the advantage of not being oxidized during seal in, and when heated in vacuum, will dissociate, leaving a clean reactive Zr surface. ZrH_2 is also used for ceramic–metal seals. It is brushed in suspension on the ceramic seal area, fired, and thereby acts as a reducing agent for less active metals (see Part 6 for more details).

TITANIUM. Titanium, which is both strong and lightweight, is a very useful metal in vacuum applications. Some of the physical characteristics of Ti are summarized in Table VIII. Titanium can be machined and formed like steel and it is nonmagnetic, therefore, it can be quite useful for structural purposes. Titanium could be an ideal constructional material but its use is generally limited to gettering and ion pump cathode use because it is a relatively rare and expensive metal such as Zr.

The property making it highly attractive for vacuum usage is its relatively high intake of the active gases such as O_2, N_2, H_2, CO, CO_2, and H_2O vapor above 650°C. This gettering action is what makes Ti widely used in uhv application such as in getter–ion pumps, where evaporating fresh Ti on the pump walls produces a highly adsorbing surface and a considerable enhancement of the pumping speed (see Part 5 for more details).

Like Al, Zr, and stainless steel, Ti has a protective oxide layer on its surface, making it corrosion resistant.

A variety of machining processes can be done on Ti with high speed steel tools. It can be welded in different ways but, as with all refractory metals, this should be done in a protective atmosphere because the adsorption of the different gases causes embrittlement and, in thin films, warping and deformation.

Ti can be incorporated into several alloys with a few percent of other metals such as Ti–Al–V, Ti–Cr–Al, and Ti–Al–Mg to enhance its hardness.

Ti should not be heated in a H_2 atmosphere due to the rapid formation

TABLE VIII. Titanium

Property	Temperature (°C) or form	Value	Units
Atomic number		22	
Atomic weight		47.90	
Density		4.51–4.54	gm cm^{-3}
Melting point		1725	°C
Specific heat	0–100	0.127	cal gm^{-1} deg^{-1}
Thermal conductivity	20	55–61	cal gm^{-1} cm^{-1} deg^{-1}
Coefficient of linear thermal expansion		8.5×10^{-6}	deg^{-1}
Brinell hardness	Soft annealed	185	kg mm^{-2}
	Cold rolled	260	
Tensile strength	Annealed	52–73	kg mm^{-2}
Young's modulus	Soft annealed	11,700	kg mm^{-2}
Electron work function		3.0	eV
Electrical resistivity	20	50	μΩ cm
Annealing temperature	1 hr	600–800	°C

of TiH. TiH itself is the active ingredient in the "active metal" method for ceramic to metal seals (see Part 6).

CADMIUM. Cadmium is common in photocell usage, and as rustproof plating. However, due to its very high vapor pressure (10^{-5} Torr at 150°C, 10^{-1} Torr at 300°C) its vacuum use should be avoided, especially in bakeable systems. Since Cd is commonly used for anticorrosion coating of nuts, bolts, and other parts, those parts should be kept separate where confusion with stainless steel parts can be prevented.

8.2.3. Refractory Metals and Alloys

Tungsten. Tungsten is the most commonly used of all the refractory metals. It has the highest melting point and correspondingly the lowest vapor pressure among them. Table IX contains some of the physical properties of W. Some of these properties are given as a function of temperature in Table IX.

Tungsten, as all the refractory metals, is made from compacted sintered powder, formed, and annealed. Tungsten is a very hard and stable ele-

TABLE IX. Tungsten

Property	Temperature (°C) or form	Value	Units
Atomic number		74	
Atomic weight		183.92	
Density	Swaged	17.6–19.2	gm cm^{-3}
	Drawn	19.2–19.4	
Melting point		3400	°C
Maximum operating temperature in vacuo		2560	°C
Specific heat	18	0.0340	cal gm^{-1} deg^{-1}
	1000	0.0365	
Thermal conductivity	0	0.399	cal gm^{-1} cm^{-1} deg^{-1}
Coefficient of thermal linear expansion	20–590	4.6×10^{-6}	deg^{-1}
Brinell hardness	Swaged	350–400	kg mm^{-2}
	Annealed	125–250	
Tensile strength	Annealed	110	kg mm^{-2}
Young's modulus	Hand drawn wire		
	0.03 mm	40,000	kg mm^{-2}
	0.3 mm	9,200	
Electrical resistivity	0	5.0	$\mu\Omega$ cm
	20	5.49	
Temperature coefficient of electrical resistivity	20–100	$(4.8 \pm .05) \times 10^{-3}$	deg^{-1}
Electron work function		4.25	eV

ment but somewhat brittle. Its mechanical strength can deteriorate at high temperatures due to recrystallization, and W wires, which are used mainly for filaments, may weaken. For such applications, delayed recrystallization is obtained by drawing polycrystalline wires, to which low vapor pressure oxides, such as ThO_2 (0.7–15%) are added. As shown in Fig. 9, thoriated tungsten also has the advantage of a greater electron emissivity at lower temperature. It should be noted that thoriated tungsten has to be heated more carefully than W and it is not self-supporting at temperatures above 2500°C. Nonsagging W filaments are made by adding alkali compounds to WO_3 powder before the reduction process. This treatment results in long interlocking grains, which add to the tensile strength of the W filaments.

Tungsten is nonmagnetic and very stable chemically. However, it oxidizes rapidly to WO_3 at high temperatures in atmospheres containing O_2 or other oxidizing gases. Tungsten is only slightly affected by most acids and bases and not at all by H_2, water, or diluted acids. It can be etched only by hot aqua regia or an equal parts mixture of HF and HNO_3. The hardness and refractoriness of W make it hard to machine and form. It is, however, readily weldable if in the form of sheet or filament.

Tungsten can be found in applications such as X-ray tube anodes, spring elements, high temperature thermocouples, boats for furnaces, and welding electrodes. They all utilize its stability and high melting point. The most common applications in vacuum are by far as a filament material for lamps, heaters, and electron emitters. Even though W has a relatively low electron emissivity (relatively high work function), this is compensated by its simplicity of operation (needs only outgassing), its resistance to gases and contaminants, to mechanical and thermal shocks and

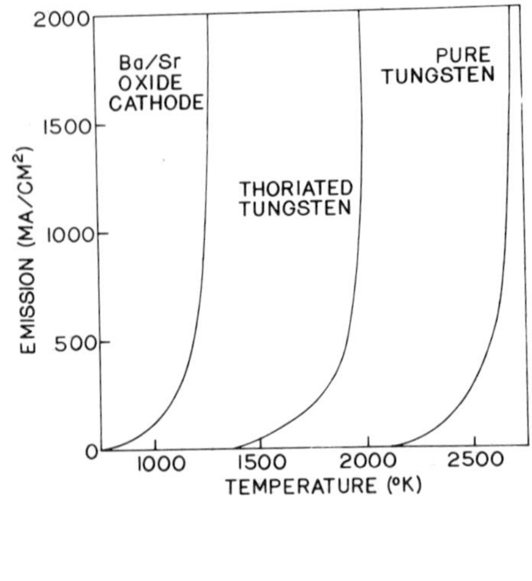

FIG. 9. Relative emission of three types of thermionic cathodes as a function of temperature.

to ion bombardment, and by its extremely high melting point and low vapor pressure.

Molybdenum. Molybdenum, whose properties are summarized in Table X, is a hard and chemically resistant refractory metal. It shows oxidation only at elevated temperatures but rapid production of MoO_3 occurs at $T > 600°C$. It is attacked only by hot dilute HCL and $HF:HNO_3$ 1:1 mixture. At room temperature there is no effect of H_2 or dry O_2.

Molybdenum, although somewhat brittle, can be worked better than W but has a tendency to chip. Its tensile strength can be enhanced by heat treatment or with alloy additives (3% Nb). Mo can be brazed to Mo or W with suitable solders and even spot welded under very clean conditions. Molybdenum is nonmagnetic.

TABLE X. Molybdenum

Property	Temperature (°C) or form	Value	Units
Atomic number		42	
Atomic weight		95.95	
Density		9–10	gm cm^{-3}
Melting point		2610	°C
Specific heat	20	0.062	cal gm^{-1} deg^{-1}
Thermal conductivity	20	0.35	cal cm^{-1} sec^{-1} deg^{-1}
Coefficient of linear thermal expansion		4.9×10^{-6}	deg^{-1}
Brinell hardness	Annealed	147	kg mm^{-2}
	Plates (0.04 mm thick)	230–250	
Tensile strength	20		
	Hard-drawn wire	140–182	kg mm^{-2}
	Annealed wire	70–100	
Young's modulus	25	17,300	kg mm^{-2}
	600	11,000	
Rigidity modulus	25	17,300–17,400	kg mm^{-2}
	600	11,000–11,206	
Electrical resistivity	27	5.78	$\mu\Omega$ cm
	727	23.9	
	1727	53.5	
Temperature coefficient of electrical resistivity	20–100	4.57×10^{-3}	deg^{-1}
	20–2620	4.79×10^{-3}	
Magnetic susceptibility		10^{-6}	(cgs)
Electron work function		4.3	eV
Annealing temperature		735	°C
H_2 solubility	400	3×10^{-1}	cm^3 (NTP)/100 gm
	600	3.5×10^{-1}	

Degassing is completely achieved only at 1670°C. Molybdenum is used for spring elements for W wires, can be made into rods, cylinders, bolts and nuts, and heater coils. It is used also for hard glass-to-metal seals, furnace boats, and electrodes for all kinds of vacuum devices.

Niobium. Niobium is a relatively strong and light refractory metal, which could potentially be very useful for vacuum structural purposes. It is currently being investigated as a potential wall material for controlled thermonuclear fusion reactors. Niobium is, however, very expensive, a fact that restricts its use mainly to brazing material for other refractory metals and for gettering. It is also used in small amounts in certain stainless steels for prevention of intergranular corrosion due to carbon.

Table XI summarizes some of the physical properties of Nb. Niobium is nonmagnetic. It is annealed in vacuum at 1070°C and then it can be more easily worked than, e.g., Ta. Its melting point is lower than Ta and its vapor pressure is higher which limits its use for a thermionic emitter although the work function of Nb is the smallest of all refractory metals. Nb resists attack by many acids (even aqua regia), alkalies, water, and Hg. It is strongly attacked by HF with or without HNO_3, and by hot concentrated H_2SO_4.

Niobium shows strong affinity to oxygen and at higher temperatures to almost all other gases except the noble gases. It is, therefore, used for

TABLE XI. Niobium

Property	Temperature (°C) or form	Value	Units
Atomic number		41	
Atomic weight		92.91	
Density		8.50	gm cm^{-3}
Melting point		2500 ± 40	°C
Specific heat		0.064	cal gm^{-1} deg^{-1}
Thermal conductivity	100	0.13	cal cm^{-1} sec^{-1} deg^{-1}
Coefficient of linear thermal expansion	20–100	7.2×10^{-6}	deg^{-1}
	20–1500	1.0×10^{-5}	
Brinell hardness	Annealed	75	kg mm^{-2}
	worked	200–250	
Tensile strength	Annealed	30–35	kg mm^{-2}
	Rolled sheet	70	
Young's modulus		8720	kg mm^{-2}
Electrical resistivity	25	0.13–0.23	Ω mm^2 m^{-1}
Temperature coefficient of electrical resistivity	20	$3–4 \times 10^{-3}$	deg^{-1}
Annealing temperature	*In vacuo*	>1050	°C

gettering, especially for high temperature applications. It is often used in sheets, which can be spot welded or electron-beam welded.

Tantalum. Tantalum, as Nb, is a light and strong refractory metal. It could be ideally suited for vacuum work because it has a very high melting point (2900°C), a consequently very low vapor pressure, and most notably it getters all active gases very well including hydrogen. Tantalum is, however, very difficult to manufacture and therefore expensive to use. Some of its characteristics are given in Table XII.

Tantalum can be machined and formed like mild steel provided it is done at room temperature because of its sensitivity to oxygen and nitrogen. It can also be spot or seam welded but this should be done under water for surface protection. For the same reason, Ta should be brazed and outgassed only in vacuum.

Tantalum withstands very well attacks by aqua regia (even boiling), chromic, nitric, sulphuric, and chloric acids. It dissolves in HF with or without HNO_3, fluoride solutions, and oxalic acid. As is the case with

TABLE XII. Tantalum

Property	Temperature (°C) or form	Value	Units
Atomic number		73	
Atomic weight		180.88	
Density		16.6–17.0	gm cm^{-3}
Melting point		2996 ± 50	°C
Specific heat	20	0.036	cal gm^{-1} deg^{-1}
Coefficient of linear	0–100	6.5×10^{-6}	deg^{-1}
thermal expansion	20–1500	8.8×10^{-6}	
Thermal conductivity	20	0.130	cal cm^{-1} sec^{-1} deg^{-1}
	1430	0.174	
	1630	0.18	
Brinell hardness	Annealed	45–125	kg mm^{-2}
	Worked	125–350	
Tensile strength	(Plate, hard 0.25 mm sheet)	100	kg mm^{-2}
	Annealed	35	
	Worked	27	
Young's modulus	0.08 mm wire	19,000	kg mm^{-2}
Electrical resistivity	18	12.4	$\mu\Omega$ cm
	1000	54	
	1630	71	
Temperature coefficient of electrical resistivity	0–100	3.8×10^{-3}	deg^{-1}
Magnetic susceptibility		$+0.93 \times 10^{-6}$	(cgs)
Electron work function		4.10	eV

Nb, Ti, and Zr, tantalum combines with hydrogen to form hydrides which destroy its metallic properties. The hydride forms even at low temperatures (100°C), causing considerable embrittlement, but the hydrogen can be released by heating above 760°C in vacuum. Other than the noble gases, Ta getters most of the regular residual gases in a vacuum system, especially at higher temperatures (700–1200°C). It is essential, however, that the getter surface be outgassed in high vacuum before operation.

Tantalum is a good electron emitter (between Mo and W). Its emissivity can be substantially increased by increasing the surface roughness mechanically or electrically.

Tantalum is used for special applications which utilize its advantages together with justifying its cost. It is found in getters, crucibles, and boats for evaporating different metals at high temperatures in high vacuum and for special structural purposes.

Rhenium. The physical properties of Re are summarized in Table XIII. Rhenium is one of the refractory metals which has an extremely high melting point (second only to W) and a low vapor pressure at elevated temperatures. Rhenium is a very strong and ductile metal after annealing. It shows however, an anomalously high surface hardness, which makes some machining operations difficult. Since it is not attacked by HF or HCl it can be evaporated onto a metal substrate (e.g., Ni) and then a Re thin film is obtained by etching the substrate.

Besides its high melting point, Re is very stable in H_2O atmospheres and is more resistant to oxidation than W. Re_2O_7 forms only above

TABLE XIII. Rhenium

Property	Temperature (°C) or form	Value	Units
Atomic number		75	
Atomic weight		186.31	
Density		20.53–21.0	gm cm^{-3}
Melting point		3176.6	°C
Coefficient of linear thermal expansion	20–500	66.6×10^{-7}	deg^{-1}
Brinell hardness		200	kg mm^{-2}
Tensile strength	Annealed	50–120	kg mm^{-2}
	Hard rolled	210–245	
Young's modulus		47,000	kg mm^{-2}
Temperature coefficient of electrical resistivity	20	1.73×10^{-3}	deg^{-1}
Electrical resistivity	20	19–21	μΩ cm
Electron work function		4.80	eV
Annealing temperature	1 hr	1500–1700	°C

600°C. Re is, therefore, useful for heating elements and filaments and electron emission sources. In connection with the latter its oxidation resistance and resistance to carbide formation are considered to be much better than those of W, but its thermionic emission efficiency is lower than that of tungsten at the same temperatures. The relatively higher cost of Re filaments is also a factor.

8.2.4. Precious Metals

Platinum. Some of the physical characteristics of Pt are given in Table XIV. Platinum is very easy to work with and is one of the most ductile metals. Its hardness can be enhanced by alloying with Ir or Ni.

Platinum is used mainly in connection with its being inert to O_2 at high temperatures. Any surface oxide that can develop, decomposes above 500°C. Above 700°C it is permeable to H_2, but not to any of the common gases. Platinum is not attacked by Hg but at higher temperatures reacts with alkalies, halogens, sulfur, and phosphorus. Platinum is not attacked by common acids except hot aqua regia.

Platinum is used in compound form in some high temperature braze alloys and in high temperature thermocouples, crucibles, and filaments, especially in cases where corrosive gases are used. Pt-coated Mo grids have long life characteristics and low secondary electron emission. Physically, the expansion coefficient of platinum makes it very suitable for metallizing glass or ceramics for glass–metal or ceramic–metal seals. It can also be sealed through glass as a fine wire for making high electrical conductivity feedthroughs. In spite of its fine ductility and inertness, the uses of Pt are limited to special cases because of its relatively high cost.

Palladium. Palladium is one of the cheapest of the noble metals, similar to platinum as hardness and strength go, but a little harder to work than Pt. A collection of its physical properties in given in Table XV. It can be resistance welded or brazed with Pt. Palladium, unlike Pt, can be oxidized by heating in air. However, the oxide decomposes at 870°C. Its major use is due to its exceptionally high permeability to H_2, especially at 300–400°C. Thus it can be found in H_2 filters and leak detectors. Palladium shows a very high solubility of hydrogen (~ 1000 times in volume).

Gold. Gold is an exceptionally malleable and ductile metal and an extremely good electrical conductor. Also it is extremely inert to gas chemisorption or dissolution, especially oxygen. Gold does amalgamate with and dissolves in Hg. Some of the physical properties of gold can be found in Table XVI. Gold can be easily welded and self-soldered with Ag. It is sometimes used as gasket material for bakeable systems, for

TABLE XIV. Platinum

Property	Temperature (°C) or form	Value	Units
Atomic number		78	
Atomic weight		195.23	
Density		21.45	gm cm^{-3}
Melting point		1773	°C
Thermal conductivity		0.17	cal cm^{-1} sec^{-1} deg^{-1}
Coefficient of linear thermal expansion	20–100	10.2×10^{-6}	deg^{-1}
Brinell hardness	1100		
	Annealed 50% pure	50	kg mm^{-2}
	Worked 50% pure	90	
	Annealed very pure	40–45	
	Cold worked very pure	97–103	
Tensile strength	Annealed	18–20	kg mm^{-2}
Young's modulus		10,000–17,500	kg mm^{-2}
Rigidity modulus	100	6000–7000	kg mm^{-2}
Electrical resistivity	0	9.8	$\mu\Omega$ cm
	20	10.6	
Temperature coefficient of electrical resistivity	0–100	0.003923	deg^{-1}
Electron work function		6.27	eV
Annealing temperature		800–1200	°C
H_2 solubility	400	3×10^{-2}	cm^3 (NTP)/100 gm

TABLE XV. Palladium

Property	Temperature (°C) or form	Value	Units
Atomic number		46	
Atomic weight		106.7	
Density		11.9–12.02	gm cm^{-3}
Melting point		1554.	°C
Thermal conductivity	18	0.168	cal cm^{-1} sec^{-1} deg^{-1}
	100	0.182	
Coefficient of linear thermal expansion	0–100	1.19×10^{-5}	deg^{-1}
	0–500	1.28×10^{-5}	
Brinell hardness	Annealed	46	kg mm^{-2}
	Cold rolled	109	
Tensile strength	Annealed	14–21	kg mm^{-2}
	Cold rolled	33	
Young's modulus		12,000	kg mm^{-2}
Electrical resistivity	0	100	$\mu\Omega$ cm
Temperature coefficient of electrical resistivity	0–100	3.8×10^{-2}	deg^{-1}
	0–350	3.1×10^{-2}	
Electron work function		4.99	eV
Annealing temperature	5 min	800	°C
H_2 solubility		≤850 times its own volume	

coating of glass–metal seals and glass surfaces, for electrical contacts, and is also a constituent in some braze alloys. The coating of glass or production of thin films is conveniently done by vaporizing gold, which is very easily done. An advantage of using gold in some applications lies in the fact that it is recoverable and retains its value.

Silver. Silver, like gold, finds a rather narrow use in vacuum technology as a construction material but is often found as a coating on other materials. Table XVII summarizes some of the physical properties of silver. Like gold, its softness renders it useful in making or coating gasket seals for bakeable systems and for bakeable uhv valve seats or noses. Unlike gold, silver shows high solubility of O_2, therefore any heating or annealing in O_2 atmosphere should be avoided. On the other hand, the high O_2 diffusion through heated silver makes it possible to use Ag as a selective filter for oxygen in a similar way to the admission of pure H_2 through Pd. Besides the applications implied by those properties, Ag is used as a brazing material and hard solder, as coating for Cu electrodes, for metallizing glass, and for electrical contacts. Ag alloys such as contact silver (Ag/Cu, 91/9) or sterling silver (Ag/Cu, 95/5) are sometimes used where enhanced strength and high thermal conductivity are desired.

8.2.5. Soft Metals

Indium. This extremely soft, low melting point metal is useful in vacuum applications due to its very low vapor pressure. Since it is not work hardenable at all, it can be used for gaskets below its melting point (156°C). Up to this temperature it is resistant to oxidation, but is attacked by most acids. A collection of physical properties of indium can be found in Table XVIII. It readily forms amalgam with Hg and forms alloys with many metals which are widely used for low resistivity contacts, solders, transducer contacts, glass-to-metal seals (In–Sn 1:1, for example, can wet glass and mica), and gasket and valve seals.

Gallium. Gallium is a soft metal whose properties can be found in Table XIX. Gallium melts at 30°C and therefore can be used as a liquid seal especially since it has a rather low vapor pressure. Since it is a liquid up to 2400°C, it can be used for high temperature thermometers.

Ga is a wetting agent for most constructional metals and is very suitable as a seal, which is easily demountable by slight heating. Since it can form alloys with many metals, it is usually used in an alloy with other low vapor pressure metals. In–Ga and In–Ga–Sn eutectics melt below room temperature. They are excellent wetting agents for many metals and glasses, will not attack them, and therefore can be used in valves, motion seals,

TABLE XVI. Gold

Property	Temperature (°C) or form	Value	Units
Atomic number		79	
Atomic weight		197.2	
Density		19.32	$gm\ cm^{-3}$
Melting point		1063	°C
Specific heat		0.031	$cal\ gm^{-1}\ deg^{-1}$
Thermal conductivity		0.71	$gm\ cm^{-1}\ sec^{-1}\ deg^{-1}$
Coefficient of linear thermal expansion	Drawn wire (0–100)	142×10^{-7}	deg^{-1}
Brinell hardness	Cast	33	$kg\ mm^{-2}$
	Annealed	25	
	Cold rolled	58	
Young's modulus	Cast	7560	$kg\ mm^{-2}$
	Annealed	8100	
	Cold rolled	8050	
Electrical resistivity	20	0.0235	$\Omega\ mm^2\ m^{-1}$
	400	0.0633	
Temperature coefficient of electrical resistivity	0–100	4×10^{-3}	deg^{-1}

TABLE XVII. Silver

Property	Temperature (°C) or form	Value	Units
Atomic number		47	
Atomic weight		107.88	
Density		10.55	$gm\ cm^{-3}$
Melting point		960.8	°C
Specific heat		0.0562	$cal\ gm^{-1}\ deg^{-1}$
Thermal conductivity		1.0	$cal\ cm^{-1}\ sec^{-1}\ deg^{-1}$
Coefficient of linear thermal expansion	0–100	196.8×10^{-7}	deg^{-1}
Brinell hardness	Annealed	15–36	$kg\ mm^{-2}$
	Drawn	75–90	
Tensile strength	Hard	29–40	$kg\ mm^{-2}$
	Annealed	13–16	
Young's modulus	Hard	8050	$kg\ mm^{-2}$
	Annealed	6000–8000	
Rigidity modulus		2600	$kg\ mm^{-2}$
Electrical resistivity		1.59	$\mu\Omega\ cm$
Temperature coefficient of electrical resistivity		4.1×10^{-3}	deg^{-1}
Electron work function		3.56–4.33	eV
Annealing temperature		400–600	°C

TABLE XVIII. Indium

Property	Temperature	Value	Units
Atomic number		49	
Atomic weight		114.76	
Density		7.31	gm cm^{-3}
Melting point		156.4	°C
Specific heat	20°C	0.057	cal gm^{-1} deg^{-1}
Thermal conductivity		0.057	cal cm^{-1} sec^{-1} deg^{-1}
Coefficient of linear thermal expansion	20°C	3.3×10^{-7}	deg^{-1}
Brinell hardness		0.9	kg mm^{-2}
Tensile strength		11	kg mm^{-2}
Young's modulus		1100	kg mm^{-2}
Electrical resistivity	20°C	9	$\mu\Omega$ cm
	156.4°C	29	
	3.38 K	Superconducting	

fluid seals (should be dust covered), and for thermocouple bonding (good thermal contact with low contact resistance). The extreme electrical and thermal conductivity anisotropy of Ga should be noted.

8.2.6. Alloys

Nickel Alloys. Most of the nickel alloys contain Mn, Fe, Cu, and carbon. Sometimes other compounds are intentionally introduced to enhance special properties. *Nickel 201* is similar to Nickel 200 (the commercially pure nickel—see Section 8.2.2) but has only a third of the carbon content of the latter. It is less hard and has a low rate of age hard-

TABLE XIX. Gallium

Property	Temperature (°C)	Value	Units
Atomic number		31	
Atomic weight		69.72	
Density		5.91	gm cm^{-3}
Melting point		29.78 ± 0.2	°C
Specific heat	0–24	0.089	cal gm^{-1} deg^{-1}
	12–200	0.095	
Coefficient of linear thermal expansion	20	180×10^{-7}	deg^{-1}
Electrical resistivity	0	53	$\mu\Omega$ cm
	20	56.8	
Magnetic susceptibility	18	-0.24×10^{-6}	(cgs)
	100	-0.04×10^{-6}	

ening. It is, therefore, easily spun and coined. It is generally preferred in the intermediate temperature range (300–600°C).

Nickel 211 (93.7% Ni) is somewhat stronger and harder than nickel 200 and therefore preferred in some structural parts subject to high temperature outgassing. Its Mn content (4–5%) offers some resistance to sulfur corrosion. It has relatively low electron emission even when coated with Ba. Its electrical resistivity is 18.3 $\mu\Omega$ cm at 0°C (equal to half the value of nickel 200).

Nickel 212 has intermediate composition and properties between nickel 200 and nickel 211.

Duranickel alloy 301 with 93% Ni includes Al (4–4.75%) and Ti (0.25–1%) which give it age-hardening properties together with lowering its thermal conductivity and its Curie temperature and increasing its electrical resistance. This is a strong, hard alloy which offers good corrosion resistance and good spring properties. It becomes magnetic after age hardening.

In the event that lower electrical resistivity and enhanced corrosion resistance and magnetic properties are needed one should use *permanickel alloy 300* (90.6% Ni, 0.5% Ti, 0.35% Mg, and 0.25% C).

Ni is also found in the composition of many alloys like *Micoro, Nioro,* etc., which are used as braze alloys and others like *Niromet, Niron,* etc., which are used for glass–metal seals.

Nichrome (60% Ni, 16% Cr, 24% Fe) is very commonly used in heating elements up to 930°C due to its corrosion resistance and convenient resistivity. Nichrome V (80% Ni, 20% Cr) is superior in performance up to 1093°C due to its higher resistivity and lower temperature coefficient of resistivity. It is nonmagnetic at room temperature and has an extremely low thermal conductivity, lower than that of 304 stainless steel.

Monel. *Monel alloy 400* is a very useful alloy due to its high corrosion resistance to most solvents, acids, and all alkalies and its easy machinability. It contains 64% Ni (+Co), 30% Cu, 2% each of Fe and Mn and 0.5% each of Al, Si, and C. Some of the physical properties of Monel 400 are given in Table XX.

Monel alloy 404 has only 54% Ni and therefore is essentially nonmagnetic for $T > 25$°C. Otherwise its properties are similar to Monel 400 as well as to Monel R-405 which has 0.5% S added to it for improved machinability.

Monel alloy K-500 and *501* are high Ni alloys, which are nonmagnetic down to −100°C. They are hard and strong and also age hardenable. The 501 has added carbon for improved machinability.

Inconel. Inconel is another corrosion resistant and widely used alloy, which comes in many varieties. They all contain about 72% Ni, 15% Cr,

6% Fe, and offer high tensile and rupture strengths, high resistance to most corrosives, and are virtually nonmagnetic. They are therefore excellent structural and spring materials, even at moderate temperatures. Some of the properties of Inconel 600 and Inconel 750 are given in Table XXI; the other varieties have some additives to enhance particular properties.

Fe Alloys. All Fe–Ni alloys are used almost exclusively in vacuum technology for glass-to-metal and ceramic-to-metal seals.

KOVAR. This is the most common alloy for glass-to-metal seals since its expansion coefficient matches those of several hard glasses. Some of its other physical characteristics are given in Table XXII. Kovar-glass seals are bakeable to 460°C. The coefficient of thermal expansion of Kovar alloy and some matching glasses as a function of temperature is shown in Fig. 10. The alloy is composed of 53% Fe, 29% Ni, 17% Co, <0.5% Mn, 0.2% Si, and <0.2% total of Al, Mg, Zr, and Ti. Kovar is commercially available in a large variety of alloys with slightly different composition for better strain-free matching with different glasses.

Kovar can be readily machined and joined to other metals by welding, brazing, or soldering. In vacuum technology the most common joining method is Cu brazing in a H_2 furnace. Kovar has fairly good corrosion resistance but not as good as that of stainless steel, and should therefore be stored in a dry place. Notably, it is inert to mercury and its vapors.

Due to its composition, Kovar is ferromagnetic up to 453°C (its Curie point). Its permeability rises as a function of the magnetic field to a maximum value of 3700 G^{-1} at 7000 G and then decreases (μ = 2000 G/Oe at 2000 G and 12,000 G).

INVAR. This alloy of 63% Fe, 36% Ni, and Mn, Si, and C additives is the result of the observation of Guillaume[2] that at this particular composition the average coefficient of thermal expansion of the alloy goes down to about one-tenth of its value at other compositions (to 1.26×10^{-6} deg^{-1} for $-18°C < T < 93°C$). It is therefore very stable structurally and geometrically in this temperature range. This is a magnetic alloy, easily welded and brazed but somewhat hard to machine and not age hardenable. It is fairly corrosion resistant.

Cu Alloys. BERYLLIUM COPPER ALLOY. This alloy consists mostly of copper with additions of beryllium and cobalt. A standard composition and its properties are given in Table XXIII. These properties can be changed somewhat according to slight changes in composition.

Beryllium copper alloy is easily machined, welded and formed and can be age hardened (3 hr at 600°C) to increase its strength and hardness con-

[2] C. E. Guillaume, *Proc. Phys. Soc.* (*London*) **32,** 374 (1920).

TABLE XX. Monel 400

Property	Temperature (°C) or form	Value	Units
Density		8.84	gm cm^{-3}
Melting point		1300–1350	°C
Thermal conductivity	0–100	0.62	cal cm^{-1} sec^{-1} deg^{-1}
Coefficient of linear	20–100	1.4×10^{-5}	deg^{-1}
thermal expansion	20–300	1.45×10^{-5}	deg^{-1}
Brinell hardness	Annealed	100–145	kg mm^{-2}
	Hard rolled	200–280	
Tensile strength	Annealed	48–60	kg mm^{-2}
	Cold rolled	65–85	
Young's modulus		18,200	kg mm^{-2}
Electrical resistivity		48.2	$\mu\Omega$ cm
Temperature coefficient of electrical resistivity		1.1×10^{-3}	deg^{-1}
Magnetic properties	Slightly magnetic		
Annealing temperature		750	°C

siderably. It shows good resistance to corrosion by dilute alkalies and cold dilute acids but is easily attacked by mercury and halogens.

The most notable characteristics of BeCu alloy is its mechanical flexibility and fatigue properties which far outdo those of phosphor bronze. The latter is not recommended for vacuum use due to the usually present impurities of volatile Zn and P. Because of its mechanical properties, BeCu alloys are very useful in making springs, bellows, diaphragms, and electrical contacts. In addition, since BeCu alloys can be made to have

TABLE XXI. Inconel

Property	Temperature (°C) or form	Inconel 600 value	Inconel x-750 value	Units
Density		8.43	8.3	gm cm^{-3}
Melting point		1395	1395–1425	°C
Thermal conductivity	25–100	0.036	0.036	cal cm^{-1} sec^{-1} deg^{-1}
Coefficient of	38–93	11.5×10^{-6}	13.7×10^{-6}	deg^{-1}
linear thermal expansion	38–760	16.1×10^{-6}		
Brinell hardness	Soft	120–170		kg mm^{-2}
	Hard	To 290		
Tensile strength	Soft	56–70	130–160	kg mm^{-2}
	Hard	91–123	220–270	
Young's modulus		21,800	21,800	kg mm^{-2}
Electrical resistivity	20	98.1	123	$\mu\Omega$ cm
Curie temperature		−40	−175	°C

TABLE XXII. Kovar

Property	Temperature (°C) or form	Value	Units
Density		8.3	gm cm^{-3}
Melting point		1450	°C
Specific heat	0	0.105	cal gm^{-1} deg^{-1}
	430	0.155	
Thermal conductivity	30	0.0395	cal cm^{-1} sec^{-1} deg^{-1}
	300	0.0485	
	400	0.053	
Coefficient of linear thermal expansion	30–200	$43–53 \times 10^{-7}$	deg^{-1}
Brinell hardness	Annealed	150	kg mm^{-2}
	Cold worked	200–250	
Tensile strength	25	63	kg mm^{-2}
	500	30	
Young's modulus		14,000	kg mm^{-2}
Electrical resistivity	25	49	$\mu\Omega$ cm
Annealing temperature	15 min	700–1100	°C

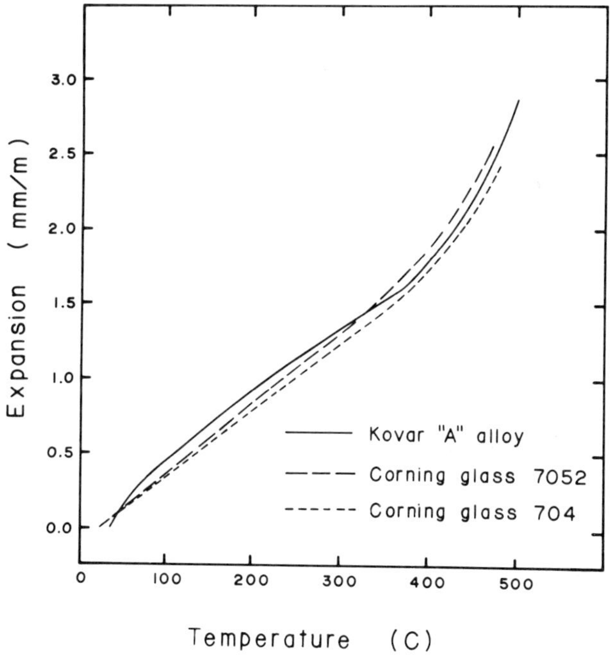

FIG. 10. Expansion of Kovar compared with two hard glasses.

TABLE XXIII. Standard Be Cu

Property	Temperature (°C) or form	Value	Units
Be content		1.8–2.5	%
Co or Ni	(Sometimes)	0.18–0.35	%
Density		8.2	gm cm^{-3}
Melting point		871–982	°C
Specific heat	30–100	0.1	cal gm^{-1} deg^{-1}
Thermal conductivity	20	0.26–0.31	cal cm^{-1} sec^{-1} deg^{-1}
	200	0.32–0.38	
Coefficient of linear thermal expansion	20–100	16.7×10^{-6}	deg^{-1}
	20–200	17.0×10^{-6}	
	20–300	17.8×10^{-6}	
Brinell hardness	Hardened	380	kg mm^{-2}
Tensile strength		100	kg mm^{-2}
Young's modulus	Hardened	12,900	kg mm^{-2}
Poisson's ratio		0.30	
Electrical resistivity	20	7.8–5.7	$\mu\Omega$ cm
	200	9.4–6.8	
Temperature coefficient of electrical resistivity		9×10^{-4}	deg^{-1}

high secondary electron emission, it is very useful as a multiplier dynode material.

Bronze (with 92–93.5% Cu and the balance Sn) and phosphor bronze (with a trace of P) are sometimes used in casting large structures since Cu and Sn have low vapor pressure and are vacuum tight. Most of the commercial bronzes, however, contain Zn and therefore should not be used when heating is involved.

BRASS. Brasses contain 66–95% Cu and the rest Zn. Due to the high vapor pressure of zinc they should not be used in bakeable systems. Their use is therefore generally limited to structural parts of demountable systems which do not have to be degassed.

8.2.7. Special Elements (Hg, C)

Mercury. Mercury is a heavy metal which is liquid at room temperature. The properties of mercury are listed in Table XXIV. Its most useful properties are fairly good electrical conductance, regular coefficient of expansion, and high surface tension. This can be made use of in switches of many kinds, thermometers (temperature range −39°C up to 800°C) and shut-off valve sealant material (will not pass through holes with $d < 15$ μm).

Its main disadvantage is its relatively high vapor pressure (several milli

TABLE XXIV. Mercury

Property	Temperature (°C)	Value	Units
Atomic number		80	
Atomic weight		200.61	
Density	20	13.55	gm cm^{-3}
	360	12.74	
Melting point		−38.89	°C
Boiling point		356.7	°C
Specific heat	20	0.0335	cal gm^{-1} deg^{-1}
	200	0.032	
Thermal conductivity	20	0.027	cal cm^{-1} sec^{-1} deg^{-1}
Coefficient of volume	20	1820×10^{-7}	deg^{-1}
thermal expansion	200	1841×10^{-7}	
Electrical resistivity	20	95.8	$\mu\Omega$ cm
	200	114	

Torr at room temperature). This necessitates dry ice traps [v.p. (Hg) at $-78°C = 10^{-9}$ Torr] or liquid nitrogen traps [v.p. (Hg) at $-180°C = 10^{-27}$ Torr] when using Hg in vacuum systems. The most common uses of Hg are in McLeod gauges, manometers, and Hg diffusion pumps. (In some of those applications Hg is replaced by low vapor pressure oils.)

Since Hg oxidizes slightly in air, it is used in a sealed environment in most of its applications. It also should be extremely pure. The common, simple, but very sensitive test is to run a drop across clean white paper or porcelain. Pure Hg should not leave a "streak" behind it. Mercury should also be sealed while in use because its evaporation rate in air (10^{-4} gm cm^{-2} sec^{-1} at room temperature) is high and its vapor is poisonous.

Hg vapor is used commonly in discharge lamps and rectifiers. Mercury will not be affected by dry S-free air and its chemical properties are much like those of the noble metals. It forms amalgams with most metals so they cannot be used as containers but stainless steel and other steels are an exception. This is also an important consideration when an Hg diffusion pump is used; if the cold trap fails, vaporized mercury can condense on and amalgamate with metal components of the vacuum system.

Carbon. The most common form of C used is electrographite. Table XXV gives some of its physical properties. Carbon is used occasionally in vacuum technology due to several useful properties. It has a high melting point and low vapor pressure. It has a high thermal conductivity, sufficient electrical conductivity, high work function for electron emission, but high thermal emissivity. It is chemically inert, highly rigid, a good getter, and relatively cheap. It can be therefore used for furnace boats and crucibles (although carbon contamination of the melts is to be expected), for heat shield coating of metals, and for arc- and

TABLE XXV. Carbon

Property	Temperature (°C) or form	Value	Units
Atomic number		6	
Atomic weight		12.010	
Density	True	2.21–2.25	gm cm^{-3}
	Apparent	1.5–1.75	
Melting point		3700 ± 100	°C
Specific heat	20–300	0.20	cal gm^{-1} deg^{-1}
	20–1500	0.40	
Coefficient of linear thermal expansion	20–100	Long. 11–22 × 10^{-7}	deg^{-1}
		Trans. 22–46 × 10^{-7}	
Brinell hardness		≥3	kg mm^{-2}
Tensile strength		0.2–0.6	kg mm^{-2}
Young's modulus	20	600–800	kg mm^{-2}
	2000	Up to 1000	
Electrical resistivity		600–1100	μΩ cm
Electron work function		4.0–4.8	eV

resistance-welding electrodes. Carbon is also used for enhancing thermal emissivity of, e.g., nickel by C coating which at the same time suppresses secondary electron emission. The primary disadvantages of carbon are low strength, high gas content, difficulty in outgassing, and poor machinability and workability. It cannot be welded or soldered.

8.3. Glasses

Glasses are made by the fusion of several inorganic oxides, the primary ingredient being silica sand (SiO_2). Glass can be very soft at high temperatures, but when the melt cools to room temperature, its viscosity increases rapidly to practically infinite values. The solidification is amorphous, with no internally ordered structure and the product does not have a specific freezing point. Glass is very hard and rigid, brittle, mostly transparent, chemically inert, and impervious to most gases.

In general, glasses can be divided into two groups. A category of soft glasses which have coefficients of thermal expansion higher than 70 × 10^{-7} mm mm^{-1} deg^{-1} and a category having values between 5–65 × 10^{-7} mm mm^{-1} deg^{-1}, which are called hard glasses. Usually, the soft glasses are also distinguished by lower working temperatures than those of the hard glasses.

The various oxides fused with silica sand give glasses a variety of characteristics and they can be classified as follows:

(a) Lead glasses contain 35–65% of silica (SiO_2) and alumina (Al_2O_3), lead oxide (~15%), and alkali oxides. This is the type of glass used for electric bulb stems and neon tubes. Lead glasses are fairly easy to work with and have fairly high electrical resistivity. A partial list of properties of lead glasses is in Table XXVI under column A. Heavy Pb glasses are usually softer and are used as radiation shields.

(b) Lime glasses contain 65–75% SiO_2 + Al_2O_3, ~15% soda (Na_2O), ~10% lime (CaO), and other oxides. Lime glasses, which are common in bulb envelopes are very easy to work with but are not to be used where high thermal resistance and chemical stability are required. A partial list of properties of lime glasses is in Table XXVI under column B. Most lime and lead glasses are considered soft glasses.

(c) Borosilicate glasses contain 75–95% silica and alumina, and most of the rest is boric oxide. They contain also some soda, which is used in reduced quantities or totally eliminated when harder borosilicate glass is needed. Some properties of borosilicate glasses are listed in Table XXVI under column C while those of the low alkali hard borosilicates are under column D. The list under column D pertains also to alkali free alumina–silicate glasses. Those glasses, such as Corning 1720, are very hard, bakeable to 700°C, and much less permeable to He than borosilicate

TABLE XXVI. Properties of Technical Glasses (See Text)

Property	A	B	C	D	Units
Transformation point	360–420	430–500	430–540	600–800	°C
Strain point	380–430	470–530	445–790		°C
Anneal point	425–460	500–570	480–890		°C
Softening point	580–660	670–750	690–1510		°C
Specific heat, 0–1000°C		0.08–0.23			cal gm^{-1} deg^{-1}
Thermal conductivity	1.7–3.7	2.5–3.0	2.5–3.0	3	10^{-3} cal cm^{-1} sec^{-1} deg^{-1}
Coefficient of linear thermal expansion	85–95	80–110	30–60	8–38	10^{-7} deg^{-1}
Thermal shock resistance	50	60–115	150	1000 (96% SiO_2)	°C
Tensile strength	3–7	4–15	4–15	5–15	kg mm^{-2}
Compressive strength	50–90	70–100	70–100	100	kg mm^{-2}
Bend strength		10–25			kg mm^{-2}
Torsional strength	9	9	9		kg mm^{-2}
Young's modulus		5000–8000			kg mm^{-2}
Poisson's ratio		Usually 0.22–0.25 often <0.20			
Electrical resistivity 30 C	10^{14}	10^{12}	10^{16}	10^{17}–10^{18}	Ω cm
Dielectric constant	8.2	7.8	4.8		
Electrical breakdown strength d = 0.2 mm, 20°C	3100	4500	4800		kV cm^{-1}
300°C	102	32	200		
Loss factor		5–250 × 10^{-4}			
		Best soft-glass dielectrics (1.0 MHz) 5–10 × 10^{-4}			
		Poor soft-glass dielectrics (1.0 MHz) 100–150 × 10^{-4}			

glasses. They are, however, much harder to work with. Borosilicates such as Pyrex, made by Corning, are the glasses usually chosen for laboratory and vacuum use. They have low coefficient of thermal expansion and their electrical resistivity, thermal shock resistance, and chemical stability are high.

The coefficient of thermal expansion can reach even lower levels in quartz glasses ($SiO_2 \geqslant 96\%$), which therefore can withstand severe thermal shocks. Quartz glasses can be used continuously at 900°C and soften only at 1200–1500°C. Quartz glasses (like Vycor made by Corning) are used only when very high temperatures or thermal shocks are encountered since they are considerably more expensive and harder to work with than borosilicates. A list of some of the important parameters of quartz glasses is given in Table XXVII.

8.3.1. Mechanical Properties

The mechanical properties of glass such as strength, hardness, elasticity, etc., have to be discussed in a manner totally different than that of metals. This is because (a) glass is not a true solid and therefore there are no shear stresses on it, and (b) the stress–strain curve of glass is linear up to the rupture point. There is no flow before fracture due to the brittleness of glass.

The inherent strength of glass is very high especially as rod diameters decrease. However, due to surface imperfections, maximum continuous loading should lie between 50–100 kg/mm^2. The strength values are essentially independent of the glass composition.

Young's modulus of elasticity of glasses has values of 5000–8000 kg/mm^2. This is important for evaluating the tensile stresses occuring under thermal or mechanical processes with constraints such as in sealing. For glass–metal seals it is important to know Poisson's ratio (of the transverse contraction to axial expansion under load). It can be taken as 0.22 for all technical glasses. It is almost never more than ± 0.03 of that value.

The mechanical hardness of glasses cannot obviously be measured by Brinell or Rockwell machines as in the case of metals. On the Moh's scale (diamond = 10), hardness values for glasses are 5–7 (quartz = 7). Glass hardness is usually specified by scratch resistance (determined by the force needed to load a diamond cone to produce a scratch 0.01 mm wide) or by impact abrasion resistance (measured under standard sandblasting conditions). Lead glasses, soda–lime, and borosilicates show increasing hardness in that order.

TABLE XXVII. Properties of Quartz Glass and Quartzware (>95% SiO_2)

Property	Value	Units
Density	2.0–2.3	gm cm^{-3}
T_g	1100–1120	°C
T_{str}	1050	°C
T_{anneal}	1150	°C
T_{soft}	1655	°C
Working range	1900–2000	°C
Max operating temperature		
Continuous	1000	°C
Transient	1300	
Specific heat		
100°C	0.20	cal gm^{-1} deg^{-1}
500°C	0.27	
1000°C	0.29	
Thermal conductivity		
20°C	0.0035	cal cm^{-1} sec^{-1} deg^{-1}
950°C	0.0064	
Average coefficient of linear thermal expansion		
20–300°C	6.27×10^{-7}	deg^{-1}
20–1000°C	5.40×10^{-7}	
Thermal shock resistance	>1250	°C
Tensile strength	7–12	kg mm^{-2}
Compressive strength	16–200	kg mm^{-2}
Bend strength	7	kg mm^{-2}
Young's modulus	6200–7200	kg mm^{-2}
Poisson's ratio	0.17	
Electrical resistivity		
20°C	10^{17}–10^{18}	Ω cm
350°C	$10^{10.5}$	
Elecrical breakdown strength		
(thin plates) 20°C	250–400	kV cm^{-1}
500°C	40–50	
Dielectric loss angle tan δ (1MHz)	$\sim 2 \times 10^{-4}$	
Dielectric constant (10^3–10^{10} Hz)	3.78	

8.3.2. Viscosity

Viscosity data are also very important for vacuum technology applications. Since there is no definite melting point, several temperature ranges with corresponding viscosities should be known when working with glasses. They are depicted schematically in Fig. 11. The brittle state of glass extends up to the strain point defined by a viscosity value of $10^{14.5}$ P. This is the extreme serviceability limit of the glass. Up to the transformation temperature T_g (which is ~450°C for "soft" glasses and >500°C for

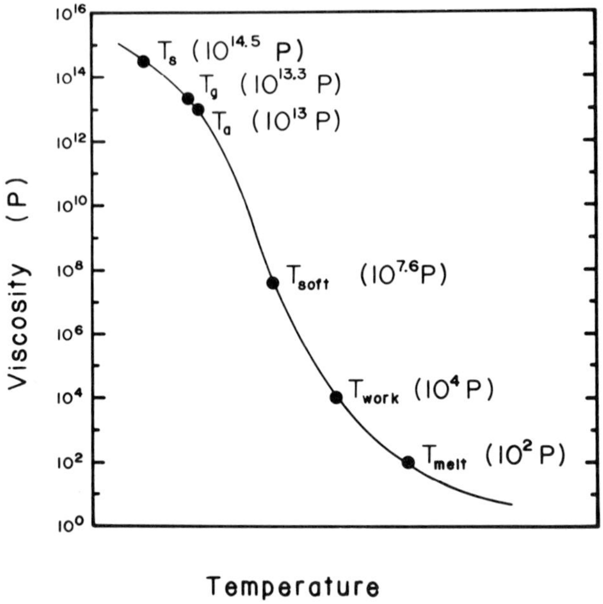

FIG. 11. Schematic viscosity diagram for glass. Circles indicate special temperatures with designated viscosity values. (Specific T values for some glasses are in the tables.)

"hard" glasses) there is hardly any stress-relieving motion and stresses can build up. The viscosity drops from $\eta > 10^{15}$ P at room temperature to $\eta \sim 10^{13.3}$ P at T_g. At T_g, stress relieving takes several hours. The temperature at which η falls below 10^{13} P is called the annealing point. At this temperature stresses can be relieved in a few minutes. Every finished glass product unless made of quartz glasses should undergo annealing to relieve internal stresses. At the softening point $-T_s$ ($\eta = 10^{7.6}$ P), the glass cannot support its own weight and the working range T_{work} is when η drops to 10^5–10^4 P making the glass workable by conventional pressing and blowing. At T_{melt}, $\eta = 10^2$ P, the glass flows as an ordinary liquid.

8.3.3. Coefficient of Thermal Expansion

Thermal expansion, as already mentioned, is a characteristic generally used to divide glasses into "hard" and "soft" categories. Its main importance is for sealing, whether with other glasses or with metals. The temperature dependence of the relative thermal expansion of several Corning glasses is shown in Fig. 12.

When glasses of different coefficient of thermal expansion α are joined

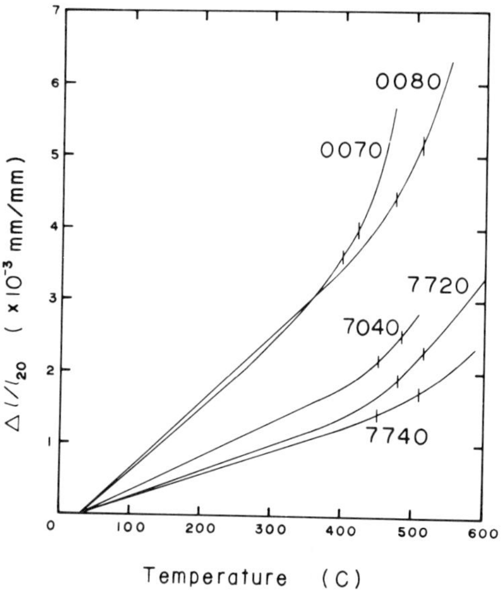

FIG. 12. Expansion curves for several soft and hard Corning glasses. The vertical lines indicate the annealing range limits ($T_a - T_a$ of Fig. 11).

together, the joint might crack under the stresses, when the seal cools below Tg. Soft glasses can be joined if their α's do not differ by more than 10% and if they have close transition temperatures. If those conditions are not satisfied, graded seals are used, where intermediate glasses keep any of the joints from exceeding the highest permissible stress. In the case of glass–metal seals, consideration has to be given to additional factors, such as shape and form of the seal, plasticity of the metal used, and the annealing process (see Part 7).

8.3.4. Thermal Shock Resistance

The coefficient of thermal expansion $-\alpha$ has a great importance in vacuum technology since it also determines the thermal shock resistance of the glass. Thermal shocks are frequent in vacuum work and, especially in the case of quick local cooling, are liable to crack the glass. The effect will be stronger the higher α is and the higher the temperature gradient is. Thermal shocks are more severe in the case of fast local cooling than heating. This is because glass has a low tensile strength, especially on the surface.

Thermal shock resistance is hard to calculate since it is momentary. After thermal equilibrium is attained, the stresses disappear. The resistance depends on the coefficient of thermal expansion, the thermal conductivity, and on geometrical factors. Thicker glass or more complicated structures are more vulnerable to thermal shocks. The resistance is determined empirically and is usually given in degrees, designating the temperature difference between two glass plate surfaces necessary to create a given stress.

8.3.5. Thermal Conductivity

The thermal conductivity of glasses is very small compared to metals, about 2 orders of magnitude lower than that of iron and nickel. The value of thermal conductivity does not depend very much on the type of glass. It varies between $1.6–3 \times 10^{-3}$ cal cm^{-1} sec^{-1} deg^{-1} with Pb glasses at the low end and borosilicates at the high end. The value for pure SiO_2 is 3.3×10^{-3} cal cm^{-1} sec^{-1} deg^{-1}. These low values of thermal conductivity may cause cracking when heat is not distributed equally over a glass part and especially over a glass-to-metal seal. The poor equalization of thermal stress has to be considered together with the coefficient of thermal expansion in determining thermal shock resistance, as well as continuous thermal gradient resistance.

8.3.6. Electrical Resistivity

Here we have to discriminate between the intrinsic bulk resistivity of glasses which can be extremely high (up to 10^{19} Ω cm) and the surface conductivity which is usually governed by adsorbed water or other contaminants on the surface and can be, therefore, many orders of magnitude lower.

The conductance in the bulk is governed by a hopping mechanism predominantly between Na^+ centers. As the temperature rises, the mobility of the charge carriers increases exponentially and, as seen in Fig. 13, the resistivity drops many orders of magnitude at only moderate temperatures. A well annealed glass will usually show higher resistivity than strained glass.

At temperatures below 100°C, the measured resistivity of glass is usually far below the intrinsic bulk value because of surface leakage. The adsorbed water layer, which is usually the cause of the surface conductance, evaporates above 150°C; above 400°C water molecules come out from deeper layers in the glass and the material is well outgassed if in vacuum. For vacuum use, therefore, an outgassed internal glass surface should not present a problem regarding insulation. The external surface, however, may have a high leakage current which will depend on the relative humidity (Fig. 14). A special surface treatment with silicone or other water repellents can considerably reduce the effect of humidity on the surface conductivity as seen from the dashed curve in Fig. 14.

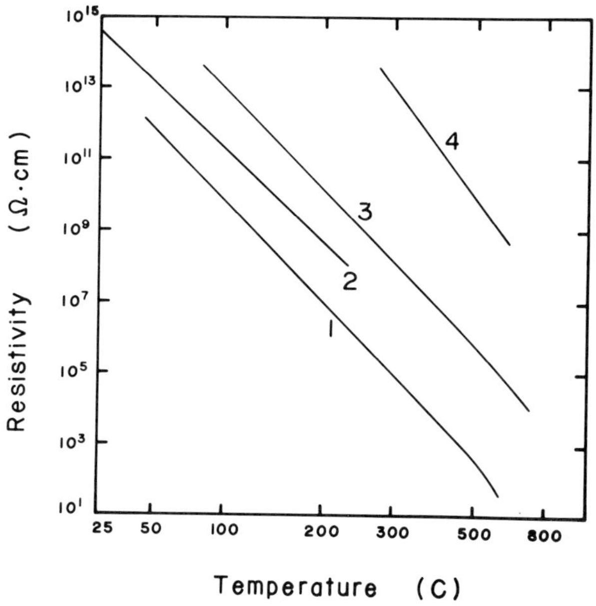

FIG. 13. Resistivity (on log scale) of several glasses as a function of temperature T (scale proportional to $1/T$). (1) Lime magnesia glass; (2) Pyrex; (3) lead glass; (4) hard aluminoborosilicate.

8.3.7. Dielectric Constant and Dielectric Loss Factor

When glass is present between two electrodes subject to ac voltage it behaves as a dielectric in a capacitor. Accordingly, there is a loss of power P in the glass that is given by

$$P = C_0 V^2 (2\pi\nu\epsilon tg\delta),$$

where C_0 is the capacitance of the array without the glass, V the applied voltage at a frequency ν, and $\epsilon tg\delta$ is called the dielectric loss factor. The dielectric power loss is an important figure to consider especially when dealing with high voltages and frequencies. This is applicable whether the glass is an insulator between two ac electrodes or a window for transmission of high frequency electromagnetic waves, e.g., in klystrons, magnetrons. In such applications, the loss of power (given mostly to optical phonons in interaction with Na^+ ions) can cause enough local heating so the glass wall caves in under the atmospheric pressure.

Empirically, ϵ at room temperature is determined from the density γ by

$$\epsilon = 2.2\gamma \qquad (\gamma, \text{gm cm}^{-3}).$$

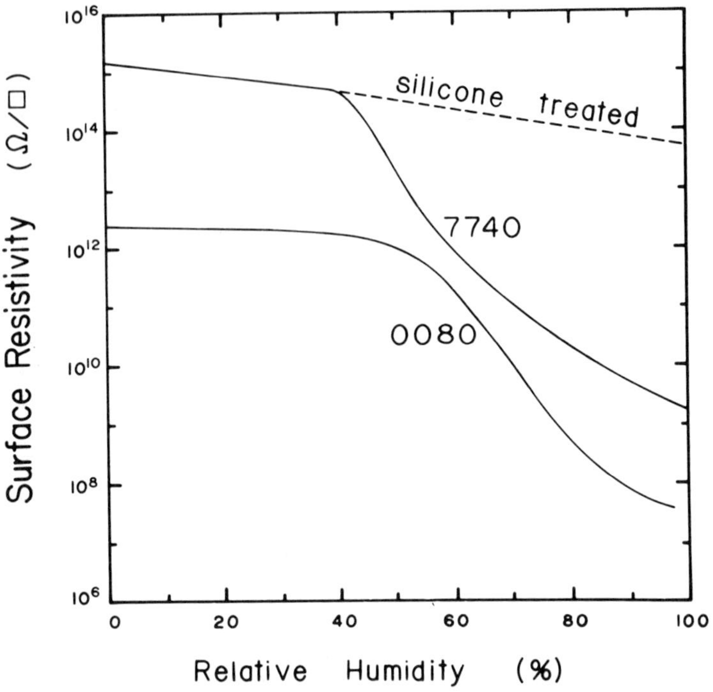

FIG. 14. Surface resistivity of soda lime glass (Corning 0080) and Pyrex glass as a function of relative humidity. Dashed curve is for silicone treated Pyrex surface.

The dielectric constant rises with temperature and, as shown in Fig. 15, it is higher and more sharply temperature dependent in the case of softer glasses.

The loss angle $tg\delta$ increases exponentially with temperature above 0°C, especially at low frequencies. Its dependence on frequency has a definite minimum for most glasses around 1–10 MHz.

For applications where dielectric loss presents a problem, glasses with low $\epsilon tg\delta$, such as borosilicate glass 7070 made by Corning, are used for windows. In the case of uhf leads, geometrical arrangements can be made to reduce the capacitance.

In extreme cases quartz and Vycor glass are used as they have the lowest values of $\epsilon tg\delta$ among glasses due to the strength of the Si–O bond.

8.3.8. Electrical Breakdown Strength

The electrical breakdown strength is measured by the field necessary to cause such breakdown. Its value for glasses at room temperature is normally much higher than that for air so arcing through air will occur at

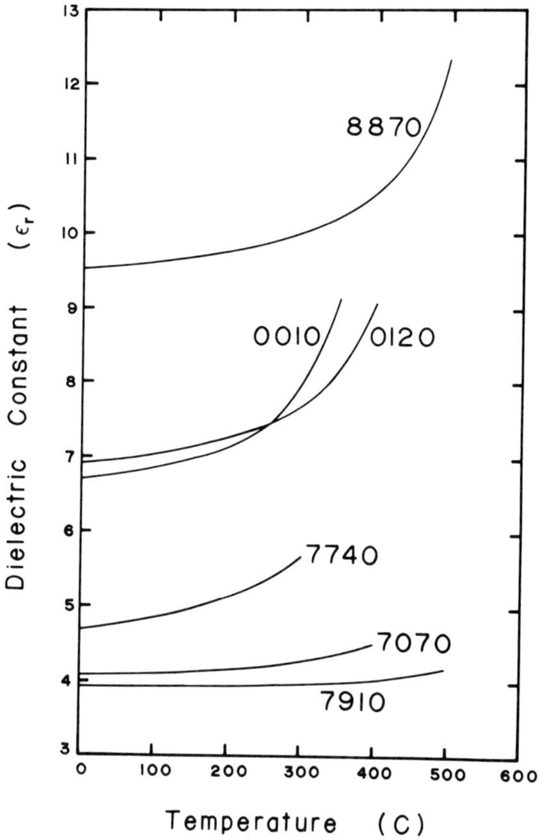

FIG. 15. Relative dielectric constants ϵ_r of several Corning glasses at 1 MHz as functions of temperature.

lower voltages. Under vacuum, glass surfaces can charge up due to the lack of the conducting water layer usually adsorbed on them. This can create substantial electric fields and accompanied stress that may bring about the evolution of pinholes. Those stresses are especially destructive in the case of gas bubbles trapped in the glass wall.

The pure electrical breakdown may be avoided by thicker glass walls. However, this can lead to thermally assisted breakdown due to the poor thermal conductivity of glass and the dependence of its resistivity on temperature. This is also the reason for the decrease in breakdown strength at higher frequencies (and higher dielectric losses).

Avoiding electrical breakdown is achieved by using borosilicate glasses (Vycor or quartz in extreme cases). The glass walls should not be too thin, cooled if possible, should have enlarged or corrugated areas, and, if possible, shielded from electron bombardment which can cause charging.

8.3.9. Permeability to Gases

Glasses have a very low vapor pressure at room temperature. Glasses are practically impervious to all gases except for helium and hydrogen. This is another factor in considering glass as a wall material for vacuum systems. especially of intricate shape. Helium can permeate through glass at room temperature, although at a very small rate (5×10^{-12} Torr liter sec^{-1} cm^{-2}), but the rate increases exponentially with temperature. Since the permeation is apparently through the lattice of structural components of glass (Si, B, P) it is greater in quartz than in soda–lime glasses where the holes are blocked by alkali atoms. Hydrogen also permeates through glass but generally at a much lower rate than He. Although the concentration of He in the atmosphere is only 5 ppm, there is a large pressure gradient on glass walls of a uhv system evacuated to, for example, 10^{-10} Torr. Helium therefore constitutes the main residual gas in uhv glass systems and vacua better than 10^{-12} Torr can be achieved only by the use of different kinds of glass such as alumino silicate glass. Hydrogen is usually found to be the second abundant residual gas in uhv glass systems.

8.3.10. Optical Properties

One of the main reasons for using glass in vacuum applications is its transparency, primarily in the visible part of the spectrum. Usually glasses transmit around 90% of the incident light, most of the losses being due to reflection at both surfaces.

Glasses can be tinted by suspension or dissolution of different metal oxides in the melt and thereby made to absorb different bands in the spectrum. Glasses can be made opaque by "clouding" the bulk (usually with cryolite Na_3AlF_6) or by "frosting."

In the ir range, glasses are also usually very good transmitters. Figure 16 shows a comparison of the transmissivity of soft and hard glasses in this range. The transmission remains the same as in the visible up to ~2.6 μm, drops to about 50% at longer wavelengths depending on the glass type, and finally reaches very low levels around 5 μm. The shape of the transmissivity spectrum will obviously be attenuated as the glass thickness increases. The ir transmission can be strongly reduced by very small FeO concentrations. In the special case of borosilicate glasses there is a very strong decrease in the transmission above 3.5 μm.

At the other end of the spectrum, most of the glasses, whether soda–lime, lead, or borosilicate glasses, have a cutoff in the transmission curve around the near uv (320–340 nm). The transmission spectrum of several typical glasses is shown in Fig. 17. However, some special glasses are

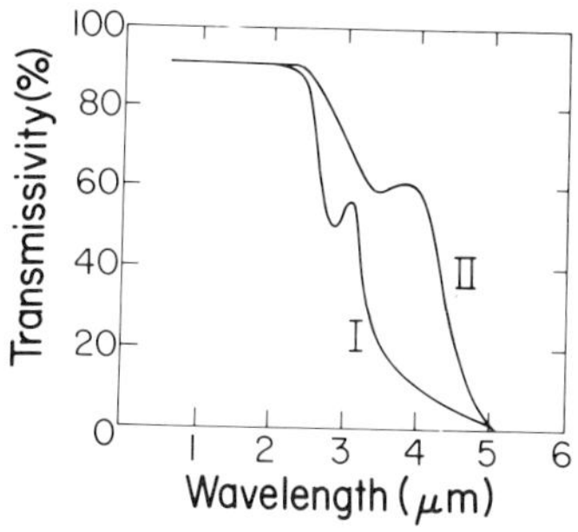

FIG. 16.

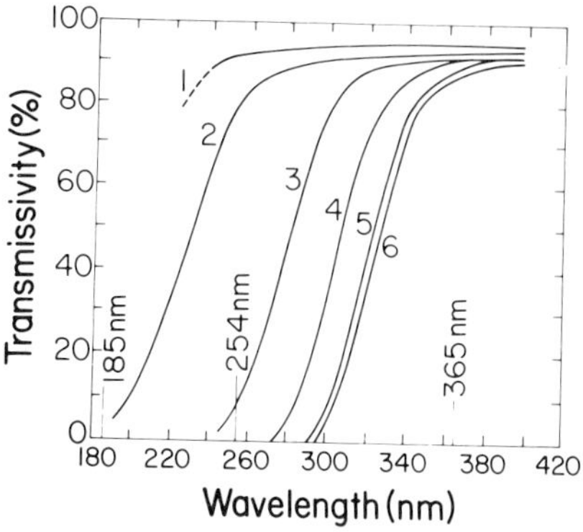

FIG. 17.

FIG. 16. Transmissivity to ir of a borosilicate hard glass (I) and a soda–lime soft glass (II) of the same thickness (1 mm) as a function of wavelength of radiation.

FIG. 17. Transmissivity to uv of some glasses of the same thickness (1 mm) as a function of wavelength of radiation. (1) Quartz glass; (2) Corning 9741; (3) Corning 9700; (4) Corning 7740 (Pyrex); (5) lead glass; (6) lime glass.

made to be transparent at shorter wavelengths. This is done by reducing the concentration of Fe_2O_3 which is the main uv absorbent. Some examples are Corning glasses 9700, 9741, and 7912 (quartz glass), which have above 50% transmission down to 290, 230, and 220 nm, respectively. For the extreme uv range (180–120 nm), quartz single crystal can be used, the uv limit of which is not matched by any glass.

In most of the good uv transmitters, exposure to strong uv irradiation results in an effect called "solarization" by which FeO converts to Fe_2O_3, resulting in a considerable reduction in transmissivity. This effect is different from the darkening of the glass surface by Hg ions in discharge lamps and sometimes can be cured by heat treatment.

In this context, some window materials other than glasses should be mentioned. The most common of them is synthetic sapphire, which is a pure form of single crystal Al_2O_3. It has high transmission in the whole spectral range from uv to ir. The transmittance falls below 50% only at wavelengths below 180 nm and on the other end of the spectrum above 6 μm. Sapphire is a very strong and hard material, it has low dielectric losses, high resistivity, resistant to wear, mechanical abuse, chemical attack, and high temperatures.

Lithium–fluoride and calcium–fluoride are also good uv transmitters which are used in vacuum technology because as sapphire they can be sealed to metal flanges.

For microwave windows, mica is often used. It has excellent chemical stability and extremely low dielectric power loss and it can be sealed to metal. However, mica is extremely hard to work. It does find use as an insulator in internal vacuum components.

Glasses which are good uv absorbers can also be made. Those containing TiO_2, for example are also very good ir transmitters.

8.3.11. Chemical Resistance of Glasses

Glass is extremely inert to almost all types of usual corrosives such as bases at room temperature, metal vapors, salts, and acids with the notable exception of HF, hot concentrated H_2SO_4 and superheated water.

At room temperature it is attacked exclusively by HF. This acid destroys the silicon–oxygen bonds and dissolves the lattice. HF-resistant glasses can be made, which do not contain silicon or boron oxides in their composition. The same type of corrosion is caused by water vapor which results in a loss of up to 0.25 in. a year at 200°C and by alkaline solution, primarily NaOH, which can cause corrosion at the same rate at 100°C. As the temperature decreases, the corrosion rates fall sharply.

8.3.12. Degassing Glass

The intrinsic vapor pressure of glass is extremely low (10^{-15}–10^{-25} Torr). However gases are trapped in the glass during preparation as composition products and are also adsorbed on the surface. Those gases consist mostly of H_2O, CO_2, and O_2. Especially when glass is exposed to steam, its silica gel surface becomes spongy and a large water vapor intake occurs. Therefore, glass in vacuum work should be thoroughly outgassed to at least the highest operating temperature.

Most of the adsorbed gases and in particular the conducting water layer on the surface usually desorb around 150–200°C from most glasses. Around 300°C most of the water comes from dehydration of the Si–OH–OH–Si compounds to create Si–O–Si + H_2O. The gas output at that temperature levels off after a few minutes. Only above 450°C does one observe a slow outgassing of the bulk, exponentially dependent on the temperature. Therefore, to achieve a thorough outgassing of the glass in vacuum work, the parts are usually heated to within a few tens of degrees from the strain point of the glass (defined as the temperature at which the viscosity is $10^{14.5}$ P).

8.4. Ceramics

Ceramics are widely used in vacuum technology applications, such as electrical insulation and for small support structures, and are especially useful when very high temperatures are involved. Most of the ceramics are very hard, brittle, and strong materials, able to withstand high temperatures and high electric fields. Their hardness is the reason for their main disadvantage, namely, the inability to machine them, except by expensive and difficult grinding. There are, however, two or three exceptions, most notably the development of machinable glass–ceramic (Macor by Corning), which will be discussed in Section 8.4.3. Otherwise, most of the ceramics are molded and formed in the green state and then fired to remove water of crystallization. After firing they are virtually nonmachinable, other than by difficult diamond or ultrasonic machining.

Almost all ceramics are somewhat porous and therefore tend to absorb water to a certain extent even after firing. The water absorption of any ceramic can be defined as the weight percent of the absorbable water at saturation referred to the dry weight of the ceramic. In cleaning ceramic surfaces, water should not be used. Clean pure solvent is recommended for degreasing followed by air drying and in some cases vacuum degassing (at 800°C for alumina). Until used, vacuum storage is recommended.

The mechanical properties of ceramics are usually only generally indicated, since they are very dependent on cross section and shape of the ceramic part. Generally, ceramics are hard and brittle and stronger in compression than in tension. Therefore, in the case of snugly fit or bonded metal–ceramic parts, the ceramic part should be fitted into an opening in the metal part, and not vice versa. Consideration should also be given to the difference in thermal expansion rates of the matched parts.

Electrically the ceramics are very good insulators and that constitutes their main use. They show a very high resistivity at room temperature with both bulk and surface resistivity decreasing with rising temperature. This behavior is shown in Fig. 18 with comparison to some glasses. The resistance of ceramic insulators, especially silicates, drops sharply as a function of the humidity. As in the case of glasses, protective silicone coating or glazing for outside walls or parts reduces the effect of moisture. The ceramics have generally higher dielectric constants and lower loss angles than most glasses and therefore better resist electrical breakdown. However, the electrical breakdown strength is a function of the shape, thickness, and type of the ceramic piece, as well as the electric field distribution and its frequency.

Generally, ceramics are optically opaque. Most ceramics are chemically inert especially in corrosive or acidic environments and they show good weather and age resistance even at high temperatures. The noncrystalline silicate ceramics, due to their high glass content, do not have a definite melting point but rather a softening temperature may be indicated. Usually this is quite high and as expected most ceramics have a very low vapor pressure and a low coefficient of thermal expansion.

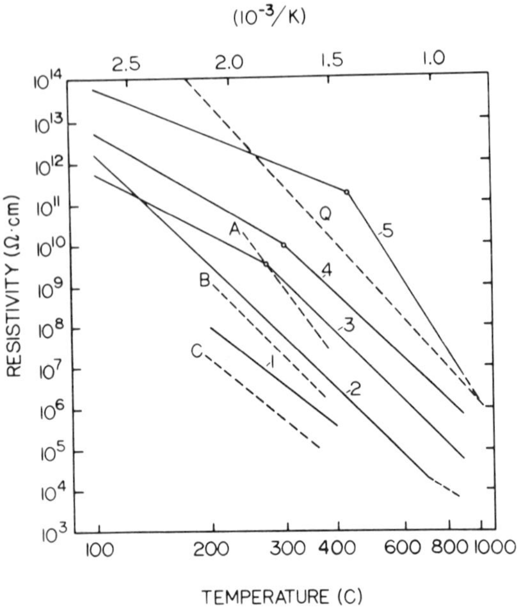

FIG. 18. Electrical resistivity of some ceramics as a function of temperature. (1) Porcelain; (2) ordinary steatite; (3) lava (grade A); (4) porous high-alumina ceramic; (5) Forsterite. For comparison: Q–quartz, A–lead glass (Corning 0120), B–Pyrex (Corning 7740), C–soda-lime glass (Corning 0080).

In the next sections, specific ceramics will be discussed in detail. They include silicate ceramics, pure oxide ceramics, and ceramics with easier machinability.

8.4.1. Silicate Ceramics

Silicate ceramics are the most commonly available ceramics for a variety of vacuum and insulation uses. Table XXVIII summarizes some of the pertinent properties of four different kinds of these ceramics. They all have a maximum operating temperature around 1000°C. They are all very good insulators, both for ac and dc operation. Figures 19 and 20 show a comparative temperature dependence of the dielectric constant and the loss angle of several silicate ceramics [compared to fused silica (quartz) in dotted line]. Silicate ceramics have very high resistivities, although they are highly humidity dependent due to the ceramics' tendency to absorb and adsorb water. The electrical breakdown strength of the silicates is very much temperature dependent above a certain "transition temperature" (between 100–200°C), where it falls rapidly. The thermal conductivity of ceramics is well below that of metals but higher than most glasses, values between $3–8 \times 10^{-3}$ cal deg^{-1} cm^{-1} sec^{-1} being typical. The thermal shock resistance depends on the type of ceramic and is higher the smaller the coefficient of thermal expansion. For

TABLE XXVIII

Property		Hard porcelain	Steatite	Forsterite	High alumina ceramics	Units
Density		2.3–2.5	2.5–2.8	2.8–3.1	3.4–3.5	gm cm^{-3}
Maximum operating temperature		1000	1000	1000	1350	°C
Tensile strength		2.8–5	5–10	7	20	kg mm^{-2}
Compressive strength		35–55	50–100	60	210	kg mm^{-2}
Bend strength		4–10	11–17	14	32	kg mm^{-2}
Impact bend strength		1.7–2.2	3.7–5	4.3	6.5–8.6	kg cm cm^{-2}
Young's modulus		4900–9800	9100–12,600	—	23,800–38,500	kg mm^{-2}
Moh's hardness		7–8	7.5	7.5	9	
Coefficient of linear thermal expansion	25–100°C	30–60	60–80	≈90	40–60	10^{-7} deg^{-1}
	25–700°C	50–70	80–90	100–110	60–90	10^{-7} deg^{-1}
Thermal conductivity		0.0003–0.0004	0.0006	0.0008	0.020	cal cm^{-1} sec^{-1} deg^{-1}
Elec. rupture strength (50 c/sec)		100–380	350–450	300–450	100–500	kV cm^{-1}
Electrical resistivity (50 c/sec)	20°C	$>10^{14}$	$>10^{14}$	$>10^{14}$	$>10^{14}$	Ω cm
	300°C	10^{7}	10^{10}–10^{11}	10^{11}–10^{12}	10^{8}–10^{9}	Ω cm
	600°C	10^{4}–10^{5}	10^{7}	10^{9}	10^{5}–10^{6}	Ω cm
	900°C	10^{3}	10^{5}–10^{6}	10^{6}–10^{7}	10^{4}	Ω cm
T_e value (10^6Ω cm)		450	820	>1000	600–700	°C
Dielectric constant ϵ		5–7	5.5–6.5	6.2–6.4	8–10	
Loss angle tan δ at 1 MHz, 20°C		60–120	3–5	≈2	6–40	10^{-4}

ceramics this coefficient is almost linear to the softening point and reversible unlike the case of glasses.

We shall discuss here only a few of the most common silicate ceramics, used in vacuum technology: hard porcelains, Mg–Al silicates, alkali-free Al ceramics, and zircon ceramics.

Hard porcelains are dense ceramics composed of silica, alumina, and Na_2O (or K_2O). It is made by sintering clay (kaolin), quartz sand, and feldspar. Hard porcelain is very similar to technical glasses, due to its high silica content. It softens around 800°C with thermal expansion coefficient in the hard glasses range ($\sim 40 \times 10^{-7}$ deg^{-1}). Therefore, porcelains show good vacuum characteristics and high thermal shock resistance. The melt of hard porcelains is relatively easy to work with and large structures can be obtained from it. This fact, combined with the electrical breakdown strength and weather resistance of porcelains, made them the most common line frequency insulators in the laboratory and outdoors.

Magnesium–aluminum silicates are well suited for and widely used in vacuum technology. They are usually sintered to specific compositions containing SiO_2, MgO, and Al_2O_3. They can be found also as natural high purity minerals such as soapstone that can be easily machined and then fired. There are several types of Mg–Al silicates.

Steatites, which are composed mostly of soapstone (with 10% each of

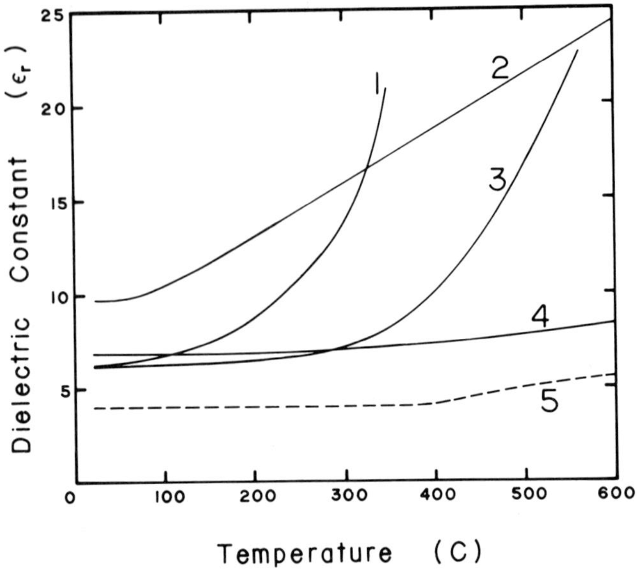

FIG. 19. Dielectric constants of some ceramics at 0.1 MHz as functions of temperature. (1) High-voltage porcelain; (2) Zirconia porcelain; (3) steatite; (4) special steatite; (5) fused silica (for comparison).

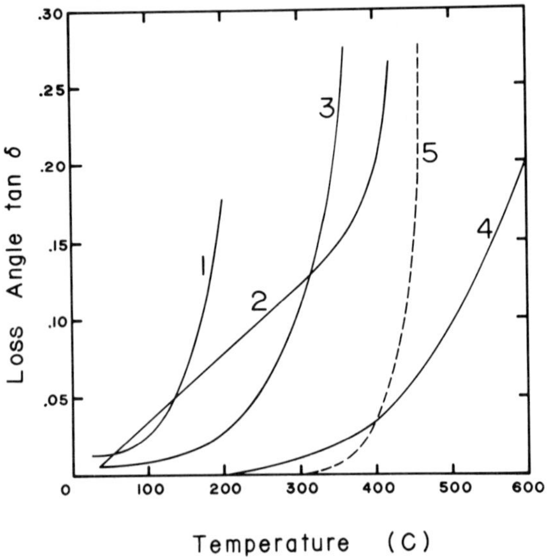

FIG. 20. Loss angle tan δ of some ceramics measured at 0.1 MHz as functions of temperature. The curves are numbered as in Fig. 19.

clay and feldspar), have a dense crystalline, and pore-free structure. This gives rise to their superiority over porcelains in both mechanical strength and insulating properties for ac and dc (see Table XXVIII). "Special steatites" which can be made by reduction of the alkali content, show even lower dielectric losses (see Figs. 18 and 19).

Forsterites are made from quality steatite batches with increased MgO content. Forsterites are strong and hard, have a higher thermal conductivity, and considerably higher resistivity at high temperatures than steatites. Their dielectric losses are also considerably lower than steatites. The higher thermal expansion of forsterites permit hard-soldered joints to Fe–Ni alloys, but results, on the other hand, in poorer thermal shock resistance.

Cordierite is another type of ceramic from this group, which has a higher alumina content than the other two. This imparts low thermal expansion, making this ceramic highly thermal shock resistant. Therefore the main use of cordierites is in applications which are exposed to sudden temperature changes. The insulation properties of cordierites, though better than porcelains, are inferior to steatites and forsterites.

The chief materials for electrical insulation in vacuum are the alumino–silicates, which have low MgO content and almost no alkalies. This last provision imparts better insulation properties and low dielectric losses, even at elevated temperatures. The alumino silicates excel in extremely high refractoriness and mechanical strength. Higher alumina content imparts extreme hardness, very low dielectric losses, and vacuum tightness. This, however, increases the firing temperature and the relative cost.

Zircon ceramics are obtained by adding $ZrO_2 \cdot SiO_2$ to steatite batches, resulting in greater ease of frabrication, especially for ceramic metal seals. Those ceramics also have low thermal expansion, and therefore higher thermal shock resistance. Zircon ceramics are dielectrically inferior to special steatites, but can rival porcelains in their mechanical and electrical properties and also ease of fabrication. Due to their higher glass content they have better vacuum tightness.

8.4.2. Pure Oxide Ceramics

Some properties of pure oxide ceramics are given in Table XXIX. The most commonly used is alumina which is discussed below. Other oxide ceramics, which are superior to alumina in some respects are used only for special requirements, where their considerably higher price is justified. The pure oxide ceramics have a crystalline structure, contrary to the silicates. Therefore, they have different characteristics. They are generally used when higher temperature durability is necessary.

TABLE XXIX. Properties of Dense-Sintered Pure-Oxide Ceramics

		Al_2O_3	MgO	$Al_2O_3 \cdot MgO$ (72:28)	ZrO_2	BeO	ThO_2
Molecular weight		102	40		123	25	264
Density	gm/cm³	3.97–3.87	3.39–3.57	3.53–3.58	5.5–5.7	2.7–3.0	9.6–10.0
Melting point	°C	2015–2050	2800 (2672)	2135	2700–2715	2550 ± 25	3000 (3300)
Maximum temperature for use							
in air	°C	1950	2400		2500	2200	2700
in vacuum	°C	1800	1600–1700	1800	1700	2000–2100	>2300
Specific heat 100°C	$\frac{cal}{gm\ °C}$	0.206	0.234	0.194	0.120	0.299	0.06
1000°C	$\frac{cal}{gm\ °C}$	0.261	0.280	0.257	0.157	(900°C:0.497)	—
Thermal conductivity	$\frac{10^{-3}\ cal}{cm\ sec\ °C}$	20°C:47–90 1000°C:12	20°C:100 1000°C:15	200°C:30 1000°C:13	Low	140–>200	100°C:20–24
Thermal shock resistance		Very good	Poor	Average	Very Poor	Very good	Poor
Coefficient of linear thermal expansion 25–1500°C	$10^{-7}\ deg^{-1}$	96	160	96	65	92–93.5	104
Tensile strength	kg/mm²	20°C:26.5	(980°C:2.2)	20°C:13.5 1300°C:0.8	20°C:15–18 1500°C:1.3	20°C:9.7 1300°C:0.45	20°C:7–10.5 1000°C:7–10
Compressive strength 20°C	kg/mm²	300	(78)	173–190	210	80	150
1600°C		5	—	6	Soft	5	(1400°C:4)
Young's modulus	kg/mm²	20°C:34,800 1250°C:24,200	20°C:3400–8900 900°C:2860–7500	20°C:22,500 1250°C:8600	20°C:17,500 1250°C:1100	20°C:31,000 1250°C:11,000	20°C:14,000 1170°C:8500
Electrical resistivity	Ω cm	1000°C:10^6–3×10^7 1600°C:$(1–5) \times 10^3$	100°C:$(1–3) \times 10^5$ 1500°C:10^3	1000°C:10^5–10^6 1600°C:10^3–10^4	1000°C:10^3–10^4 2000°C:0.6	1000°C:10^7–10^8 2000°C:10^3–10^4	1050°C:10^4
Dielectric constant ϵ	—	9.5–12	8–10	9		7.5	17
Dielectric loss angle tan δ	10^{-4}	5×10^7 c/sec: 10–20	2×10^3 c/sec: 800	2×10^9 c/sec: 8		2×10^9 c/sec: 3	2×10^9 c/sec: 15

Aluminum oxide is also called alumina, alundum, or sapphire. This commercially available material consists mainly of Al_2O_3. It has very high resistivity and good thermal conductivity together with chemical inertness even at high temperatures. Alumina is very hard (9.0 on Moh's scale). It is not as strong as typical metals at room temperature but keeps its strength at high temperatures (see Fig. 21). Alumina has a low vapor pressure and it is vacuum tight to 1950°C. It resists reduction and corrosion by all gases except fluorine.

Alumina is very useful in vacuum technology, mainly for insulation and supports or spacers and for ceramic to metal seals. It can be used at high temperature and in highly corrosive environments (even HF). Its main disadvantage is due to its extreme hardness. It can be machined only by grinding or with diamond tools. Pure MgO ceramic can operate at even higher temperatures than alumina, but has a vacuum limitation due to its vapor pressure (10^{-5} Torr at 1000°C). Contrary to alumina, pure MgO ceramic is considerably less hard, strong, and shows poor thermal shock resistance. Its insulating characteristics, especially ac, are also much less impressive. This ceramic is, however, relatively cheap and is used mainly for furnace linings.

Beryllia (BeO) is another useful ceramic, which is superior to alumina in many respects. As alumina, it has a high operating temperature (to 2000°C), very low vapor pressure, and thermal expansion. It is very hard but light in weight. It is superior to alumina in having an excellent thermal shock resistance and a much higher heat conductivity, which is also higher than steel. Although it has high dielectric strength, its electrical resistivity drops to 100 Ω cm at 1000°C, while alumina retains a value of 1 MΩ cm at that temperature.

Pure ZrO_2 ceramics are hard to process by usual ceramic methods due to their high shrinkage and a transformation point at 1100°C. They can be partially stabilized by adding CaO and MgO. The properties of stabilized ZrO_2 are given in the appropriate column in Table XXIX. Pure ZrO_2 ceramics are used only for very high temperature applications, where alu-

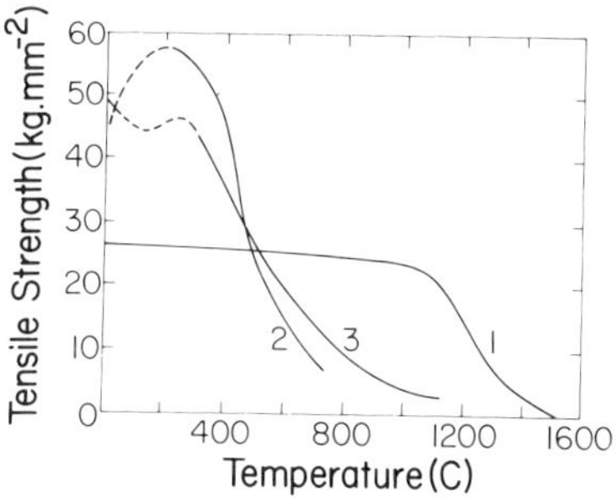

FIG. 21. Tensile strength of pure alumina ceramic (1) as a function of temperature compared with annealed mild steel (2) and nickel (3).

mina is not serviceable. They show high thermal expansion and poor thermal shock resistance.

Pure ThO_2 ceramics are the most refractory of all ceramics and have an extremely low vapor pressure. They are used in cases of extreme high temperature and oxidizing conditions. The main disadvantage of this type of ceramic is its very high price and its tendency to give off radioactive gases, as do all thorium products.

8.4.3. Machinable Ceramics (Macor, Lavas, BN)

Macor. This is a brand name for a new machinable glass–ceramic, manufactured by Corning, which was recently discovered by Beall.[3] This material, which can be described as a real breakthrough in ceramic technology, consists of fluorophlogopite phase mica, which is controllably crystallized in a boro alumino silicate matrix. The mica crystallites form randomly oriented blocks rather than sheets, resulting in a fracture-arrest mechanism, which imparts both thermal shock resistance and machinability to close tolerances with conventional metal working tools. Macor still retains the usual hardness and strength of ordinary ceramics without the need of firing after machining.

In addition to machinability, Macor shows no porosity or water absorption, very high resistivity (10^{14} Ω cm) and dielectric strength (1.4 MV/cm at 25°C), and low loss angle for good insulation for both ac and dc operations. Macor shows a good chemical durability, low helium permeability, no outgassing, and it is easily sealed to metal or glass by solder glass, metalizing, or bonding. Although its thermal expansion is as high as soft glasses, which enables easy sealing to metals and glasses, it has good thermal shock resistance and a maximum operating temperature of 1000°C.

Due to the possibility of easily machining Macor to precision tolerances, it is currently used for fabricating accurate fixtures of complicated shapes. Together with its other excellent properties it will definitely find a wide use as a ceramic material in all vacuum technology uses.

Lavas. Lavas, a brand name of the American Lava Corporation, are the classical "machinable" ceramics. They are supplied in the unfired "green" state, in which they can be machined by conventional methods like brass. The only precaution that should be taken is for the dust, which is abrasive. High temperature firing changes the Lava into a very hard material by driving out absorbed water. This process, together with crystalline changes, induces volume shrinking or growing of about 2% de-

[3] G. H. Beall, *Adv. Nucleation Crystallization Glasses* (L. L. Herch and S. W. Frieman, eds.), *Amer. Ceram. Soc. Spec. Publ.* 5, p. 251. Frieman, Columbus, Ohio, 1972.

pending on the type of Lava. Therefore, for close tolerances, the fired part has to be ground to precision.

The firing has to proceed very carefully to prevent cracks and major distortions. The Lava pieces should be protected from direct flame, and the heating rate should not exceed 150°C/hr for pieces thicker than 1 cm. A maximum temperature of 1015–1093°C should be held for 30–45 min for pieces of thickness 0.6–1.8 cm, respectively. Then, cooling is obtained by shutting off the heat without opening the oven door. Fired Lava should not be taken out before its temperature drops below 90°C.

Lavas come in two grades: grade A is hydrous aluminum silicate and grade 1136 is hydrous magnesium silicate, both natural stone products. The first expands on heating (above 650°C), its volume changing by about 2% at temperatures above 870°C, and if it is to be used in vacuum, should be fired in a hydrogen atmosphere.

Some properties of Lavas are given in Table XXX. Their main differences after firing are in their coefficient of thermal expansion and their dielectric loss angle. Grade A Lava is therefore considerably less insulating at higher frequencies.

Boron Nitride—BN. An exceptional member of the III–V compounds, BN still retains the typical crystal structure but has ceramic-like proper-

TABLE XXX. Properties of Fired Lava[a]

Property	Grade A	Grade 1136	Unit
Density	2.35	2.82	gm cm^{-3}
Water absorption	2.5	2.5	%
Softening temperature	1600	1475	°C
Maximum operating temperature	1100	1250	°C
Hardness (Moh's scale)	6	6	
Coefficient of linear thermal expansion:			
25–100°C	2.9×10^{-6}	11.3×10^{-6}	deg^{-1}
25–600°C	3.4×10^{-6}	11.9×10^{-6}	deg^{-1}
Electrical resistivity			
25°C	$>10^{14}$	$>10^{14}$	Ω cm
100°C	6×10^{11}	9×10^{12}	Ω cm
900°C	5×10^{4}	4.2×10^{5}	Ω cm
Dielectric constant (1000 kHz 25°C)	5.3	5.8	
Loss factor (1000 kHz 25°C)	0.0530	0.002	

[a] All data on Lava are from Bull. No. 576 of the American Lava Corp., subsidiary of Minnesota Mining, Chattanooga, Tenn. By permission.

ties. BN has an extremely high resistivity and dielectric strength coupled with an excellent thermal conductivity and thermal shock resistance. Its dielectric strength is more than 500 kV/cm at high frequencies and does not change at high temperatures. These properties make it highly suitable for all kinds of electrical insulating parts without losing the ability to transfer heat. It also shows excellent machining to close tolerances and to high surface finishes which makes it superior to other ceramics.

Structurally BN is also very stable even at high temperatures. Its thermal expansion is low. It will stay mechanically and chemically stable in vacuum up to 1650°C. BN, unlike most ceramics, does not have to be fired after machining and is also less brittle.

Boron nitride is extremely resistant to many kinds of corrosion by molten metals and salts, corrosive acids and bases, and to oxidation up to 700°C. It is, however, affected by water (hydrolysis) and by humidity which tends to reduce its resistivity due to water adsorption. It should therefore be kept under strictly dry conditions after cleaning or vacuum firing. This restriction, together with the somewhat inferior mechanical properties and higher price of BN, limit its use relative to ordinary ceramics.

Pyrolytic BN is also available. It is a harder material, excellent insulator, and impervious to water.

8.5. Elastomers and Plastomers

The high elasticity organic polymers or elastomers are used in vacuum technology, mainly in gaskets for static and motion seals in demountable systems and for flexible tubing. Their main disadvantages for vacuum uses are their high vapor pressure and gas permeability, low operating temperature (inability to fully degas), and the general need for greasing the elastomer gasket. However, elastomers have many useful properties such as their flexibility, their resilience and reusability, and their low thermal and electrical conductivity. Those properties, together with development of new low vapor pressure and low cost elastomers, make those materials very attractive for use in dynamic systems which do not need high temperature bakeouts.

When working with elastomers, one should note several important characteristics. They include vapor pressure, gas permeability and outgassing, working range of temperature, and elastic properties such as compressibility and permanent deformation after compression. The operating temperature is extremely important because most of the properties of elastomers are sharp functions of temperature. Most notably, the elastic properties of the organic elastomers are changed drastically from embrittlement at low temperature to softening and decomposition at higher temperatures. A comparative survey of some general properties of several elastomers is given in Table XXXI.

When elastomers are used for gaskets, several rules should be noted. Rubber is fairly incompressible and subject to plastic flow, especially when the temperature goes up. Therefore, flange grooves should be designed to allow for deformation, the flanges should not be pulled too tight, and the gasket should be heat shielded. The latter holds especially when the gasket is greased because temperature range and vapor pressure of the grease may be the limiting factors. Generally speaking, the groove design should be such that there is enough room for deformation of the gasket without filling the groove totally, that the part of the gasket ex-

TABLE XXXI. General Properties of Various Elastomers

Chemical and physical Properties	Natural rubber	Nitrile rubber or Buna N	Neoprene	Silicone	Viton A
Composition		Butadiene acrylonitrile	Chloroprene	Polysiloxane polymer	Copolymer of vinylidene fluoride and hexafluoro-propylene
Tensile ($lb/in.^2$)					
Pure gum	Over 3000	Below 1000	Over 3000	Below 1500	Over 2000
Black	Over 3000	Over 2000	Over 3000		Over 2000
Practical hardness range					
Shore Duro A	30–90	40–95	40–95	40–85	60–90
Rebound					
Cold	Excellent	Good	Very good	Excellent	Good
Hot	Excellent	Good	Very good	Excellent	Excellent
Tear resistance	Good	Fair	Fair to good	Poor	Fair
Abrasion resistance	Excellent	Good	Excellent	Poor	
Ozone resistance	Fair	Fair	Very good	Very good	Very good
Sunlight aging	Poor	Poor	Very good	Excellent	Excellent
Oxidation resistance	Good	Good	Excellent	Excellent	Excellent
Heat resistance	Good	Excellent	Excellent	Outstanding	Outstanding
Solvent resistance					
Aliphatic Hydrocarbons	Poor	Excellent	Good	Poor	Excellent
Aromatic Hydrocarbons	Poor	Good	Fair	Poor	Excellent
Oxygenated solvents, alcohols (ketones, etc.)	Good	Poor	Poor	Fair	Poor
Oil and gasoline	Poor	Excellent	Good	Fair	Excellent
Acid resistance					
Dilute	Fair to good	Good	Excellent	Excellent	Excellent
Concentrated	Fair to good	Good	Good	Fair	Good
Flame resistance	Poor	Poor	Good	Fair	Good
Permeability to gases	Fair	Fair	Low	Fair	
Electrical insulation	Good	Poor	Fair	Excellent	Excellent
Water swell resistance	Fair	Excellent	Fair	Excellent	Excellent

posed to the vacuum side be minimal, and that the groove be smooth with no sharp edges. The greasing, which should be used only in grooved flanges, is to be used in small amounts, just enough to make the elastomer shiny. If possible, the best quality greases and oils should be used. (Some of them are referenced in Chapter 8.6.) For small loading of flanges a design, different than groove-to-flat mating, is sometimes preferred. These designs are similar to the step or knife edge seals used for metal gaskets. Naturally, no sharp edges are allowed in this case.

In case of corrosive atmospheres, or other dangers to the elastomer gasket from bombardment heating or chemicals, the elastomers can sometimes be clad with metal or Teflon sheath.

8.5.1. Natural and Synthetic Rubbers

Rubbers are available commercially in three general categories: vulcanized rubbers made of natural latex, synthetic rubbers or neoprene, and silicone rubbers.

Soft rubber is made of natural crude rubber which is usually mixed with fillers and additives, vulcanized and sulphurized. These treatments obviously cause rubber to give off large amounts of gases, usually hydrogen and hydrocarbons, and of volatile vapors, besides making its surface rather porous. Natural rubber has a high vapor pressure at room temperature ($\sim 10^{-4}$ Torr) and it is also highly permeable to most of the gases, mainly H_2 and CO_2. The gas permeability of rubbers, compared with other materials, for certain gases is given in Fig. 22. Other mechanical and electrical properties of natural rubbers depend very much on their preparation and average values are given in Table XXXII. A survey of the main properties of soft rubbers makes it clear that their use in vacuum technology is limited only to systems operating at or around room temperature, under constant pumping, and restricted to very moderate vacuum. The common uses of rubber have included gaskets and tubing. For these uses it is generally being replaced by synthetic materials, which will be surveyed below.

Neoprene is a brand name for a chlorobutadiene polymer (manufactured by E.I. duPont) which has a variety of formulations, all resulting in a rubberlike, elastic, and resilient material. Some of the physical properties of Neoprene are given in Table XXXIII; some others have been summarized in Table XXXI.

Although Neoprene shows many qualities very similar to natural rubber, such as toughness, resilience, and elasticity, it is superior to natural rubbers in many other aspects. Neoprene has a much better resistance to oils, greases, sunlight, ozone, weathering, and high temperatures

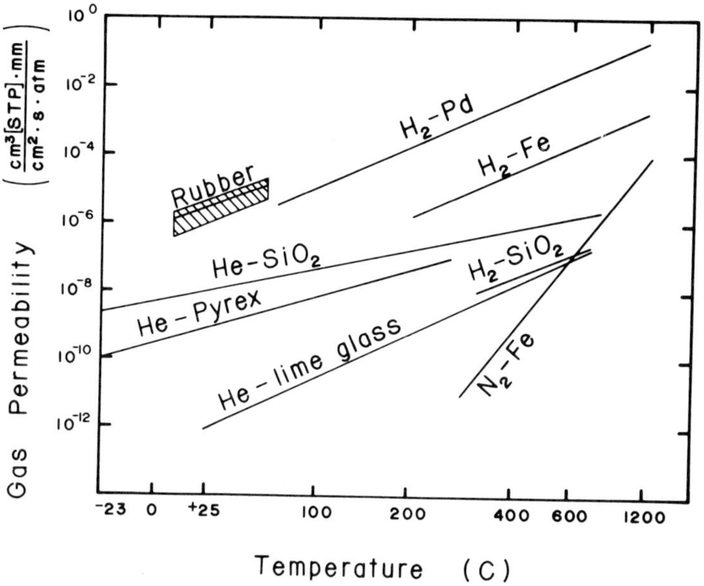

FIG. 22. Gas permeability of rubber (shaded area) as compared to that of several metals, glass and quartz for different gases as functions of temperature.

TABLE XXXII. Properties of Soft Rubber

Density		0.93	gm cm^{-3}
Working temperature	Continuous	−30– +60	°C
	Transient	to +75	°C
Specific heat		0.50	cal gm^{-1} deg^{-1}
Thermal conductivity		3–5 × 10^{-4}	cal cm^{-1} sec^{-1} deg^{-1}
Coefficient of linear thermal expansion	20–120°C	1.4–2 × 10^{-4}	deg^{-1}
Tear strength		80–250	kg cm^{-2}
Rupture elongation		350–1000	%
Young's modulus		10–70	kg cm^{-2}
Permanent elongation	After 22 hr of 100% extension at 20°C	14	%
	at 100°C	40	%
Electrical resistivity		10^{13}–10^{16}	Ω cm
Electrical breakdown strength	25°C	16–24	kV mm^{-1}
Dielectric constant		2–4	
Loss angle tan δ	10^3 Hz	~10^{-2}	

(up to 150°C). Neoprene is much less permeable to different gases than is natural rubber (10–25% of natural rubber values). It gives off considerably less dissociable hydrocarbons, but this gas liberation, consisting mainly of butane, limits its vacuum use. The liberation of butane is attributed to cleaning procedures and a fresh Neoprene which is degassed

TABLE XXXIII. Properties of Neoprene

Density		1.15–1.24	gm cm^{-3}
Working temperature	Continuous	−40–+90	°C
	Transient	to +150	°C
Specific heat	15°C	0.4	cal gm^{-1} deg^{-1}
Thermal conductivity	25°C	5×10^{-4}	cal cm^{-1} sec^{-1} deg^{-1}
Coefficient of linear thermal expansion		7×10^{-4}	deg^{-1}
Tear strength	20°C	150	kg cm^{-2}
	90°C	55	kg cm^{-2}
Rupture elongation		250–400	%
Permanent elongation	After 22 hr of 100% extension at 20°C	65	%
	at 70°C	Destroyed	
Electrical resistivity		10^{12}	Ω cm
Dielectric constant		6.7–7.5	
Loss angle tan δ		250	10^{-4}

for several hours at 90°C can be pumped down to the 10^{-7} Torr range. Such degassing can also remove water vapor which can be absorbed by Neoprene and cause some swelling.

Neoprene resists attacks by most dilute acids and alkali solutions as well as alcohols. It is, however, severely damaged by nitric or chromic acids, acetone, and by electron bombardment. Neoprene is used widely as gaskets with different cross sections, most commonly the round type called O-rings. More recent and better types of materials for O rings are discussed below.

Silicone rubbers are synthetic high polymers made by hydrolysis of pure dimethyl dichlorosilane. The polymers cannot be sulfurized as natural rubber and its vapor pressure is about 1×10^{-6} Torr at room temperature. A summary of some physical properties of silicone rubbers is given in Table XXXIV (see also Table XXXI). Properties can vary according to formulation.

Silicone rubber has some superior qualities relative to other rubbers. Its working temperature range extends from −80°C to above 300°C (momentarily) and it can continuously withstand temperatures up to 180°C. Silicone rubbers are also superior to other types regarding electrical insulation, resistance to oxidation, ozone, uv irradiation, and, especially in some types, resistance to permanent set throughout the temperature range. The latter property enables excellent reusability of gaskets and O rings made of silicone rubber. Although the heat, weather, and permanent set resistance of silicone rubbers is relatively high, they have disadvantages relative to other rubbers. They have a relatively low tear and

TABLE XXXIV. Properties of Technical Silicone Rubber

Density		1.1–2.1	gm cm^{-3}
Working temperature	Continuous	−40–180	°C
	Transient	to 250	°C
Thermal conductivity		1×10^{-3}	cal cm^{-1} sec^{-1} deg^{-1}
Tensile strength		40–56	kg cm^{-2}
Rupture elongation		70–350	%
Permanent set	After 100% extension for 22 hr at 150°C	50–60	%
Electrical resistivity	To 50% relative humidity	7×10^{14}	Ω cm
Electrical breakdown strength		30	kV mm^{-1}
Dielectric constant		3–10	
Loss angle tan δ	10^2–10^8 Hz	12–280	10^{-4}

abrasion resistance, especially at room temperature. The porosity of silicone rubbers is higher than Neoprene and although their vapor pressure is low, their high permeability and formation of capillary leaks, generally prevents reaching ultimate vacua better than 10^{-6} Torr in systems using silicone rubber gaskets. In spite of the good chemical resistance of silicone rubbers, especially to mineral oils, it should not be exposed to strong acids or alkalies and should not be cleaned with any solvents. If necessary, cleaning should be restricted to mild detergent and thorough rinsing. Silicone rubbers have no corrosive effects on metals and no physiological effects.

Viton. Viton is a brand name by E. I. duPont for a synthetic elastomer very useful as gasket material even for ultrahigh vacuum systems. Viton gaskets are resilient and strong, reusable and serviceable up to 270°C (continuous). Viton gaskets are superior to Neoprene because, while maintaining the same elasticity, they exhibit considerably lower hydrocarbon gas evolution, mainly butane, and higher serviceable temperature. With continuous pumping vacua better than 1×10^{-9} Torr can be obtained with Viton gaskets. Viton is a long chain polymer (molecular weight 60,000) which is resistant to hydrogen peroxide, oxygen, ozone, many solvent oils, fuels, and lubricants.

8.5.2. Plastomers

Teflon. Teflon is a trade name by E. I. duPont for tetrafluoroethylene made by polymerization of chloroform and hydrofluoric acid under high pressure. Molding and sintering the resulting powder at 370°C produces a tough resin which is not thermoplastic and not porous. Teflon, though mostly crystalline, does not have a melting point, but turns into nonflowing amorphous jelly above 327°C and can then be molded and sintered. Above 400°C it decomposes, releasing poisonous fluorine volatiles.

Some properties of Teflon are given in Table XXXV. Teflon is a strong and tough polymer which is notable for low friction coefficient to itself or to steel. Teflon is highly resilient and will return to its original dimensions even after 300% elongation, the extent of elastic recovery increasing with increased temperature. However, Teflon is not as elastic and compressible as rubbers, and tends to flow and even split under high loads. It should therefore be used as gasket material only in grooved flanges and not loaded to more than 35 kg/cm^2.

Teflon is superior to rubbers, being nonporous and having considerably lower vapor pressure as well as degassing rate. The permeability to gases of fluorocarbon resins is comparable to Neoprene. Teflon is serviceable and elastic from −80°C to a maximum of 200°C, and does not lose strength even at 300°C. Machining Teflon can be done with ordinary tools. Drilling should be done preferably in a lathe with special drills used for plastics. Teflon should be stress relieved before machining by annealing at 300°C.

Electrically, Teflon is a remarkable insulator exhibiting extremely high resistivity and low dielectric loss. Since it does not adsorb or absorb

TABLE XXXV. Properties of Teflon (PTFE)

Structure	To 327°C	Crystalline	
	Above 327°C	Amorphous jelly	
	Above 400°C	Decomposes	
Density		2.1–2.2	$g\ cm^{-3}$
Working temperature		−50–+250	°C
Specific heat		0.25	$cal\ gm^{-1}\ deg^{-1}$
Thermal conductivity		6×10^{-4}	$cal\ cm^{-1}\ sec^{-1}\ deg^{-1}$
Coefficient of linear thermal expansion		1500×10^{-7}	deg^{-1}
Tensile strength	90°C	140–350	$kg\ cm^{-2}$
Young's modulus		4300–5200	$kg\ cm^{-2}$
Rupture elongation	20°C	200–400	%
Permanent set			
Tested on rings	200 $kg\ cm^{-2}$ for 24 hr		
55 × 75 × 2 mm	at 20°C	2	%
	at 100°C	12	%
Electrical resistivity		$>10^{19}$	Ω cm
Electrical breakdown strength		20	$kV\ mm^{-1}$
Dielectric constant	60–10^8 Hz	2	
Loss angle tan δ	25°C	5	10^{-4}

water vapor, it maintains its extremely high surface resistivity even at 100% relative humidity. This property together with the arcing resistance of Teflon make it extremely useful in all kinds of insulation needs.

Chemically, Teflon is extremely inert, more so than any other elastoplastic material. It resists all known acids or alkalies and all weather effects. It cannot be wetted by water and has no known solvents, but dissolves in molten alkali metals. Teflon is nonflammable and nontoxic but will give off considerably poisonous vapors if heated above 315°C. Its extreme inertness makes it impossible to bond or cement without using special techniques. In case of Teflon joint usage, attention should be given to the maximum operating temperature of the adhesive.

It should be noted that the indicated characteristics of Teflon, primarily mechanical and thermal, can be considerably modified by addition of fillers, such as graphite, asbestos, or glass fibers. Teflon is used in vacuum technology for gaskets, seals for permanent or motion feedthroughs (e.g., vee rings), insulation parts, and for low friction moving parts.

Tygon. Tygon is a brand name for a vinyl plastomer widely used for flexible tubing of various sizes and bores and is manufactured by Norton Company. Tygon is superior to rubber with regard to porosity, cleanliness, vapor pressure (10^{-7} Torr at room temperature), and temperature and weathering resistance. The type of tubing designated for vacuum work (R-3603) has extra thick walls to prevent collapsing and according to the manufacturer has the following properties:

Tensile strength	1500 psi
Rupture elongation	400%
Brittle at	−46°C
Flexible at	−29°C
Serviceable up to	74°C
High temperature resistance:	
dry heat	93°C
steam	Fair

Tygon is clear plastic, tasteless, and nontoxic. It has a slight odor, burns in flame, but has excellent weather and oxidation resistance. Tygon resists corrosive atmospheres, oils, weak acids, and alkalies. It is attacked by ketones and esters.

Polyethylene and Polypropylene. Polyethylene and polypropylene are thermoplastic high molecular weight polymers of ethylene and propylene, respectively. They are distinguished by extreme inertness, and will withstand almost any chemical at room temperature at exposures up to 24 h. Only at higher temperature are they subject to oxidation by gases or acids. This effect can be enhanced by uv illumination. Halogens and organic acids can diffuse through or be absorbed by these thermoplastics and some swelling can be caused by some hydrocarbons such as carbon-tetrachloride and trichloroethylene. Although their vapor pressure is in the 10^{-9} Torr range, the use of these materials in vacuum technology is very limited because of their strength and temperature limitations.

Nylon. Nylon, like Zytel V by E. I. duPont, is a thermoplastic polymer that is notable for a rather sharp transition point (about 230°C), above which it loses its rigidity and flows as oil. Nylon should be used therefore below 100°C. In this range it exhibits high machinability, strength and toughness, low vapor pressure, high resistance to most solvents, acids, oils, and other chemicals. Zytel is dissolved in phenols and formic acids. It absorbs moisture, and swelling (to 0.6%) should be considered when close tolerances are required. Mylar is a durable and strong form of transparent nylon film.

Phenolics. Thermosetting plastomers which should generally not be used in vacuum technology because of high vapor pressure.

Acrylics. Acrylics such as Plexiglas (by Rohm and Haas Co.) or Lucite (by duPont) are light, transparent, very easily molded, and machinable thermoplastics. They exhibit low water absorption, high dielectric strength, and are chemically resistant, but their vapor pressure can at best be reduced to 5×10^{-7} Torr after suitable baking.

8.6. Sealants, Waxes, Greases, and Oils

8.6.1. Sealants and Lubricants

Greases and glues are often used for sealing purposes in vacuum technology. The use of greases is most common in demountable systems, for improving elastomer gasket seals, whether in fixed or motion seals, and in glass systems for lubricating stopcocks, valves, and ground joints.

Glues such as picein or epoxy resins are used for permanent seals of leaks or joints or for joining different materials such as in metal–mica seals.

Oils are rarely used as sealants. Oils, such as Apiezon J or K, can be found mostly in specific applications, such as on large valves or rotating shaft seals. Table XXXVI contains some pertinent properties of commonly used sealing greases. As all tables in this subsection it should be regarded as indicative only and the user should look into the details of the types of sealants and lubricants available locally at that time.

The main problem in using sealants and lubricants such as greases is their relatively high vapor pressure, especially when they are fresh. The

TABLE XXXVI. Properties of Some Sealing Greases

Name	Supplier[a]	Vapor pressure (25°C)	Maximum working temperature (°C)
Apiezon L	I	$10^{-9}-10^{-10}$	30
Apiezon T	I	10^{-8}	110
dc high vacuum grease	II	10^{-7}	(−40 to) 200
Leybold R	III	Fresh; 10^{-6} Degassed at 90°C; 10^{-8}	30
Vacuseal	IV	10^{-5}	40

[a] Suppliers: (I) James G. Biddle Co., USA; (II) Dow Corning Co., USA; (III) E. Leybold, W. Germany; (IV) Central Scientific Co., USA.

temperature dependence of the vapor pressure of some common sealants and greases is given in Fig. 23. Compounds such as Apiezon Q, Ramsay, picein, and some epoxy cements have vapor pressures in the $10^{-3}-10^{-4}$ Torr range. Greases can have much lower vapor pressure especially after degassing in vacuum. Apiezon L, for example, is rated below 10^{-9} Torr. Sealing oils such as Apiezon J and K have even lower vapor pressures.

Another point to be aware of is the temperature working range of the greases, which usually does not exceed room temperature by much. The limits of the temperature range for usage of grease are determined by the viscosity. This is another very important parameter of greases. The greases should not be too viscous, so that free motion of valves is assured. On the other hand, a low viscosity will cause leaking of the grease into the system under the external atmospheric pressure.

The common type of greases are the ones based on lanolin, on natural

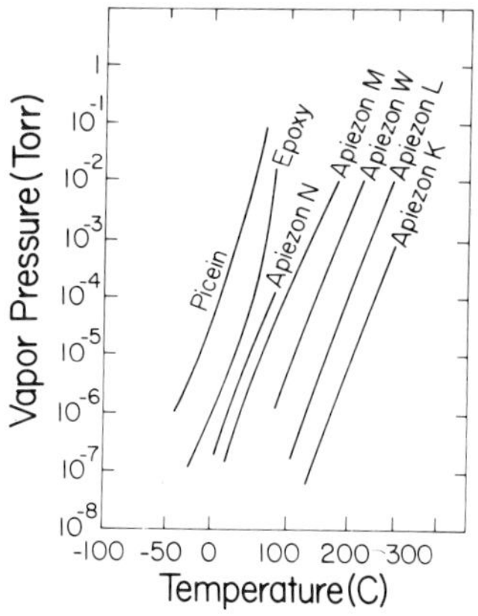

FIG. 23. Saturation vapor pressure as a function of temperature for several degassed oils (Apiezon K), greases (Apiezon N, M, and L), waxes (Apiezon W), and cements (picein, epoxy).

vaseline (e.g., Ramsay grease with vapor pressure around 10^{-4} Torr), on vacuum-distilled hydrocarbons (such as Apiezon or Celvacene having low vapor pressure but very limited working range), and on silicone greases, which have a low vapor pressure and keep a constant viscosity over a fairly large temperature range (up to 150°C).

For high and ultrahigh vacuum and especially for higher temperature applications, the use of molybdenum disulfide (MoS_2) as lubricant is quite common. Very clean metal surfaces in vacuum tend to "stick" when contacted. The layered structure of MoS_2 gives metals coated with it a smaller sliding friction coefficient than graphite (0.05–0.1). (Silver should also be mentioned as a lubricant in this context.) Molybdenum disulfide has a very low vapor pressure, is useful to above 400°C and has very high corrosion resistance.

For temporary and permanent seals different kinds of waxes, cements, varnishes, and resins are available. Properties of some commercially available sealants are shown in Table XXXVII. Some of them are very soft and easily removable such as adhesive wax, picein, and Apiezon Q compound. They are used as temporary seals usually for leak checking. Others can be hardened irreversibly such as epoxy resins (e.g., Araldite, see Table XXXVIII), Apiezon W, and hard wax. They are used mainly for vacuum-tight cementing of windows, connector, or permanent joints in vacuum systems. These glues have relatively low vapor pressure and can be heated to 90°C (Apiezon W) or higher (epoxy glues to 200°C). Silver chloride is an example of an inorganic glue that can be used for vac-

TABLE XXXVII. Properties of Some Sealants

Name	Supplier[a]	Vapor pressure −25°C	Vapor pressure +25°C	Softening point (°C)	Maximum working temperature (°C)
Picein	I	10^{-5}	8×10^{-4}	80	60
DeKhotinsky		—	7×10^{-3}	50–70	40–50
Apiezon Q (soft)	II	—	10^{-4}	—	30
Apiezon wax W-100 (medium hard)	II	—	$<10^{-7}$	80	50
Celvacene heavy	III	—	$<10^{-6}$	130	—
Silver chloride		—	$<10^{-8}$	455	300
Glyptal varnish		$<10^{-6}$	2×10^{-4}	—	100
Araldite	IV	—	$<10^{-6}$	—	60

[a] Suppliers: (I) E. Leybold, W. Germany; (II) James G. Biddle Co., USA; (III) Consolidated Vacuum Corp., USA; (IV) Ciba A.G., Switzerland.

TABLE XXXVIII. Properties of Araldite I

Curing temperature		120	°C
Tensile strength		7–8	kg mm^{-2}
Bend strength		12–13	kg mm^{-2}
Young's modulus		300–310	kg mm^{-2}
Electrical resistivity	20°C	10^{16}	Ω cm
Electrical breakdown strength	1 mm thick	60	kV mm^{-1}
Dielectric constant	20°C	6	
Loss angle tan δ	2 MHz	60	10^{-4}
	82 MHz	200	10^{-4}
Solubility	Water, benzene, alcohols		Insoluble
Chemical resistance	Acids and bases (30%)		Excellent

uum tight joints up to 300°C. Varnishes such as shellac and glyptal (by General Electric) are also used in vacuum technology for leak sealing or for preventing H_2 diffusion through metal walls. After drying and hardening they withstand temperatures to 200°C and have fairly low vapor pressure (around 1×10^{-6} Torr) together with good corrosion resistance. Varnishes used for sealing should not be applied to systems under vacuum or they will be forced inside by the atmospheric pressure.

8.6.2. Pump and Manometer Fluids

Organic fluids ("oils") used in pumps such as mechanical, diffusion or ejector pumps, or in manometers affect the ultimate pressure attainable in the vacuum system while they are used. The pressure cannot be lower than the vapor pressure of the oil at the coolest spot of the last stage of the device. This normally requires a good cold trap between the device and the system, generally employing liquid nitrogen. In other cases, oils with low enough vapor pressure are sufficient.

Oils for mechanical roughing pumps usually do not have to have vapor pressures below 10^{-3} Torr. With cold traps, better vacua are achieved. These oils are available commercially from pump manufacturers. Such mechanical pump oils are distilled from their low boiling point fractions and roughly outgassed in vacuum.

Diffusion pumps require oils of much better grade. They have to have vapor pressures below 10^{-6} Torr at room temperature. In good pumps, which are very well trapped, pressures below 10^{-10} Torr can be obtained.

Diffusion pump oils must also have as high a vapor pressure as possible at the boiler temperature for higher efficiency of pumping together with the lowest possible cracking tendency. Another additional requirement is a good oxidation resistance because a complete elimination of oxygen backflow generally cannot be achieved. This problem can be helped by a lower backing pressure. An excellent oxidation resistance is characteristic of silicone oils which also have relatively low vapor pressure (DC 703, DC 704, and DC 705 by Dow Corning). Although silicone oils are very adequate for vacuum use, particularly in systems which are frequently vented, their strong tendency for surface migration into the vacuum chamber limits their use to systems where cleanliness is not of utmost importance.

Diffusion pump fluids should also be nonhygroscopic and should not contain any other high vapor pressure impurities. Therefore they have to be very carefully vacuum distilled and kept under vacuum. Apiezon oils, especially Apiezon C, are examples of oils produced in that manner.

The oils used in diffusion pumps can also be used in manometers sometimes together with Hg. Their low vapor pressure together with their lower density make it possible to read lower pressures more accurately. They can enable reading accuracy of 0.1 Torr in U manometers and below 10^{-5} Torr for McLeod gauges. The only marked disadvantage of oils, which sometimes will not need any cold trapping, is their tendency to dissolve gases. They should therefore be evacuated and degassed slowly, especially when used for the first time. Care should also be taken for gases diffusing from the high pressure side to the low one. Those problems have been successfully encountered by structural designs published in the literature.

8.7. Gases—Preparation, Properties, and Uses

Gases can play a significant role in many processes of vacuum technology (and not always as something which should be pumped out). Of the most common uses, one can mention (a) processing of vacuum materials in inert, reducing, or oxidizing atmospheres, (b) conducting current in discharge lamps, (c) conducting heat from hot electrodes and for uhv techniques: ion bombardment and sputtering and production of controlled environments. Some of the physical properties of several important inert and active gases are given in Table XXXIX.

For most uses, ordinary tank gas is sufficient. Gas tanks of various specifications are commercially available and are used with suitable pressure reduction valves and, if necessary, with special precautions. In uhv techniques, pure gases can be obtained in special glass containers with various mechanisms for backfilling the gas into the uhv system, which should be well degassed previously.

8.7.1. Inert Gases

The inert gases are generally obtained by condensation and fractionation of air. Liquefaction of air and rectification, with additional chemical processes if necessary, can produce any of the inert gases with sufficient purity, 1000 m^3 air contains about 9 m^3 Ar, 15 liters Ne, 5 liters He, 1 liter Kr, and 80 ml Xe. He can also be obtained from natural gas wells in the United States.

A collection of several important properties of the inert, or rare gases, is shown in Table XXXIX. The rare gases are colorless, odorless, nonflammable, and nontoxic, and will normally not combine with other elements or with themselves, with the exception of C or silica gel. Besides He, they are not known to diffuse through glass or metal walls even at elevated temperatures. Helium diffuses through all types of glasses, especially at high temperatures.

Inert gases are used mainly for many types of discharge and hot filament lamps and for protective atmospheres for preparation and joining vacuum materials. Specific uses will be discussed individually.

Helium is the lightest rare gas. As all other rare gases it is inert and nontoxic. It is used as a hydrogen substitute in many industrial processes. Helium is slightly soluble in water (about 10 cm^3 $liter^{-1}$) and slightly absorbed by Pt.

The main use of He in vacuum technology is for leak detection by mass spectrometry. The very small radius of He and its inertness makes this a very convenient and sensitive technique. Helium is also used with Ne in He–Ne lasers and in the liquid state (below 4.2 K) as a refrigerant. It is less often used for visible light production because of the high voltage needed.

Neon is used mainly in all the varieties of neon lamps giving them the distinct red glow. Neon lamps with different additives can produce other shades and colors. Ne/Ar (75/25) mixture is well known for deep blue emission.

Argon is the third constituent in the earth's atmosphere (0.94% by volume). It is therefore the cheapest inert gas and often used for flushing and drying gas systems. Argon is soluble in water to a certain extent, slightly more so than oxygen. Argon is also known for a distinct violet light sometimes used in decorative discharge devices. It is used for filling fluorescent lamps, general purpose filament lamps, and in the famous argon laser. Argon is commonly used for argon-arc welding which is mostly used for light metal welding. Argon is also the most common backfilling gas for ion bombardment of solid surfaces for sputtering and cleaning.

TABLE XXXIX. Properties of

Property	Units	Inert gases				
		He	Ne	Ar	Kr	Xe
Molecular weight	gm/mole	4.003	20.183	39.944	83.7	131.3
Molecular (atomic) mass	10^{-24} gm	6.66	33.6	66.5	139	218
Gas–kinetical radius	Å	0.95	1.15	1.4	1.6	1.75
Weight per standard liter	gm	0.1786	0.9002	1.7838	3.708	5.845
Boiling point (760 Torr)	°C	−269	−246	−186	−153	−109
Mean free path λ_g of molecule (1 Torr 293 K)	10^{-3} cm	13.32	9.4	4.73	3.63	2.62
Specific heat c_p (760 Torr, 293 K)	cal/gm deg	1.25	(0.246)	0.125	(0.06)	(0.038)
Ratio of specific heats c_p/c_v (293°K)	—	1.63	1.64	1.648	1.689	1.666
Thermal conductivity						
0°C	$\frac{10^{-4}\text{ cal}}{\text{cm sec deg}}$	3.43	1.09	0.39	0.21	0.12
20°C		3.61	1.14	0.42	(0.23)	0.13
100°C		4.08	1.33	0.52	—	0.16
500°C		—	—	—	—	—
Ionization potential	V	24.58	21.56	15.76	13.99	12.12

Kr and Xe are the heaviest inert gases. As Ar, they are sometimes used for heavy ion bombardment. Their common use is for illumination and discharge devices, such as high and super high pressure lamps for uv light, photoflash, and general-purpose lamps.

8.7.2. Reactive Gases

Hydrogen. The lightest element, hydrogen, is highly reactive, combines with many elements, and is especially common in organic materials. It is flammable and explosive when mixed in air (above 4% by volume at atmospheric pressure) and therefore precautions for adequate venting should be taken when it is used. Hydrogen is not soluble in water. It has the highest ionization potential among the reactive gases and can be used in discharge devices, such as continuous spectrum uv sources. The breakdown voltage for H_2 (compared with air and CO_2) as a function of the gap between the electrodes multiplied by the gas pressure is given in Fig. 24.

Hydrogen is a widely used gas, especially due to its reducing action. It is very reactive and tends to attack metal oxides, producing pure metal and water vapor. This is utilized in many types of heat treatments such as firing, brazing, welding, annealing, and filament "forming," where protection from oxidation is needed. For general use, tank hydrogen, which

Important Gases in Vacuum Technology

Reactive gases							
H_2	N_2	O_2	Air	CO_2	CO	NH_3	H_2O
2.016	28.016	32.00	28.96	44.010	28.010	17.032	18.016
3.35	46.6	53.3	—	73.2	46.6	28.3	30.0
1.15	1.55	1.45	—	1.6	1.6	0.8	1.3
0.08987	1.2507	1.4290	1.2928	1.9768	1.2504	0.7708	—
−253	−196	−183	−193	−78.5	−191.5	−334	100
8.81	4.5	4.82	4.56	2.96	4.48	3.4	2.96
3.42	0.247	0.219	0.240	0.201	0.249	0.500	0.444
1.407	1.401	1.396	1.402	1.293	1.40	1.317	(1.33)
4.19	0.57	0.58	0.576	0.34	0.53	0.52	
4.45	0.61	0.62	0.61	0.38	—	0.57	—
5.47	0.73	0.76	0.74	0.50	—	0.80	0.60
6.80	1.15	—	1.84	0.68	—	—	0.78
15.8	15.7	12.1		14.4	14.1	11.2	13.0

is a by-product of benzene synthesis is sufficient. For high purity, H_2 obtained by water electrolysis, is preferred. The latter contains mostly water vapor and oxygen and can be purified by Pd or Pt filters, suitable drying tubes and/or cold traps.

Oxygen. Oxygen is a highly reactive gas and, contrary to hydrogen, is known for its oxidizing action. It forms oxides by contact with most

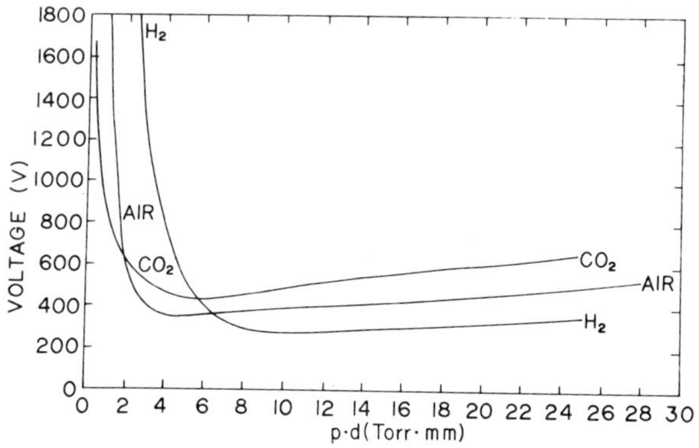

FIG. 24. Electrical breakdown voltage of H_2 (as compared with air and CO_2) as a function of electrode gap in millimeters multiplied by the gas pressure in torr.

metals, especially at elevated temperatures, and they should be protected if oxide formation is to be avoided.

Oxygen is one of the mjaor constituents of the earth's atmosphere (~21%) and is essential for combustion and flame production. In the liquid phase, it is explosive and should be used with serious precautions.

In most cases tank oxygen made from liquefied air or by water electrolysis is of sufficient purity. It can be further purified and dried. Small quantities of pure O_2 can be produced by heating of pure $KMnO_4$ to 350°C in vacuum. The same as true for barium peroxide and silver oxide. The oxygen produced can be spectrally purified by diffusion through a heated silver filter.

N_2. Nitrogen, the major constituent of the earth's atmosphere (78%) is an odorless, colorless, and generally a chemically inert gas. Only at elevated temperatures can it form some stable nitrides with some metals. Inertness and low cost make nitrogen useful for purging and flushing systems, for protecting and storing water vapor and oxygen sensitive items, and in torchbrazing or glassblowing of parts which should be protected from oxygen. In the liquid phase it is a common refrigerant. A major use of N_2 is in neutralizing the explosion hazard of H_2, which is achieved in a N_2:H_2 mixture with less than 67% H_2. This "forming gas" has the reducing action needed in many heat treatments, without the explosion risks of pure H_2.

N_2 is available in tanks, made from liquefied air. Most of the contaminants are O_2, H_2O, and CO_2, which should be removed, when higher purity is needed. Very pure N_2 can be obtained in the laboratory by heating a dried alkali–azide in vacuum.

CO_2. Carbon dioxide is an odorless, colorless gas which constitutes 0.3% of the earth's atmosphere. It is a relatively inert gas and used sometimes as a protective atmosphere usually bubbled through isopropyl alcohol. CO_2 is highly soluble in water, alcohol, or acetone. From its solid state (dry ice) at −78.5°C, CO_2 sublimes directly into the gaseous phase at atmospheric pressure. A mixture of dry ice and acetone or alcohol is a common intermediate temperature refrigerant for cold traps. CO_2 is also used sometimes for coarse leak detecting with the use of a Geissler tube.

Tank CO_2 is usually not more than 99% pure. It can be further purified or obtained in better grades. It can be produced in the laboratory from heating a clean dried $MgCO_3$ in vacuum. Another convenient way is to obtain CO_2 from dry ice, which is oxygen free.

CO. Carbon monoxide is odorless and colorless but a very toxic gas, which is a common product of combustion processes. Precautions should be taken against excessive inhalation of CO. Carbon monoxide is a common residual gas in uhv systems, as it is a product of hot filaments and its pumping speed in ion pumps is not high. It can be efficiently pumped off by a Ti sublimator. It attacks many metals through reducing and carburizing, especially at high temperatures.

8.8. Conclusion

This chapter attempts to summarize some of the important properties that should be considered when choosing materials for use in vacuum technology. For additional information, the reader is referred to some of the reviews on materials for vacuum technology.[4–8]

The reader who intends to use materials for vacuum applications should be alert to new developments in the fields of materials science and vacuum technology.

[4] W. Espe, "Materials of High Vacuum Technology." Pergamon, Oxford, 1966. Vol. 1: Metals and Metalloids. Vol. 2: Silicates. Vol. 3: Auxiliary Materials.

[5] F. Rosebury, "Handbook of Electron Tube and Vacuum Techniques." Addison–Wesley, Reading, Mass., 1965.

[6] S. Dushman, "Scientific Foundations of Vacuum Technique." Wiley, New York, 1962.

[7] A. Roth, "Vacuum Sealing Techniques." Pergamon, Oxford, 1966.

[8] R. M. Mobley, *Meth. Exp. Phys.* **4B,** 318 (1967). Academic Press, Inc.

9. GUIDELINES FOR THE FABRICATION OF VACUUM SYSTEMS AND COMPONENTS*

One must avoid the unnecessary costs of following recipes blindly; accordingly, these guidelines must be weighed carefully in each application. We emphasize the choice of materials and techniques including those of cleaning, joining, installation, and operation primarily to avoid real and virtual leaks initially and during the lifetime of a specific vacuum system.

Having said this we must be alert to what we cannot see with our eye. For example, we must not be misled by appearances, e.g., metals baked out in oxygen (air) and thereby made discolored are less gassy than the same type of metal polished to high brightness. Consider also the relative importance of invisible microcrack structure to macrocracks that we can see with our eye and the contribution of gross cracks and voids that are lined with microstructure. Figure 1 illustrates the point here. As an example, suppose we avoid deep, narrow cracks and trapped volumes that we can see. Have we done our job in this regard? The answer is, probably not. Examine the cross section of a microcrack versus a macrocrack. Let the aspect ratios of these cracks be the same, that is, the depth of the crack divided by the width of the crack equals the same number. As we let the width of the crack become narrower and narrower approaching molecular dimensions, the significance of the crack can become greater regarding the retention of, for example, water molecules. At room temperature water molecules can enter such a crack by two dimensional diffusion processes. Notice that the amount of work necessary to remove

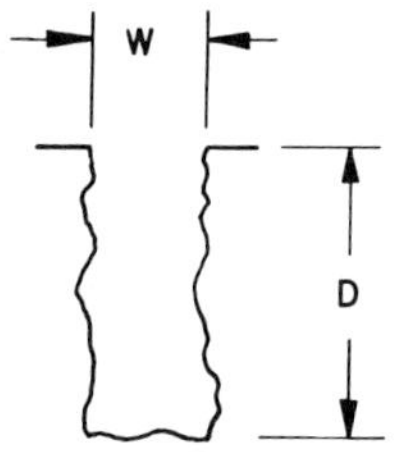

FIG. 1. Crack of width W, depth D, aspect D/W = constant.

* Part 9 is by N. Milleron and R. C. Wolgast.

ISBN 0-12-475914-9

all of the water from pores in zeolite can be as high as 100 kcal/mole. This may be compared to the heat of vaporization of water (10 kcal/mole) and the heat of desorption of submonolayer water molecules on a solid geometric plane (14 kcal/mole).

9.1. Soft Soldering

Is it necessary to avoid soft soldering in every case? In appropriate cases soft soldering can and must be used. The major problems with soft soldering concern getting the solder to stick, the low mechanical strength of the joint, and the service temperature of the joint. First and foremost, it is difficult to get low melting point solders to wet the surfaces to be joined. All chemical fluxes present difficulties from two points of view. First of all, it is difficult, if not impossible, to remove a chemical flux after it has been used to insure the wetting of the solder. The residual amounts of flux can contaminate the system and, perhaps of equal importance, can promote corrosion effects over long-term operation within the joint itself. Considerable effort has been made to use ultrasonic soldering techniques to wet the low melting point solder to the solid substrate. Using ultrasonic soldering techniques and solders of, e.g., pure indium and the indium alloys, soft soldering has been successfully used to join parts together and to fix leaks. However, even with the most scrupulous care, such joints or repairs can subsequently leak. In short, ultrasonic soldering techniques without flux are practical provided that the area of the joint is not overly large or that there are not a large number of joints.

No reliance should be placed on the mechanical strength of soft soldered joints. Aptly designed and supported, soft solder joints can serve under severe mechanical vibration. In general, soft solders are limited by the corrosive action of the solder at high temperatures. Volatile components in the solders must be avoided and, in rare instances, tin bearing solders, >70% tin, for example, may not be usable at very low service temperatures. At temperatures of 300°C, well above its melting point, pure indium is corrosive on stainless steel. However, at lower temperatures, indium, together with ultrasonic soldering techniques, can be used to real advantage. For example, indium remains ductile and malleable at least to liquid helium temperatures.

9.2. Hard Soldering

Hard soldering can be used in many instances. Once again, the problem is not so much with the solder as with the methods used to apply it. Vacuum furnace brazing that may involve the use of a flux, such as hydrogen as a step in a sequence, can yield excellent results. On the other hand, hand-held torch techniques, the use of cadmium bearing solders, and/or paste fluxes must be employed with the greatest care in order to achieve acceptable results. The following references[1-3] should be consulted for a discussion of both furnace brazing and hand-held torch techniques, the latter being almost impossible to avoid completely.

9.3. Welding

To be acceptable, welding must be under strict supervision, control, and inspection. Arc welding using flux coated rods must be avoided. This blanket rule is made as much to avoid the rough and ready approach to system fabrication as to avoid any problems with the welding per se. Various forms of shielded arc welding, on the other hand, lend themselves to excellent quality control and inspection procedures where the most rigorous clean room procedures can be followed. Wherever possible, automatic and semiautomatic welding must be preferred to hand-held procedures. Even with the use of plasma needle arcs and other modern developments, a skillful welder can make the difference between leak-free, trouble-free smooth operation and time and money wasting iterations involving welding, testing for leaks, patch-up, retest, and reweld. As a rule of thumb, encourage the welder to take time to set the job up carefully. For example, in the author's personal experience, a welder working full time in the vacuum laboratory, working primarily on the welding of type 304 stainless steel parts, produced leak-free results during a five-year period. The maxim, "there's never time to do it right but always time to do it over," should be considered.

9.3.1. Welding Design Details

Welding is a very useful, common, inexpensive, and reliable fabrication technique, to the point that it is largely taken for granted. One should however "remember that welding is the casting of metals under the most difficult circumstances," and realize that the success of welding is due to good design and preparation, as well as the skill of the welder. Normal good joint design and welding practice should be followed whenever possible. In spite of this, special equipment needs will still arise to challenge the welder. Joint material should be clean and free of embedded foreign matter that can produce porosity or affect the metallurgy of the weld.

[1] W. Espe, "Materials of High Vacuum Technology," 3 Vols., Pergamon, Oxford, 1968.

[2] F. Rosebury, "Handbook of Electron Tube and Vacuum Techniques," Addison-Wesley, Reading, Massachusetts, 1965.

[3] W. H. Kohl, "Handbook of Materials and Techniques for Vacuum Devices," Van Nostrand-Reinhold, New York, 1967.

Machining should be done with sharp tools and good machine shop practice to avoid excessive cutting forces which could tear the metal surface and create many fine cracks. Burnishing or smearing the surface with a dull tool should be avoided. In machining stainless steel, in particular, the use of sharp tools with narrow cutting faces and cutting lubricant is best. An attempt to reduce the outgassing rate of the steel by not using any cutting lubricant failed, presumably due to the extensive tearing of the surface, observed in the microscope.[4]

For best vacuum quality it is generally desirable to have the weld bead on the vacuum side, but there are exceptions. For example, when vacuum quality is not critical, it may be less expensive to weld on the outside. Also outside welding may be best for large and complicated assemblies where inside weld repair is difficult.

Design considerations particular to vacuum applications include: the advantage of eliminating internal recesses or cracks where possible, the frequent joining of thin walls to heavier sections, and the frequent material defects in thick sections of normally produced steel and other metals, especially heavy bar stock. Internal cracks resulting from weld–joint geometry are undesirable because they collect dirt during fabrication or operation which may be impossible to remove. These cracks, however, are large compared to molecular diameters. If it is kept clean, the effect of the crack in limiting gas and surface phase molecular transfer out of the crack is usually small unless it is very tight and deep, or there is a significant trapped volume that communicates to the vacuum through a tightly pressed joint. Examples are shown in Fig. 2. The situation occurs more frequently when screwed joints are used inside the chamber, as is shown in Fig. 3.

The welding considerations in this paragraph apply to the AISI 300 series austenitic stainless steels (except type 303, which is a high sulfur content-free machining grade which is usually not acceptable in high temperature vacuum systems and cannot be welded satisfactorily). Although these steels are relatively easy to weld, the weld shrinkage is high, which can cause distortion of the component. Some typical distortion shapes are shown in Fig. 4. Rupture cracks in or near the weld due to welding are unusual. Stress relief is not usually required in vacuum work but is important in avoiding stress corrosion. Mild relief of the weld stresses by heating to about 400°C is adequate to provide dimensional stability with little effect on the austenitic structure of the base metal. The austenitic structure is partly modified by welding in the fused and heat affected

[4] R. S. Barton and R. P. Govier, "Some Observations on the Outgassing of Stainless Steel following Different Methods of Cleaning," Report CLM-R 93, United Kingdom Atomic Energy Authority (1968).

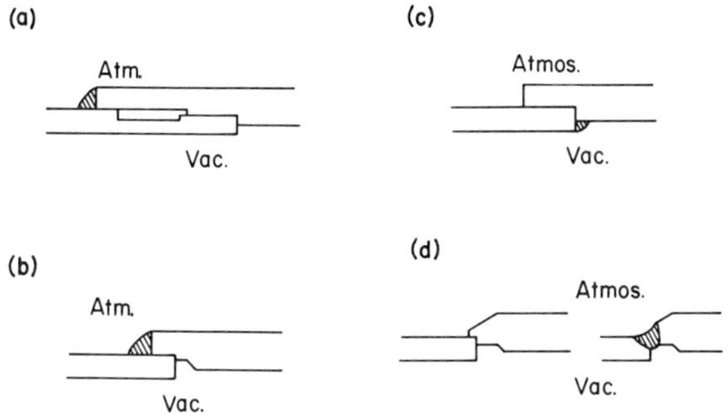

FIG. 2. Good and bad design for welding an aligned tubing joint. (a) Bad design. Trapped gaseous and adsorbed molecules migrate slowly from the trapped region and can contribute an undesirable gas load for very long time periods. (b) Undesirable. If this joint is kept clean and dry there is little bad effect on vacuum performance, but dirt, atmospheric contaminants, moisture, or solvents accumulated during assembly or operation will have a bad effect. (c) Best. The inside weld joint is best for uhv applications. (d) Good. If an outside weld must be made, as on small tubes, a full penetration weld may be made.

zone. A complete recovery is possible only by high temperature annealing and very rapid uniform cooling, usually impossible except in simple shapes. Magnetic effects can be minimized by using stabilized type 316, making narrow welds and using special filler rod that remains austenitic. The low carbon grades, designated 304 L, for example, experience less chromium carbide precipitation which makes the weld region susceptible to intergranular corrosion. Well developed automatic welding techniques produce welded stainless tubing from sheet stock that is as good strengthwise and vacuumwise as seamless; the surface finish is better and the cost is lower.

The design details shown in the following figures are appropriate to TIG (tungsten inert gas) and MIG (metal inert gas) welding methods. These methods are also called gas tungsten arc welding and gas metal arc welding. TIG uses an electric arc with a permanent tungsten electrode and inert gas shielding. The weld may be made by simply melting together the edges of the pieces to be joined, or by adding filler metal to

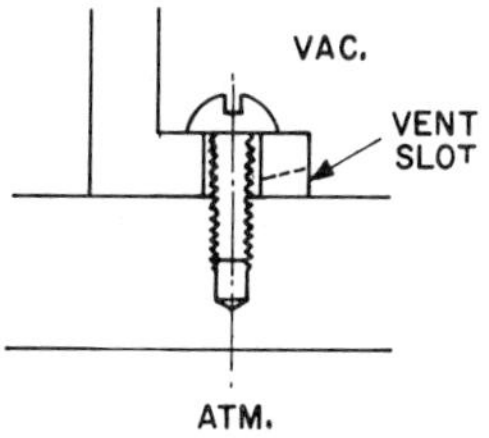

FIG. 3. Screwed joint in the vacuum chamber. A vent slot should be provided to pump out the trapped volume in the screw hole. Locating the vent in the bracket allows use of standard replacement screws.

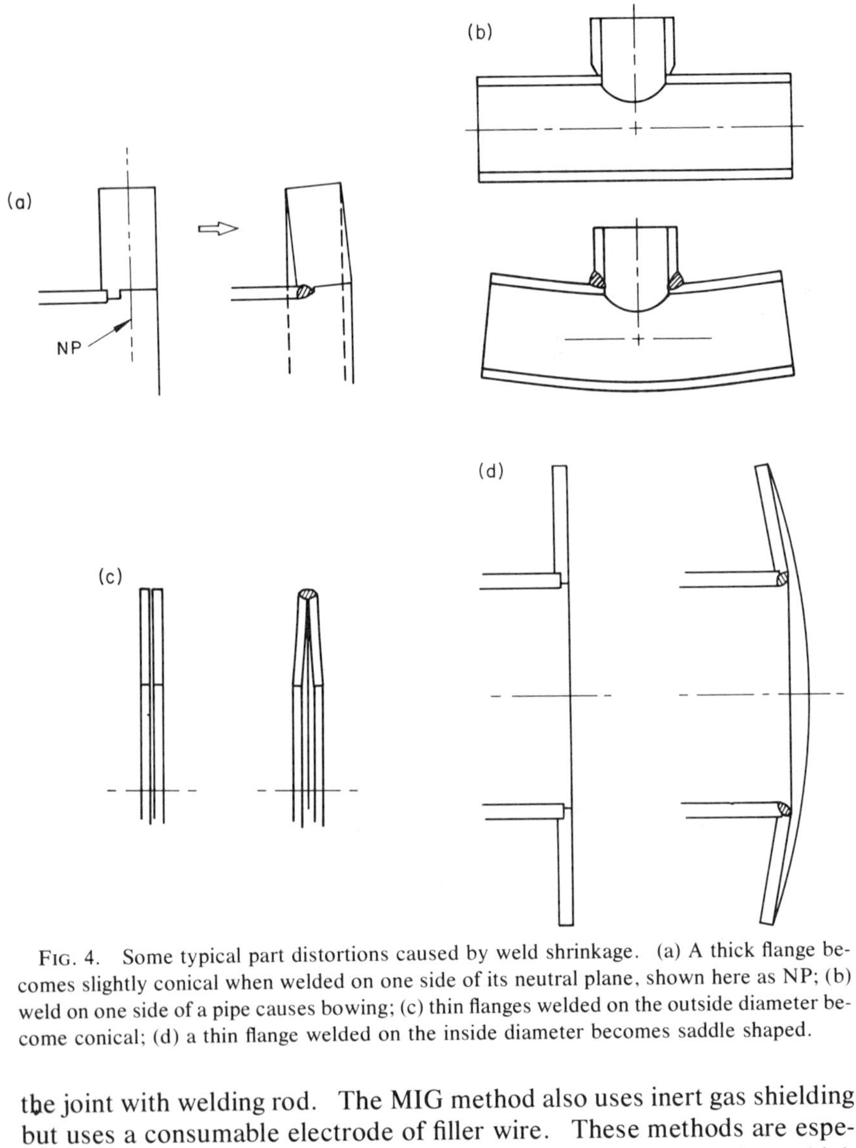

FIG. 4. Some typical part distortions caused by weld shrinkage. (a) A thick flange becomes slightly conical when welded on one side of its neutral plane, shown here as NP; (b) weld on one side of a pipe causes bowing; (c) thin flanges welded on the outside diameter become conical; (d) a thin flange welded on the inside diameter becomes saddle shaped.

the joint with welding rod. The MIG method also uses inert gas shielding but uses a consumable electrode of filler wire. These methods are especially suited to vacuum work because they do not require a flux which would need to be removed. The slight surface oxidation resulting from inert gas welding does not have a bad effect on the vacuum.

It is best to design the weldment so machining or filing the weld will not be required to complete the part. If such machining is necessary after welding, the weldment should be leak-checked after machining. If multipass welds are required, the root-pass weld should be leak-checked before continuing.

Clean stainless steel wire brushes may be used to clean the surfaces to be welded. Abrasive grinding compounds such as alumina, silicon carbide, or steel grit should be avoided if possible, or followed by an acid cleaning to minimize porosity in the weld.

Figure 5 shows some welds where the material thicknesses are comparable. Locating shoulders may be machined to make alignment easier. When the parts are large and the alignment is not critical, the extra cost of machining aligning shoulders may not be justified.

Figure 6 shows some outside welds where an internal crack is avoided by making a full penetration weld. The back side of the weld should be flushed with inert gas (without creating a difference in pressure across the weld).

Figure 7 shows thin tubes or sheet welded to heavy flanges or plates. If greater strength is needed an intermittent weld can be made outside—the outside weld must not be continuous to avoid an undectable virtual leak. The internal radius on the weld side of the groove is desirable to avoid high stress concentration and cracking due to stress from the weld shrinkage. If a broad-faced tool which tears the metal is used to cut the radius, the tearing may be worse than a square corner, cut with a sharp pointed tool. Figure 8(a) shows an example of bad vacuum practice to be avoided where possible. This case may occur often in the use of socket–weld tube fittings for small diameter tubing. It may be avoided by using tubing that fits the outside diameter of the fitting and making full penetration butt welds as shown in Fig. 8(b).

Thick stainless steel materials, such as bar stock, plate thicker than 6 mm and heavy wall seamless tubing, very likely contain filamentary leak paths along the forming dimension. Vacuum quality flanges are made from cross or ring forged blanks or from specially remelted steel to avoid these faults. If nonvacuum quality materials must be used, a departure from normal welding design may be required. Figure 9 shows some examples.

Figure 10 shows alternate details for welding intersecting tubes. The bevel for the second weld detail follows a saddle shape and is usually ground manually. A pull-formed tee (requiring special tooling or available commercially) can avoid this difficulty.

A joint for thin-wall bellows less than 0.25 mm thick approximately, is shown in Fig. 11(a), which unfortunately involves an internal crack. Plasma–arc welding equipment provides easier control in welding thin pieces.

For greater thicknesses the joints in Fig. 11(b) and (c) may be used.

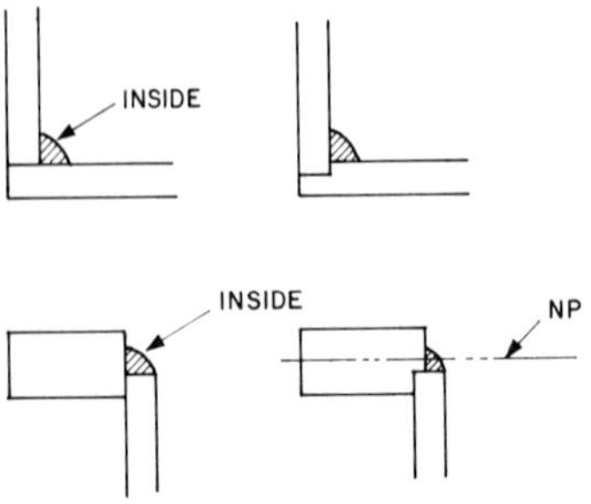

FIG. 5.

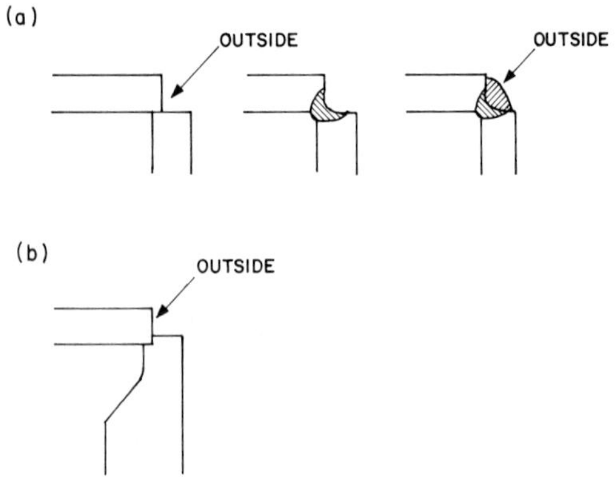

FIG. 6.

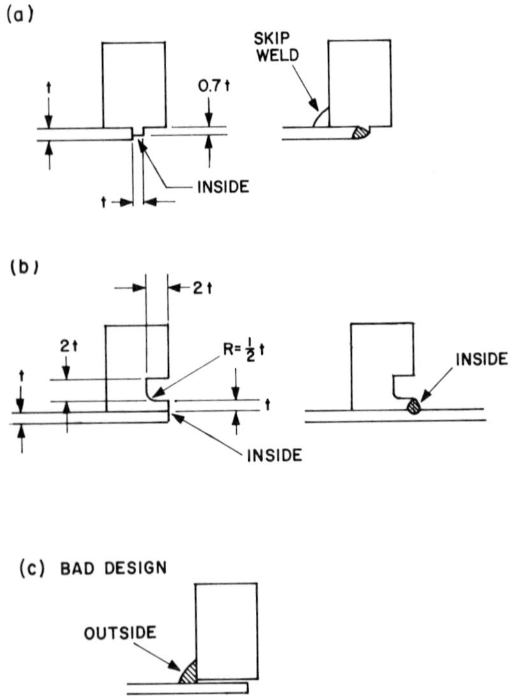

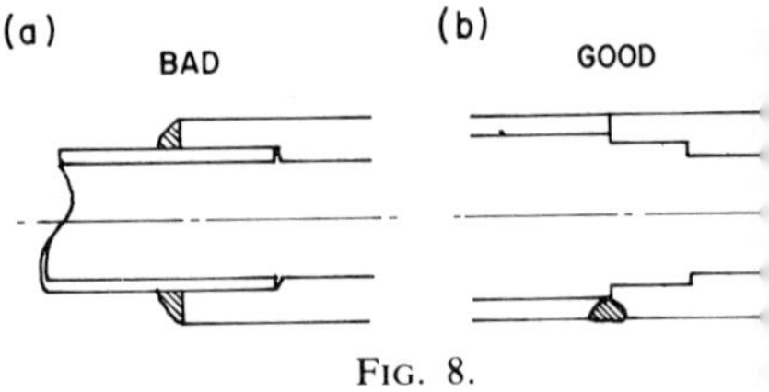

FIG. 8.

FIG. 7.

FIG. 5. Weld details for materials of comparable thickness. Locating the weld on the neutral plane of the flange, shown as NP, minimizes the distortion of the flange.

FIG. 6. Full penetration outside welds useful where inside welds are not practical. (a) The parts have comparable thickness. A second filler pass may be added for strength. (b) A similar weld technique may be used with locating shoulders for parts of unequal thickness.

FIG. 7. Thin tubes or sheet welded to heavy flanges or plates.

FIG. 8. (a) Typical tube socket weld for pressure service which is bad vacuum practice. (b) Good vacuum use of a socket-weld fitting by making a full-penetration butt weld.

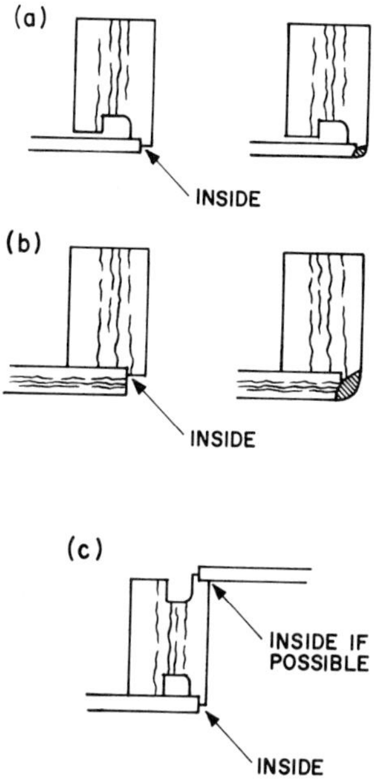

FIG. 9.

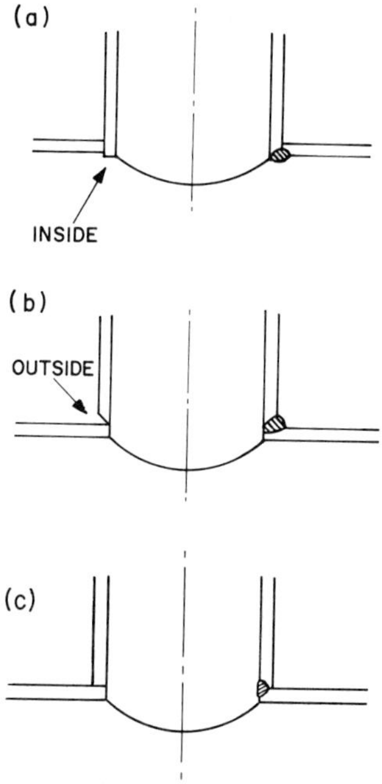

FIG. 10.

FIG. 9. Some welding examples which circumvent the filamentary leaks which may occur in thick sections.

FIG. 10. Alternate details for welding intersecting tubes when one tube is smaller than the other. Detail (a) is useful when there is accessibility for welding through the larger tube. The full penetration weld of detail (b) is useful when the welding must be done from outside. Detail (c) is useful when there is accessibility for welding through the smaller tube. This detail is also useful for making a brazed joint; the hole may be machined after brazing, taking advantage of the stiffening of the tube to make the machining easier.

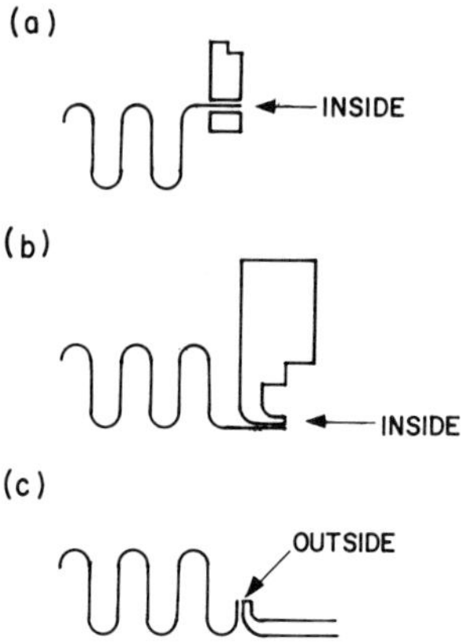

FIG. 11. Details for welding thin-wall bellows. (a) An inner backup ring is recommended for bellows with wall thickness less than 0.25 mm. The parts should fit snugly. Such a thin bellows is often welded to a transition piece rather than directly to the flange or part. (b) For bellows wall thickness in the range from 0.25 to 0.8 mm, the welding lip may be thicker than the bellows to make it practical to machine. (c) The tube end is flared and trimmed to provide a welding lip.

Kovar may be welded by the same techniques as stainless steel. Rewelding or intersecting welds may be less successful. Kovar welds satisfactorily to stainless steel. The same considerations apply to Invar. Welding properties of Invar should be checked before planning extensive use of the material.[5] Invar and Kovar are not corrosion resistant, and are magnetic.

Aluminum and its alloys weld reliably with relative ease, but not so easily as stainless steel generally. The MIG or TIG method with filler rod is preferred. For best welds, especially fusion only welds, the joint surfaces must be freshly scraped to remove the naturally built-up oxide layer.

Copper is relatively difficult to weld reliably. Higher power is required because of the high thermal conductivity. Copper is "hot short," i.e., weak at high temperature, and shrinkage stress as the weld cools often causes cracks. The use of oxygen-free, high conductivity (OFHC) copper instead of the more common electrolytic tough pitch (ETP) does not eliminate the problem. Filler metal welds with a copper alloy filler wire are best (0.75% Sn, 0.25% Si, 0.2% Mn). Copper can be TIG welded to stainless steel. An extra length of copper melt-down lip should be provided. The TIG "torch" has been used with brazing alloy rod to make a fluxless copper-stainless joint.

Metals such as titanium and molybdenum, which oxidize rapidly, require special inert gas shielding. They are often welded in an inert gas enclosure fitted with glove ports to allow manual operation from outside. The enclosure should be evacuated to about 10^{-5} Torr before backfilling with inert gas to insure a low contaminant level. Adsorbed water vapor is probably the chief offender. It dissociates at high temperatures and releases oxygen to attack the metal.

9.3.2. Cut–Weld Joints

Welded joints in stainless steel and in aluminum alloys provide a higher degree of leak-tight reliability and less cost than metal gasket sealed flanges. This is especially true when the joint is subject to high or low temperature cycles. In order to have the advantage of welding and still provide the required separability, "cut–weld" joints have been used. In the simplest form, the joint may be a butt weld in a length of tube, which may be cut with a standard tubing cutter. In larger sizes the tubing may lack sufficient rigidity to be cut by a standard tubing cutter and joints like those in Fig. 12 have been used.[6] The flanges were press formed from

[5] D. A. Wigley, "Mechanical Properties of Materials at Low Temperature," Plenum. New York, 1971.

[6] N. Milleron, *IEEE Trans. Nuc. Sci.* **14**, 794 (1967).

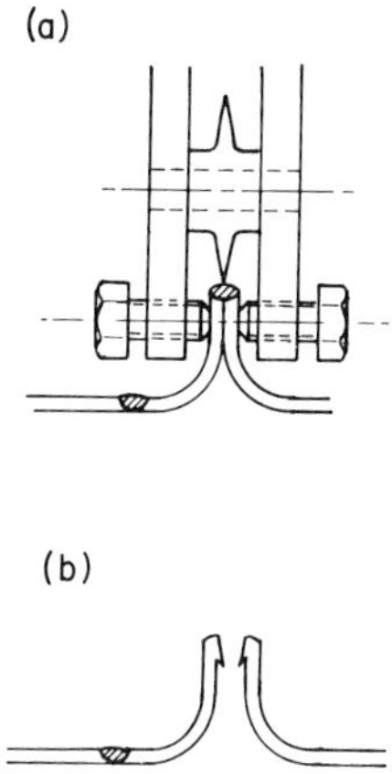

FIG. 12.

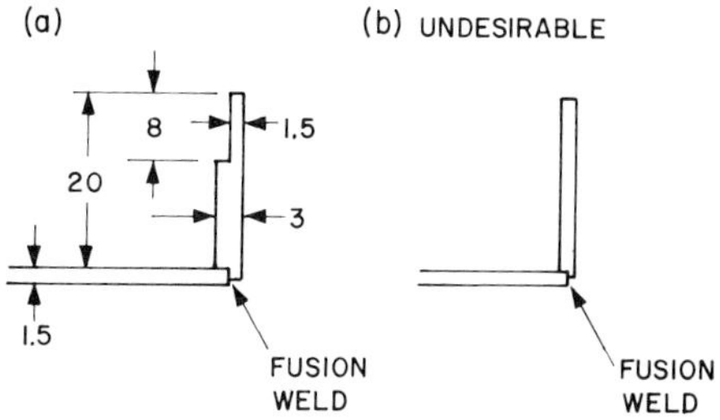

FIG. 13.

FIG. 12. (a) Cutting technique used for a cut–weld tubing joint. The cutter was held in a clamshell fixture that was manually driven. The alignment screws were retracted after the initial cutting groove was established. (b) The flanges after the completed cut. The remaining burr was removed with a broad flat file before rewelding.

FIG. 13. Flange used for cut–weld tubing joint. (a) The flange is thick enough to adequately limit distortion caused by the inner weld, up to sizes of 16 cm diameter. (b) The plain thin flange is undesirable because of saddle-shape distortion from the inner weld.

sheet metal 1.5 mm thick. A plain, thin, flat flange is not desirable because the shrinkage of the inside weld to the tube will make the flange saddle shaped. The added thickness of the flange in Fig. 13 provides enough stiffness to control this distortion for 16 cm diam or smaller flanges. The weld can be made manually and cut open with a hand-held high speed carbide cutter. When they were welded with a roller-guided electrically driven welding fixture, the shape remained very regular, and the flanges have been cut apart with a tubing cutter wheel held and guided in a clamshell fixture, manually driven. Some dress-up filing was needed to remove the cutting burr before rewelding. Prototype welding and cutting fixtures for tube diameters up to 50 cm have been used. The cutting is not easy and requires some skill and strength from the operator, especially in larger sizes. As many as 15 cuts and rewelds have been made on the same 14 cm diam pair without the appearance of leaks or other difficulties that would limit the number of cut–weld cycles.

The welding has been done by the tungsten electrode inert gas shielded (TIG) fusion method. A very steady drive motion of the welding fixture makes regular welds with controlled penetration depth, which makes the cutting operation as easy as possible. Plasma–arc welding equipment would make the welding control easier and more reliable. Orbiting welding equipment is available commercially. Some of these units re-

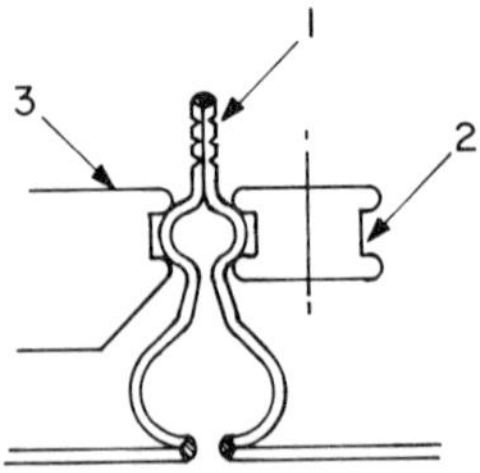

FIG. 14.

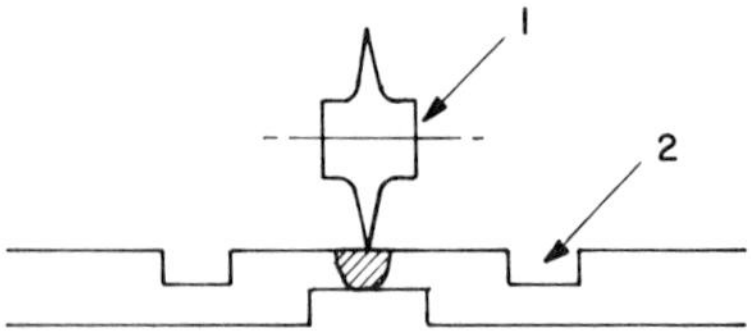

FIG. 15.

FIG. 14. Cut–weld tubing joint opened by tearing. (1) Premachined grooves permit a number of tear-off opening and rewelding cycles before replacing the flange. (2) Guide rollers for the welding fixture. (3) Fixed clamshell support ring.

FIG. 15. Cut–weld joint used on the main ring of the Fermilab synchrotron. (1) Cutter mounted in clamshell ring. (2) Guide grooves for the cutter ring and for the aligning jig for rewelding.

quire a minimum of space, and stock sizes of equipment can weld tubes or flanges up to 10 cm diameter. A large number of required welds may justify the cost of such equipment.

High speed carbide cutting tools can be mounted in an orbiting fixture to remove the weld bead. Commercial "weld shavers" are possibilities for this application. The throwing of chips and vibration are problems to be designed for. Even the tubing cutter knife generates fine abraded particles which may need careful removal. There is still room for the development of cutting techniques. Cutting of aluminum flanges may be easier.

Another method for removing the weld bead is shown in Fig. 14. A number of grooves are premachined in the flange. The weld bead may then be cut and torn off at the groove. A number of "tear-off" openings are provided before a new flange pair is needed.[7]

Cut–weld joints were used for the two mile long beam tube of the Stanford linear accelerator.[8] The flanges were formed of stainless steel sheet 0.635 mm thick. One tear-off groove was provided to remove the temporary blank-off cover welded to the flange for preinstallation bakeout. Installation welds and removal cuts were done with hand-held equipment.

The main ring vacuum beam tube of the 6 km proton accelerator at the Fermi National Accelerator Laboratory is assembled by cut–weld joints in the 15 cm diameter stainless steel tubing. The design of the joint is illustrated in Fig. 15. The tubing cutter is mounted in a clamshell ring

[7] J. M. Voss, "Automatically and Remotely Welded and Removable Weld Flange Vacuum Joint," Report ISR-VAC/67-23, Organization Européenne pour la Recherche Nucléaire (CERN) (1967).

[8] R. B. Neal, Ed., "The Stanford Two Mile Accelerator," Benjamin, New York, 1968.

which is guided by grooves premachined in the tubing. It is driven manually. Another clamshell jig aligns the parts for manual TIG rewelding. Some dressing of the cutting burr is required before rewelding.

9.3.3. Bonded Aluminum to Stainless Steel Joints

Aluminum alloys are commercially bonded to stainless steel by friction, forge, and explosive welding. Large plates of aluminum alloy and stainless steel, each 5 cm thick, are explosively bonded with a thin silver interlayer about 0.3 mm thick. Heavy roll-bonded plate without any interlayer is also available. Strong and reliable transition pieces for vacuum and cryogenic service of tubular or other shapes may be machined from the bonded plates. One vacuum use for particle accelerator beam tubes has been to provide stainless steel metal-gasketed flanges at the ends of long aluminum extrusions, by welding the aluminum side of the transition piece to the extrusion. Another application is to weld low heat conduction stainless tubing fill lines to aluminum cryogenic dewars, or to cryopump panels. The bonding techniques mentioned are applicable to other metal combinations.

9.3.4. Electron Beam Welding

In electron beam welding the welding heat is supplied by sharply focused, high speed electron beams.[9] The electron beam quickly vaporizes a hole to the entire depth of the joint. The walls of the hole are molten, and, as the hole is moved along the joint, the metal flows from the advancing side around the hole and solidifies on the rear side of the hole to make the weld. The source of the electrons is called the electron gun, which is an assembly consisting of a hot filament to supply electrons, a field-shaping electrode, a high voltage insulator, the shaped accelerating anode, and a focusing coil. The filaments and electron beam require a vacuum of about 10^{-5} Torr. There can be no obstruction to the electron beam, and as a result the electron gun and the work piece are typically mounted in a vacuum chamber. Rotating and traversing tables, which hold the workpiece, move the weld joint under the electron beam; the electron gun may also be movable. The welding is observed by the operator through an optical system or by closed circuit television. The movements may be manually controlled by the welder, or preprogrammed or controlled by seam-following feedback equipment. Separate vacuum systems may be provided for the electron gun and the workpiece

[9] S. T. Walker, Ed., "Welding Handbook, Section 3A; Welding, Cutting and Related Processes," 6th Ed., American Welding Society, New York, 1970.

chamber. The process may be used with the workpiece in partial vacuum or in atmosphere by providing adequate differential pumping along the electron beam exit channel. This mode of operation is useful (1) when one wants high production rates; (2) in avoiding the size or handling limitations imposed by the workpiece vacuum chamber; and (3) when the full capabilities of the high vacuum mode are not needed.

A variety of electron beam welding equipment permits the welding of delicate electronic components (viewed through microscopes) and heavy structural parts. Single pass welds may be made with penetrations of 5 cm in copper and steel and 8 cm in aluminum and magnesium alloys. A single pass square butt weld has been demonstrated in aluminum plate 22 cm thick. Welding equipment usually falls into two classes from practical design considerations: (1) beam voltages up to 60 kV, and (2) beam voltages 60–150 kV, with beam power up to 50 kW and 30 kW, respectively.

The particular safety hazards of electron beam welding result from the high voltage, the X rays generated when the high speed electrons strike the work piece, and, in the case of the atmospheric mode of operation, the generation of ozone by the electron beam passing through atmosphere. In addition to a substantial vacuum tank wall, lead shielding is required in higher voltage equipment to reduce the X-ray flux. Safety standards are prepared by the American Welding Society (New York).

The characteristic advantages of electron beam welding stem from the high rate of energy deposition per unit area, about twice that of shielded arc welding processes, and the vacuum environment. Deep narrow welds are possible in a single pass. For example, a square butt weld of 12.5-mm-thick stainless steel plates may have a weld width of as little as 1.5 mm. The heat affected zone is narrow, and the grain sizes in both the heat affected and the fusion zones are small, with less deterioration of the metallurgical properties of most materials. Distortion is reduced, and it is easier to weld thick to thin sections than with other methods. Gears have been welded together after final machining while maintaining an eccentricity tolerance of 0.04 mm.

The contamination level of a modest vacuum of 3.8×10^{-2} Torr is equivalent to that of 99.995% pure argon at atmospheric pressure (assuming that the contaminant species are the same). Besides eliminating most of the gas from the weld area, the high vacuum mode removes vapors that may result from melting the metal. The process produces high weld purity and minimum porosity. Metals that react with atmospheric gases as well as a greater range of dissimilar metals may be welded. Oxidation of the weld and heat affected zone is eliminated.

The parts to be joined should be fitted carefully. Gaps should not exceed 0.13 mm for the best results. A light press fit is recommended for circular parts. As in all high quality welding, cleanliness is very important. A final cleaning with acetone is recommended. Surface oxides should be removed either chemically or by scraping. Fuller wire may be added during welding to avoid underfill, and filler shims sandwiched into the joint may be used to improve the metallurgy of the welds or to aid in the welding of dissimilar metals.

The features in this process of special interest to vacuum technology are as follows: (1) the welds are inherently sound and leak tight; (2) full penetration welds may be made with a single pass, eliminating both internal and external cracks even though the welding may be done externally; and (3) thick and thin sections are easily joined. The fact that tanks and enclosures may be welded shut, sealing in the work chamber vacuum, is another particular feature of electron beam welding in the vacuum mode. This feature has an important application in producing composite superconductors. Superconductor rods and copper shapes are densely fitted into a copper can, which is welded shut. The absence of air and surface contamination permit the superconductor and the copper to bond metallurgically in the subsequent extrusion and wire-drawing operations. This process also may be used for hermetic sealing. The quality of the sealed-in vacuum depends upon the gases released during welding as well as the quality of the initial vacuum.

For the fabrication of vacuum components, electron beam welding complements other welding methods. The excellent results and, in most cases, lower cost of the more readily available Heliarc (gas tungsten arc and gas metal arc) welding methods indicate that it will be used for most work.

9.4. Bakeable and Coolable Vacuum Joints

If a joint is made of one material and heated or cooled uniformly throughout the part, no substantial changes in the stresses or deflection in the joint will occur, provided the temperature excursion does not include creep, stress-relieving, or transformation temperatures. Low heating rates may be required to maintain sufficient uniformity of temperature. When the assembly includes materials with differing thermal expansion properties or when the temperature changes are not uniform, stresses can be created that will cause yielding or rupture of the material or bond.

The problems of creep and the unequal thermal expansion between stainless steel and copper (though small) in regard to bakeout temperature cycles has been discussed in Section 6.2.3, Vacuum Flanges. Problems resulting from nonuniformity of temperature also occur in flanged joints.

Suppose, for example, that the flanges are heated and expanded more rapidly than the bolts. The bolts, which were initially highly stressed to make up the joint at room temperature, are forced to stretch further and may elongate in the plastic flow range. When the joint is subsequently cooled to room temperature, the elongated bolts may no longer maintain adequate seal force. Another instance occurs where a tube flange is bolted to a massive blank-off flange. With its higher thermal mass, the blank-off flange may heat and expand more slowly than the tube flange and disturb the seal.

The problem of unequal thermal expansion is most acute in the case of metal-to-ceramic joints, where the typical metals used have a higher coefficient of expansion than the ceramic. Let us consider the case of a ring-shaped metal cap brazed to an alumina ceramic tube, starting at the moment it is in the brazing furnace after the braze material has solidified and the joint is still near the braze temperature. When the joint starts to cool, the metal first stretches elastically until the yield point is reached. As cooling progresses toward room temperature, the metal must continue to stetch plastically if the braze bond remains unbroken. The joint is often designed with the metal around the outside diameter of the ceramic to avoid high tensile stress on the bond during this cooling period. When the joint is finally cooled to room temperature, the metal has been work hardened to a higher yield point material and is under tensile stress.

Suppose that the joint is now to be heated for a bakeout cycle. As the joint materials expand, the tensile stress in the metal relaxes to zero. If the heating continues, the metal goes into compression and the stress on the bond reverses. With still more heating the metal again may be forced to flow plastically. If the temperature cycle is too high, the joint may fail from fatigue due to repeated plastic flow cycles or from too large a tensile stress on the bond. This upper temperature limit is usually substantially below the initial braze temperature.

Let us suppose that, instead of a bakeout cycle, the cooling of the joint was continued toward cryogenic temperatures (such as that of liquid nitrogen at − 196°C). The plastic flow of the metal would continue to a higher degree, and the subsequent high temperature cycle limit would be lower than that for a joint cooled only to room temperature. Some typical operating ranges for commercial ceramic–metal joints (such as those made for electrical feedthroughs) are 0–600°C (maximum continuous temperature in air, 450°C), − 100–300°C, and − 200–450°C, with rates of temperature change limited to 25°C per minute.

Flanged joints that must be cycled to low temperatures have design requirements of the same kind that bakeable joints have, which is the problem of maintaining sealing force during the various operating conditions. In cooling stainless steel from room temperature to liquid nitrogen temperature (77 K), for example, the fractional thermal contraction is about 3×10^{-3}. In a flanged tube joint in a cryogenic liquid pipe the flanges will cool more rapidly than the bolts when the cryogenic flow starts. Leakages may occur until the bolts become cold, restoring the sealing force. This problem has been observed with a 12-mm-diameter tubular coupling. The coupling nut on the outside cooled more slowly than the inner tubular parts. The solution is to provide sufficient elastic springback to overcome the relaxation of seal load due to unequal thermal contraction. Thus far this is accomplished most reliably by the principle of the Batzer flange (discussed in Section 6.2.3, All-Metal Flanges), or some other spring-loading method, in conjunction with limited plastic flow of the gasket. The problem at low temperatures is less difficult than that at high temperatures because there is no high temperature creep and mechanical properties improve at low temperatures. Mild steel is an exception, undergoing a brittle transition at low temperatures, and should not be used in a critical situation, nor should it be relied upon to carry substantial stresses or possible shock loads.

A flanged joint for low temperatures using a thin Mylar or Kapton sheet gasket was described near the end of Section 6.2.2, Elastomer Gasketed Flanges. Elastic springback can be provided by using extra length bolts with tubular standoffs under the head or nut. The standoff should have about the same cross-sectional area as the bolt. Belleville spring washers may also be used; the spring force may not be adequate to make the initial seal, but may be adequate to maintain the seal.

9.5. Cleaning Techniques

Before exposing anyone directly or indirectly to procedures that may be injurious to them, up to date information should be acted upon to reduce (and inform concerning) likely risks and hazards.

Cleaning parts before assembly and cleaning assembled systems entails removing contamination without allowing significant residues to lodge out of sight or to be formed in cracks, joints, or bulk materials. In other words, the aim of cleaning is to improve the operational stability of the surfaces of all walls and other components under operating conditions, which include elevated temperatures, cryogenic temperatures, and bombardment by electrons or photons or more massive particles. The aim of cleaning is not, in general, to obtain an atomically clean surface, but rather to ensure that all surfaces be as free as possible of microstructure and that the species of molecules bound to surfaces be held there tightly, without significant additions of chemicals to the bulk phase. In special cases, the cleaning is done under sufficiently controlled environmental conditions so that atomically clean surfaces are obtained on specific areas for a significant interval of time, for example, by techniques such as evaporation, flash filament, sputtering, and cutting.

There are good surface treatment techniques available that provide outgassing rates after pump down that are sufficiently low to be time consuming to measure accurately at room temperature. The measuring system, gauges, molecular analyzer, temperature, length, and type of air exposure are a few of many factors that affect such an outgassing measurement. Often the outgassing rate from only a single sample of a given type has been measured by an experimenter and observed for only a limited time under vacuum. Negative net outgassing rates have been observed, thus indicating that, under the test circumstances the sample, the measuring system, gauges, etc., or combinations thereof have become a net absorber of outgassing. The information on cleaning methods presented here is offered as consistent with plausible understanding and experience and is not significantly in conflict with published data. The reader must choose methods that fit specific requirements. A continuous, tightly bound oxide layer on stainless steel, for example, is desirable because it inhibits the outgassing rate of H_2, and probably CO.

For metals, the most effective cleaning processes are those that remove a controlled amount of the base metal, without adding significant amounts of chemicals to the bulk or surface phase, for example, chemical polish, electropolish, and chemical etch. In this discussion the word polish means that the process results in a more level surface on the small scale; the surfaces become more mirror-like rather than matte, but a bright shiny surface may not be lower in outgassing than a baked dull one. It is desirable to remove enough base metal to remove the microcracks and foreign materials imbedded in the surface by the handling, assembly, manufacturing, and machining processes. A microlevel, polished surface is better than a microrough, matte surface because it will pick up and retain smaller amounts of foreign materials from air exposure and handling, and it can be cleaned better by solvent touch-up cleaning, which may become necessary after inadvertent contamination. Seal surfaces are generally improved by polishing and degraded by etching.

Going down the scale in cleaning strength are the pickling processes, generally understood to be those processes that are strong enough to remove metal oxide scales. Next are the chemical cleaning processes that remove almost any substance except oxides and base metals.

9.5.1. Stainless Steel and Mild Steel Acid Pickle

These techniques are usually multistep processes. For example, the process of pickling stainless steel is as follows: (1) a vapor degreasing to remove the bulk of oil and grease; (2) a hot alkaline bath (possibly with electrolytic effect) to remove stubborn greases; (3) a potassium permanganate soak to condition the metal oxide scale; (4) a hydrochloric acid dip to sensitize the surface (to remove the natural oxide surface passivation); and (5) the pickling solution, consisting typically of 30% by volume nitric acid and 3% by volume hydroflouric acid, at room temperature for 3–30 min.[10]

Carbon and low alloy steels are pickled by a hydrochloric acid bath 8–12% by weight, at 100–105°F for 5–15 min.[11]

9.5.2. Mechanical Methods

Mechanical polishing accomplished by rubbing or blasting with abrasive particles is not in itself a good preparative technique for vacuum because of the creation of microscratches and the imbedding of gassy grinding compounds into the surface. However, on large steel structures, where chemical descaling and cleaning is very expensive, sandblasting may be the best alternative. Small scale sandblasting may be used to advantage to remove carburized deposits and scale on units composed of several materials where chemical cleaning methods are excluded. Glass bead blasting may be useful as a milder blasting method. The SiO_2 grit can be removed by pickling solutions that contain hydroflouric acid.

9.5.3. Electropolishing

Electropolishing is applicable to most metals. This is the process of electroplating in reverse. The workpiece is made the anode (electrically positive) and is "unplated." Metal removal occurs more rapidly at the high points because of field concentrations, resulting in a more level surface. Some hydrogen from the acids used in associated cleaning steps may become dissolved in the workpiece, adding to the outgassing rate in operation. High temperature vacuum bakeout (900°C for stainless steel uhv applications) before final assembly will remove most of the hydrogen. Lower temperatures are adequate to reduce the hydrogen content to the normal level as supplied by the manufacturer of the stainless steel.

[10] American Society for Testing Materials, "1975 Book of ASTM Standards," Stainless Steel: Part 3; Copper: Part 6 ASTM, Philadelphia, 1975.

[11] American Society for Metals, "Metals Handbook," Vol. 2, 8th Ed. Am. Soc. Metals, Park, Ohio, 1964.

9.5.4. Chemical Polish for Stainless Steel, Kovar, Invar, and Nickel and Cobalt Alloys

Stainless steels type AISI series 300 (except type 303, which is unacceptable because it is a sulfur rich grade) can be chemically polished by the DS-9* process, which also applies to Kovar, Invar, and other nickel and cobalt alloys. This process includes a number of steps, which clean, pretreat scale, sensitize, and polish the surface. The bath DS-314* does the polishing. It contains strong acids (no hydroflouric acid is used in the process) and an organic filming agent, which, by surface tension, stretches thinly over the high points and thickly over the valleys. The acids must diffuse through this film before attacking the steel. The diffusion and metal removal is more rapid at the high points, where the film is thin, resulting in a leveling action. About 0.013 mm of metal is removed. The Technical and Operating Data* for the process states:

> Silver and brass solder are cleaned in the DS-9-314 bath without detrimental effect. The process should not be used however on tin and lead soldered work.

After the DS-9 process, a rinse in dilute nitric or other oxidizing acid may be used to remove remnant organic material and to retain the passivity of the steel surface. The DS-9 process should not be used on parts that have narrow cracks, where the chemicals cannot be easily rinsed clean, even though these cracks are external to the vacuum system (for example, between tubing and flange where an internal weld has been made). The retention of chemicals is the suspected cause of leaks occuring in welds after weeks of system operation. Weld material is typically 5% ferrite, which does not have the anticorrosive properties of the austenitic phase. There also may be some chromium carbide precipitation at the grain boundaries. Full penetration welds may be cleaned by the DS-9 process. This is an example of the need to integrate the cleaning methods with the fabrication and installation procedure. The DS-314 bath degenerates with use and must be monitored and maintained by replenishing chemicals in order to get the best results.

The acid pickle for stainless steel described earlier is a good method for vacuum service when the added cost of polishing is not justified, especially when large parts are involved, and other factors may determine vacuum limitations. In order to remove more base metal, the acid pickle has been strengthened to 33% nitric acid and 33% hydroflouric acid. Etching by this solution will leave a rough surface.

* Diversey Chemical Co., Industrial Division Issue No. 70–73, Technical and Operating Data DS-9-314, Div. Chem. Co., Chicago, Illinois, 1970.

9.5.5. Nonferrous Metals

Copper is chemically polished by the common copper bright dip process. A typical bright dip consists of a solution of 60% by volume sulfuric acid, 15% nitric, and $\frac{1}{2}$ oz per 1 gal hydrochloric acid, at (20° Baumé) room temperature, for 5–45 sec, followed by an alkaline rinse and a sodium cyanide bath to remove stains.

The process leaves a finish on brass that is matte in appearance but not rough enough to interfere with elastomer seals.

Aluminum and aluminum alloys are well prepared by a cleaning process that uses a sodium hydroxide bath to remove base metal. The finish is white matte in appearance but not rough enough to interfere with elastomer seals. Chemical polishing processes using phosphoric and nitric acids are available. The improvement in vacuum applications of the aluminum polish compared with the hydroxide etch is not nearrly so definite as that with stainless steel.

The processes described above, with the exception of the chemical polishing, are available in most plating shops and are not especially expensive. Less strong chemical cleaning process (as distinct from etching and polishing) using acid and alkaline baths are available. Chemical cleaning is useful as a preinstallation cleaning of completed welded parts to remove soil accumulated from final machining operations, handling, or long-term storage.

9.5.6. Solvent Cleaning

NOTE: Human health must come first. Check out up to date procedures for handling specific chemicals, especially chlorinated hydrocarbons.

Vapor degreasing is accomplished by suspending the piece to be cleaned in a vapor degreaser so that solvent vapor condenses on the piece and drips off. The advantage of this process is that only fresh, warm, distilled solvent touches the piece, with continuous rinsing. Trichlorethylene or tetrachlorethylene (also called perchlorethylene, which is nonflammable), and sometimes Freon (type TF), is usually used with this equipment. "Trichlor" and "Perchlor" are especially good for removing oils and greases. They are heavier, less volatile compounds that tend to remain on the surface in significant amounts; a subsequent cleaning with Freon TF or ethyl or methyl alcohol will remove these compounds. Methyl alcohol is a somewhat stronger solvent than ethyl alcohol, but is more toxic. Methyl ethyl ketone (MEK) is a strong solvent, and may be useful in cases when the chlorine and fluorine content of trichlor and the freons are undesirable for certain applications. Acetone is a strong solvent, but acetone remnants often appear in the residual gas, and this substance therefore is often not used for final cleaning. As wiping solvents, acetone and methyl alcohol tend to dry too rapidly; ethyl alcohol is often preferred. Small parts may be suspended by a wire and sprayed with Freon TF from spray cans. Freon or other solvents can be used in ultrasonically vibrated baths to increase the cleaning effect.

Detergent cleaning as a final step is not desirable because it tends to leave a thin adherent film. However, it may be the only available method for cleaning some plastics that are attacked by the usual solvents. A very thorough rinsing is required.

9.5.7. Sequence of Fabrication and Cleaning Steps

In many cases the most effective cleaning methods may not be permitted on a finished part. Examples include the chemical polish of stainless steel when the part has narrow cracks or thin-wall bellows, or on parts that are assemblies of different materials requiring incompatible cleaning chemicals. To achieve the best results the cleaning steps should be scheduled along with the fabrication, leak-checking, and installation procedure. When possible, the vacuum furnace braze or the high temperature vacuum oven degassing of H_2 serves as an excellent final cleaning step. An example of a part that requires machining after high temperature vacuum degassing is shown in Fig. 16. The cleaning and fabrication sequence is given in the figure legend.

9.5.8. Effect of Air Exposure

The molecular bombardment rate on a surface exposed to ambient air is so high that, if only a minute fraction of the impinging molecules are contaminant species, contamination cannot be avoided. Fortunately, the air layer near the surface becomes depleted of contaminants and the rate of contamination is then limited by diffusion. To minimize the contamination by particulates, the part can be wrapped promptly and stored in a dust-tight enclosure. Aluminum foil is convenient for wrapping. Brown paper is better than white paper, which contains remnants of bleaching and whitening agents. Polyethylene remnants, which could have come from polyethylene bagging, have appeared in residual gas analyses. Generally the part emerges hot from the last distilled or deionized water rinse of the cleaning process and will dry quite quickly in "still" air except for the last drop at the drip edge, which can be picked off with a lint-free paper or cloth. An air blast to hasten drying is probably not useful unless

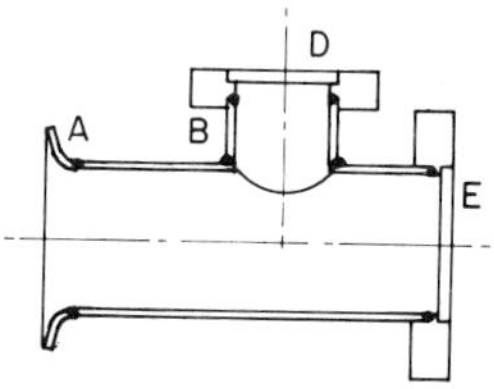

FIG. 16. An example of a planned sequence of fabrication and cleaning steps. The fabrication and cleaning steps of a typical stainless steel part for ultimate vacuum performance, where in this case flange E must be accurately perpendicular to D. (1) Make full penetration welds A and B using TIG welding; (2) Diversey DS-9 polish the incomplete weldment and flanges; (3) weld on the flanges using TIG welding; (4) leak check; (5) bake in vacuum furnace at 900°C for 4 hr at less than 10^{-5} Torr. (6) second leak check; (7) complete precision machining of the flange face E using a sharp tool and cutting lubricant; (8) clean finished part by degreasing and acid cleaning.

the air is very clean and does not increase the circulation of contaminated air over the part.

9.5.9. Cleanup Methods *in Situ*

Some methods for the removal of contamination from a system after it has been under vacuum and usually when no further air exposure is planned before the next operating period are given in this section.

Bakeout is the term applied to the technique of temporarily raising the temperature of the entire vacuum system (or all parts above the pump trap) in order to accelerate the rate of thermal desorption of molecules and to speed their movement to the pump. (See also Sections 5.14.4, 7.3.1, and 14.1.) The effects of time and temperature and procedures for bakeout with diffusion pumps and traps are discussed by Santeler *et al.*[12]

High temperature bake out is to be avoided if at all possible. Alternatives include careful cleaning and handling plus low temperature heating, internal glow discharge scrubbing, and electron desorption cleaning, which is of more limited usefulness. Sources of contamination can be blocked off by suitably placed and cycled cryopumping and, to a lesser degree, getter pumping. This use of pumping surfaces to block off sources of contamination is limited to reasonably low rates of contamination for reasonable periods of time; the blocking off approach is not applicable for long time periods at extremely low levels of contamination.

Electron bombardment desorption can be accomplished by using a hot

[12] D. J. Santeler, D. H. Holkeboer, D. W. Jones, and R. Pagano, "Vacuum Technology and Space Simulation," NASA SP 105, N66-36129, NTIS, Springfield, Virginia (1966); since published as: Aerovac Corp., "Vacuum Engineering," McGrath, 1967.

filament as an electron source, electrically biased negatively with respect to the surfaces to be cleaned. The cross section for desorption by electrons increases rapidly above the energy to overcome the binding energies of the molecules to the surfaces, reaching a practical maximum value above 500 eV. The cleaning effectiveness depends on the geometry of filaments and their relationship to surfaces, that is, it depends on the number of electrons per unit area of surface, on the electron energy, and hence on the power supply available. The necessity for a source of electrons in the vacuum may preclude use of this technique.

Electrical glow discharge cleaning is accomplished by introducing an optimum partial pressure of inert gas, typically argon, and the application of a high positive voltage to a suitable electrode in the chamber.[13] The inert gas becomes a conducting plasma, which bombards the inside of the chamber. Most of the voltage drop occurs in the "cathode sheath" at the chamber walls. Positive ions leave the plasma and bombard the surfaces. The geometry of electrode and walls affect the cleaning rate and result. The molecular cross section for ion desorption is higher than that for electron desorption. Better cleanup has been obtained using 10% oxygen in the discharge gas, which presumably oxidizes some of the hydrocarbons to more volatile gases. Some experimenters have used 100% oxygen. High purity Ar and O_2 are required if a continuous flow of gas is used to remove desorbed gases. Some argon is imbedded in the walls and later outgasses into the system. The imbedded argon can be removed by bakeout. In an aluminum alloy chamber a bake temperature of 280–300°C was sufficient to remove the argon.[14]

The subsequent outgassing of hydrogen may result from the dissociation of water and the formation of atomic hydrogen in the bulk phase. The dominant outgassing species change from H_2O, CO_2, CO and H_2 to Ar and H_2, which may be less detrimental to a desired operation. The glow-discharge treatment can substantially reduce the total outgassing rate, and can be especially useful when the chamber wall is subjected to ion bombardment during operation.

A less explored possibility for *in situ* cleanup involves the use of chemically active gases such as ozone, in which absorbed gases are oxidized to more volatile compounds.

How much benefit from *in situ* cleanup methods is retained after subsequent air exposure will depend on the particular circumstances.

[13] R. P. Govier and E. M. McCracken, *J. Vac. Sci. Technol.* **7,** 552 (1970).
[14] H. J. Halama and J. C. Herrera, *J. Vac. Sci. Technol.* **13,** 463 (1976).

10. PROTECTIVE DEVICES FOR VACUUM SYSTEMS*

Protective devices for vacuum systems may be conveniently divided into two categories; namely, those which protect the system itself, and those which protect the system operator, maintainence personnel, or others in sufficiently close proximity that the system may present a hazard. Of course, it commonly happens that devices that primarily serve to protect the system also protect personnel due to the failure mode of the system. The purpose of this part is to point out some of the more important hazards to personnel and system integrity.

10.1. Safety Considerations and Protective Devices

This aspect of good vacuum practice has been largely ignored by system design engineers and a great deal of reliance has been placed upon the skill and experience of the operator. Such a design philosophy is fraught with dangers for all concerned since the assumption that all operators of the system are aware of the possible hazards is often invalid. One need only recall one's own early days as a graduate student should a concrete example be needed!

It is somewhat difficult for manufacturers of a system to provide the needed information on safety, since to do so is generally considered to imply completeness. That is, if one operates the system in the stipulated manner, it is assumed that no *additional* significant hazards will be encountered. At this time, no manufacturer seems willing to assert that he has found all the dangerous modes of (mis)operation, and one finds, at most, general references to "established well-known safety practices" and personnel "skilled in the art" of vacuum system operation. However, one excellent source of information of a general sort on the subject of safety does exist. It is the *Vacuum Hazards Manual*[1] produced under the

[1] L. C. Beavis, V. J. Harwood, and M. T. Thomas, "Vacuum Hazards Manual," American Vacuum Society, 1975. Available from The American Vacuum Society, 335 East 45th Street, New York, NY, 10007.

* Part 10 is by Lawrence T. Lamont, Jr.

METHODS OF EXPERIMENTAL PHYSICS, VOL. 14

ISBN 0-12-475914-9

auspices of the Education Subcommittee of the American Vacuum Society. This monograph is highly recommended to all users of vacuum equipment.

10.1.1. Mechanical Hazards

It is somewhat disconcerting to note that many large vacuum systems are not mechanically stable—particularly when physically opened—and require external braces to prevent tipping over. Any competant physics freshman could solve the problem, so it will not be dwelt upon here. We merely note that if the system is not stable against reasonable perturbations, then appropriate braces *must* be designed. The best sort of stability criterion would simply be that the system will slide before tipping over. Where this is not possible, it may be reasonable to anchor the system to the floor.

The belt and chain and drive mechanism should always be shielded from physical access during normal operation. Such shields are commercially available for all current mechanical vacuum pumps and are mandatory in this writer's opinion. Simple shields for custom drives are also required by common sense and are easily fabricated. When such shielding is removed, as for routine maintenance, it is critically important that the mechanism cannot be energized by a remote operator. This is considered further in Chapter 10.2.

10.1.2. Explosion and Implosion Hazards

Explosion hazards are generally thought to be rare in vacuum technology. This is not strictly correct, however, and there are several worthy of consideration here. One rather obvious danger is associated with the use of pyrophoric gases such as silane (SiH_4) and oxygen at (or near) stoichiometric concentrations (for the production of SiO_2). The danger of these gases in combination for storage pumps such as the new cryosorption pumps for uhv application is obvious (see Section 5.4.5), but one must also be aware of the great potential danger of the use of these gases without special precautions in ordinary oil sealed rotary pumps. The hazard is simply that the solubility of SiH_4 in common mechanical pump oil is high enough that explosive concentrations may build up within the pump oil reservoir. Should the pump explode under these conditions, the resulting shrapnel from the reservoir casting can cause severe injury. The remedy for this is to provide for complete conversion of the gas prior to the mechanical pump. For this example, a simple pyrolytic converter would suffice. A second explosion hazard is related to water cooling in vacuum. Obviously, all cooling circuits must be able to withstand the forces generated by the normal operating water pressure (with an appropriate safety factor). This is particularly difficult to do with copper brazements since they are usually fully annealed when the braze is made and tend to deform slowly under pressure. However, no reasonable coolant circuit will tolerate power on the cooled device and the coolant valved off at inlet and outlet. This happens when "quick disconnect" water lines are used and inadvertantly disconnected (or, of course, when valves are closed). For this problem, one must interlock against coolant *flow*, not pressure (see Chapter 10.2).

By contrast to the explosion hazard, the threat of implosion is generally well recognized. All glass portions of the system, *including windows* in "all metal systems," must be protected from impact and workers from flying glass should they fail. Coarse mesh bell jar implosion shields are almost universal, but one occasionally encounters a system without any shielding or with only a portion of the jar shielded. Many laboratories are large enough that overhead walkways are provided. Tools dropped from such walkways onto the top of a vacuum system can result in a serious accident if the top of a bell jar is not protected.

10.1.3. Magnet Hazards

The large permanent magnets used on sputter–ion pumps and magnetic sector spectrometers can be a serious safety problem when they are removed from the apparatus. The danger is not so much from flying wrenches or other ferromagnetic material, although many painful injuries have resulted from such carelessness, but from *other magnets*. Magnetic assemblies must be separated by robust separators (wooden blocks are adequate) and in no case should the disassembly of a sputter–ion pump to this level be attempted by the inexperienced.

10.1.4. Electrical Hazards

Most commercially available high voltage feedthroughs and their matching connectors are provided with adequate protection and grounding provisions and it remains to the user not to defeat their purpose. However, difficulties can arise when such assemblies are not properly grounded. It is neither safe nor (usually) legal to use the cable shield (when one is provided) and the connector shell as the safety ground. This is particularly true for large power supplies, and one *must* provide grounding which is independent of the cable shields.

The entire question of grounding a system is often inadequately treated or even sidestepped entirely in instruction manuals. However, the increasing demand for systems under microprocessor control seems likely to alter this unacceptable state of affairs since the noise sensitivity of such devices requires considerable attention be paid to grounding points and return currents under pulse conditions. It is somewhat sad commentary upon human nature that this has been the sequence of events. Nonetheless, let us examine the requisite conditions.

(1) Each system should have its own grounding rod in multiple system installations.

(2) The grounding rods should be copper rod stock at least 1.5 cm diameter and 2.5 m long driven into the ground as near to the system as possible. If the conductivity of the earth is too low, multiple grounds may be required, or soil treatment with an electrolyte, such as $CuSO_4$, may be appropriate where allowable. A benchmark in this regard is that the resistance of a 2 cm diameter rod buried 7 m in earth of conductivity 10^{-4} mho/cm is approximately 20 Ω. Grounding to water pipes is very poor technique due to the high resistance of most pipe thread joints.

(3) For grounding up to power line frequencies only, copper wire at least 4 mm in diameter (AWG B&S 6 gauge) is adequate. Braided wire should *not* be used. The wire should be silver soldered to the grounding rod and not merely clamped.

(4) For grounding of systems incorporating radio-frequency equipment such as rf sputtering, induction heaters, rf ion sources, and the like or for systems using electron beam gun evaporation apparatus, or any other devices with significant stored energy, the ground must be adequate for rf. This is extremely difficult to accomplish in any meaningful way since the limit here is the inductance of the ground path and not the dc resistance. For reference, almost any reasonable ground return path has an inductance of at least 1.5 μH/m. In general, the following may be helpful.

(a) The *maximum allowable* system ground point to the ground rod is 2 m, with shorter distances being strongly preferred.

(b) The ground strap should be at least a 5 cm wide copper strap 0.1 mm thick. It should have no significant bends or kinks.

(c) Power supplies must not be "grounded" to the system frame. Usually, it is acceptable to connect them to the same grounding rod as the system but, in extreme cases, they may require separate grounding rods.

As a general commentary, we may note that systems which are subject to arcing are likely to present a more severe grounding problem than rf excited apparatus due to the broad spectrum of Fourier components in the arc signal. These arcs may not present an electrical shock hazard in themselves, but may create a dangerous situation by actuating valves or other mechanisms outside of their normal sequence.

For more detailed requirements concerning grounding and shielding techniques, we may turn to the electronics reference texts on the subject.[2]

Many feedthroughs are purchased without matching connectors—and indeed there are feedthrough assemblies available for which there are not standard matching connectors other than the special units specific to the feedthrough manufacturer. Fortunately, the trend is away from nonstandard feedthroughs, and so the problem should be less of a practical concern in the future. Potentials in excess of a few tens of volts must be isolated from accidental contact by personnel as must all constant current power supplies due to the stored energy that may exist in system inductances or the fact that the open circuit voltage of the supply may far exceed the normal operating potential (recall that the "idle" condition for a constant current supply is a short circuit whereas a fault condition is an open circuit).

Cables must also be protected from thermal stress and mechanical abuse. Most common errors of this sort are improper cable restraints—or no restraints at all—in the vicinity of heaters of diffusion pumps, cables draped carelessly across the laboratory floor, and cables of insufficient length to permit the bell jar, on such systems, to be fully raised. Should a lift truck, dewar, dollie, etc., be run over an unprotected cable, one may have the misfortune to generate an intermittant short between wires within the cable. This may cause dangerous intermittant malfunctions, or may put hazardous potentials on unshielded (normally low voltage) terminals. The solutions here are obvious.

10.1.5. Thermal Hazards to Personnel

Once again, the problems are obvious. Hot surfaces must not be directly accessible to accidental contact by the operator. Even if the temperature of the object is not itself high enough to constitute a hazard, the involuntary withdrawal of the hand or other burned portion of the anatomy may precipitate other problems should the person accidentally strike another hazardous portion of the system. The worst offenders are typically diffusion pumps. Only in recent times have manufacturers been

[2] See, for example, R. Morrison, "Grounding and Shielding Techniques in Instrumentation." Wiley, New York, 1967; or H. W. Ott, "Noise Reduction Techniques in Electronic Systems." Wiley, New York, 1976.

shielding these pumps in more than a cosmetic manner. There is clearly much room for improvement.

Similar conclusions hold for bakeout shrouds when they are installed, and the entire system is a hazard to personnel when they are not used. At the very least, some highly visible warning must be posted on the system that it is under bakeout and must not be touched. For an obscure reason, presumably having to do with human nature, handwritten signs seem significantly more effective than standard approved "commercial" signs. Hopefully, as safety practices are more universally improved within vacuum technology, this distinction will become less significant.

10.1.6. Radio Frequency and Microwave Radiation Hazards

Most rf sources are of such low frequency—generally tens of megahertz—that *radiated* energy is not the true danger. Rather, since rf applications involve power coupled into a plasma load, the true hazard is from voltage drop across conductors carrying large reactive currents. This voltage drop is determined, of course, primarily by the inductance of the conductor. For example, a 25 cm long rf buss strap has an inductance of about 0.1 μH. If this buss must carry a reactive current of 50 A, as it commonly must in rf sputtering systems, then for the frequency of 13.560 MHz usually used in such applications, the voltage drop will be over 400 V! Obviously, the "trick" of designing such systems is to never permit such currents to flow on the outside surfaces of the vacuum system. This design problem involves a great deal of "art"—which is to say, we really don't know all that much about how to do it—and the details of construction for successful systems are usually proprietary to the manufacturers who have developed them for their product line. Most "solutions" are quite specific, which again points to out general lack of fundamental technology.

Microwave power is quite another matter. Here, radiated power *is* the danger, and the mechanism of damage to living tissue is frequently internal heating. Since the body requires extremely precise control of certain temperatures, one *must not* ever assume that the hazard will be "felt" in time. Microwave safety, like the techniques for the safe handling of radionuclides, is a highly specialized topic which is far too complex to be covered meaningfully here. Those without special knowledge in this area should either acquire the requisite knowledge or simply avoid the use of microwave power above the level of a few milliwatts. It is also worth noting that rf induced gas discharges may generate such microwave power even though they are excited by far lower frequency.[3] If this is suspected, it is advisable to have a radiated power survey performed by someone skilled in that area to be certain that no significant personnel hazards exist. Small, relatively inexpensive survey meters have recently become available commercially.* These may be of use in large installations.

10.1.7. X Rays

X rays may be produced either by a bremsstrahlung process or as the characteristic x rays of the target material (if this is energetically possible) when electrons bombard a target. For potentials less than a few keV, the minimum wavelength of the bremsstrahlung produced x rays is the order of the interatomic lattice spacings and they have very little penetrating power. For this reason, such radiation is not generally a danger to personnel. However, should the bombarding electrons have energies much in excess of a few tens of keV, appropriate shielding may have to be designed for such items as viewports. Again, this is a highly specialized topic, and the reader is referred to someone experienced in that field.

10.2. Interlocks and System Protective Devices

The function of system protective devices is to return the apparatus to a safe status in the event of failure of some component or components in a way that minimizes damage to the system itself and the process underway. The priority of these two may be altered by the value of the items under process and the cost of damage to the system that may result. These judgments are far beyond the scope of this text and are specific to the process, so their existence as possible considerations is merely mentioned.

Failures with which the system must deal may be conveniently divided into internal and external. The former classification includes component failure, degradation of some component outside tolerance limits, and process failure. The latter class includes such "outside utilities" as

[3] L. T. Lamont, Jr., and J. J. DeLeone, Jr., *J. Vac. Sci. Technol.* **7**, 155 (1970).

* For example, the RAHAM Model 2 radiation hazard meter produced by General Microwave Corp., Farmingdale, New York, is sensitive to radiated power in the frequency range from 10 MHz to 3 GHz and can detect a power flux as low as 0.05 mW/cm^2. Other units with similar capability may be available.

power failure, coolant failure (overtemperature or inadequate supply), and inadequate cryogen for chilled traps.

Systems pumped by diffusion pumps and turbomolecular pumps should generally respond to any serious external failure or to a failure threatening vacuum integrity by closing all valves while leaving all normally functioning pumps running. Any mechanical pumps which have drive failures or which are vented to atmosphere should be turned OFF. The latter condition is due to the motor size which is generally not adequate to long term pumping against atmosphere. Overheating of both the motor and the pump itself may result. Diffusion pumps can be left ON even though they are not connected to a backing pump for rather long periods of time. However, if there is a leak in the pump area, its heater power should be turned OFF (but the cooling water left ON!). Turbomolecular pumps are particularly susceptable to overheating under high gas flow conditions. Most current pumps, however, are adequately protected against this problem, assuming the user doesn't override the protective device! It is particularly important to close the valve between the mechanical pump and a cold diffusion pump. This is because the mechanical pump oil will migrate slowly into the diffusion pump. When the diffusion pump heaters are turned on, the mechanical pump oil will decompose, contaminating the pump walls with a "varnish" which can be quite difficult to remove. The diffusion pump oil, of course, would have to be changed.

Systems with sputter-ion pumps are fail-safe with respect to power failure, and one needs only to close off any significant leaks into the chamber. Of course, the pressure in such a system will rise slowly due to normal outgassing, but the titanium in the pump may continue to getter the active gases for a significant period of time. Naturally, the nongetterable gases, such as the inert gases and saturated hydrocarbons like methane (CH_4), will not be gettered and their partial pressures will rise considerably faster than the getterable gases.

Systems pumped by uhv closed-loop cryopumps should be treated in the same manner as diffusion pumped systems in the event of power failure. As covered in Section 5.4.5, we assume the presence of a functional pressure relief valve for when the pumping element warms up!

Internal failures which threaten vacuum integrity should be treated in the same manner as external failures. Systems response to process threatening failures should be at the discretion of the operator, or "preprogrammed" into the system as appropriate.

All interlock overrides must be located at or very near the hazardous area and *nowhere else*. This is absolutely necessary to avoid injury to maintenance personnel.

11. DESIGN OF HIGH VACUUM SYSTEMS*

11.1. General Considerations

There are two basic ways of reducing the pressure or time needed for evacuation of a chamber: one can increase the pumping speed or decrease the outgassing rate. In some cases, it may be less expensive to provide additional pumps rather than attempt to reduce gas evolution. This depends on the degree of vacuum and other process requirements. The following check points will be useful before specifying or designing a vacuum installation. The gas handling capacity of the pumping system at a given pressure is usually more important than the ultimate pressure; the net pumping speed at the chamber is more significant than the nominal speed of the pump; increasing the number of pumps in high vacuum region does not necessarily decrease the evacuation time in the same proportion; well chosen pumps, working fluids, baffles, and traps can reduce or eliminate contamination problems; rapid pumping requirements increase system cost; long exposure of the chamber to atmospheric conditions will increase the subsequent outgassing rate; temperature history and distribution in the chamber before and during evacuation should not be neglected; the amount of surface area present in the vacuum chamber is more significant (at high vacuum) than the volume.

11.1.1. Surface-to-Volume Ratio

Generally, at high vacuum conditions, there is more gas adsorbed on the walls of the vacuum chamber than present in the space. This, of course, depends on the size of the chamber. For small devices, the surface-to-volume ratio is higher and there is a tendency for small vacuum chambers to have a higher ratio of surface area to pumping speed. Therefore, generally 5-cm-diameter diffusion pump systems, for example, do not produce low ultimate pressure as easily as 10- or 20-cm-diameter pumping systems.

A similar comment may be made about other sources of gas. Pumping speed grows approximately with the square of the pump diameter but the

* Part 11 is by M. H. Hablanian.

METHODS OF EXPERIMENTAL PHYSICS, VOL. 14

ISBN 0-12-475914-9

exposed area of gaskets increases almost at a linear rate. Thus, it is relatively easy to produce high vacuum in large chambers (as high as 10 m diameter). The rough vacuum system in such cases may be the expensive part because of its dependence on volume.

In laboratory systems, pumps of 15–20 cm diameter are usually more convenient for low ultimate pressure work even if the high speed requirements are not necessary.

11.1.2. Pump Choice

The selection of diffusion pumps is not as straightforward as selection of mechanical pumps. The considerations for reaching a given pressure in a short time may not prove to be an adequate measure of the system's capability to handle steady gas load. Characteristics of pumping curves of the diffusion pumps will indicate the proper components. More frequently, gas loads are not known and the selection of the pump must be based on experience with similar applications.

The vacuum equipment suppliers generally will offer performance based on pumpdown characteristics of an empty chamber. This can be misleading if it is construed to be a measure of pumping efficiency under full load conditions. The vacuum equipment designers must relate performance to a set of conditions which are known and can be defined and duplicated, which perhaps has misled some users to construe the pumpdown time to be the true indicator of process time.

The selection of diffusion pumps should be concerned with all factors involved in the system. These factors normally consist of pumping speed, forepressure tolerance, backstreaming rates, pressure at which maximum speed is reached, protection devices, ease of maintenance, backing pump capacity, roughing time, ultimate pressure of the roughing pump, baffles or cold traps employed, valve actuation, etc.

Unless these factors are carefully weighed in the system design, the system may not perform satisfactorily as a production system. It is to the advantage of both the supplier and the user to carefully consider these factors before finalizing system design.

It is obvious from the previous discussion that jet breakdown should be avoided in system operation. As a general rule, diffusion pumps should not be used at inlet pressures above the "knee" for prolonged periods of time. In a properly designed system with a conventional roughing and high vacuum valve sequence, the period of operation above the knee should be measured in seconds. As a "rule of thumb" it may be said that if this period exceeds half a minute, the pump is being overloaded (Section 4.3.3).

It is clear then that in applications requiring only 10^{-4} Torr process pressure, the mechanical pump roughing period is normally much longer than the diffusion pump portion of the pumpdown cycle, or at least it should be for economic reasons.

11.1.3. Evacuation or Process Pumping

Vacuum pumps are used for two somewhat distinct applications. The first involves evacuation of gas (usually air) from a vessel, the other maintaining process pressure with a given gas load. It is difficult to choose the correct pump size with any degree of precision for both applications when required pressure is in the high vacuum region. In the first case, this is due to the uncertainty of surface outgassing rates which depend on temperature, humidity during atmospheric exposure, and the presence of gas adsorbing deposits on the chamber walls. The water vapor is the usual cause of concern. In the second case, the process gas evolution is uncertain due to variation of gas content in materials and the variations in the process parameters. For example, when electron beams are used for evaporation the power distribution in the focal spot may not be exactly repeatable. This may produce variations in the degree of heating of the surfaces in the vicinity of the electron gun.

As a very rough rule for evaporation, melting, or welding or sputtering of metals, at least 100 liters/sec of net pumping speed should be provided for every kilowatt of power used for the process.

Generally, one can hardly make an error in the direction of using too large a pump. There is a tendency of using 25 cm diffusion pumps in a large bell jar system to produce a net system speed exceeding 1000 liters/sec. A few years ago, the net pumping speed at the base plate was only 500 liters/sec. These differences are not important in improving system pumpdown time but they do help to maintain lower process pressures. As far as evacuation time is concerned, the differences between a 15 or 25 cm pump is hardly noticeable. When outgassing is taken into account, the evacuation time is not inversely proportional to the pumping speed as might be expected from the experience at higher pressures. However, during rapid process gas evolution or introduction, there often exists an inverse relationship between the process pressure and the pumping speed.

In some cases, such as sputtering, the maximum throughput becomes an important item because it may be desirable from the process point of view to have a certain flow of argon through the system. A high pumping speed for the initial evacuation of the system will be of lesser importance.

11.2. Metal Systems

Several design considerations for vacuum systems cannot be easily put into a single set of requirements. A practice which can be easily tolerated in one case must be completely forbidden in another. The requirements span over 3 or 4 orders of magnitude in pressure range and surface area. For a small device which must be rapidly evacuated to higher vacuum with a system of limited pumping speed, any trapped air pockets must be carefully eliminated. For example, a screw or bolt in a threaded hole can be a source of virtual gas leakage. It is a good practice to provide an evacuation passage for the trapped gas. A hole can be drilled through the bolt, threads can be filed flat on one side, or the blind holes can be generally avoided. However, all this will be of no significance in a system which has a pumping speed of 50,000 liters/sec and is operated only at 10^{-5} Torr. Obviously, vacuum level, pumping speed, size of chamber, and the desired time of evacuation must be considered to provide a balanced design.

In large vacuum systems, structural distortions during evacuation may be high enough to be troublesome. To preserve precise alignments of optical equipment, mechanical drives, etc., it may be necessary to support the various components independent of the external vacuum vessel.

For a 2.5-m-diameter door used in an industrial coating system or furnace, the total force due to one atmosphere pressure difference is nearly 50 T. This is certainly sufficient to seal an elastomer gasket without the use of any clamping devices. However, if the chamber is "air released" with nitrogen or argon, it is advisable to prevent an internal overpressure. A small overpressure may suddenly open the door if the rubber gasket has a tendency to stick. A spring-loaded catch can be used to prevent this.

Another common malfunction due to distortion occurs in large high vacuum valves. The pneumatic pistons used to actuate such valves are usually not strong enough to provide the vacuum seal. The pressure from the atmospheric side is relied upon to provide the tight closure. Thus, such valves cannot be opened against the atmospheric pressure. A malfunction can occur during the roughing cycle. As the atmospheric pressure in the chamber is reduced, the valve seat may lift slightly and begin to leak. If the valve is used above a diffusion pump system, the leak may be large enough to overload the diffusion pump and cause backstreaming and pumping fluid loss problems.

11.2.1 Pumping System Design

High vacuum pumps are used, together with mechanical pumps, in applications where system operating pressures of 10^{-3} Torr and below are desired. The physical arrangement of system components depends on the characteristics of the process to be carried out, such as the pressure level, cycle time, cleanliness, etc. To some extent the availability and compatibility of components influences the system design. In some instances, the economic aspects of component selection may determine the system layout. The following paragraphs illustrate briefly the most common component arrangements, referred to as valved and unvalved systems, and outline their major respective advantages and disadvantages. A recommended operating procedure to insure minimum backstreaming and work chamber contamination is outlined for each type.

To furnish maximum effective pumping speed at the processing chamber, it is generally desirable to make the interconnecting ducting between the chamber and the pump inlet as large in diameter and as short in length as practical. The amount of baffling and trapping required depends on the desired level of cleanliness in the chamber, the necessary reduction of the inherent backstreaming characteristics of the pump, and the migration of the pumping fluid. This phenomenon has been previously discussed and its effects can be significantly reduced by correct component selection and operating procedures.

In regard to diffusion pumps, baffles, traps, and valving must be selected for minimum obstruction to gas flow, without sacrifice of efficiency of their specific function.

The size of roughing and foreline manifolding is governed by the capacity of the mechanical pump, the length of the line, and the lowest pressure region in which it is expected to function effectively. Additionally, the size of the foreline is influenced by the forepressure and backing requirements of the diffusion pump under full load operating conditions.

The need for a holding pump is determined by several factors including the forepressure characteristics of the diffusion pump, time cycle requirements, manifold configuration, and leak tightness.

The selection of the type and capacity of mechanical pump will depend on the desired operating cycle, the throughput and forepressure requirements of the diffusion pump, and the proposed operating procedures for a system pumpdown.

The general system operation and equipment layout will affect the optimum selection and location of vacuum gauges, safety devices, and interlocks.

11.2.1.1. Valved Pumping Systems. For applications involving rapid recycling, a fully valved pumping system is essential. This type of system is shown schematically in Fig. 1. It permits isolation of the diffusion pump from the work chamber at the conclusion of a pumpdown and prior to air admittance. The pump can, therefore, remain at operating

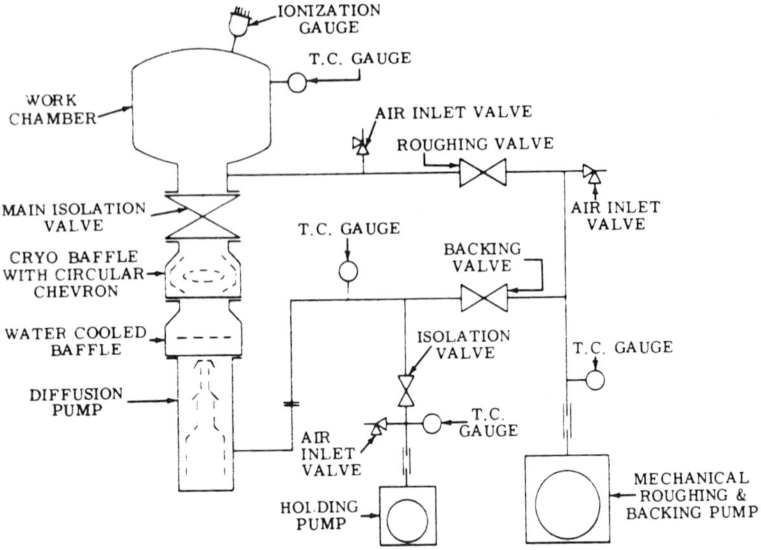

FIG. 1. A schematic arrangement of a typical vacuum system with a high vacuum valve.

temperature and pressure during periods when the chamber is at atmosphere and during the rough pumping portion of the cycle. The length of these periods may indicate the need for a holding pump. The main isolation valve also permits the continuous operation of the cryobaffle between it and the diffusion pump inlet. Neither this nor rapid cycling can be realized without the valving indicated, in view of the cool down and heat up time lapse inherent in diffusion pump operation, and cool down and reheat time of the cryobaffle. Judicious operation of the main valve at the changeover phase from roughing to diffusion pumping can significantly reduce the backstreaming of oil vapors to the work chamber. Valved systems are generally confined to operating pressures in the 10^{-8} Torr range and above. Most large commercially available valves contribute too high a gas load to the system to allow operation at pressures lower than 10^{-8} Torr.

Leak testing and leak hunting are considerably easier in valved systems, and repair procedures are also generally less time consuming than in unvalved systems. However, the following disadvantages are noted.

Valved systems are initially more expensive, especially when large size valves are involved. In addition, the use of valves inevitably adds to the system complexity and generally results in lower effective pumping speed at the chamber.

For operation below 10^{-8}–10^{-9} Torr and for use with baked ultrahigh vacuum chambers, the availability and/or cost of valves may make their use prohibitive.

11.2.1.2. Valveless Systems. Valveless vacuum pumping systems are generally considered for applications where the length of the pumpdown is of less importance, and process or test cycles are of extended duration. They are found on baked ultrahigh vacuum chambers using diffusion pumps, due to lack of availability and/or prohibitive cost of valves compatible with the proposed operating pressure levels. The valveless system (Fig. 2) generally offers higher effective pumping speed at the chamber and, in view of its prolonged operation at very low throughputs, lends itself to the use of a smaller holding pump. This pump is sized to handle the continuous throughput of the diffusion pump at low inlet pressures, 10^{-6} Torr and below, and allows economy of operation by permitting isolation and shut off of the considerably larger main roughing and backing pump. The valveless system, however, also has a number of disadvantages. More complex operating procedures are necessary to insure maximum cleanliness in the work chamber and minimum backstreaming from the diffusion pump.

To minimize chamber contamination it is recommended to keep at least one of the baffles or traps operative while the vacuum chamber is being baked. When the inlet ducts are at the elevated temperature the rate of oil migration into the system is accelerated. At temperatures near 200°C the oil vapors are not condensed in the chamber and are subsequently pumped by the diffusion pump. A very thin (invisible) film of oil will con-

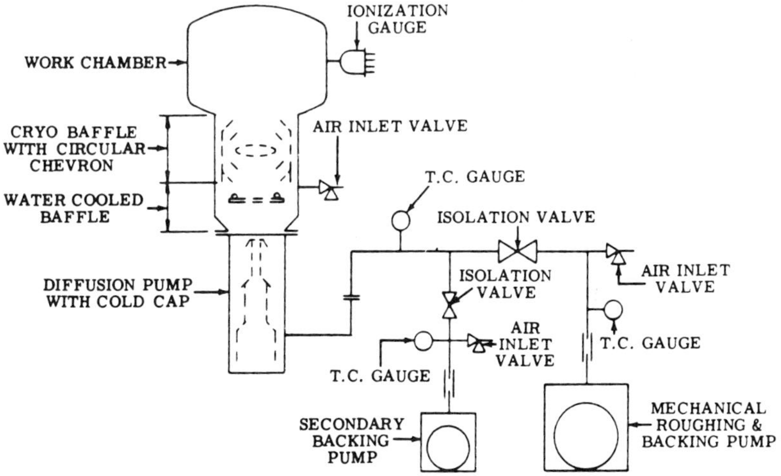

FIG. 2. A vacuum system arrangement without a high vacuum valve.

dense in the chamber during the cooldown. This film formation is minimized by continuous operation of at least one of the baffles.

The diffusion pump must be cooled down to a safe level before the chamber can be repressurized. This makes servicing cumbersome and time consuming. Failure of utilities necessitates immediate shut down of the system to protect the work load and equipment. Operating costs of cryogenic baffles are higher, leak testing and leak hunting are less convenient and more time consuming.

12. OPERATING AND MAINTAINING HIGH VACUUM SYSTEMS

12.1. General Considerations*

Operation of high vacuum systems requires certain care and attention to several items. General cleanliness can be extremely important, especially in small systems. It should be remembered that a droplet of a substance which slowly evaporates producing a pressure of 10^{-7} Torr may take an entire day to evaporate if the available pumping speed is 100 liters/sec. Humidity and temperature can be important in view of constant presence of water vapor in the atmosphere. When vacuum systems are back filled or air released and then repumped, it will make a significant difference whether the air was dry or humid. The time of exposure is also significant. If necessary, the back filling can be done with nitrogen or argon. For short exposures (30 min or less) this appears to reduce the amount of water vapor adsorption in the vacuum system.

It is extremely important to develop good habits in sequences of valve operations, especially in systems with manual valves. Graphic panels showing valve locations and functions are very useful. A single wrong operation may require costly maintenance procedures. It is well to pause a few seconds before operating any valves in a high vacuum system unless a routine procedure is followed. Automatic process control sequences have been widely used in recent years to avoid operational errors.

* Chapter 12.1 is by M. H. Hablanian.

METHODS OF EXPERIMENTAL PHYSICS, VOL. 14

ISBN 0-12-475914-9

12.2. High Vacuum Valve Control*

Usually, in a high vacuum system which is expected to be pumped to 10^{-6} Torr in less than half an hour, the outgassing will be negligible (compared to the maximum throughput of the pump) during the initial period after the high vacuum valve is opened. In practice, due to lack of precise valve control, the period between the time when the valve is opened and the time when inlet pressure of 1 mTorr is reached can be shorter than that computed using Eq. (5), Section 4.1.3.2. This is due to the expansion of air across the high vacuum valve into the higher vacuum space downstream. This downstream space (part of the valve, trap, baffle, upper part of the pump) can be significant compared to the chamber volume. On the other hand, when outgassing is severe, the pump may be overloaded for longer periods of time. The same is true if the pump is too small for the chamber volume. This can be encountered in furnaces having porous insulating materials and in coaters where large drums of thin plastic film are present in the vacuum chamber. In such cases, it may be better to elongate the rough pumping period before returning to the diffusion pump. When throughput is nearly constant, the pumping time will be the same regardless whether the high vacuum valve is only slightly cracked or fully opened. In general, the high vacuum valves should be opened slowly; very slowly at the start. The motion of ordinary pneumatic valves can be controlled to some extent with the air inlet and exhaust adjustments. Special valve controls can be made either to maintain approximate constant throughput during initial opening or to have a two position interrupted operation. When the valve is almost closed, it serves as a baffle. This prevents the possibility of pumping vapor getting into the chamber due to the turbulent flow of air entering the pump. Figure 1 shows an example of a large valve operation.

12.2.1. When to Open High Vacuum Valve

During the evacuation of a vessel, the question arises regarding the proper time to switch from rough pumping to the diffusion pump or, in other words, when should the high vacuum valve be opened. No direct answer can be given because it depends on the volume and the gas load of the chamber. Ideally, the answer should be: when the gas flow into the

* Chapter 12.2 is by M. H. Hablanian.

ISBN 0-12-475914-9

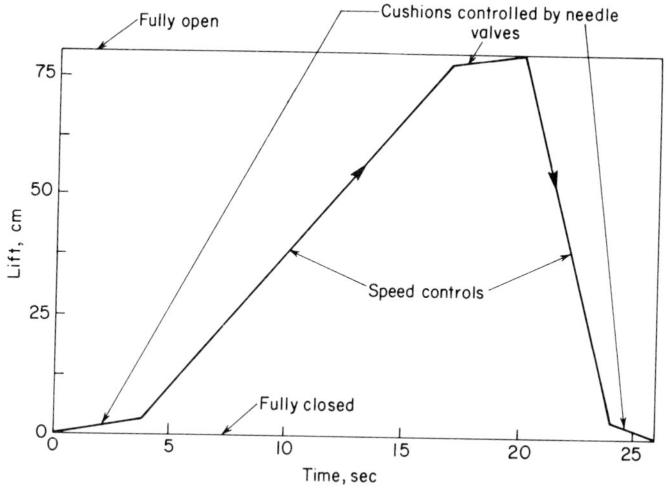

FIG. 1. An example of slow operation of a large high vacuum valve.

high vacuum system is below the maximum throughput of the diffusion pump. This would imply an inlet pressure near 1 mTorr. However, it is not necessary to obtain this pressure with mechanical roughing pumps or even with mechanical booster pumps (Roots type). In practice, the transfer from roughing to diffusion pumping is normally made between 50 and 150 mTorr. Below this pressure region, the mechanical pumps rapidly lose their pumping effectiveness and the possibility of mechanical pump oil backstreaming increases. Although the throughput of the diffusion pump is nearly constant in the region of inlet pressures of 1–100 mTorr, the initial surge of air into the pump, when the high vacuum valve is opened, will overload the diffusion pump temporarily.

The general recommendation is to keep the period of inlet pressure exceeding approximately 1 mTorr (0.5 mTorr for large pumps) very short, a few seconds, if possible. Consider the following example. For constant throughput, the evacuation time (between 100 and 1 mTorr) is

$$t = V/Q(P_1 - P_2)$$

[see Eq. (5), Section 4.1.3.2].

For a common 45 cm diameter bell jar the volume is about 120 liters and using a 15-cm-diameter diffusion pump with a maximum throughput of about 3 Torr liters/sec, we obtain

$$t = 120 \text{ liters} \times 0.1 \text{ Torr}/3 \text{ Torr liters/sec} = 4 \text{ sec.}$$

If the high vacuum valve were to open slowly to admit the gas into the diffusion pump at the maximum throughput rate, it would take only 4 sec to reach the stable pumping region.

12.3. Mechanical Pumps*

Routine maintenance for mechanical pumps is usually limited to oil changes, clean up of oil reservoir, and to drive adjustments. The need for oil changes depends on service and can be as often as once a day or as rare as once in several years. A most common period is once every six months in reasonably clean applications, such as roughing and backing of systems involving diffusion pumps.

Lubricants used are usually hydraulic or turbine grade mineral oils with a viscosity of 300 SSU at 100°F (SAE 20 W). A 100 SSU change in oil viscosity is an ample reason for an oil change as is inability to produce the desired operating pressure level. Contamination pressure can be measured at zero flow by subtracting the partial pressure of air (McLeod gauge) from the total pressure gauge reading and correcting for apparent vapor pressure of the lubricating oil. In the absence of identifiable sources of contamination, a 100 mTorr increase in vapor pressure over that of a fresh oil sample indicates undesirable oil degradation.

If gas ballast is used for oil clean up, the end point is determined when there is no further improvement in pressure with time or when the difference between total pressure and partial pressure of air is less than 10 mTorr for compound pumps and less than 25 mTorr for single stage pumps when using recommended lubricants. In the absence of suitable instrumentation, if oil looks clean and has no unusual odor, it is likely to be good and needs no replacement. Traces of free water, the most common contaminant, can be detected by heating an oil sample with a match flame on a piece of aluminum foil. Crackling noise with small vapor explosions indicates presence of free water; smoke only indicates dry oil.

Handling of large water vapor load can result in the formation of oil emulsion. This presents no mechanical problems since oil emulsion is a good lubricant. Operation with a 50/50 mixture of oil and water is possible, but the pump can fail mechanically almost immediately when lubricated by pure water instead of oil or oil emulsion. Such a situation occurs if a heavily contaminated pump is stopped and water separation is allowed to occur, so that the oil intake port becomes immersed in water. When restarted, water enters the pump and a rapid failure can occur.

* Chapter 12.3 is by Z. C. Dobrowolski.

ISBN 0-12-475914-9

This can be prevented by draining the pump prior to starting or by using a lubricant heavier than water.

While gas ballast is very effective in handling water vapor and other vapors not soluble in oil, it is not very effective in handling soluble vapors (toluene, turpentine) particularly those with boiling temperature above 80°C. With such contaminants pumps may require frequent oil changes. High boiling point contaminants can be removed by oil changes only. "Contamination" of pump oil by helium can be cleaned up quite effectively by the use of gas ballast or by operating the pump briefly with a high air throughput. Gas ballast is also effective in suppression of the hydraulic noise associated with the low pressure operation of rotary piston pumps.

The lowest point in the oil reservoir serves as a depository for sludge, solids, and other debris heavier than the lubricant. This area should be cleaned about once a year in rough applications and once in several years when the process is clean.

The need for other maintenance is subject to diagnostic observations. Presence of oil in the vicinity of the drive shaft indicates necessity for shaft seal replacement (small shaft seal leaks are difficult to confirm with a helium leak detector). Unusually warm pump temperature combined with high power requirements and deteriorated blank off suggests defective discharge valves. Hot zones around shaft supports mean defective bearings. Unusually noisy operation of the pump, particularly if it persists in the presence of gas ballast flow, indicates either a failure of the lubricating system (oil lines if externally accessible will be cold) or general mechanical wear. When mechanical distress is observed or suspected, the use of a noise meter scanning for highest noise intensity in the close vicinity of the pump surface is helpful in pinpointing the location of the trouble source. Wear can be confirmed by the presence of cast iron sediment in the separator or by the cast iron discoloration of the lubricant. Inspection of discharge valves, which are normally quite accessible, can be used also as a diagnostic tool. Excessive wear of the valve seat and springs suggest the presence of abrasive materials and hence potential internal wear. Discharge valves are first to fail in the presence of abrasives. If internal wear is suspected, gas ballast test can be used in place of pumping speed test as a quick and sensitive check on the mechanical clearances in the pump.

For example, a 5% gas ballast flow would produce a pressure of $5 \times 760 \times 0.01 = 38$ Torr if admitted at pump suction. In a tight single stage pump such gas ballast flow admitted in the normal fashion during the compression stroke would produce an inlet pressure of about 2 Torr for a compression ratio of $38/2 = 19$. A pump which can produce a compression ratio of 4:1 (9.5 Torr blank off with 5% gas ballast flow) will provide still about 80% of the nominal pumping speed, particularly at low pressures. With a compression ratio of 2:1 pumping speed is reduced to about $\frac{1}{2}$ nominal and can be considered uneconomical. Excessive clearances have relatively little effect on the pump blank off. External pump leaks, provided these do not interfere with intended use, can be tolerated at the expense of reduced pumping speed. If the base pressure due to leaks is less than 20% of the operating pressure, leaks need not be taken seriously as the sacrifice in pumping speed is not significant. In the absence of a leak detector, the presence and size of external leaks can be determined by the use of an oil or water manometer located at hermetically sealed pump discharge. A length of clear plastic tubing can be used as a U-tube manometer and leaks as small as 10^{-3} atm cm^3/sec can be measured. The use of a McLeod gauge together with a total pressure gauge can facilitate diagnostic decisions. Low McLeod gauge reading with a high total pressure reading indicates absence of leaks in the presence of contaminated oil. Elevated pressure indication by both gauges confirms the presence of leaks. Handling of very aggressive gases (Cl_2, HCl) results in a rapid formation of varnish deposits. It is best to operate the pump continuously with frequent oil changes with the pump running (dilution method). Once stopped, the pump may bind up and has to be cleaned mechanically or by flushing with a varnish removing compressor fluid. The varnishing problem can often be avoided by the choice of a lubricant resistant to the contaminant. In this case, a suitable choice would be a fluorocarbon. Pure oxygen can be handled with the help of phosphate fluids, while polyglycols are useful with food derivatives.

It is sometimes necessary to clean the discharge gases for analysis or reuse. A good exhaust filter can remove all particulate matter (zero transmission for particles with a diameter of 0.3 μm) but it cannot remove oil vapor. Should that be necessary, activated charcoal or molecular sieves can be helpful.

Cold starting of pumps at temperatures below 5°C can be difficult. It is a frequent difficulty in winter when cooling water is not turned off after a water-cooled pump is stopped. Occasionally it is necessary to start a pump in cold environment. Complications of pump heating can be avoided by the use of low pour point (−40°C) high viscosity index oil (lubricating silicones).

A well operated maintenance and diagnostic service program needs to be supported by the following supplies and equipment.

(a) Ample supply of lubricating oil, sufficient for five complete oil changes;

(b) drive shaft seal;
(c) discharge valves;
(d) bearings;
(e) sealer;
(f) drive belt or coupling;
(g) filter elements;
(h) total pressure gauge with a good resolution in the 10–500 mTorr range;
(i) McLeod gauge;
(j) viscometer;
(k) clamp-on ammeter;
(l) surface and immersion thermometers;
(m) gasketing material;
(n) leak detector;
(o) noise meter.

12.4. Diffusion Pump Maintenance*

The diffusion pump is basically a relatively trouble-free device as it has no moving parts and requires only proper water flow, proper power input to the heaters, and the correct charge of clean pumping fluid. Extensive damage to the internal jets or incorrect assembly of the parts can have an effect on pumping speed and backstreaming. A routine check should be made whenever a problem is encountered to be sure the heaters are producing the rated power. The outlet water temperature should be usually 40–55°C. Water flow should be adjusted to maintain outlet water temperature in this range. If a durable pump fluid is used such as the Dow Corning 700 series or Monsanto Santovac, a check of the oil level is sufficient at this point. If, however, one of the lower cost fluids is used, it may have been decomposed and require cleaning of the pump and a fresh change of fluid. The life of the fluid is dependent on the exposure when hot to relatively high pressures. Excessive loss of pump fluid suggests misoperation of the valves, that is, opening either the foreline valve or the high vacuum valve at pressures in excess of the tolerable pressures specified by the pump manufacturer.

A slow pumpdown is normally associated with high gas loads. This is evident by high foreline pressures if the diffusion pump is operating correctly. The diffusion pump will be pumping a relatively large mass of gas when pumping at inlet pressure of 100–1 mTorr. It is normal for the foreline pressure to rise to 200 or 250 mTorr during this part of the pumpdown, and the pressure will decrease rapidly as the chamber pressure is reduced to below 5×10^{-4} Torr. Previous records should indicate "normal" operation and the foreline pressure can be a key to diagnosis. If the foreline pressure is below normal, the condition of the diffusion pump is suspect—if the foreline pressure is higher than normal, high gas loads in the chamber or leaks in any part of the system are suspected.

High gas loads in the system can come from two sources, leakage and outgassing. Outgassing is any gas that is adsorbed in the chamber, work fixtures, substrate, or component surfaces. The gas evolved is usually predominantly water vapor but may also be volatile materials coming out of plastics or lubricants. As both leaks and outgassing have the same ef-

* Chapter 12.4 is by M. H. Hablanian.

METHODS OF EXPERIMENTAL PHYSICS, VOL. 14

ISBN 0-12-475914-9

fect on pumpdown or on the minimum pressure achieved, a check should be made to determine the nature of the problem. Checks should be made both with an empty chamber and with fixtures and the product inserted.

Diffusion pumps generally require little attention when correctly operated. However, it is advisable to perform some periodic checks to insure continued trouble-free operation.

By simple preventive maintenance, costly down time and cleaning procedures can be avoided. A day-to-day log of pump and system performance will indicate the condition of the pump and marked variations will show the need for corrective action.

The frequency of inspection will depend on the type of system, its operation, and utilization. The maximum interval between inspections is established on the basis of experience.

Complete cleaning of the pump may be periodically required because of gradual deterioration of some pump fluids. Removal of the pump from the system is then necessary.

12.4.1. Power Variations

Heat transfer and fluid mechanics devices, such as diffusion pumps, cannot be designed and manufactured with high precision. Some variation in performance is to be expected from pump to pump and for the same pump under varying conditions. The major performance parameters of well designed pumps are not strongly affected by small changes in heat input, pumping fluid level, cooling water flow rate, and temperature. For small changes in heater voltage, the various effects can be considered to be linearly dependent on power. Thus, the same heater can be used for 220 and 240 V operation regardless of the nominal rating of the heater. The only exception is the maximum pressure ratio which the pump can sustain which is highly dependent on the density of the vapor jets (see Section 4.1.5.1, Fig. 14).

The power of the electric heater depends on heat transfer conditions. The heater manufacturers can only specify the electric resistance when the heater is cold. The heat transfer conditions will depend on the degree of contact between the heater and the boiler plate, the cleanliness of the internal boiler plate surface, the heat loss from the jet assembly, etc. Generally, $\pm 10\%$ variations of major specifications are possible. High vacuum systems should be designed with some safety factors to avoid marginal performance.

The performance variation may also be affected by changes in experimental conditions and instrumentation. It is possible in some pumps that the sharp increase of the helium compression ratio may occur within the $\pm 10\%$ band shown in the figure around the nominal power value (100%) which may become noticeable in some applications.

12.4.2. Safety, Hazards, and Protection

Diffusion pumps perform very well within their normal operating range. They have been widely used for many years as a simple and efficient means for attaining high vacuum. Knowledge of their limitations—particularly at the higher pressure end of their operating range—can be extremely helpful in achieving best performance.

All diffusion pump fluids, with the exception of mercury, are to some extent heat sensitive. In general, if the pressure in a diffusion pump rises above a few torr near the operating temperature the power should be turned off. Hydrocarbon fluids when hot may be damaged even by a brief exposure to air. Silicone fluids are much more resistant to oxidation, but even they can be damaged when the pressure and temperature are significantly higher than normal for a period of time. It is possible under certain conditions of temperature and pressure for most diffusion pump fluids to ignite and cause an explosion. Such explosions are not strong enough to burst the vacuum chamber but damage to high vacuum valves and sudden opening of doors can occur. This has been reported in a few cases when an overheated pump was exposed to air.

Diffusion pumps may be operated for brief periods of time at inlet pressures as high as 500 mTorr without damage to the fluid. But a steady gas flow at pressures above the normal operating range of the pump may result in a rapid depletion of pump fluid. This may result in overheating. Thermostatic protection, fail-safe valve arrangement, and means of checking fluid level, particularly for large pumps, should be provided.

12.4.3. Dos and Donots Regarding Diffusion Pumps

The following recommendations should be generally followed unless some compelling circumstances require deliberate departures.

The maximum permissible discharge pressure (the forepressure tolerance) should never be exceeded under any circumstances when the pump is operating. Strict observance of this most basic rule is mandatory and should be clearly explained to all equipment operators. This will eliminate most of the gross backstreaming difficulties encountered in diffusion pump systems.

The maximum gas load capacity of the pump (maximum throughput) should not be exceeded in steady state operation. The transient overload period immediately after the high vacuum valve is opened should be kept as short as possible. If this period is longer than a few seconds, the roughing should be continued for a longer period to a low pressure before returning to pumping with the diffusion pump.

An operating diffusion pump should normally not be air released to atmospheric pressure to prevent oxygen exposure of the hot pumping fluid. If necessary, in systems without valves, air release should not be done from the discharge side. The high pressure will suppress the boiling, but the vapor in transit between the nozzle and the pump wall will be carried into the chamber.

High vacuum valves upstream of diffusion pumps should be opened slowly to reduce the degree of overload.

If the system or process tends to operate mostly near and sometimes above the overload point, the conductance of inlet ducts (baffles, traps, valves, etc.) should be reduced. This can be done by choosing baffles of lower conductance and better baffling capacity. In sputtering work, throttle valves are introduced to allow chamber pressures in the 10^{-2}–10^{-1} Torr range without exceeding the maximum throughput of the pump.

The baffles and traps or, in their absence, the diffusion pump itself, should be the coldest part of the vacuum system. Otherwise, the pumping fluid vapor will tend to migrate to the colder area.

Liquid nitrogen traps should be periodically defrosted to avoid heavy pumping fluid deposit which may break up when the trap is refilled. Radiation of thermal energy directly into the cryogenic surfaces of the trap should be avoided.

13. DESIGN AND PERFORMANCE OF BAKEABLE ULTRAHIGH VACUUM SYSTEMS*†

For the purposes of this Part, we shall quite loosely define "ultrahigh vacuum" as covering the pressure range below 10^{-5} Pa. The reader is referred to Chapter 14.1 for a discussion of the appropriate bakeout procedures.

13.1. Small Glass Uhv Systems

The details of system design and performance for such systems are covered in Part 7 and will not be repeated here. However, there are some aspects of the choice between glass and metal systems which are worthy of some consideration here, although we may observe at the outset that properly constructed and operated, there is little to choose between the two approaches for most applications.

Certainly, any researcher contemplating the construction of a small uhv system would do well to review the construction details of glass and metal systems and compare them to the available techniques in his institution. Beyond this, certain technical considerations exist which may or may not be of significance. First, at the limits of performance, the residual gas composition in glass and metal systems is generally quite different. If a large fraction of the chamber wall area is constructed of glass, the base pressure of the system may be limited by helium permeation.[1] To a lesser degree, hydrogen and neon permeation may be of concern. By contrast, the extreme high vacuum gas composition in a stainless steel system is generally predominantly hydrogen. A second consideration involves requisite bakeout temperatures. When a system is air exposed, a considerable quantity of atmospheric water vapor is adsorbed. The outgassing of this water vapor almost always constitutes the major gas load in the initial portion of the high vacuum pumpdown. Removal of this water by

[1] P. A. Redhead, J. P. Hobson, and E. V. Kormelsen, "The Physical Basis of Ultra-high Vacuum." Chapman & Hall, London, 1968.

† See also Vol. 4B in this series, Chapter 5.1.

* Part 13 is by Lawrence T. Lamont, Jr.

ISBN 0-12-475914-9

bakeout requires somewhat higher temperatures for glass systems than for metal due, apparently, to the formation of surface hydrates.[2] This question is considered from the viewpoint of practical techniques in Chapter 14.1. A final consideration may be safety requirements. Any sizeable glass portion of a vacuum system *must* be provided with an adequate implosion shield. Such considerations notwithstanding, the literature abounds with photographs of systems with no shielding whatever! Increased sensitivity to safety as a part of the system design, however, seems certain to alter this in most laboratories.

13.2. Small "All Metal" Uhv Systems

13.2.1. Build or Buy?

As a practical matter, this question must be faced by any person seeking acquisition of a uhv system. Most industrial users find that it is best to buy a system from a reputable manufacturer which is complete or nearly so. By contrast, the university researcher nearly always begins with the idea of building his own system—particularly if the system is quite small and specialized.

If the decision is to buy the system, the user's primary task will be the generation of a meaningful set of specifications. There are two presumptions here which may not hold: namely, (1) that the user knows his situation well enough to translate his *scientific* objectives into *technological* prerequisites, and (2) that the user is aware of what the manufacturer can actually accomplish. The current state of affairs is such that the first point is often not carefully thought out, and the second is complicated by the disparity between specification sheets and actual performance. In fairness, it must be pointed out that many system performance parameters are quite subtle and exceedingly difficult to properly specify—and that they are even more difficult to unambiguously measure. If some parameter is truly crucial and is specified, the user would be well advised to specify the *technique for measuring* as well. This is now standard practice, for example, in large procurements of sputter–ion pumps for high energy particle accelerators and storage rings.

13.2.2. Applications and General Considerations

Small "all metal" uhv systems find considerable application as the "typical" laboratory table top system. As such, they are required to be

[2] W. Espe, "Materials of High Vacuum Technology," Vol. 2. Pergamon, Oxford, 1968.

appropriate for a broad range of projects—and concomitant specifications. Hence, the prospective user should attempt to delineate the most critical applications.

One decision which will have to be made at an early point in the design phase centers about the question of access time. If the system must be cycled to air frequently, it may be desirable to use an elastomer seal on the main access port. This is not a trivial point, since the outgassing from such a seal will almost always dominate the system performance at uhv pressures. On the other hand, large metal gasket seals are time consuming to break and reseal and add considerably to the cost of the system and its operation (due to the cost of the gasket which must be renewed each time the seal is remade). Of the elastomer seals, two types are common; namely, O ring and L gasket. The L gasket is much easier to make and is usually encouraged by manufacturers for that reason. O-ring main seals, however, expose far less elastomer to the uhv chamber and are therefore strongly recommended. For either case, the best design is that which minimizes the exposure of elastomer, so if L gaskets must be used, they should be trimmed internal to the system and the excess discarded. O-ring design is determined by the properties of the gasket material. Design data is available from most suppliers,* or Chapter 6.2 of this text.

Another design question centers about the number and location of flanged penetrations. Many of these will be determined by the specific requirements of the experiments envisaged, but others will be allotted for general purposes. It is wise to include as many of the "general purpose" ports as reasonably possible. They are relatively inexpensive to have, impose few limitations on performance if metal gasket sealed, and somehow always seem to end up being needed! Location of a port is determined by the use to which it will be put in cases where there are special functional requirements, but is generally otherwise not critical. Pressure gauges must not be located in any region where the distribution of the gas is not isotropic in momentum space. Generally speaking, this requires that gauges be located well away from any pump or pump port which has a high capture coefficient. Gas inlets for highly reactive gases should be located as far as practical from ionization gauges owing to interconversion of gas species which is known to occur quite efficiently on hot filaments and in the discharge space of all Penning and magnetron discharge gauges. In addition, if sputter–ion pumps are to be used, gas inlets should be located as far from *them* as well for the same reason. A similar effect is ob-

* See, for example, design data available from Parker Seal Co., Culver City, Calif., or other manufacturers.

served for titanium sublimation pumps, but only during active sublimation, so they may be more tolerable in close proximity to gas inlet flanges.

13.2.3. Acceptable Materials and Methods of Construction

By far, the preferred material for the vacuum envelope and associated flanges and hardware is 304L stainless steel. The "L" signifies "low carbon," a fact of considerable importance to us here since it is much easier to obtain leak-free welds with this alloy rather than common 304 stainless steel. Other stainless steel alloys may be suitable for specific applications, such as the "400" series for magnetic feedthrough sections, but extreme caution is advised since some of these contain significant amounts of sulphur and/or other high vapor pressure elements. 304L stainless steel, despite its nearly ideal characteristics for vacuum chamber fabrication, has two related limitations which are often overlooked; namely, it has poor electrical conductivity, and extremely poor thermal conductivity (far worse than any of its constituents). The former is of concern if any radio frequency or short duration pulses must use the vacuum vessel as a return path, and the latter places severe conditions on the design of any bakeout oven and its related cycle. This question is covered in detail in Chapter 14.1.

Aluminum is occasionally suggested as an alternate to stainless steel since it seems to possess all the properties of that alloy (albeit welding is somewhat more difficult) and adding the advantages of extreme ease of machining, low weight, and excellent electrical and thermal conductivity. Unfortunately, aluminum has two serious limitations in that metal gasketed seals are essentially out of the question since its yield strength is quite low and, more seriously, aluminum is surprisingly difficult to clean. It also happens that the very "protective" oxide which is responsible for the utility of such an active (chemically) metal makes it difficult—sometimes impossible—to make a good electrical connection to the metal. Thus, while aluminum is often used in "medium vacuum" chambers where weight is at a premium and performance is not, its use in small uhv chambers is rare.

OFHC copper and certain of its alloys are acceptable materials. The designation "OFHC" means "oxygen free, high conductivity," a specification which is directly related to the brazeability of the material in hydrogen furnaces. The use of this grade of copper is generally recommended even if braze joints are not immediately contemplated, since a great deal of use of this material has uncovered no serious shortcomings. Numerous alloys of copper are available, but only AmZirc™, a high strength alloy of OFHC copper and zirconium, has been extensively used.

It seems to be entirely satisfactory from a vacuum standpoint and can be brazed with CuSil™, a copper/silver eutectic braze alloy, without losing its strength or creep properties. Other alloys, such as brass, bronze, AmSulf™ (a sulfur bearing "free machining" alloy), and AmTel™ (a tellurium/copper alloy used for electrical contactors) are not acceptable due to their high vapor pressure constituents. This judgment holds even if the part is electroplated with an acceptable metal such as silver or pure copper since even slight flexing under high vacuum conditions can permit diffusion of the offensive material from the substrate through the coating and into the vacuum chamber.

A final caution regards screws. Cadmium plated screws are not acceptable in uhv systems of any size. Such contamination, once introduced to a vacuum chamber, can only be removed by acid etching the entire inside surface of the system. Plasma etching will not work in this case, since the metal will merely be transported around the system and become even more difficult to remove due to enhanced diffusion along the grain boundaries of the walls of the system.

Of the nonmetals that might be used in uhv systems, only glass and the various ceramics should be given serious consideration. A recent addition to this list at this writing is MaCor™, a machinable glass ceramic, which promises to be of considerable utility in the laboratory. Teflon, including PTFE, should be avoided if possible, or at least its use minimized due to its permeability to helium and its unfavorable outgassing characteristics. Of the elastomers, only Viton™ appears acceptable in bakeable systems, although butyl rubber may be acceptable for unbaked systems with less stringent requirements.

13.2.4. Surface Area and Volume Considerations

For small systems, the volume is almost entirely irrelevant and the uhv performance is determined by the *area* of the outgassing surfaces and, of course, their outgassing rates and the system pumps, etc. Since the outgassing rates of materials vary widely according to the material, its configuration, and past history, it is wise to study the system as contemplated to determine what the limiting item will be. A trivial example already mentioned is the use of an elastomer main seal. Other likely problem areas are Teflon, any sintered or hot pressed part, large quantities of granular or powdered samples in analysis systems (due to their enormous surface area per unit mass), or, of course, even microscopic quantities of molecular sieve material from a cryosorption pump!

Since many innocuous seeming materials can behave like huge virtual leaks, time spent in this simple exercise is particularly well invested.

13.2.5. Special Pumping Requirements

Since even trace quantities of mechanical pump oil represent gross contamination in a small uhv system, *any mechanical pump ever connected to the system must be adequately trapped.* This includes leak detecting operations and initial testing of the system—situations where operating considerations seem secondary, but most assuredly are not! Three main trap designs are in common use: (1) simple cryogenic (usually with liquid nitrogen), (2) sorption traps utilizing Zeolites (metal–aluminosilicates) which may be cryogenic or operate at ambient temperature, and (3) copper foil traps. The choice is largely up to the system designer, since all three appear to do the job equally well. However, it is worth noting that the copper foil trap does not require cryogens, nor does it need to be rejuvinated periodically as do the other configurations.

The problem of oil contamination can be circumvented entirely by using cryosorption pumping for the roughing system. This technique is particularly well adapted to larger laboratories where liquid nitrogen (a required commodity) is readily available and small systems where the consumption is small enough to be of little economic consequence. As mentioned in Section 13.2.4, however, great care must be exercised so that none of the molecular sieve material ever finds its way into the system. All major manufacturers of these pumps are aware of this problem and most have taken steps to prevent its occurrence by installing appropriately sized screens in the pump throat and placing appropriate cautions in the instruction manual. If the designer of a system should "inherit" an older model pump without a screen, he should contact the manufacturer for his current recommendations.

The high vacuum pumps may be diffusion, sputter–ion, sublimation, or cryogenic, or any combination of the above depending largely on the specific requirements of the process and, more often than not, the personal preferences of the designer. Most commonly, however, one finds sputter–ion pumps in combination with sublimation pumps. This combination is easily capable of performance to the 10^{-11} Pa range given sufficient time and attention to good vacuum practice. Such systems are extremely simple to operate and are essentially fail-safe in that power loss of any sort does not result in lost vacuum. At low pressures, the saturation time of the sublimed layer is sufficiently long that protracted intervals may pass between sublimations and on the average very little energy is used. It should be realized, of course, that these pumps are essentially residual gas garbage cans, so that all the gas evolved by the system walls is *still inside the system and can be reevolved by improper operation.* For a discussion of these effects, see the Sections appropriate to the pump of interest. Contrary to common belief, diffusion pumps are easily capable of ultraclean uhv operation. In practice, their performance on "medium vacuum" systems is limited not by the pump but by the other aspects of vacuum practice such as copious quantities of Buna-N O rings, electroplated brass hardware, and other techniques which would never be acceptable in a uhv system regardless of the pump used. In fact, it is a tribute to these pumps that they do so well under such crude circumstances. Cryogenic pumping is only now beginning to become common thanks to the recent advent of several commercial pumps. Although they are usually used in combination with sputter–ion pumps, they can be used alone in some cases if desired.

Turbomolecular pumps are a separate consideration here since they are limited by their nature to *true base pressures* which range from 10^{-5} to about 10^{-7} Pa, depending on the design and speed of the rotor. This limit arises from the small dynamic compression ratio of the pump for low molecular weight gases such as hydrogen—the most common residual gas in all metal systems. If turbomolecular pumps are desired for some reason peculiar to the application, their base pressure can be greatly improved by backing the turbo with a small diffusion pump. Of course, this makes the system somewhat more complex, and it is not fail-safe (due both to the diffusion pump and the turbomolecular pump) but the technique has been demonstrated to work well.

13.2.6. Conductance Considerations

The idea of "conductance" in a chamber can be seriously misleading. The system walls have widely varying sticking coefficients and gas flow through the system is far from amenable to such simple interpretations. Monte Carlo techniques have been applied to some of the more common large components such as traps and valves, but the more general problem has not received much attention as yet.

Nonetheless, some general recommendations are possible which may be of use to the system designer. First, identify the major sources of gas within the system and arrange the system so that these sources "look" at the pumps with as little in the way as possible. Remember that while most surface sources emit according to Knudsen's cosine law, a simple gas inlet system such as that seen in Fig. 1(a) will be much more directed. As much can be said for the gas flow isotropy in any high aspect ratio geometry such as long tubulations or the narrow channel between two shields in a vacuum furnace. This "beaming" effect arises from the preferential transmission of gas atoms initially aligned so that they do not intercept the system walls. As the gas flow continues down the path, this

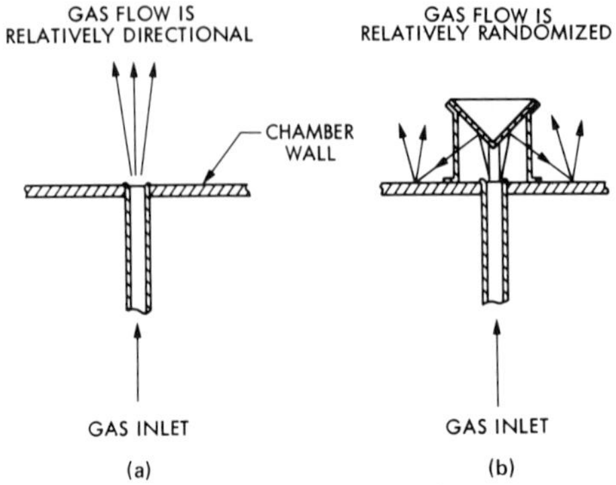

FIG. 1. (a) Simple tube gas inlet. The gas flow is relatively directed since the population of gas molecules directed along the axis of the tube becomes enhanced along the straight sections of the inlet. This sort of inlet may be of use when anisotropic gas distribution within the vacuum chamber is desired. (b) Gas inlet with a simple diffuser cone. Gas flow is randomized to some extent by requiring molecules to make at least two collisions before entering the main population within the chamber. Such an inlet system may be desired when the inlet gas is reactive, as in reactive sputtering, reactive ion etch, or various plasma–chemical reaction systems.

class of gas atoms becomes increasingly populated, and the flow is therefore increasingly more directed. If, on the other hand, a randomized flow is *desired,* as in the gas inlet system mentioned previously, a small diffuser can be added as seen in Fig. 1(b).

Normally, one desires to maximize the pumping speed delivered to the chamber. To accomplish this, the conductances between the pumps and the chamber must be large compared to the rated speed of the pumps. It is *not* sufficient that they merely be of the same order of magnitude. For example, if the limiting conductance equals the rated speed of the pump, the speed delivered to the chamber will be halved. This can be a very important consideration for sublimation and cryogenic pumps where the intrinsic speed at low pressures can be very high.

The other possibility is that it may be desirable to severely conductance limit the pump to reduce "memory effects" or the natural selectivity of the pump. This is often done in gas sampling systems where the object is to present an atmosphere to the residual gas analyzer which is simply relatable to the atmosphere being sampled.

13.2.7. Ultimate Pressure Performance

The ultimate pressure performance of any properly designed small uhv system is well below the limits of measurement of all the commonly available total pressure gauges provided no elastomer seals are used. The pressure reached by the system in a given period is quite another matter, being determined by the available pumping speed and the outgassing characteristics of the materials exposed to vacuum. For comparison, the pressure as a function of time is shown in Fig. 2 for two small commercially available systems. One has all metal seals and the other uses a Viton-A main seal and a Pyrex glass bell jar. Notice that the performance to the 10^{-7} Pa range is essentially the same for these two systems, while the all-metal sealed system is far more limited regarding ease of access.

The residual gas composition at extreme low pressures is dominated by hydrogen, except for cases where a significant fraction of the vacuum envelope is constructed of glass. In that case, helium may become an important constituent due to permeation through the glass, and is present roughly in proportion to the fraction of the envelope which is fabricated of glass.

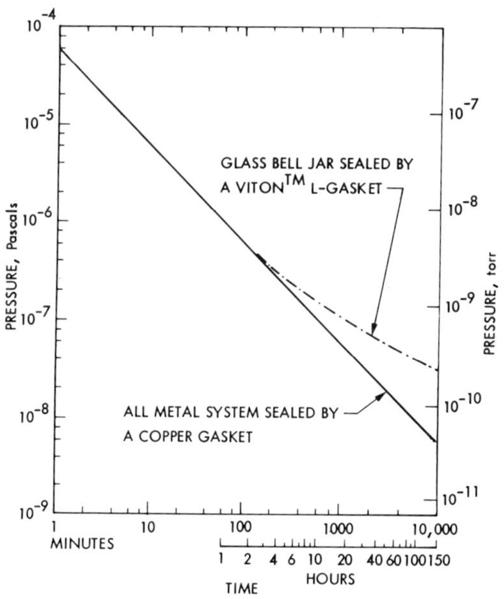

FIG. 2. Pumpdown of a small uhv system with two different main chambers of nominally the same surface area, but with one being a stainless steel chamber sealed with a copper wire gasket of the ConFlat™ configuration and the other being a glass bell jar sealed with a trimmed Viton™ L gasket. Note that there is very little difference between the two pumpdown curves for the systems in question until quite low pressures are attained. Time $t = 0$ corresponds to the point at which the system was valved into the uhv pumps.

13.3. Intermediate Size Uhv Systems

13.3.1. Applications and General Considerations

Uhv systems of "intermediate" size—say, with chambers roughly the proportions of a typical 45-cm-diam by 75-cm-high bell jar—are used for the same sorts of purposes as their smaller relatives, and few of the conclusions of the preceding sections are significantly altered.

One consideration which is strongly effected, however, is the matter of access time. Due to the sheer physical size of an average "intermediate" uhv system, access through a metal gasketed main seal can be very time consuming. Moreover, the cost of such a seal and its gaskets—which MUST be renewed each time the seal is remade—begins to be a factor of some importance. For example, a pair of 18 in. Wheeler™ Flanges can cost on the order of $2000 installed, the clamp assembly several hundred dollars more, and the required metal gaskets over $10 apiece. One way to approach this problem is to use the flanges on the system but install the optional Viton-A gasket. This allows the flexibility often required in a more "general purpose" uhv system. Other aspects of the use of large metal gasket sealable flanges include their weight and thermal mass. These are considered separately in Section 13.3.7 and Chapter 14.1, respectively.

13.3.2. Acceptable Materials and Methods of Construction

Again, most of the considerations in this regard as developed previously in Section 13.2.3 apply. TIG welded stainless steel is the preferred material for the chamber as well as all flanges and peripherals with aluminum to be used only when its limitations are specifically tolerable. The size and weight of many of the components and the chamber usually require that some consideration be given to mechanical design. This is covered in more detail in Section 13.3.7.

13.3.3. Special Pumping Requirements

Usually, the same pumps used on small uhv systems find application in similar combinations on the "intermediate" size systems with the only difference being size. This is because the emphasis is still on total performance, and economic considerations which come into play on very large uhv systems are not yet significant.

An exception to this concerns the roughing pumps. Since the roughing portion of the cycle is controlled by pumping of the *volume* gas rather than normal surface area controlled outgassing, the fact that the system volume has increased by perhaps a factor of 20 or more alters the roughing time considerably for most situations. This is because the typical rough pump for a small system delivers between 150 and 300 liters/min of speed to the chamber. To obtain the same time constant for the system, then, one would have to use 3000–5000 liters/min roughing pumps: Usually, the order of 500 liters/min is used, and even a 1500 liters/min mechanical pump is considered very large—as, indeed, it *is* physically. Rootes-type blowers have been used with some success, but the added complexity is not usually warranted, and they are subject to the same considerations regarding oil backstreaming as are mechanical roughing pumps, and so require traps. High conductance ambient temperature traps are difficult to obtain commercially, and so the application of these pumps is somewhat restricted.

13.3.4. Pressure vs Time Performance

Generally speaking, the pressure in any normally functioning uhv system falls in inverse proportion to time as discussed in Section 15.1.1. Any significant departures from this behavior should be taken to indicate a system fault. These may be either related to the presence of a material in the vacuum chamber which is not truly uhv compatable, a virtual leak, or a true leak. These are discussed fully in Part 15 along with methods for their detection, diagnosis, and repair.

13.3.5. Residual Gas Composition

As with small systems, the residual gas composition prior to bake is largely water vapor, carbon monoxide, and methane. As the pressure falls, hydrogen becomes an increasingly significant fraction of the background, and eventually dominates after bakeout, or an extremely protracted period of pumping. The residual gas "fingerprint" of a system is a very useful monitor of the general condition of the system. To this end, the inclusion of a residual gas analyzer on a system is highly desirable.

13.3.6. Conductance Considerations

As the systems get larger and the pumps increase in size, the difficulty of efficient coupling of the pumping mechanisms to the chamber increases considerably. A benchmark in this regard is that a careful look at the manufacturer's specifications for such pumps on systems of his own design usually quote "intrinsic" speed! The general utility of this parameter is open to question, and is certainly of little use to the researcher attempting to define system performance.

The same considerations discussed in Section 13.2.6 apply here.

13.3.7. Mechanical Limits

Since we are here dealing with a rather sizeable system which, in total, may weigh in excess of 750 kg, mechanical considerations become important from several viewpoints. The first of these is simply the opening of the system. Small chambers present no particular difficulties in this regard since in most cases the chamber is light enough to remove manually. The larger chambers used here require the use of a carefully designed hoist. The "ball–screw" hoist design is preferred for reasons of reliability and safety.

Another matter is the stability of the system when the chamber is open (and, of course, also when it is closed!). Since this is a problem virtually any competent physics freshman could solve, it is a bit disconcerting that one still occasionally sees chambers in the field which require special braces when opened to keep from tipping over.

Finally, it is necessary to allow for the effect of all that weight on the various components of the system and its frame and their net effect on the laboratory floor. Most buildings have maximum allowable floor loading specifications and the user would be ill advised to discount their importance.

13.4. Large Uhv Systems

13.4.1. Applications

Large uhv systems are used primarily as space simulation chambers. As there is relatively little activity in that area today, few such systems are being built. They are always specialized in their requirements and require close cooperation between the manufacturer and the user for a reasonable measure of success to be attained.

13.4.2. Special Pumping Requirements

In small systems, pumps could be combined largely at the whim of the system designer to get the desired effect. As the systems grew in size, so did the pumps, but the combinations—albeit more costly and subtle—remained essentially unchanged. For the very large systems we are considering here, much begins to change drastically. Sputter–ion pumps slowly but inexorably become impractical as the system grows due to many factors such as starting problems, magnetic fields, and sheer cost and weight. A single order of pumps for a very large system can represent a major fraction of a manufacturer's business for a year—a matter not to be taken lightly by either party to the transaction! There are some very large turbomolecular pumps which can be used here, but their base pressure performance is generally inferior to their smaller relatives and hence they are not truly adequate. Properly trapped diffusion pumps are used for many such systems, but eventually even they become impractical. The largest standard diffusion pumps have 48 in. diameter apertures and consume considerable amounts of energy. The last resort is cryogenic pumping for such systems, and a great deal of work has been done in that direction.

The considerations that ultimately drive the user to cryopumping are simple (!) economics. However, the benefits of (nearly) fail-safe operation accrue as well.

14. SPECIAL REQUIREMENTS IN THE DESIGN, OPERATION, AND MAINTENANCE OF ULTRAHIGH VACUUM SYSTEMS*

14.1. Bakeout of Uhv Systems: Theory and Practice

14.1.1. Basic Considerations

The term "bakeout" refers to elevating the temperature of a vacuum system during some portion of the pumpdown cycle and subsequently returning it to ambient temperature. The fundamental idea is that the adsorbed gas will be desorbed from the heated surfaces at significantly higher rates than at ambient temperature and thus can be removed from active circulation within the system by the pumps. The benefits of this are either a lower pressure attained within a given pumping time, or a shorter total pumping time to reach a given pressure. In some cases, only certain small sections of the system are subjected to bakeout. Common examples of this include the degassing of ionization gauges prior to using them as more than rough indicators of the system condition, and the outgassing of samples and sample holding fixtures for surface science studies. Here, the total system pressure is generally only negligibly reduced by the bakeout—and may even be slightly increased—but the fundamental reasoning remains; namely, that the lower binding energy components of the adsorbed gases are selectively removed from the surfaces of interest and the relevant portions of the vacuum environment are "cleaner."

It is important to understand that under virtually all uhv conditions, the surfaces have not equilibrated with their environment and the walls are heavily loaded with condensables such as water vapor in quantities far in excess of their equilibrium values. This is the fundamental reason behind the experimental fact that wall outgassing rates in the uhv regime depend almost exclusively upon the pumping time and past history of the surface and not upon the system total pressure. This is fortunate indeed, for if

* Part 14 is by Lawrence T. Lamont, Jr.

METHODS OF EXPERIMENTAL PHYSICS, VOL. 14

ISBN 0-12-475914-9

this were not the case, control of temperature and temperature uniformity during bakeout would be far more critical than it actually is.

Of course, it is true that when the colder surface is both very clean and chemically reactive, chemisorption of desorbed gas molecules may become a significant process. However, it is also true that this does not present any special difficulties since the activation energy for desorption of a chemisorbed gas atom is quite large and hence the mean residence time is long compared to times of interest to a vacuum physicist. One obvious exception to this would be the surface of a sample in a surface research system where care must be taken to ensure that the sample cannot act as a pump for gases which are desorbed from other parts of the system. This may be easily accomplished in most cases by simply performing the general system bakeout to attain an adequate total pressure prior to outgassing the sample surface.

Some final considerations remain which are related to the system pumps and their properties. Recall that most uhv systems are pumped by sputter–ion pumps, titanium sublimation pumps, cryogenic pumps, or some combination of the three. These pumps all share some important characteristics: namely, (1) they all have walls which do not contribute significantly to the pumping speed, and (2) the gas that is pumped is trapped and stored within the pump and hence *within the system*. The first of these considerations means that some special care will have to be taken to either bakeout the pumps themselves or to minimize their contribution to the gas load. When bakeout of the pumps is impractical or, in extreme cases, impossible, outgassing from them may be minimized by either careful cleaning and protection from recontamination—difficult to do in any meaningful way—or by reducing their *fractional* contribution to the system surface area—a parameter which lies almost totally under control of the pump manufacturer. Obviously, it is desirable to select pumps which can be baked along with the rest of the system! The second consideration means that care must be taken in the normal operation of the system as well as during the bakeout portion of the pumpdown cycle to avoid reevolution of previously pumped gas, insofar as is reasonably possible.

14.1.2. Temperatures Required for Bakeout

As recently as 10 years ago, bakeout to extreme temperatures—450°C and above—was the rule. This is because the uhv technology associated with the invention of the sputter-ion pump evolved via the manufacturing techniques associated with high vacuum microwave tubes. The vacuum tube, however, is a special situation since it is not usually dynamically pumped except in the very largest devices. Being a sealed "static" environment, the tube was generally subjected to rigorous processing to ensure a reasonable lifetime. Since the outgassing rate is a very strong function of absolute temperature, it was necessary to bake the tube at a temperature in excess of the maximum temperature it could reasonably be expected to reach in normal operation prior to sealing. Notice two crucial differences between this situation and the typical uhv system: (1) the tube is "static," but the system is dynamically pumped, and (2) the tube is an isolated system of which the pumps are not a part after sealing—from the outgassing viewpoint—whereas outgassing from the pumps in the uhv system may play a crucial role in determining the vacuum environment by virtue of their own outgassing, memory effects, or interconversion of gas species.

Outgassing studies due largely to Strausser[1] demonstrated that a far less severe bakeout was generally adequate and that temperature uniformity was of much greater importance than the temperature used. Based on that work and subsequent experience, a temperature of 150°C is recommended along with a caution to be certain that the entire system is uniformly baked. The effect of leaving various fractional surface areas unbaked is shown in Fig. 1, where the outgassing is taken to be initially uniform throughout the system. This latter assumption is quite well followed in most conventional all-metal sealed uhv systems. Note that only a few percent of the surface remaining unbaked has a pronounced effect on the system performance.

Certain conditions may make it advisable to either increase or decrease the above recommended temperature. For example, it may be that some material within the system cannot tolerate 150°C or the simple thermal expansion of some critical part may cause difficulty. Should that be the case and a lower temperature selected, the operator should be certain that all parts of the system are baked as uniformly as possible and the time for the bakeout increased as discussed in Section 14.1.3. On the other hand, it may be that a higher temperature is required due possibly to the inclusion of some substance that requires a higher temperature to outgas at a reasonable rate due to high binding energy components on the surface or perhaps even to the knowledge that a uniform bake *cannot* be achieved. This latter case is all too common due to the inclusion of thermally massive flanges or thermally isolated devices within the chamber that are difficult to bake and which must transfer heat primarily via radiation. Of

[1] Y. E. Strausser, *Proc. Int. Vac. Congr.*, **4**, p. 469. Inst. Phys. London (1968). Also, see Y. E. Strausser, *J. Vac. Sci. Technol.* **4**, 337 (1967), or reprint "VR-51, Review of Outgassing Results," by the same author, available from Varian Associates, 611 Hansen Way, Palo Alto, Calif. 94304.

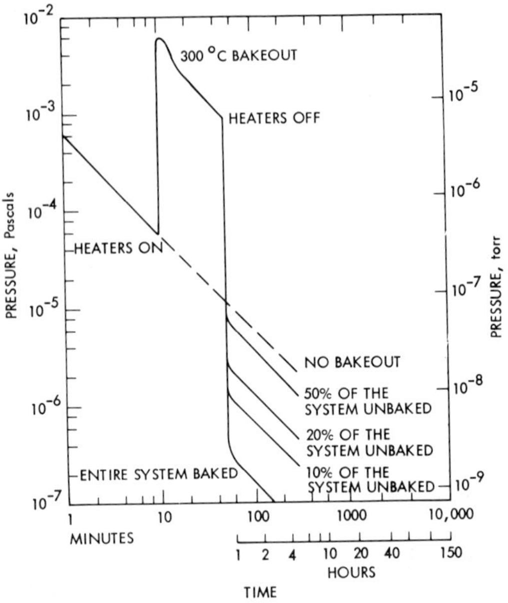

FIG. 1. The effect of leaving various fractions of the system unbaked is shown for a small all-metal system baked uniformly at a temperature of 300°C for 40 min. The entire system was baked in a small oven which totally enclosed the system. These curves should be taken as a general indication of behavior *only*, since the outgassing in vacuum systems is quite variable. Nonetheless, it is instructive to notice that even a small fraction of the system being unbaked can have a significant effect.

course, at the temperatures involved, radiative heat transfer is a rather slow process. Parenthetically, we should note that it is generally not possible to achieve good thermal transfer between two "contacting" surfaces where the surfaces are held together by a clamping arrangement. This is due to the fact that the heat transfer between such surfaces is dominated by conduction through a thin layer of gas between the surfaces under normal conditions and the absence of this gas under uhv conditions leaves only radiation and physical contact transfer—usually a maximum of three points unless significant distortion of the surfaces occurs.

14.1.3. Time/Temperature Relationships in a Typical Bakeout Cycle

Most operators of uhv systems are aware, at least in a qualitative way, that increasing either the bakeout temperature or the bakeout time or both will generally result in a lower pressure being reached after termination of bakeout. However, very little quantitative—or even semiquantitative—data exist to aid the user of vacuum systems in determining the appropriate tradeoff relationships. As a result, most operators either determine the bakeout cycle experimentally or, more commonly, follow some generally unrelated "cookbook." More serious than the likely less-than-optimized bake cycle is the fact that one cannot judge the adequacy of the bakeout in reported experiments in the open literature. "The system was baked at 350°C for 4 hr and pumped to a pressure of 2×10^{-10} Torr" does *not* necessarily give the reader an unambiguous point of judgement unless, for example, he knows how the system was baked, how it was pumped, and how the pressure was measured.

Referring back to Fig. 1, notice that both before, during, and after termination of the bakeout, the pressure falls with an inverse time dependence. This being the case, it is easy to see that there is an "excess gas removal" during the bakeout itself which, for the baked portions of the system *only*, must equal the amount of gas that had to be removed to attain the lower pressure achieved after termination of the bakeout. As seen in Fig. 2, this simply means that the quantity of gas removed during the bakeout, Region I, must equal the quantity of gas in Region II, assuming that the pumping speed remains constant throughout the pumpdown. Formally, we may write

$$\int_{t_1}^{t_2} S(P, t) \left[P_{\mathrm{B}}(T, t) - P_{\mathrm{U}}(t)\right] dt = \int_{t_2}^{t_3} S(P, t) \left[P_{\mathrm{U}}(t) - P_{\mathrm{BO}}(T, t)\right] dt,$$

where $S(P, t)$ is the pumping speed as a function of time and pressure; $P_{\mathrm{B}}(T, t)$ is the pressure during the bakeout as a function of temperature and time; $P_{\mathrm{U}}(t)$ is the function which describes the pressure in the unbaked system as a function of time. This is taken to be valid to describe the behavior of the system pressure after the bake had the bakeout not actually taken place; and $P_{\mathrm{BO}}(T, t)$ is the pressure reached immediately after the system has returned to ambient temperature as a function of bakeout temperature and time. These quantities and the limits of integration may be more easily visualized by referring to Fig. 2.

In actual practice, the system does not usually attain the pressure after termination of bakeout, which is predicted by the prior analysis. This is almost invariably due to the fact that not all of the system has been baked, and clearly reemphasizes the necessity of baking the entire system uniformly.

14.1.4. Miscellaneous Thermal Considerations

In practical vacuum systems, there are many significant limitations placed on the objectives outlined in the preceding sections. The most

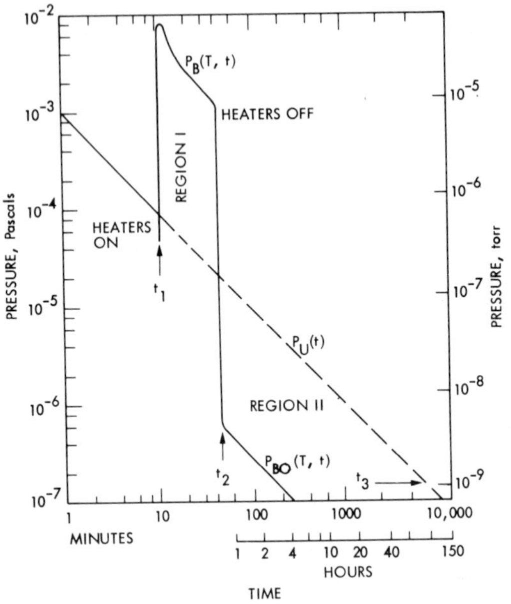

FIG. 2. Schematic representation of the bakeout process. Notice that the gas removed in Region I in excess of that which would have been removed had the bakeout not taken place must equal the gas removed in Region II under the same condition. Clearly, the gas in Region II is unbounded as shown, which implies that the curve $P_{BO}(T, t)$ must approach the curve $P_U(t)$ as a limit as $t \to \infty$. The most likely candidate for the general shape of both these curves is an outgassing characteristic limited by gas diffusing from the bulk of the chamber wall. It is anticipated that this process would be extremely variable, so no attempt has been made to indicate the detailed termination of Region II for this figure.

severe of these is associated with the poor thermal conductivity of the most common material of construction—304-L stainless steel. The thermal conductivity of this alloy is only about 0.16 W/cm °C as compared to 3.94 W/cm °C for copper. This means that some care must be taken to ensure that nonuniform heating of the system chamber and flanges will not result in either poor temperature uniformity during the bakeout or, in extreme cases, thermally induced distortion of the chamber or its peripherals. Beyond these considerations, and the possibility of "swamping" the uhv pumps due to excessive outgassing, there do not appear to be any constraints upon the rate of rise of the system temperature.

Differential thermal expansion of the materials used in flanges and their gaskets may be a source of difficulty under extreme conditions or with custom designed flanges using nonstandard seal configurations. For this reason, most users are well advised to avoid the use of nonstandard flange configurations where possible unless they are experienced in this special design area. For standard flanges, one need only keep the system temperatures within the range specified by the manufacturer.

14.1.5. Techniques for Heating the System

The most obvious way to uniformly heat a uhv system would seem to be by means of a bakeout oven. However, remembering that the system is fabricated of a poor thermal conductor and of many separate pieces with widely varying thermal masses, it is evident that auxiliary heating of large flanges and the like is mandatory for optimum performance. Thus, the primary recommendation is to use a bakeout oven with extra heat supplied to thermally massive parts of the system by directly mounted heating strips as appropriate. The power required for the supplementary heating can be estimated using standard elementary physics, but will probably have to be determined experimentally for best results.

The least desirable—and most common, unfortunately, for obvious economic reasons—way to supply heat to a system for bakeout is by directly bolting some heating strips to the system main chamber. This is wasteful of energy and poor technique and should therefore be avoided. When such a system has been acquired, possibly by default or for economics, the situation can be greatly improved by fabricating a simple bakeout shroud for the system. Even without insulation, such a shroud does much to improve the system performance and the addition of, for example, fiberglass batting as insulation in the shroud begins to approach the ideal described above. Another step that can be taken is to simply space the heaters off the walls of the system. Even a gap the thickness of a few washers is adequate to significantly alleviate most of the more severe nonuniformity problems on thin chamber walls.

There is little one can do, in a practical sense, about the cooling rate of the system parts. Of course, it is possible, in principle, to add heat during the cooldown to the thermally *less* massive parts of the system to achieve more uniform cooldown. Fortunately, however, this is rarely necessary and so the technique is not recommended unless a real need arises.

14.1.6. Cycled Systems

A general purpose uhv system with heat is commonly cycled back to air for each new process or experiment to be installed. This generally is not a problem, since the bakeout portion of the pumpdown sequence can be controlled by simple timers and thermostats overnight. However, analytical systems, which must effectively be frequently cycled, present special problems in this regard. One of the best techniques for such systems is to introduce the samples to be analyzed through a vacuum transfer lock and avoid the problems of exposing the entire system to laboratory ambient. However, the design of transfer locks of acceptable reliability is surprisingly difficult and adequate valves have only recently become available.

An alternate to the transfer lock is to load the samples in batches of a size appropriate to the process throughput desired. Again, this is a special design problem, although it is far simpler than that of designing a transfer lock.

Beyond these special systems, there are a few obvious elements of technique that can help considerably. The key fact to realize is that the bulk of the gas adsorbed on the system walls while exposed to the laboratory environment is water vapor. Thus, the following recommendations are made: (1) the system should be vented to dry nitrogen, or minimally, dry air, (2) the system should be physically opened for an absolute minimum of time, preferrably less than 2 min, and (3) large systems should be provided with a positive flow of dry gas out of the system to inhibit back migration of atmospheric water vapor into the system. Where systems must be physically open for periods between a few minutes and an hour or so, the system walls can be warmed to reduce condensation. These techniques are obviously of far greater significance in high humidity environments than low, but are of significant value in most operating circumstances.

14.2. Routine Maintenance and Cleaning

With few exceptions, laboratory scale uhv systems require periodic attention to what amount to "housekeeping" details if trouble-free operation is expected. While many details of this complex topic are so specific to particular systems that their inclusion here is not warranted, some are quite general. It is to this latter class that we now direct our attention.

14.2.1. Cleanliness and Cleaning

Two aspects of this topic are obvious; namely, *preventive* and *curative* cleaning. The preventive classification involves the avoidance of introduction of offensive material to the vacuum environment—often in connection with curative cleaning!—while curative cleaning refers to the removal of contaminants from the system. Since the former is usually much simpler than the latter, effort in that direction is generally well invested.

14.2.1.1. Proper Pumpdown Technique. Surprisingly, many otherwise cautious investigators introduce serious contamination to their uhv systems simply by improper pumpdown technique. The following list is by no means exhaustive, but includes the more common errors.

(a) Excessive exposure of the system to an untrapped mechanical pump;
(b) allowing cold traps to warm up overnight or over long weekends in diffusion pumped systems;
(c) allowing molecular sieve material from a cryosorption pump to enter the system;
(d) allowing diode sputter–ion pumps to operate in the unconfined glow regime ("starting" mode) for long periods of time when organics are present in the system.

14.2.1.2. Use and Abuse of Organic Solvents. Few organic solvents are appropriate to use in uhv systems, and in any case, the quantities should be minimized. Remember, cleaning is a two stage process: (1) solvation, which is followed by (2) dilution. This points out that one must not be deluded into thinking that the offensive material has been removed when, in fact, the quantity has simply been reduced and the remainder distributed thinly around the system along with traces of the solvent. These remaining quantities of solvent can be particularly troublesome for sputter–ion and titanium sublimation pumps if they are allowed to remain in the system to be "pumped away." The following are useful as indicated.

(a) 1,1,1-Trichloroethane is a useful degreasing agent, particularly in a vapor degreaser, but must not be used to clean nickel parts since it forms HCl on contact with the adsorbed water vapor and etches the part. This solvent should not be used as the final cleaning step.
(b) Acetone is useful, but tends to leave a residue which can be difficult to remove. It may seriously degrade Viton seals and certain plastic materials used around (but not in!) uhv systems, and so its use is generally discouraged, as are most other ketones
(c) Various alcohols are of considerable utility for general small-scale cleaning since they may be readily removed by water, for which they have a great affinity. Methanol is excellent as a solvent, but is highly toxic and cannot be recommended unless appropriate safety precautions are taken.[2]

[2] N. I. Sax, "Dangerous Properties of Industrial Materials." Van Nostrand-Reinhold, New York, 1968. This reference work is strongly advised for all researchers working with potentially hazardous materials.

2-Propanol is a good substitute for methanol, and is particularly useful in a vapor degreaser. Other alcohols are not generally recommended for various reasons.

It is recommended that any cleaning procedure be as mild as possible, and in most cases, that the next to the last solvent be 2-propanol and the final rinse be in ultrapure water. This may be surprising in light of earlier comments that water vapor is the most common residual gas in uhv systems and that care should be taken to avoid excess water vapor adsorption while the system is exposed to laboratory ambient. But, the facts are simply that of all the reasonable choices for a final rinse, water appears to be the best. It outgasses readily from the system walls and is pumped easily and generally without serious side effects by most uhv pumps. Blowing the rinsed part dry with filtered dry nitrogen or air is usually adequate, or it may be heated mildly with a "heat gun" or in a *clean* laboratory oven prior to introduction to the system if desired. In any case, bear in mind that if the cleaned part is recontaminated, the purpose of the cleaning will be defeated. It is also quite difficult to store "ultra clean" parts or samples since most containers—including the ubiquitous plastic bags—introduce significant contamination due to their own lack of cleanliness and the presence of such undesirable substances as mold release agents and the like on their surfaces.

14.2.1.3. Abrasives and Abrasive Etch. "Bead blasting," similar to "sand blasting" but utilizing glass beads instead of sand, is another acceptable technique for many cleaning problems. Small-scale laboratory bead blast cleaning systems are commercially available.

As with solvent cleaning, one should be as conservative as possible in its use and it must be recognized that microscopic quantities of the glass beads will remain embedded in the surface along with traces of all prior materials and contaminants. This is particularly true with softer materials, such as aluminum. Thus, the bead blast cleaner should not be used both for the removal of gross contamination and for final cleaning unless the beads are changed and the system decontaminated, if necessary. In addition, the insulating nature of the residue means that *soft metal parts which are to be used near electrical gas discharges or in experiments where the contact potential is significant must not be cleaned in this manner!* Obviously, this includes gauge parts (excepting the insulators themselves, which are readily cleaned in this manner), ion source parts (again, with the exception of the insulators), and virtually all the conducting parts and shields of sputter and electron beam thin film sources. The new low voltage "magnetron" sputter sources are particularly susceptible to this problem.

14.2.1.4. Removal of Thick Layers of Deposited Material. This problem is of great concern for vacuum coaters, but is also important for Auger spectroscopy systems with depth profiling and, of course, for the refurbishing of sputter-ion and titanium sublimation pumps.

For a great number of cases, the net adhesion of thick deposited films is so poor that they can be simply peeled off. This arises from the fact that net adhesion is the difference between the "intrinsic" adhesion bond strength and the stress in the film. Since the latter quantity is roughly proportional to film thickness, the film will eventually be torn from its substrate—the system walls—by its own stresses. This is a mixed blessing on several points. On the positive side, the contaminant is easily removed and the surface left behind may be extremely clean. However, such films and their release is generally far from uniform and so much of the offending material will probably have to be removed by other techniques. In addition, some of the film may release and foul other parts of the system prior to the cleaning exercise and become a serious particulate contamination (see Section 14.2.1.5). A further negative aspect of this situation is that many of the common materials to be removed are dangerously pyrophoric when extremely clean—as the delaminated surface is—and exposed to air. Examples of this latter problem are Ti, Ta, and Nb films. Removal of Ti films in particular must be approached with great caution as the danger of fire is considerable.

Some deposited films must be removed by chemical etch techniques. Some obvious examples include the removal of aluminum from stainless steel by an alkali bath, such as NaOH or KOH solutions, or by acid etchants when metallic ion impurities must be avoided, as in the semiconductor industry. Again, the problem must be approached with care since the chemical properties of the deposited film are likely to be quite different from the bulk properties of the material. That is, etch rates may be either significantly reduced or greatly enhanced over bulk material rates, the latter case posing a possible safety hazard.

Removal of insulating films, such as glass from a chamber wall, presents special difficulties. It is generally advisable to use the bead blast technique when possible and resort to severe etchants such as HF only as a last resort.

Removal of conducting or insulating films from insulators such as ceramics is also best approached by bead blasting, but care must be taken to not introduce leaks in feedthrough insulators by overly zealous cleaning. The same comment can be applied to the chemical cleaning of such parts, since most of the braze materials used are preferentially attacked by the common etchants.

In all the foregoing techniques, care should be exercised to avoid serious safety hazards, and in no case should chemical etch cleaning be attempted by the inexperienced or without proper equipment such as fume hoods and the *proper* fire extinguishers and emergency equipment.

14.2.1.5. Particulate Contamination. Particulate contamination may be introduced to the system either from the spontaneous self-destruction of highly stressed films or as dust or lint from the outside world.

Should the source be stressed films, then the only viable techniques are system redesign to avoid the accumulation, modification of technique to reduce stressing, or a more frequent cleaning schedule. Sometimes the "system alterations" consist simply of the relocation of a bit of shielding or the addition of a bit of heat or cooling at some critical location, so this possibility should not be dismissed as "too difficult" without first considering the nature and degree of the problem.

Particulates entering the system from the laboratory environment may be avoided by improved technique and "housekeeping." Aside from dust, lint from laboratory coats, and exfoliated human skin, a common source is the use of ordinary (dust laden) laboratory air for venting the system. This is trivial to prevent with a simple filter on the system vent line. As mentioned previously, venting the system with dry filtered nitrogen or dry filtered air is excellent technique since it avoids both the particulate problem and the excessive adsorption of water vapor on the system walls.

14.2.1.6. Cleaning of Pumps. Cleaning of sputter–ion pumps and titanium sublimation pump modules should be restricted to the removal of excessive deposits on the system and pump module walls. This may be accomplished using the standard techniques discussed in the preceding sections, except that the use of organic solvents must be avoided in sputter–ion pumps and titanium sublimation pump modules. Since the titanium films in the latter two pumps are unlikely to be saturated chemically, particularly for titanium sublimation pump modules, attention is again drawn to the potential fire hazard (see Section 14.2.1.4).

Cryogenic pumps rarely need this sort of cleaning unless fouled by some other system component. The pump manufacturer should be contacted for his recommendations and procedures appropriate to the situation.

Diffusion pumps (occasionally, but not commonly, used on uhv systems) do not require any special cleaning if they are properly operated beyond routine changes of oil and removal of system induced contaminants (rare). The manufacturer should be contacted for specific recommendations.

Turbomolecular pumps, like diffusion pumps are occasionally, but not commonly, used on uhv systems and generally require no special cleaning beyond the removal of system introduced contaminants. However, since these pumps are only rarely used with cryogenic traps, they are more subject to the problem than diffusion pumps which *must* be trapped for use in uhv applications.

14.2.2. Routine Maintenance

Beyond the cleaning considerations covered in the preceding sections, certain maintenance problems must be considered.

When mechanical or diffusion pumps are used, their pumping fluids must be regularly changed. This is due to the deterioration of the fluids caused by heat and by contamination from the gas being pumped. When this begins to become significant, the pumps will appear to function normally, but the low pressure performance will deteriorate. Mechanical pump oil should probably be discarded, but the diffusion pump oils used in uhv systems are sufficiently expensive that their recovery and reprocessing is generally warranted. This service is commercially available as of this writing and is highly recommended from the environmental as well as the economic viewpoint.

Traps of all sorts in the system will require periodic regeneration with the single possible exception of the "copper foil" traps frequently used in forelines and on roughing carts. In the case of cryogenic and cryosorption traps, this is easily accomplished by allowing the trap to warm up to laboratory ambient followed by a bakeout of the trap as appropriate. Cryosorption traps, of course, require more extensive bakeout cycles than do simple cryotraps without sorbants.

When new filaments are fitted to ionization gauges, ion sources, and electron guns, a period of outgassing must be allowed to avoid problems with surface generated ions and the like. Following the initial "burn-in," the device should be thoroughly outgassed.

Elastomers, particularly those subjected to mechanical motions or the presence of an electrical gas discharge, require periodic inspection. Their degradation, except where gross and therefore obvious, is best detected by observing the pressure as a function of time for a normal pump-down cycle. This technique is described and interpretation discussed in Section 15.1.2.2. Should the elastomer have failed in this manner, the only option is replacement.

15. THE FINE ART OF LEAK DETECTION AND REPAIR*

15.1. Basic Considerations

15.1.1. Review of Basic Equations

The response of an evacuated chamber to sources and sinks of gas is properly analyzed by the pumping equation

$$Q = SP + V\frac{dP}{dt},$$

where Q must be taken to include all significant sources of gas and S is the effective pumping speed delivered by all significant sinks. A very considerable degree of complication arises from the fact that there are generally several sorts of sinks in addition to the system pumps, such as "wall pumping" and "gauge pumping," and numerous sources of gas, such as normal outgassing, permeation, virtual leaks, "diffusion leaks," and simple real leaks. In general, the behavior of these quantities as a function of time and past history allows the experimenter to separate and identify them, a fact of paramount importance to us here since, before attempting to *find* a leak, one ought to determine that one *has* a leak. We therefore approach this complex situation by considering the response of the system to the various sources of gas as a function of time taken singly.

Physically, the simplest case is that for which $Q = 0$. The system pressure then decays exponentially as

$$P = P_0 e^{-(S/V)t}.$$

This is simply the pumping of "volume gas," and hence the equation describes such situations as the "roughing" portion of the normal pumpdown cycle, or the decay of the partial pressure of a gas species which *does not form a significant surface phase with the chamber walls* and for which *no other significant sources exist within the system.*

Normal outgassing rates from most materials considered acceptable for use in uhv systems are found experimentally to decrease with time in a

* Part 15 is by Lawrence T. Lamont, Jr.

ISBN 0-12-475914-9

simple inverse proportion in the high and ultrahigh vacuum regimes. This is thought to result from the desorbing species having a broad distribution of activation energies for desorption. After a sufficient period of time, the exponential decay term associated with volume gas pumping may be neglected and we approach an equilibrium situation for which the pumping equation reduces to $Q = SP$. Thus, for normal outgassing

$$P(t_2) = P(t_1)[t_1/t_2].$$

Certain types of leaks, and the normal outgassing rates from some materials occasionally used in high vacuum systems exhibit a diffusion limited time dependence. That is to say, they lead to a source term Q which is proportional to $t^{-1/2}$. Thus, as with the preceding case, we approach a (different!) equilibrium situation described by

$$P(t_2) = P(t_1)[t_1/t_2]^{1/2}.$$

Finally, we consider the case of a constant source term Q_0. The system will approach, as a limit, a *true base pressure* P_0 given by

$$P_0 = Q_0/S.$$

Parenthetically, we should note that this situation, or one for which the total effective pumping speed for the system approaches *zero*, constitute the *only* sense in which one ought to use the term "base pressure". The common usage of the term to denote the pressure reached after an (arbitrary!) "appropriate" period of pumping is, at best, misleading.

15.1.2. Establishing the Presence and Nature of a Leak

Barring truly gross leaks (known in the jargon as "hissers" for physically obvious reasons!), the first evidence of a leak in a vacuum system is an unexpectedly protracted pumpdown. However, the pumpdown characteristics of a system may be altered significantly by many means, and the fact that a system appears to be taking too long to pump down, or even that it appears to have "based out" at a given pressure, is not generally sufficient basis for the determination that a leak is present. However, the analysis of Section 15.1.1 allows us to establish the presence (or absence) of a leak with much more certainty and avoid the waste and frustration of searching for nonexistent leaks.

15.1.2.1. Leaks in the Roughing Portion of the Pumpdown Cycle. We expect that the initial portion of the rough pumping cycle will be a simple exponential decay with time, so that a plot of log P versus time should yield a straight line of slope $-2.3\ (S/V)$. However, as the system pressure decays, significant departures from this prediction are observed as seen in Fig. 1. Many factors contribute to this degradation including the falloff of performance of the mechanical pump as its "blankoff" pressure is approached, the vapor pressure of the mechanical pump oil (which may be considerably higher than that of fresh oil if more than a few months of normal operation have elapsed), and, of course, the normal outgassing of the system itself.

Nonetheless, leaks the order of 10^{-1} Pa liters/sec or larger may be easily and unambiguously detected as seen in Fig. 2 for a typical bell jar system with volume the order of 100 liters pumped with a 500 liters/min (nominal free air displacement) mechanical pump. Two cautions are in order here; namely, (1) the change of slope from -2.3 (S/V) to *zero* is quite abrupt and characteristic, and (2) exposure of the system to an untrapped mechanical pump under conditions of low pressure and gas flow will likely lead to problems of contamination with mechanical pump oil. Therefore, this technique is *not recommended* for establishing the presence of leaks which give equilibrium pressures lower than 10 Pa with an untrapped mechanical pump or lower than 1 mTorr with a trapped mechanical pump.

15.1.2.2. Leaks in the High Vacuum Portion of the Pumpdown Cycle. Upon crossover to the high vacuum pumps, the time constant of the system is decreased in proportion to the ratio of the speed of the high vacuum pump to the speed of the mechanical pump. This ratio is typically the order of several hundred, and so the volume gas source is rapidly depleted and becomes a negligible part of the total system gas load. Thus, as discussed in Section 15.1.1, the pressure is expected to decrease in inverse proportion to time.

We can conveniently test for this behavior by generating a graph of log P versus log t, as in Fig. 3, where the pumpdown of a properly functioning uhv system is displayed. If, as in Fig. 3, a slope of -1 is observed, it is quite pointless to search for a leak, since the system pressure, no matter how unsatisfactory it may be, is determined by normal outgassing.

If a straight line of slope $-\frac{1}{2}$ is observed for a plot of this sort, then the presence of a diffusion limited gas source is indicated. This may or may not signify the presence of a leak since certain types of outgassing, such as that from "spongy" layers of deposited material in vacuum coaters and the normal outgassing from most plastics, exhibit a similar time dependence. An example of this sort of behavior is seen in Fig. 4. This particular case was found to be caused by a faulty butyl O ring in the main gate valve of the system.

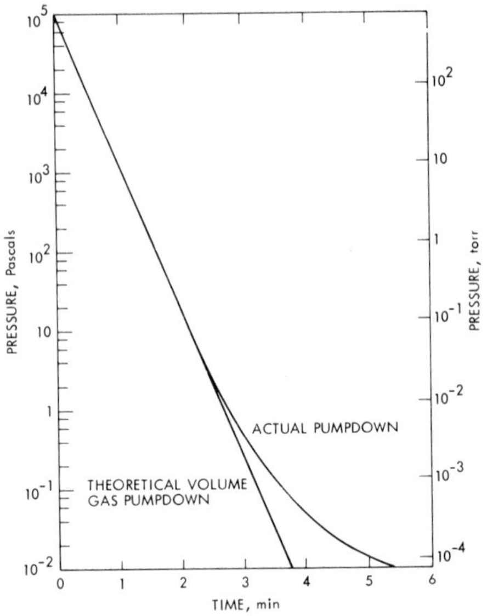

FIG. 1. Simple exponential pumpdown of volume gas compared to an actual pumpdown. Factors contributing to the difference between the two curves include normal outgassing and variations of the speed of the pump with pressure. The chamber volume is 100 liters and the pumping speed (free air displacement) is 500 liters/min. Speed variations with pressure are for a *new* pump and may be far more severe as the pump ages.

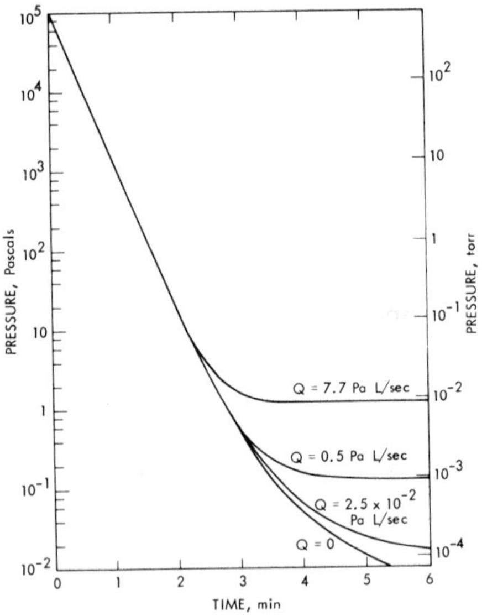

FIG. 2. Pumpdown of a 100 liter chamber by a 500 liters/min (free air displacement) pump with various size leaks. A leak-free pumpdown with normal outgassing and pumping speed variation with pressure is given for comparison. Speed variations with pressure are for a new pump and may be far more severe as the pump ages.

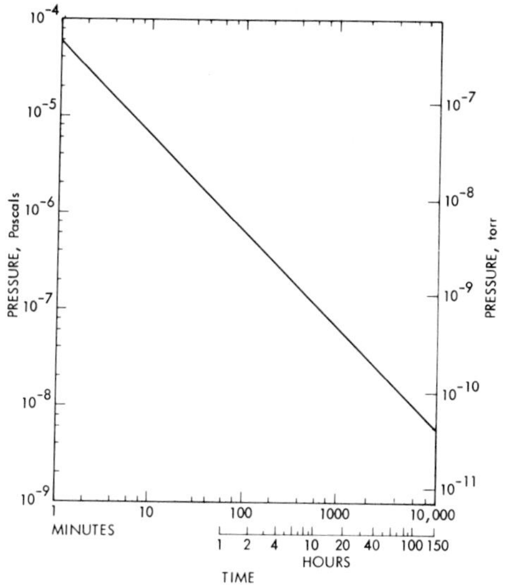

FIG. 3. Pumpdown in the free-molecular flow condition with normal outgassing. Time $t = 0$ corresponds to the point at which the uhv pumps were valved in to the system.

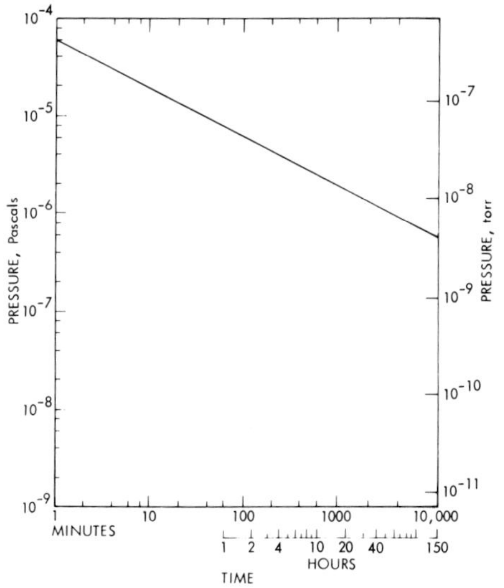

FIG. 4. Pumpdown in the free-molecular flow condition with the gas load dominated by a diffusion or diffusionlike process. This behavior is characteristic of faulty O rings *or* the outgassing from certain materials such as plastics, fluorocarbons such as PTFE, or the outgassing from "spongy" surfaces.

15.1.3. Specifying Acceptable Leak Rates

The specification "zero leak rate" is not physically meaningful, since any real vacuum system will certainly manifest all of the properties under appropriate circumstances. The proper definition of a vacuum is most certainly environmental, so that any specification for leak rate—in fact, any specification for the system at all—must depend in detail upon what the operator desires to accomplish.

For the simple case where only the total pressure is significant, the maximum tolerable leak rate may be fixed at a point such that the system total pressure will not be leak limited above that specified point. Of course, some leeway should be allowed for the fact that the various uhv total pressure gauges do not have equal sensitivity for all gases, and hydrogen, for which the variation in sensitivity is generally considerable, is usually the dominant gas species in all metal uhv systems at extremely low pressures which are of interest to surface science researchers.

In addition to a simple total pressure specification, it is frequently necessary to stipulate that the partial pressure of some particular gas or class of gases must be considerably below the total pressure limit. A common example of this would be limits on certain highly reactive gases in vacuum deposition systems or in surface science systems such as AES or LEED instruments. Not infrequently, one of the critical gases is oxygen and another nitrogen, so that an air leak would become a major source of the offending gas. To monitor such a specification would require the inclusion of a partial pressure analyzer on the system. This instrument would obviously be capable of monitoring *any* change in residual gas composition within its range and is a powerful diagnostic tool for any of the more sophisticated applications. An alternative to a partial pressure analyzer for more routine applications could be an instrument sensitive to total pressure and one of the constituent partial pressures. Such an instrument based on the analysis of light emitted from a discharge gauge was demonstrated by Milleron[1] and is just now commercially available.*

Frequently, some concern is expressed over the anticipated increase in pumpdown time associated with a leak. It should be obvious from the analysis of the preceding sections that unless a major diffusion limited source is present, the pumpdown time to a pressure only half a decade higher than the leak limited base pressure of the system will be negligibly increased. In this regard, diffusion limited gas sources of *any* sort are much more important than simple true leaks and it would be wise to eliminate them if at all possible.

[1] N. Milleron, *Trans. Natl. Vac. Symp.* **10**, p. 283 (1963). Macmillan, New York, 1963.

* Smart™ Gauge, Varian Associates, Palo Alto, Calif.

A final consideration is the possibility of a detrimental effect of ambient atmospheric gases on the system pumps. In most cases, this is not a problem, but special combinations of pumps and perhaps corrosive atmospheres present exterior to the vacuum chamber, may alter this conclusion.

15.2. Types of Leak Detecting Equipment

15.2.1. Leak Detection, Locating, and Measuring

Virtually all equipment sold in this category is referred to as a "leak detector," a custom to which we shall conform here. However, it is worth distinguishing between the three possible functions of these units as mentioned in the title for this section. The *detection* of a leak—that is, establishing that a leak is present and significant to the system performance—can almost always be accomplished using only the basic instrumentation on any system and the graphical analysis discussed in Sections 15.1.2.1 and 15.1.2.2. The power of this simple exercise cannot be overemphasized! It can take hours to "leak check" a typical vacuum system. Leak *location* is the major function of most units both in principle and usage. They do very little to establish the presence of a leak without first locating it. The single exception to this is the LI gauge mentioned previously* and it does not have leak detection as a primary function, being intended as a "system readiness monitor." Leak *measuring,* in more than a rough sense of the word, is a function none of the available techniques or units perform very well. Fortunately, this function is rarely needed by the "troubleshooter" of a vacuum system. "Standard" leaks for calibration of leak detectors are available, but, with the exception of those operating on the principle of permeation, their accuracy is subject to question at the lower leak rates which would be of interest to a researcher testing a uhv system.

15.2.2. Leak Detection Equipment and Methods

In this section, we shall consider the various types of leak detection units and methods for using them in the laboratory. Specific manufacturers' detectors will not be discussed. For that material, the reader is referred either to an excellent book on the subject by Holland *et al.*,[2] or to the manufacturers themselves.

When the system is leak limited to a pressure sufficiently high that an electrical gas discharge can be supported, a small hand-held Tesla coil can

[2] L. Holland, W. Stecklemacher, and J. Yarwood, "Vacuum Manual," Spon, London, 1974.

be used to search for leaks through insulating surfaces on the vacuum envelope. This technique is far less common than it used to be since the Tesla coils are no longer common laboratory equipment, and most uhv systems made today are all metal construction. There is still the possibility of using the method to check the various windows and feedthrough for leaks, presuming a Tesla coil can be located, but a word of caution is in order: A very intense spark can *make* a leak in a thin layer of glass or ceramic. The use of Tesla coils as leak locators is therefore not recommended.

Another popular, but questionable, technique is the "rate of rise" method. Here, the system is isolated from its pumps and a pressure rise linear with time is taken to indicate the presence of a leak. Unfortunately, there are many problems with this simple method. The gauges are not passive elements and both pump at speeds which may be several lamberts per second or higher in extreme cases and rapidly convert gas species. This, combined with the fact that the pressure will certainly rise due to the normal outgassing of the chamber walls, makes the observed pressure rise difficult to interpret. At best, this technique can suggest that a leak is present unless another gas is squirted about in an attempt to alter the rate of rise when the leak is found. This method is not recommended.

Bubble detection may be used when the chamber under test has a gross leak and can be pressurized. The suspected regions can then be squirted with a weak soap or detergent solution and inspected for bubble formation. In practice, the sensitivity is limited by the time it takes to form a bubble with the leak rate present. For a 1 mm diameter bubble, a leak rate of about 10^{-1} Pa liters/sec will yield bubbles at the rate of about 1/sec. Most vacuum systems are not amenible to pressurization, however, and the utility of the method is quite restricted.

Thermocouple gauges respond dramatically when acetone is squirted on a system leak. This fact is used and abused in laboratories throughout the world. The abuse is related to the fact that acetone can seriously degrade O-ring seals and certain insulating materials occasionally used on vacuum systems such as glass-filled Noryl. Generally, the method may be used on any metal sealed component or joint in the system so long as great care is taken to prevent the acetone from contacting any organic material used on the system. The acetone should be used sparingly. A final caution regarding acetone concerns the nature of the chemical: Acetone is a powerful solvent, dangerously flammable, and mildly toxic. It must be used with care, and indiscriminate squirting of acetone around a vacuum system is ill advised—and possibly illegal in industrial installations due to OSHA regulations.

Ion gauge and ion pump leak detectors operate on the same fundamental principle; namely, the pressure signal is electronically nulled and an attempt is made to alter the leak rate either up or down and detect the change in the pressure signal. Since these instruments measure "nitrogen equivalent total pressure" and have considerable variations in sensitivity for the various gas species without being able to distinguish between them, response to a probe gas entering the system through a leak may be much less than anticipated. In addition, most real leaks have a conductance which is *not* a simple known function of the atomic mass of the leaking gas. This technique works best when the system is at its leak-limited pressure. Under these circumstances, the sensitivity may be sufficient to locate leaks as small as 10^{-9} Pa liters/sec. If, however, the system pressure is high and/or limited by other sources of gas—including other small leaks—sensitivity is greatly reduced and may be as low as 10^{-3} Pa liters/sec.

Halogen leak detectors, like the bubble detection technique, require that the system be pressurized. In this case, the system must be pressurized with a gas containing an organic halide such as one of the freons. The exterior of the system is then probed with a "sniffer" probe sensitive to traces of the halogen bearing gas. The most common detector for these units is the platinum wire diode detector in which an axially located platinum wire is heated to form positive ions which are collected by the cylindrical cathode which surrounds it. This ion current is extremely sensitive to traces of organic halides. A recent refinement to this method is the application of the electron capture detector (ECD) developed as a detector cell for gas chromatography. This cell is a cylindrical ionization chamber containing a tritium bearing foil. The electrons emitted by the natural decay of tritium have maximum energies the order of 18 keV. By collisions with gas atoms and molecules within the cell, these electrons are rapidly thermalized. The free electrons are swept from the cell by applying a positive pulse to the anode. Immediately following the charge collection pulse, ionization causes the electron population to increase until the rate of production is exactly balanced by the rate of recombination. When an electronegative gas is present in the cell, this balance will occur at a lower electron density since some of the thermalized electrons will be captured by the electronegative gas. The ECD detector can be used to locate leaks as small as 10^{-6} Pa liters/sec, whereas the more conventional platinum wire diode detector is about a factor of 10 less sensitive. This method is very useful in situations where the chamber under test can be pressurized.

The helium mass spectrometer leak detector (MSLD) is the unit most often visualized when the topic of leak detectors comes up. At the outset, it is important to realize just what this unit is: The helium MSLD is a helium partial pressure gauge affixed (usually) to a small dedicated vacuum system. The conventional way to connect a MSLD to a vacuum system is shown in Fig. 4. Sensitivities the order of 10^{-8} Pa liters/sec can be obtained in this manner but the sensitivity can be a great deal less if severe throttling is necessitated by high leak-limited pressures in the system. An alternate method is to connect the MSLD spectrometer tube directly to the system chamber as shown in Fig. 5. This may be done if the system pressure is below about 10^{-2} Pa and results in an increased sensitivity to about 10^{-9} Pa liters/sec or better. Finally, a new technique has been introduced in which the helium is introduced into the *foreline* of the MSLD vacuum system rather than into the normal pump inlet. This connection, shown in Fig. 6, makes use of the fact that a diffusion pump has a far lower compression ratio for helium than for most of the other gases in the system (except hydrogen!) and helium is therefore enriched in the detector tube. This technique allows a high sensitivity to be obtained even at test port pressures as high as 50 Pa.

15.3. Special Techniques and Problems

15.3.1. "Bagging"

When a complicated feedthrough or other system peripheral is suspect, a good technique with a helium MSLD is to seal a plastic bag around the item and fill it with helium. In this manner, the test gas can penetrate leaks which would be otherwise difficult to get at with a stream of gas from a needle. Of course, the location of the leak *within* the item cannot be pinpointed in this manner. It is a good idea to separately test all accessible welds and joints prior to bagging an object.

15.3.2. "Bombing"

"Bombing" is a technique for testing hermetic enclosures such as transistors, diodes, and integrated circuits by subjecting the enclosure to helium at an elevated pressure and then testing the item in the vacuum chamber of a helium MSLD for the reemission of the bombed helium. The indicated leak rate Q_i is surprisingly close to the true leak rate Q_t for most packages encountered in semiconductor manufacturing. In general, for a noncondensable gas

$$Q_1 = P_b[-e^{-(Lt_b/V)}][e^{-(Lt/V)}]L,$$

where P_b is the bombing pressure, L the bombing leak rate per unit pres-

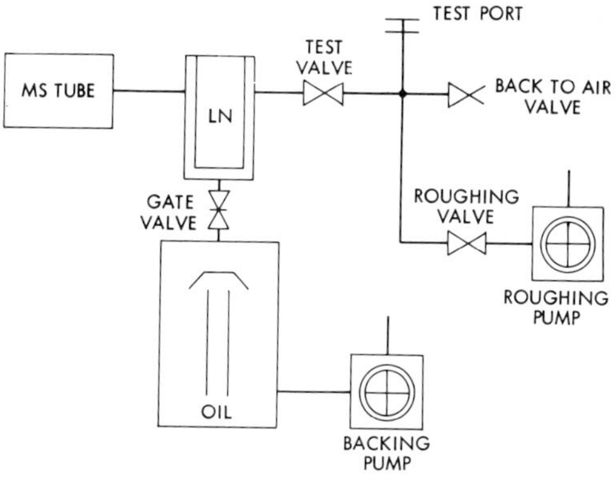

FIG. 5.

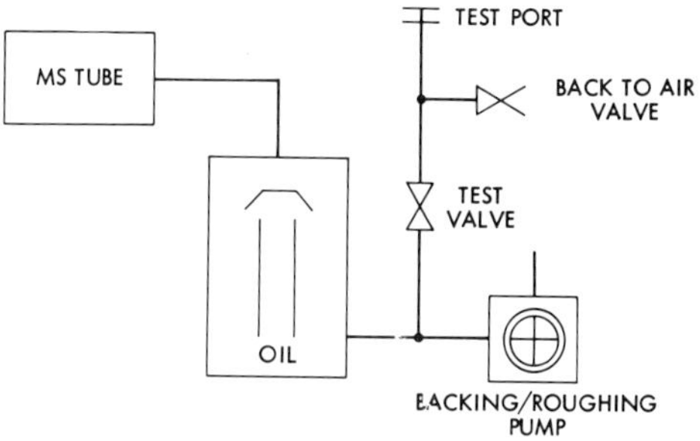

FIG. 6.

FIG. 5. MSLD connected to the system in the conventional manner. This technique is still required in the leak checking of uhv systems not pumped by diffusion pumps. The system pumps must be able to hold a pressure of at least the order of 10^{-2} Pa for this arrangement to be used.

FIG. 6. MSLD connected to the chamber in the "counterflow" configuration. Improved sensitivity may be obtained due to the selective compression of higher molecular weight gases (relative to He) by the diffusion pump. This effect could be made even more dramatic by using a turbomolecular pump for the mass spectrometer tube ("MS TUBE").

sure, t_b the bombing time, V the enclosure volume, and t the time under vacuum.

15.3.3. Helium Permeation

Obviously, if there is a permeation leak into the system with a long characteristic permeation time, a large background of helium can build up which will seriously reduce the effective sensitivity of the MSLD. The red vacuum hose commonly used to connect mechanical pumps to the system is subject to this effect as is the Teflon tape used to seal the pipe thread joints used with most thermocouple gauges. Special care should be taken to avoid saturating these items with helium.

An additional complication arises when the ambient laboratory atmosphere has a high background of helium due to an experiment being performed, or overly zealous use of helium in an enclosed area with a helium MSLD. Under these circumstances, the usefullness of a helium MSLD may be sharply curtailed.

15.3.4. Finding the Virtual Leak

Virtual leaks, by their very nature, cannot be located by any of the external locating techniques discussed here. The most common form of the virtual leak is a trapped volume within the system—that is to say, poor vacuum practice! Examples of this are such things as blind tapped holes and screws not properly vented and double welds. A double seam weld is NEVER acceptable vacuum practice from this standpoint since leaks in such a weld cannot be located. Essentially the only way to *detect* the virtual leak is from the pumpdown curve as discussed earlier. These sources can act like a true fixed leak if the trapped volume is large and the leak rate small, or a rapidly depletable diffusion limited source is the reverse is true. The only viable *location* technique is careful analysis of the interior of the vacuum system itself.

15.3.5. Response Time

The response of a leak detector to flooding a leak with the probe gas is limited by the response time. Sources of this time delay include the instrumental response time, the time required for the probe gas to penetrate the leak, and the time constant of the system itself. If a leak is flooded with a probe gas for a time t' and gas leaks into the system at a rate Q, the partial pressure of the probe gas will rise within the system according to the equation

$$P = (Q/S)[1 - e^{-(S/V)t}]$$

for time t less than t'. After the source of probe gas is removed, the pressure in the system will decay exponentially from the above value in accordance with the equation

$$P = (Q/S)[1 - e^{-(S/V)t}][e^{-(S/V)(t - t')}].$$

Thus, for small leaks in very large systems, one must not only contend with the time delay associated with the time constant of the system, but, in addition, the attendant loss of sensitivity must be considered.

The time required for the gas to flow through a leak may be a significant factor in system response, since the leaks one encounters in real vacuum systems are rarely simple apertures. Rather, the leak typically follows a tortuous path. The limiting case for such behavior is a permeation leak such as one might find if porosity or "pipes" in a metal wall of the chamber should occur.

Instrumental response time is amenable to improved electronics and is rarely the limiting factor in the response time of modern leak detectors.

15.4. Repair Techniques

15.4.1. General Considerations

Once the leak (or leaks) have been located, the problem of what to do about them remains. As pointed out in Section 15.1.3, it may be that the leak is at least marginally acceptable and repair may be so difficult, costly, or time consuming that it may not be desirable to attempt the repair. However, let us assume the repair is to be made. Techniques for repair depend primarily on the following considerations:

(1) location of the leak (foreline, main uhv chamber, gas inlet system, etc.);
(2) nature of the leak (real, virtual, crack, porosity, "pipes," etc.);
(3) what are the materials of construction? and
(4) what properties must the repair have? (vacuum related, mechanical, vibration, thermal).

Let us take the general locations one at a time, enumerate the leaks, and identify possible repair techniques.

15.4.2. Leaks in the Roughing Line

Leaks to atmosphere in the roughing lines, unless large enough to preclude reaching the pressure necessary to "cross over" to the uhv pumps, are seldom a problem of great significance. In fact, a small air leak in the roughing line may serve to inhibit the backstreaming of mechanical pump oil into the main chamber should an untrapped mechanical pump be exposed to the chamber under conditions of what would otherwise be low flow and pressure!

Leaks in rubber or Tygon tubing may arise from splits at the ends or excessive porosity (rare with Tygon). The effected end must be trimmed and discarded. In the case that excessive porosity is detected, some workers like to try to coat the effected region with vacuum grease or other sealant. This generally has the effect of producing a virtual leak and is therefore not recommended.

Another frequently overlooked problem area is the use of PTFE tape on the pipe thread seals commonly used on thermocouple gauges. As mentioned previously, this is very poor technique since leaks are very common and the joint cannot be leak checked with a helium MSLD due to the permeability of Teflon to helium. If the use of PTFE tape cannot be avoided, the offending joint will have to be redone as many times as necessary to meet the leak limit.

Soldered or brazed joints in copper roughing lines sometimes fail due to the vibration of the mechanical pump. Such joints are easily repaired by careful cleaning and resoldering of the leaking area in most cases. A good solder to use for such repairs is a Sn/Ag 5% alloy. This solder is much stronger than the usual Sn/Pb solders and maintains its strength over long periods of time far better.

Finally, leaks in the roughing valve itself must be considered. With a through valve, leaks may be either to atmosphere or through the valve itself. Leaks through the valve are usually caused by dirt on the sealing surface or a faulty gasket. The valve should be cleaned and the gasket replaced as necessary. Excessive use of vacuum grease in an attempt to seal small leaks associated with small scratches on the sealing surface should be avoided since it tends to replace a small real leak with a large virtual leak. Other causes of valve leaks, such as bellows failure or major distortions of the sealing surfaces, are generally beyond the average user, and the valve should either be returned to the manufacturer for repair or replaced. Repeated failure of the O ring in a valve reflects a design flaw in the valve itself and should be taken up with the manufacturer.

15.4.3. Leaks in the Uhv Chamber

Leaks in the main chamber fall into three main categories; namely, (1) leaks at demountable seals, (2) leaks in joints in the chamber or peripherals, and (3) leaks *through* the materials of the main chamber or peripherals.

Demountable seals may be either all metal or elastomer sealed. Each has its own set of problems and solutions. Metal gasket seals can leak for the following reasons and with the following associated solutions.

(1) There may be deep scratches across the sealing surfaces. (Superficial scratches are generally insignificant.) Replacement of the faulty sealing surface is generally the only reasonable course of action, although in some cases remachining is practical. Such repairs are usually beyond the average user.

(2) The mating surfaces may have become distorted due to excessive thermal or mechanical stresses. Again, replacement, or sometimes remachining, are the only viable alternatives.

(3) The joint fasteners may not have been properly torqued. The solution here is to replace the gasket and retorque properly. Attempts to seal the joint merely by tightening the fasteners *in situ* are to be strongly discouraged as broken or distorted fasteners can be the result.

(4) The gasket itself may be faulty. With modern seals of the ConflatTM configuration and its derivatives, this is a rare occurrence. The oxides and chemical stains sometimes seen on new gaskets are of no consequence for the sealing properties of the item.

Leaks in joints in the chamber or peripherals are occasionally found. Typical locations are the welds, brazes, metal-to-glass, and ceramic-to-metal seals. Aside from replacement, leaks in glass or ceramic parts can be repaired only by means of one of the low vapor pressure epoxy resins sold for that purpose. The various paints, lacquers, and waxes that may be used for temporary repairs in the foreline are totally unacceptable for use in the uhv system. Rewelding a leaky seam weld is generally not very successful, and in *no case* should a full double seam weld be used on a vacuum chamber. Brazes can sometimes be rebrazed using a lower melting point braze alloy, or they can be soft soldered using the Sn/Ag 5% alloy mentioned previously, or they can be sealed using low vapor pressure epoxy.

Leaks *through* the materials of a chamber can also occur. These may be the result of porosity in the metal or "pipes." ("Pipes" are voids in a billet which have been greatly elongated in the rolling process and then machined through inadvertantly in the process of manufacturing.) Both of these faults are the result of careless manufacturing processes and quality control and should be taken up with the manufacturer.

15.4.4. Leaks in a Gas Inlet System

A major problem area which is almost universally ignored concerns the problem of leaks in any source and/or delivery system for gas used in the vacuum process. These should also be carefully leak tested! The only caution in this regard concerns the leak testing of gas regulators. The common variety of gas regulator is not designed to have a vacuum on the delivery side. Special regulators designed for vacuum use should always be used in these applications. Most regulator manufacturers will supply leak test certification for these units for a nominal fee.

16. VACUUM DEPOSITION TECHNIQUES*

16.1. Introduction

Many areas of research require the fabrication of thin films.† Films are called "thin" if they have a thickness less than 10,000 Å (10^{-4} cm). Requirements of the research generally dictate the material to be used for the film, the substrate material, the dimensions of the film, and the handling to which the film will be exposed. There are many thin film parameters and deposition parameters which must be considered in each particular situation. The purpose of this part is to describe the two basic vacuum techniques for producing thin films: vacuum evaporation and sputter deposition. Emphasis will be on the experimental parameters available to the researcher and the relation of these parameters to thin film properties.

16.1.1. Thin Film Deposition Parameters

There are approximately a dozen parameters used to characterize thin films. The experimenter must decide which parameters are to be controlled for his application; some of the more important are thickness, density, spatial and compositional uniformity, impurity concentration, crystallinity, orientation, resistivity, grain size, adhesion to substrate, internal stress, surface morphology, thermal coefficients of resistivity, and Hall coefficient. These parameters are not all independent. For example, the resistivity of a film is a function of the thickness (for very thin films), the impurity concentration, and the grain size. It should be emphasized that all of the parameters can be controlled to some degree by applying the appropriate experimental techniques discussed below. It should also be emphasized that for a particular application only one, or at most, a few of these film parameters are important and thus will dictate the choice of experimental procedures. Therefore, the first task before setting up an apparatus to make a thin film or requesting a service laboratory to supply a film is to itemize those characteristics that are required

† See for instance Vol 2A in this series, Chapter 2.3, and Vol. 11, Part 12.

* Part 16 is by M. T. Thomas.

ISBN 0-12-475914-9

for the experiment. This simple first step will avoid a great deal of extra work and expense.

16.1.2. Experimental Options

Thermal vapor deposition and sputtering are the two experimental techniques for thin film deposition. The advantages and disadvantages of each technique should be considered before deciding which will be used in a given application. Briefly, thermal vapor deposition consists of increasing the temperature of the material for deposit until its vapor pressure is high enough ($>10^{-2}$ Torr) for evaporation. If the ambient pressure is sufficiently low, the vapor from the hot source will deposit on the cooler surrounding structures of the system. Details for application of this phenomenon for controlled film formation are discussed in Chapter 16.2.

Sputtering utilizes a different physical principle for the removal of material from a source to form a thin film. The sputtering phenomenon involves accelerating a gas ion and allowing it to collide with the film source material. During the collision the gas ion transfers its momentum and energy to the atoms and lattice of the source material. One of the many processes that occurs after such an ion–lattice collision is the ejection of a lattice atom. This sputtered atom is collected by other structures of the vacuum system to form a thin film. Chapter 16.3 outlines the experimental details required for producing sputtered films. Chapter 16.4 compares the thermal evaporation and sputtering techniques and itemizes the advantages and disadvantages of each.

16.2. Vacuum Evaporation Techniques

16.2.1. Simple Theory

Vacuum evaporation is accomplished by increasing the temperature of the source material until its vapor pressure is equal to or greater than 10^{-2} Torr. In a vacuum environment the source atoms or molecules in the vapor phase will leave the source and condense on cooler surfaces. The rate of evaporation can be obtained from the kinetic theory of gases as first derived by Langmuir[1]

$$G = 5.83 \times 10^{-2}\, \alpha P(M/T)^{1/2}, \tag{16.2.1}$$

where G is the rate of evaporation in grams/cm²/sec, P is the vapor pressure in Torr at temperature T, M is the moleular weight, T is the source

[1] I. Langmuir, *Phys. Rev.* **2**, 329 (1913).

temperature in degrees K, and α is the sticking or condensation coefficient. For most metals α equals unity if the substrate temperature is not too high. It should be remembered that the vapor pressure of a material is a function of absolute temperature. For vapor pressures less than 1 Torr,

$$\ln P = A - B/T, \tag{16.2.2}$$

where A and B are constants that are generally determined from experimental data. The evaporation rate is determined primarily by the source temperature. For a more detailed discussion of rates of evaporation and vapor pressure the reader is referred to Chapters 1 and 10 of Dushman.[2] Any of the many techniques used to heat material in a vacuum system could be considered for vaporizing the source for deposition.

16.2.2. Experimental Parameters

16.2.2.1. Source Temperature and Evaporation Rate. Source temperature is the main parameter controlling the evaporation rate and hence the deposition rate for a given system geometry. However oxides, carbides, and nitrides on the source generally reduce the vapor pressure and thus influence evaporation rate. In addition, source contamination can cause other film growth problems and is discussed in a later section. So assuming a pure source, the source temperature directly determines the film deposition rate. Besides determining the length of time required to grow a film of given thickness, deposition rate is important in controlling many other film parameters. For example, the amount of impurities incorporated in film from the ambient gas is reduced as the deposition rate increases. Film texture or morphology, crystallinity, grain size, orientation, resistivity, and intrinsic stress are all functions of the deposition rate. Although other experimental parameters also help to determine some of the film properties, the deposition rate is one of the easiest to control and vary. Deposition rates in thermal evaporation systems can be varied from 10^{-2} Å/sec to 10^{6} Å/min.

16.2.2.2. Substrate Temperature. Substrate temperature is one of the variables controlling surface atom mobility and thus is important in studies of single crystal films and epitaxial films. Substrate temperature influences the crystallinity, grain size, resistivity, and internal stress of the film. The initial growth of thin films tends to be in the form of very small, 20–30 Å size, disconnected islands or crystallites, even for room temperature substrates. If an amorphous film is required for a given ap-

[2] S. Dushman, "Scientific Foundations of Vacuum Technology" (J. M. Lafferty, ed.), 2nd ed. Wiley, New York, 1962.

plication, then the substrate must be cooled, perhaps to as low as liquid nitrogen temperature (77 K). Besides increasing film grain size, increasing the temperature will also tend to improve substrate film bonding.

16.2.2.3. Physical Geometry of Source and Substrate. The distance between the source and substrate is an important factor in controlling film uniformity. In general, the greater this distance the more spatially uniform the film will be. However, as the source–substrate distance is increased the deposition rate is decreased for a given source temperature. In such cases, film uniformity can be improved by using multiple sources or special source geometries. If nonplanar objects have to be coated, a cylinder, for example, it may be necessary to rotate or move the substrate during deposition.

16.2.2.4. Substrate's Surface Condition. The fact that thin films replicate the substrate has been used for many years for studying surface topography in transmission electron microscopes. Every scratch, dust particle, and surface imperfection on the substrate will show in the film. Dust particles also tend to produce pinholes in the film which may be detrimental for certain applications. Thus great care must be taken to use substrates that are as smooth and clean as possible. Not only are the macroscopic features of the substrate important, but for many applications so are the microscopic conditions. For instance, in epitaxial film studies the exposed crystalline plane of the substrate has a known atomic spacing which is one of the parameters that can be varied. Also, the number and density of surface imperfections such as atomic steps or impurities are important in single crystal film studies since they can act as nucleation sites thus controlling the film growth.

16.2.2.5. Ambient Gas Environment. The total pressure as well as the partial pressures of the ambient gas in a vacuum system during deposition can affect the purity and the structure of the film. If the partial pressures of oxygen, water vapor, carbon monoxide, or hydrocarbons are high, the vaporizing atoms have a high probability of chemically reacting with the ambient gas pressure. The depositing film would then be a mixture of the pure element and chemical compounds. Even if compounds are not formed, a high ambient gas pressure will cause a large quantity of gas to be trapped in film. On the other hand, if the purpose is to form an oxide film, it may be advantageous to do the deposition in a high oxygen partial pressure. In many vacuum evaporators it is not possible to control the ambient gas, thus making it difficult to control the film parameters. However, even under these circumstances, if correct vacuum procedures are used and if the system and pumps are cleaned, usable films can be produced.

16.2.3. Materials Used in Vapor Deposition

16.2.3.1. Materials That Can Be Evaporated. Essentially all solid elements can be vacuum deposited if the appropriate techniques are used. Table I lists the most common elements along with their melting point, temperature at which their vapor pressure equals 10^{-2} Torr, materials generally used to support the element, and important evaporation characteristics that should be noted. For example, some elements such as aluminum are very reactive and form an oxide readily, and others, like nickel, alloy with the refractory metals that are used as evaporation retainers. Holland[3] presents a good discussion of some of the problems encountered in film deposition of the more common elements along with the generally accepted solution.

In contrast to the relative ease of forming elemental films, alloys and compounds are very difficult to deposit directly. For a compound to be vacuum evaporated, it must be vacuum compatible and be chemically stable up to the temperature required for evaporation. Many compounds begin to decompose when exposed to a vacuum environment or begin to dissociate at elevated temperatures. Relatively few compounds can be vapor deposited directly: the easiest compounds to work with are SiO and MgF_2. Table 10 in the chapter by Glang[4] lists the most readily deposited oxides and other compounds. To obtain stoichiometric thin films of a compound, control of the deposition temperature is critical. For example, suboxides can easily be formed and the resulting film can be a mixture of oxides, suboxides, and pure elements. Classical thermodynamics can determine the temperature at which dissociation will occur but not all of the relevant thermodynamic parameters are known to predict the composition of the resulting film. Not only can complex chemistry occur at the vapor source, it can also occur in the vapor and on the substrate. The refractory metals often used as source supports are chemically active and can combine with dissociated oxygen from oxide compounds to form volatile refractory oxides which become incorporated in the depositing film. High levels of impurities can be trapped in films by this mechanism. Even if the individual constituent did not interact with the support material, the vapor pressures of each component would have to be the same in order to deposit stoichiometric compounds. Also, certain requirements must be satisfied at the substrate if stoichiometry is to be achieved when the elements of the compound volatize from the source separately. For

[3] L. Holland, "Vacuum Deposition of Thin Films." Wiley, New York, 1961.

[4] R. Glang, "Handbook of Thin Film Technology" (L. I. Maissel and R. Glang, eds.), Chapter 1. McGraw-Hill, New York, 1970.

TABLE I. Temperature and Support Materials Used in the Vaporation of the Elements[a]

Element and predominant vapor species	Temperature (°C)		Support materials		Remarks
	mp	$p^* = 10^{-2}$ Torr[b]	Wire, foil	Crucible	
Aluminum (Al)	659	1220	W	C, BN, TiB_2–BN	Wets all materials readily and tends to creep out of containers. Alloys with W and reacts with carbon. Nitride crucibles preferred
Antimony (Sb_4, Sb_2)	630	530	Mo, Ta, Ni	Oxides, BN, metals, C	Polyatomic vapor, $\alpha_v = 0.2$.[c] Requires temperatures above mp. Toxic
Arsenic (As_4, As_2)	820	~300	—	Oxides, C	Polyatomic vapor, $\alpha_v = 5.10^{-5}$–5.10^{-2}. Sublimates but requires temperatures above 300°C. Toxic
Barium (Ba)	710	610	W, Mo, Ta, Ni, Fe	Metals	Wets refractory metals without alloying. Reacts with most oxides at elevated temperatures
Beryllium (Be)	1283	1230	W, Mo, Ta	C, refractory oxides	Wets refractory metals. Toxic, particularly BeO dust
Bismuth (Bi, Bi_2)	271	670	W, Mo, Ta, Ni	Oxides, C, metals	Vapors are toxic
Boron (B)	2100 ±100	2000	—	C	Deposits from carbon supports are probably not pure boron
Cadmium (Cd)	321	265	W, Mo, Ta, Fe, Ni	Oxides, metals	Film condensation requires high supersaturation. Sublimates. Wall deposits of Cd spoil vacuum system
Calcium (Ca)	850	600	W	Al_2O_3	
Carbon (C_3, C, C_2)	~3700	~2600	—	—	Carbon–arc or electron–bombardment evaporation. $\alpha_v < 1$

Chromium (Cr)	~1900	1400	W, Ta	—	High evaporation rates without melting. Sublimination from radiation-heated Cr rods preferred. Cr electrodeposits are likely to release hydrogen
Cobalt (Co)	1495	1520	W	Al_2O_3, BeO	Alloys with W, charge should not weigh more than 30% of filament to limit destruction. Small sublimation rates possible
Copper (Cu)	1084	1260	W, Mo, Ta	Mo, C, Al_2O_3	Practically no interaction with refractory materials. Mo preferred for crucibles because it can be machined and conducts heat well
Gallum (Ga)	30	1130	—	BeO, Al_2O_3	Alloys with refractory metals. The oxides are attacked above 1000°C
Germanium (Ge)	940	1400	W, Mo, Ta	W, C, Al_2O_3	Wets refractory metals but low solubility in W. Purest films by electron-gun evaporation
Gold (Au)	1063	1400	W, Mo	Mo, C	Reacts with Ta, wets W and Mo. Mo crucibles last for several evaporations
Indium (In)	156	950	W, Mo	Mo, C	Mo boats preferred
Iron (Fe)	1536	1480	W	BeO, Al_2O_3 ZrO_2	Alloys with all refractory metals. Charge should not weigh more than 30% of W filament to limit destruction. Small sublimation rates possible
Lead (Pb)	328	715	W, Mo, Ni, Fe	Metals	Does not wet refractory metals. Toxic
Magnesium (Mg)	650	440	W, Mo, Ta, Ni	Fe, C	Sublimates
Manganese (Mn)	1244	940	W, Mo, Ta	Al_2O_3	Wets refractory metals
Molybdenum (Mo)	2620	2530	—	—	Small rates by sublimation from Mo foils. Electron-gun evaporation preferred

(*continued*)

TABLE I (*continued*)

Element and predominant vapor species	Temperature (°)		Support materials		Remarks
	mp	$p^* = 10^{-2}$ Torr[b]	Wire, foil	Crucible	
Nickel (Ni)	1450	1530	W, W foil lined with Al_2O_3	Refractory oxides	Alloys with refractory metals; hence charge must be limited. Small rates by sublimation from Ni foil or wire. Electron-gun evaporation preferred
Palladium (Pd)	1550	1460	W, W foil lined with Al_2O_3	Al_2O_3	Alloys with refractory metals. Small sublimation rates possible
Platinum (Pt)	1770	2100	W	ThO_2, ZrO_2	Alloys with refractory metals. Multistrand W wire offers short evaporation times. Electron-gun evaporation preferred
Rhodium (Rh)	1966	2040	W	ThO_2, ZrO_2	Small rates by sublimation from Rh foils. Electron-gun evaporation preferred
Selenium (Se_2, Se_n: $n = 1 - 8$)[d]	217	240	Mo, Ta, stainless steel 304	Mo, Ta, C, Al_2O_3	Wets all support materials. Wall deposits spoil vacuum system. Toxic. $\alpha_v = 1$
Silicon (Si)	1410	1350	—	BeO, ZrO_2, ThO_2, C	Refractory oxide crucibles are attached by molten Si and films are contaminated by SiO. Small rates by sublimation from Si filaments. Electron-gun evaporation gives purest films
Silver (Ag)	961	1030	Mo, Ta	Mo, C	Does not wet W. Mo crucibles are very durable sources
Strontium (Sr)	770	540	W, Mo, Ta	Mo, Ta, C	Wets all refractory metals without alloying
Tantalum (Ta)	3000	3060	—	—	Evaporation by resistance heating of touching Ta wires, or by drawing an arc between Ta rods. Electron-gun evaporation preferred

Tellurium (Te_2)	450	375	W, Mo, Ta	Mo, Ta, C, Al_2O_3	Wets all refractory metals without alloying. Contaminates vacuum system. Toxic. $\alpha_v = 0.4$
Tin (Sn)	232	1250	W, Ta	C, Al_2O_3	Wets and attacks Mo
Titanium (Ti)	1700	1750	W, Ta	C, ThO_2	Reacts with refractory metals. Small sublimation rates from resistance-heated rods or wires. Electron-gun evaporation preferred
Tungsten (W)	3380	3230	—	—	Evaporation by resistance heating of touching W wires, or by drawing an arc between W rods. Electron-gun evaporation preferred
Vanadium (V)	1920	1850	Mo, W	Mo	Wets Mo without alloying. Alloys slightly with W. Small sublimation rates possible
Zinc (Zn)	420	345	W, Ta, Ni	Fe, Al_2O_3, C, Mo	High sublimation rates. Wets refractory metals without alloying. Wall deposits spoil vacuum system
Zirconium (Zr)	1850	2400	W	—	Wets and slightly alloys with W. Electron-gun evaporation preferred

[a] Reproduced from R. Glang, "Handbook of Thin Film Technology," Vacuum Evaporation (L. I. Maissel and R. Glang, eds.), Chapter 1, McGraw-Hill, 1970, with permission from McGraw-Hill.

[b] Temperature at which material has vapor pressure $p^* = 10^{-2}$ Torr.

[c] α_v is the evaporation coefficient.

[d] R. Yamdagni and R. F. Porter, *Electrochem. Soc.* **115,** 601 (1968).

instance, the sticking coefficient of each element must be the same and the mobility on the substrate must be such that recombination can occur.

Formation of thin film alloys with the same composition as the parent source material has many of the same problems as the evaporation of compounds. The main difficulty at the source is that each element composing the alloy will evaporate independently of each other. Each essentially behaves like a pure metal and its rate of evaporation is governed by its vapor pressure at the source temperature. The problems at the substrate are the same as for compounds. Very few alloys can be evaporated directly; permalloy (85% Ni and 15% Fe) is one of the rare exceptions.

A number of techniques have been developed to overcome some of the problems of vacuum depositing compounds and alloys: namely, flash evaporation, multiple sources, and reactive evaporation. The flash evaporation and multiple source techniques are discussed in the next sections. Reactive evaporation involves maintaining the ambient environment in the evaporation chamber at a high partial pressure ($\sim 10^{-3}$–10^{-2} Torr) of the gas necessary to form a stoichiometric film. For example, if one is trying to deposit an oxide or a nitride compound, the resulting film will probably be deficient in oxygen or nitrogen. To compensate for this deficiency a gas source is connected to the deposition chamber by a variable leak and the partial pressure of oxygen or nitrogen in the chamber is adjusted to allow deposition of stoichiometric compounds. The partial pressure required for any particular film will depend on the deposition parameters and will have to be determined by experimentation.

Elements that sublime are generally treated separately when discussing evaporation techniques since they can be handled differently than other materials. A few such materials are chromium, rhodium, silicon, and titanium. A solid sublimes when its vapor pressure is sufficiently high ($\sim 10^{-2}$ Torr) that a reasonable deposition rate can be achieved before the solid melts. The advantage of sublimation is that there is little or no reaction with the source of support. In general, source–support reactions occur at or above the source's melting point.

16.2.3.2. **Source Heaters.** The main requirements of a source heater are to supply thermal energy so the source material can reach and maintain the vapor pressure needed for the desired deposition rate, to be nonreactive with the source, and to have a low vapor pressure at its operating temperature. There are many different types of source heaters. Choice of a particular heater is controlled by the material to be deposited, the quantity of material to be deposited, and the thickness distribution required in the deposit. In general, the geometry of a heater assembly is dictated by the geometry of the substrate and the thickness uniformity required in the film.

Resistivity heated refractory metals, such as tungsten, tantalum, and molybdenum in the form of wires and foils are probably the most widely used source heaters. Figure 1 shows a few of the wide variety of shapes that are available. The wire filaments, cone baskets, and coils can be either single or multiple strands of wire with a total thickness of about 0.5 mm. Dimpled foils and boats are about 0.13 mm thick and about 10 mm wide. Table I lists which refractory metals can be used as source heaters for the various elements.

It is important that the moltant element wet the filament to improve thermal contact and thus temperature uniformity. However, as the evaporant melts and begins to wet the filament, the filament temperature is decreased. Temperature control at this stage is important for a number of reasons. First, if the temperature is increased too rapidly, wetting may not occur fast enough. Unless wetting occurs, the evaporant forms small spheres and falls from the wire basket. Second, if the evaporant is very gassy and the temperature is increased too rapidly, the rapid evolution of the trapped gas can cause a bubbling or "spattering" of the source re-

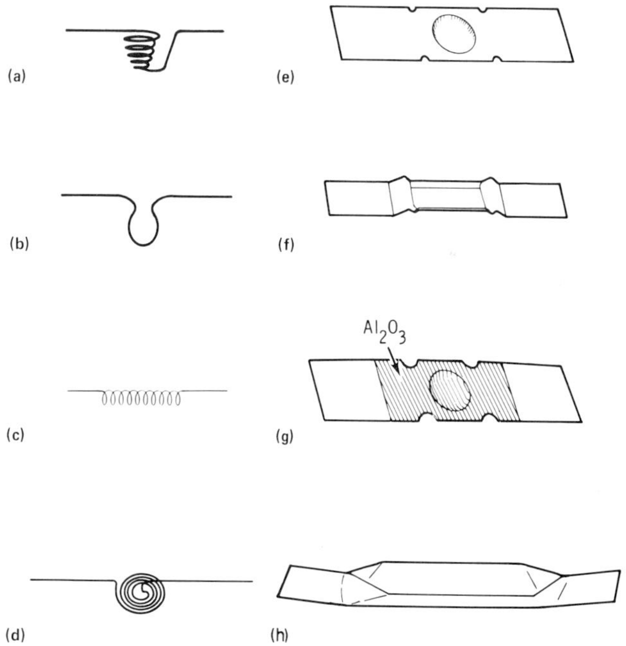

FIG. 1. Wire and foil source heaters. There are many different shapes of wire and foil source heaters. Some common ones are: (a) conical wire basket, (b) hairpin, (c) ten turn wire helix, (d) flat spiral, (e) dimpled foil, (f) shaped foil (to minimize warping), (g) Al_2O_3 coated dimpled foil, (h) boat.

sulting in small globules striking the substrate. This problem is not unique to wire filaments and will be discussed in "Deposition Procedures" (16.2.6). One of the major limitations of the wire baskets, cones, or spirals is the limited quantity of evaporant they can support. The foil has a large capacity, but it too has limitations. As the amount of evaporant is decreased, the temperature of the filament tends to increase, which increases the rate of deposition. This is not a problem if only the total film thickness is important, and not the deposition rate. However, if deposition rate is important, then rate control must be incorporated in the deposition circuitry. Crucibles made from refractory metal, refractory oxide, boron nitride, or carbon can hold up to a few grams of source material. These are particularly convenient for large batch applications or if very thick films are required. The deposition rate can be kept constant very easily since it does not depend on the quantity of material in the crucible.

Generally, crucibles are heated indirectly. Often metal crucibles are supported by a refractory wire filament basket [Fig. 2(a)]. The crucible is maintained at the desired temperature eliminating evaporant contact with the higher temperature filaments. Refractory oxide or carbon crucibles can be supported in the same way. However, some oxide and ceramic crucibles are formed around filament wires [Fig. 2(b)]. It is also possible to use rf heating of the evaporant in a refractory compound crucible. By

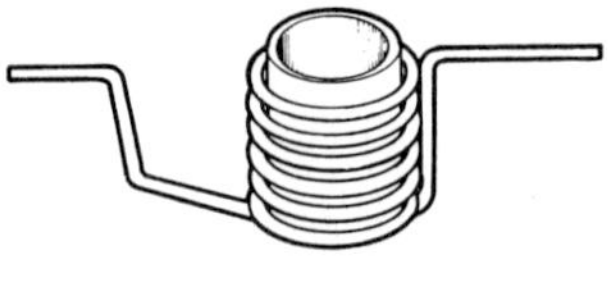

(a)

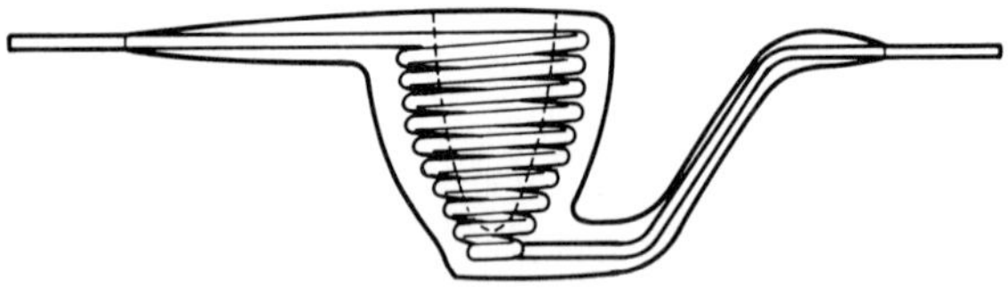

(b)

FIG. 2. Crucible sources. Crucible sources can hold up to a few grams of the source material and are convenient for repetitive, batch or thick film applications. (a) Crucible supported by a refractory wire filament basket. (b) Ceramic crucible formed around a refractory wire basket.

proper source design[5] it is possible to control temperatures to obtain high deposition rates without the problem of droplet splatter. As with wire filament and foil source heaters, it is important to use the proper crucible material to minimize reaction with the evaporant. This is particularly true when using refractory compound crucibles. Table I gives some of those crucible materials that are most useful for different elements. Glang[4] and Kohl[6] present a review of the properties of different types of refractory oxides that can be used to make crucibles.

Crucible sources are generally used for materials that have high sublimation rates. This is particularly true for those materials and compounds in the form of powders. The two major exceptions are for those materials that can be formed into a solid rod, such as chromium, and into wire, such as titanium. The rod is heated indirectly with a tungsten or tantalum heater until the required vapor pressure is reached. By proper shielding, large surface areas can be made to sublime producing a large capacity source with a high deposition rate. The deposition rate can easily be controlled by adjusting the power to the indirect heater and since the surface area does not change, the rate is constant for large periods of time. This type of source is ideal for large batch applications when rate control is important. Wire sources can be resistively heated to the desired temperature.

One of the most universal deposition sources is the electron beam source. There are three major types of electron beam sources, classified by the way the electrons are accelerated to the source material, but each type can have many different geometries. Figure 3 shows one geometrical configuration of each of the major classifications. The three types have been classified (16.4) as work accelerated [Fig. 3(a)], self-accelerated [Fig. 3(b)], and bent beam [Fig. 3(c)]. There are also various ways of supporting the source material but the most widely used is a water-cooled copper crucible. There is essentially no interaction between the evaporant and the support in this technique. The electron beam energy is almost completely converted to heat in the impingement volume. With sufficient energy this volume becomes liquid, while the portion of the source material in contact with the water-cooled crucible remains a solid. It is possible to produce very pure films in this manner. With presently available power supplies it is possible to obtain melting temperatures greater than 3000°C. This means that electron beam techniques can be used to deposit the refractory metals and carbon which are

[5] I. Ames, L. H. Kaplan, and P. A. Roland, *Rev. Sci. Instr.* **37,** 1737 (1966).

[6] W. H. Kohl, "Handbook of Materials and Techniques for Vacuum Devices." Reinhold New York, 1967.

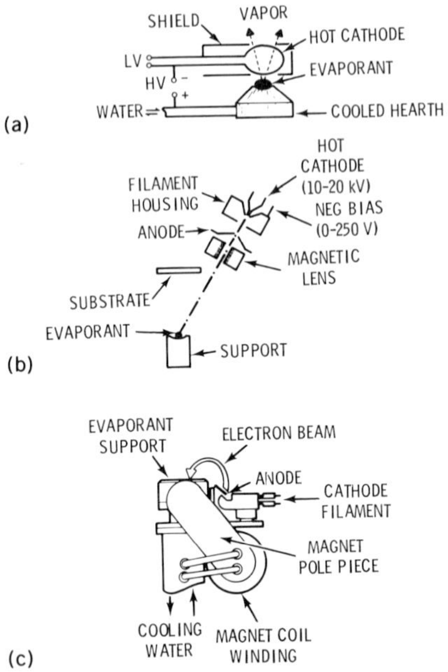

FIG. 3. Electron beam sources. Examples of geometrical configurations for each of the three major classifications of electron beam sources, (a) work accelerated, (b) self-accelerated, and (c) bent beam. (Reproduced with permission from McGraw-Hill. See Ref. 4.)

very difficult to evaporate by other methods. Although electron beam thin film evaporators are universal since they can be used with all evaporative materials, they are not as widely used as refractory wire, foil, or crucible sources. The primary reasons are that they are more complex, more expensive, and difficult to control. For example, much more experience is required to control and change deposition rates with an electron beam source than with a crucible source.

Many compounds and alloys cannot be evaporated from the source heaters discussed above because their composite elements have different vapor pressures. Flash evaporation is a technique that does allow multi-element material to be vacuum evaporated. The source consists of a refractory wire, strip foil, boat, or crucible at such a temperature as to volatize the constituent with the lowest vapor pressure. Only small amounts of the evaporant, which is in the form of a fine powder, granules, or fine wire, are placed in contact with the high temperature strip. Each particle is completely evaporated so the effects of fractionation are minimized. Some of the devices sprinkle the granules or feed the evaporant continuously onto the hot source and so become quite complicated. In addition to the complex mechanism required for flash evaporation sources, the gas released during evaporation further complicates this type of film deposition. It is not possible to outgas and process the source material before evaporation, thus impurities are incorporated in the film. However, even with these problems, flash evaporation is used because it is the only way some materials can be vapor deposited directly. Glang[4] presents a good review of different flash evaporation techniques.

16.2.3.3. Substrate and Substrate Holders. The word substrate is the term used to identify the material on which a film is deposited. In general, a substrate can be anything that is compatible with the deposition process. For vacuum evaporation this means that the substrate material has to be vacuum compatible; that is, it should not decompose or volatize during vacuum processing. To achieve desired film parameters it may be necessary to elevate the substrate temperature, so the substrate must also be stable at the required temperature. Other substrate requirements depend on the application of the film.

The most common problems encountered in vacuum deposition are film growth, film adhesion, and film appearance. These film parameters are closely related to substrate cleanliness and smoothness. Features on a substrate such as scratches, hillocks, pits, irregularities, and fingerprints stand out clearly after film deposition. If the film is for a passivation layer, such features may not matter; but if the film is the first step in a procedure to fabricate a device, then irregularities can influence the device's performance. Oils, greases, fingerprints, airborne particulates, and dust must be removed from a substrate for most film applications. Particulates and dust will cause pinholes (micrometer size and larger) which can strongly influence successful use of the film, especially in device fabrication. Oils and grease will affect and control film adhesion, growth, and reproducibility. How clean is clean enough depends on the application, and various techniques for cleaning the substrate might have to be tried to find the optimum technique for the given application. Cleaning techniques vary depending on the type of substrate used. Glass substrates can be cleaned by one of a number of glass cleaners,[7,8] solvent cleaning with ultrasonic agitation, or heating. The state of the substrate prior to use determines which methods are used. One cleaning sequence that has been found satisfactory is the following: if there are no metallic impurities to be removed, wash the substrates in a strong detergent and rinse with water. Using ultrasonic agitation, rinse the substrates in one bath of triclorethylene and two baths each of acetone, methanol, and ethanol. The time in each bath should be 2–5 min. Metal substrates might require solvent cleaning and/or electropolishing. Single crystal substrates can

[7] R. Brown, "Handbook of Thin Film Technology" (L.I. Maissel and R. Glang, eds.), Chapter 6. McGraw-Hill, New York, 1970).

[8] F. Rosebury, "Handbook of Electron Tube and Vacuum Techniques." Addison-Wesley, Reading, Mass. 1965.

be prepared in a number of ways. Some, such as NaCl, can be cleaved either in air or in the vacuum system. Metal single crystal substrates are often cleaned in a vacuum system by either high temperature heating or a combination of ion sputter cleaning and high temperature annealing. Many expitaxial growth studies are done on single crystal surfaces prepared in these ways.

Once substrates are clean, they normally have to be handled during mounting into the deposition chamber or stored for a while before use. Extreme care and "clean room" techniques should be used in the handling and mounting of substrates even though a clean room may not be available. In general, this means that clean gloves and clean tools (tweezers) are used to handle the substrates. Also, substrates are stored under controlled conditions such as in a laminar clean hood, in clean polyethylene bags, or in a dust-free dessicator. Keeping the substrate dust free is probably most difficult when loading in the deposition chamber. Keeping dry nitrogen gas flowing through the system will help prevent particulates in the air, dandruff, etc., from entering the system. The vacuum system itself can be the cause of particles on substrates. If a deposition system is used frequently and seldom cleaned, accumulated film on supports and base plates can flake. The pumpdown and backfill of the vacuum system can move these flakes randomly over the system. It is thus important to stress cleanliness and careful handling during all phases of a film deposition process.

Many other substrate characteristics might be important for a particular application. A good review of these is given by Brown.[7] Only a few will be mentioned here. If the film application requires large temperature cycling, then the thermal coefficient of expansion of the film and substrate must match; otherwise, the film might separate from the substrate. They must also match if the film is to be deposited on heated or cooled substrates. In this case it is important that the substrates have good thermal conductivity so the surface exposed to the vapor will reach the desired temperature. Finally, the chemical properties of substrates can be very important, not only for cleaning procedures but also if patterns must be etched in film after deposition. Many device fabrication techniques require chemical processing, etching, anodization, etc., and the substrate must be inert to the chemical used in these techniques.

To deposit films at ambient temperature only a pedestal support for the substrates is necessary. However, if the substrate must be heated or cooled during deposition, the substrate holder becomes more complex. It is fairly easy to design a suitably sized substrate holder that incorporates heater wire or cooling channels connected to the appropriate feedthrough for cooling. Special care should be made to have the side holding the substrate as smooth and flat as possible to attain good contact between the holder and substrate surfaces. It may be necessary to incorporate spring clips or epoxy to hold the substrate in contact with the holder. Unless good thermal contact is achieved and maintained, it is difficult to know the substrate surface temperature by monitoring the holder temperature. Monitoring substrate surface temperature is often impossible or extremely difficult.

16.2.4. Geometrical, Mechanical, and Electrical Configurations of Vacuum Evaporation Systems

16.2.4.1. Geometrical Configurations. The geometry of a vacuum evaporation system, which is the relative locations of sources, substrates, shutters, and thickness monitors, can vary over a wide range of possibilities. However, there are three basic geometrical configurations: source above the substrate, source below the substrate, and special substrate geometries in which the source is arranged to take advantage of special symmetry. Wire filaments and conical baskets allow the substrates to be either above or below the source. Figure 4 shows a typical bell jar arrangement with the source above the substrate. In such geometry the substrate needs only to rest on the base plate or pedestal if deposition can be performed at ambient temperature. Crucibles, metal foils, and electron beam sources employing water-cooled hearths normally require the substrate above the source. Figure 5 shows a typical arrangement with a simple substrate support. This type of substrate holder is easy to fabricate and works well if the deposition can be done at ambient temperature. If elevated temperatures are needed, then there is a heater element in back of the support plate and some type of retaining clips or spring to hold the substrate. Figure 6 shows a source which takes advantage of the special geometry. In this case, the problem was to gold coat the inside of a cylinder. A tungsten wire filament was fabricated to hold the gold along the length to produce a fairly uniform film. For very odd shaped substrates that must be coated on all sides, it is often necessary to rotate the object in the vapor stream.

The size of the substrate and the required film thickness uniformity will determine the details of the type and number of sources required for a particular application. There are many shapes of wire conical baskets, foil boats, and crucibles, each of which produces a different spatial distribution on film. Of course, the source–substrate distance will affect the thickness uniformity. The greater the distance between source and substrate, the more the source looks like a point source which has a spherically symmetric spatial distribution. Holland[3] and Glang[4] discuss various

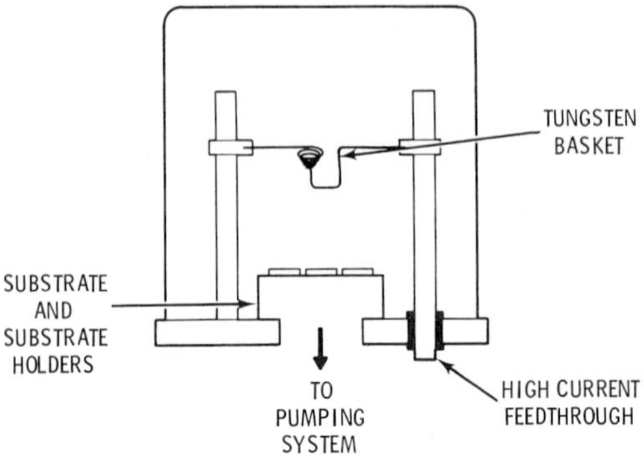

FIG. 4.

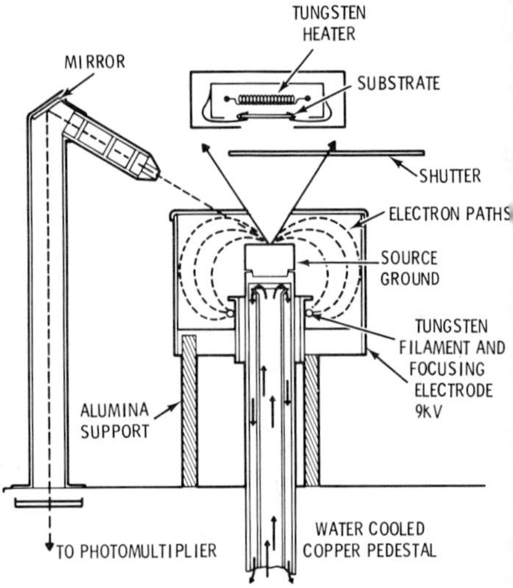

FIG. 5.

FIG. 4. Bell jar evaporation system. Typical geometrical arrangement in a simple bell jar evaporation system. The source is above the substrate which rests on a pedestal on the base plate.

FIG. 5. Electron beam source with water cooled hearth. Schematic of an electron beam source showing the water cooled hearth, shutter, and substrate support with a simple heater assembly. Note that the substrate is above the source and requires special supports. The focusing electrode acts as a shield to limit the direction of film deposition. The mirror is used to monitor the source temperature. [After B. A. Unvala, *Le Vide*, **104,** 110 (1963).]

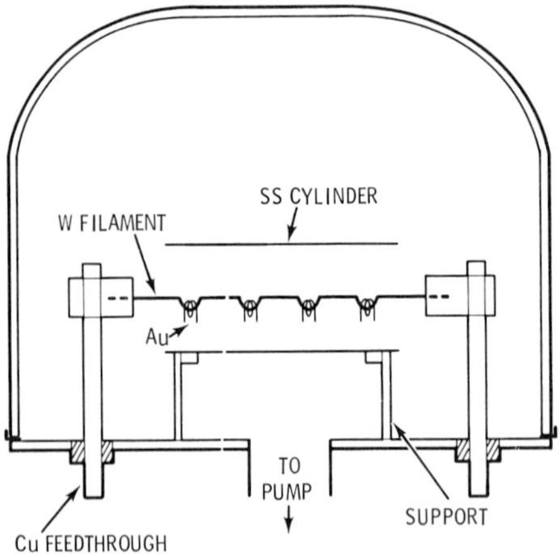

FIG. 6. Film deposition in a cylindrical geometry. Schematic of a system to deposit a uniform gold film on the inside of a cylinder. A series of hairpin sources were formed along a tungsten filament to take advantage of the special geometry.

theoretical approaches to determining what thickness distribution is expected from different sources. Some manufacturers of evaporation sources have measured thickness distributions with different source geometries. If thickness uniformity is important, then the source geometry that will produce the desired result must be considered.

There are situations when it is necessary to restrict the deposition to a portion of a system, to deposit on a portion of a substrate, or to deposit a pattern on a substrate. The first two of these objectives can be met by placing a shield with apertures around the source. The third can be accomplished by placing a metal mask with openings of a desired shape on the substrate. A mask need not be complicated but could consist of a wire above the substrate to produce a slit in the film. An example of shields used to control the area of deposit is shown in Fig. 7. The area exposed to deposition is the sample and deposition monitor only.

A special source geometry occurs when multiple sources are used to deposit alloy films. Only binary sources are discussed here but, in principle, more than two sources can be used. The basic idea is to produce a binary alloy film using two sources each of pure elemental material. The temperature of each source can be controlled separately to produce a film of the desired composition. Figure 8 shows one such arrangement that produces a graded alloy over the length of the substrate. The position of the sources could be set to produce the same composition over the entire substrate. Each source must be calibrated and controlled separately for reproducible results. It is possible for each source to have its own deposition monitor so that the source parameters can be controlled separately during the film growth. One of the main problems with dual sources is to place shielding in a location such that the substrates are coated uniformly but do not contaminate each other.

16.2.4.2. Mechanical Configurations—Shutters. A shutter is a shield that is placed between the source and the substrate to allow processing of the source, *in situ* cleaning of the substrate or both procedures without contaminating the other. In general shutters are used whenever ultrapure films are required. The source can be thoroughly outgassed and the high vapor pressure source impurities evaporated prior to exposure of the substrate. Some materials tend to bubble and spatter if they are very gassy or if heated too rapidly. A shuttered substrate can be protected during initial source heating. It is also possible to adjust the source temperature and hence the evaporation rate with the shutter closed so that the film can be grown with preadjusted deposition parameters. Keeping the shutter open for a given period of time with a known deposition rate is a good way of producing a film with desired thickness. Use of a shutter makes *in situ* substrate cleaning, either by heating or ion

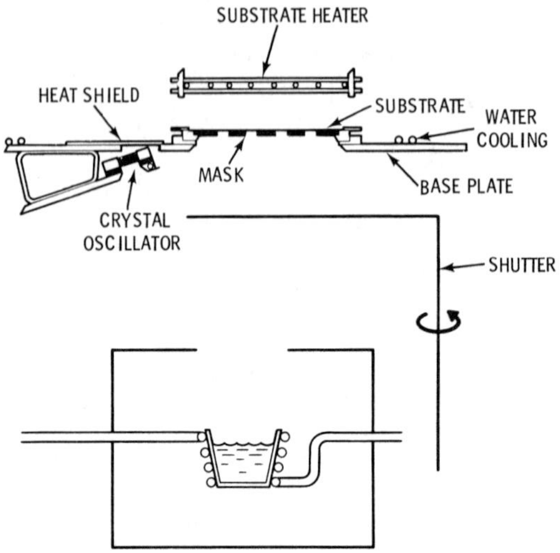

FIG. 7.

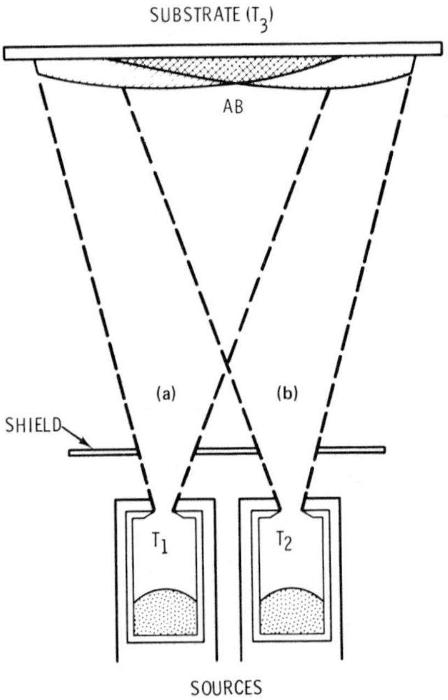

FIG. 8.

FIG. 7. Simple shield and mask arrangement. Schematic of a crucible source surrounded by a shield to limit the area of deposits to the crystal oscillator thickness monitor and the substrate. The shutter allows processing of the source and adjusting of the deposition rate prior to depositing on the substrate. Note that the shutter does not cover the thickness monitor. The substrate is shown with a mask to obtain a film of a specific geometry. Also shown is a radiant substrate heater and a heat shield to protect the thickness monitor.

FIG. 8. Binary evaporation source. Schematic of a binary source arrangement to produce a graded alloy over the length of the substrate. The temperature of each source T_1 and T_2 should be controlled separately. In addition the substrate temperature may need to be controlled and maintained at temperature T_3. (Reproduced with permission from McGraw-Hill. See Reference 4.)

cleaning, possible without contaminating the source. It should be noted that a shutter requires a vacuum motion feedthrough, either linear or rotary, in the system. Since this adds complexity and cost to a system, a shutter should be added only if the process demands it.

16.2.4.3. Electrical Configurations. All source heaters except electron beam and rf type rely on resistive heating of a filament or foil. Therefore, the only electronics needed are a high-current–low-voltage source, which may be either ac or dc, of sufficient current capacity to reach the required temperature. For single strand filaments the current required is 5–10 A; for foil boats the required current may be as high as 200 A. It is important that the electrical feedthroughs have the correct current ratings. If the current rating is too low, the feedthrough can overheat, cracking the ceramic-to-metal seal, and thus causing a severe vacuum leak and possible damage to personnel and apparatus.[9]

The electronics required for electron-gun devices are much more complex than for resistance heated sources. Voltage requirements for the different types of electron guns are shown in Fig. 3. These sources tend to be more difficult to operate and control than filament-type sources. The complexity of operation increases with the number of electrodes in the gun. Some guns have power ratings as high as 10 kW, so it is important to use electrical vacuum feedthroughs with the correct specifications to avoid failure.

16.2.5. Thickness Monitoring*

16.2.5.1. Source Precalibration. Source precalibration is the method of determining film thickness that requires the least equipment. Only a good balance, one that can measure 0.1 or 0.01 gm, is required. The procedure consists of weighing a substrate of known size before and after the deposition, which determines the mass Δm deposited on a known area A. If bulk density is assumed, then film thickness t is given by

$$t = \Delta m/\rho A, \tag{16.2.3}$$

where ρ is the density.

The calibration of the source can proceed in two ways. First, with the source–substrate geometry fixed, the source is loaded with a measured amount of material for deposition. Evaporation is allowed to continue until the source material is exhausted. Δm and thus t is obtained for a number of different source masses which produces a calibration curve. As long as the geometry of the system does not change and the source material is completely evaporated, a film of predetermined thickness can be grown. Of course, the accuracy of this monitoring method depends on producing films with bulk density. Also, it may be difficult to produce very thin films this way. Second, with a fixed source–substrate geometry and with the source operated at a given temperature, films can be made by depositing for different lengths of time. This will produce a calibration curve of thickness versus time. Besides having to assume that

[9] L. C. Beavis, V. I. Harwood, and M. T. Thomas, "Vacuum Hazards Mnual." American Institute of Physics, New York, 1975.

* See also Vol 11 of this series, Chapter 12.2 on thin film measurements.

bulk density films are produced, care must be taken to insure that the amount of evaporant in the source does not decrease too much. For example, as the material in a conical tungsten basket decreases, the temperature generally increases, thus changing the deposition rate. In both of these techniques it is important to clean the source of impurities before calibration or film growth is started.

16.2.5.2. *In Situ* Weighing. There are three devices for determining the mass of a deposited film in situ: a quartz fiber microbalance, an electromagnetic or current meter movement balance, and a quartz crystal oscillator. These devices can also be used to measure film deposition rate and thus can be used for process control. Vacuum quartz fiber microbalances have been used for mass measurement in many types of vacuum experiments and processes besides thin film deposition applications.[10,11] There are many different designs, and although some have a sensitivity of about 2×10^{-8} gm, most have a sensitivity in the range of 10^{-7} gm. Typically the microbalance vane (balance pan) has an area of 1 cm^2, which implies thickness sensitivites of $< 10^{-7}$ gm/cm^2 or < 1 Å for materials with a density of 10 gm/cm^3. Despite this high sensitivity and no significant limitation on the maximum mass that can be measured, the quartz fiber microbalance is not generally used for routine thin film applications. Because of its construction and high sensitivity, the quartz microbalance is very sensitive to external vibration and thus requires considerable experience to use effectively.

Galvanometer or current meter movement microbalances have a sensitivity of 10^{-7}–10^{-6} gm and are much easier to use than quartz fiber microbalances. The usual construction of these balances does not allow their use in ultrahigh vacuum systems. However, for many thin film deposition applications these types of microbalances are convenient and can be readily adapted to process control. Process control consists of monitoring and adjusting deposition rate and film mass. The deposition process can be stopped when some predetermined mass has been deposited. Figure 9 shows schematics of a galvanometer type microbalance and a quartz fiber microbalance.

The AT cut quartz crystal oscillator is the easiest and most convenient vacuum microbalance to use. An AT cut quartz crystal can be used as a microbalance because the resonant frequency of the crystal plate is changed when mass is added to the plate. It can be shown[12] that the reso-

[10] "Vacuum Microbalance Techniques," Vols. 1–5. Plenum Press, New York, 1961–1965.

[11] "Progress in Vacuum Microbalance Techniques" Vol. 1. (T. Gast and E. Robens, eds.). Heyden and Son, London, 1972.

[12] R. A. Heising, "Quartz Crystals for Electrical Circuits. Van Nostrand, New York, 1946.

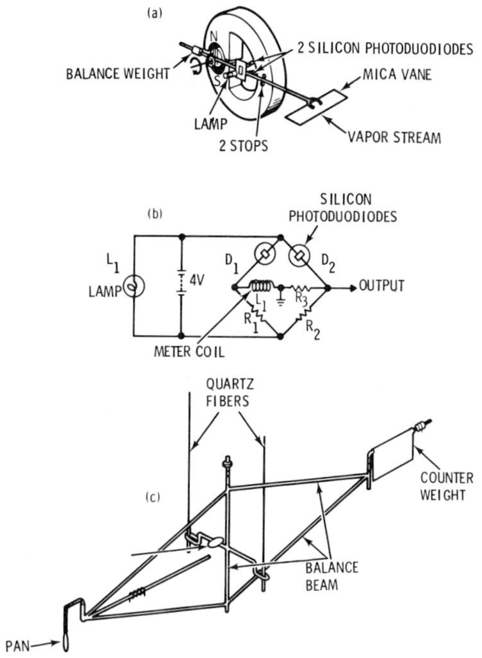

FIG. 9. Vacuum microbalances. Schematic diagrams of galvanometer type and quartz fiber type vacuum microbalances. (a) Drawing showing the coil, magnetic and collector vane, (b) the circuit diagram of the galvanometer movement [after R. E. Hayes and A. R. V. Roberts, *J. Sci. Instr.* **39,** 428 (1962)], (c) drawing of a quartz fiber microbalance showing the quartz fibers, balance beam, counter weight, mirror, and collector pan. (After W. Krüger and A. Scharmanns[11].)

nant frequency of an infinitely large quartz plate vibrating in the thickness shear mode is

$$f = v_{tr}/2t = N/t, \tag{16.2.4}$$

where v_{tr} is the transverse velocity in the plate, t the thickness of the plate, and $N = 1670$ kHz mm for AT cut crystals. This equation holds very well for plates of finite size if the ratio of the diameter of the plate to the thickness is greater than 20.[13] Using the definition of density $\delta_q = M/V = m/At$ and Eq. (16.2.4), it follows that $df/f = -dt/t = -dM/M$, and

$$df = -(f_0^2/\delta_q N)(dM/A) = -\omega\Psi, \tag{16.2.5}$$

where δ_q is the density of the quartz, f_0 is the resonant frequency before mass loading, ω is the sensitivity for mass determination, and Ψ is the mass coverage in grams per square centimeter. The minus sign means that the frequency decreases as the mass increases. Equation (16.2.5) also shows that Δf is linear with ΔM. Since in this derivation only the parameters of the quartz plate have been used, ω depends only on the properties of the quartz. This is clear once it is realized that there is an

[13] R. Beckman, *J. Sci. Instr.* **29,** 73 (1952).

antinode at the surface of a plate vibrating in the thickness shear mode. This means that material on the surface of the vibrating plate affects the resonant frequency only by virtue of its mass and not by its elastic properties. It follows that a thin layer of a foreign material on the surface of the vibrating plate should change the resonant frequency in the same way as an equivalent mass of quartz. The experimental sensitivities have been measured by many investigators[14,15] and have been found to agree with the theoretical sensitivity ω to better than a few percent. As an example, the theoretical sensitivity of a 10.7-MHz quartz crystal is 3.8×10^{-9} gm/cm^2/Hz and a 6-MHz crystal is 1.25×10^{-8} gm/cm^2/Hz. Figure 10 shows the frequency change of a 10.7-MHz crystal as a function of time when germanium was deposited on it at a constant deposition rate. The linearity of Δf with Δm is clearly demonstrated. The total frequency change during this deposition was 8800 Hz and corresponds to 3.4×10^{-5} gm/cm^2 of film. The slope determines the deposition rate which was 2.7×10^{-7} gm/cm^2/sec. Commercial quartz microbalances are available, but these balances are also simple to fabricate. A block diagram of the electronics required is shown in Fig. 11. Examples of crystal holders are shown in Fig. 12. It is important that the holders keep the quartz plate from moving, place little or no stress on the crystal, and allow easy assembly and disassembly.

There are two general areas that might be a problem when using a quartz crystal microbalance and should be considered before choosing a crystal of a given resonant frequency or a holder design. The first one is: over what range is the frequency change linear with mass loading. Behrndt[16] shows that if a deviation of 1% from linearity is permissible, then the change in frequency should not exceed 0.5% of the resonant frequency. For the 10.7-MHz crystal discussed above, this corresponds to a maximum frequency change of 53.5 kHz. For a material that has a mass of about 4×10^{-8} gm/cm^2, then one monolayer would produce 10 Hz frequency change and about 5000 monolayers can be measured within 1%.

The second area becomes important if only a few monolayers are to be measured. Thermal shock and temperature changes also cause a frequency change of an AT cut quartz crystal. The frequency change due to a thermal shock can be in the opposite direction than that due to mass loading. During any deposition a quartz microbalance will be exposed to the radiation from the source. Figure 13 shows the frequency variations

[14] G. Sauerbrey, *Z. Physik* **155,** 200 (1959).

[15] H. L. Eachbach and E. W. Kruidhof, "Vacuum Microbalance Techniques, Vol. 5, p. 207. Plenum Press, New York, 1966.

[16] K. H. Behrndt, in "Physics of Thin Films" (G. Hass and R. F. Thun, eds.), Vol. 3, p. 23. Academic Press, New York, 1966.

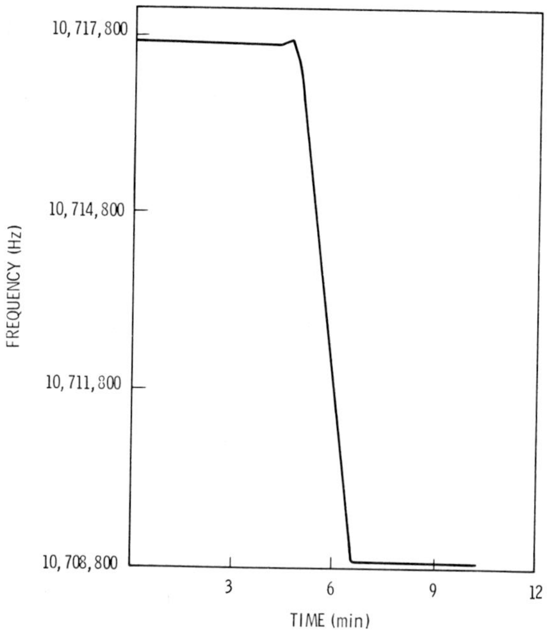

FIG. 10.

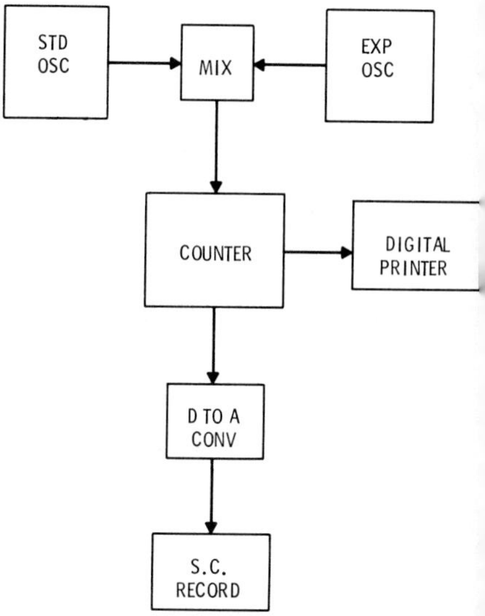

FIG. 11.

FIG. 10. Quartz crystal microbalance: frequency vs deposition time. The frequency of a 10.7 MHz quartz crystal was measured as a function of time during the vacuum deposition of germanium. These data show that when the source temperature is constant, constant deposition rate, the change in frequency is linear with time, and therefore linear with mass change. The total frequency change in this example was 8800 Hz which corresponds to a film thickness of 3.4×10^{-5} gm/cm^2.

FIG. 11. Block diagram of vibrating quartz crystal microbalance. The electronics necessary to operate a quartz crystal microbalance is fairly simple. As shown in the block diagram, two quartz crystal oscillators are required with a mixing circuit and a frequency counter. The stability of the oscillator circuits will depend on the application requirements.

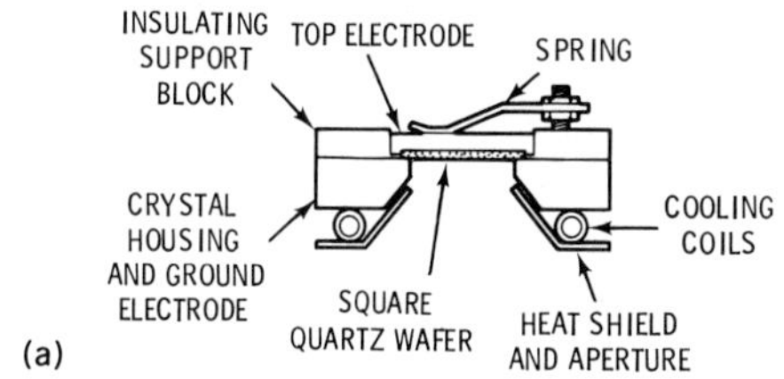

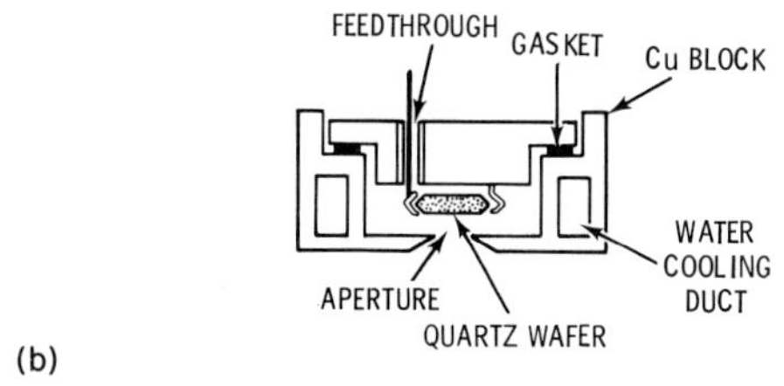

FIG. 12. Quartz crystal holders. Schematic drawings showing two types of holders for quartz crystal. It is important that the holders place as little stress as possible on the quartz plates, hold the plates stationary, and yet allow easy assembly and disassembly. For many applications it is necessary to maintain a constant temperature. In these holders this is accomplished by water cooling and shields. (a) Holder for a square quartz wafer, (b) holder for a round quartz wafer. (After R. Glang.[4])

that occur when a 10.7-MHz crystal is exposed to a hot, empty spiral filament; that is, no material was being deposited. The frequency excursion was 120 Hz in this experiment. Notice that after 5 min the frequency stabilized and the reverse frequency shifted when the hot filament was turned off. This problem can be minimized in a number of ways. One way is to place the quartz plate against a water-cooled plate which will help keep the temperature constant. A second way is to expose two quartz crystals to the same thermal shock treatment but deposit on only one of the crystals.[17] Figure 14 shows how the beat frequency changed during a deposition using this two-crystal technique. There is no frequency excursion at the onset of deposition, therefore very small (0.1 of a monolayer) deposits can be made. The frequency variations at the end of the deposition are thought to be due to the thermal energy carried to one of the crystals by the evaporant. Another technique to avoid thermal problems with quartz crystal microbalances is to shutter the substrates and deposit on the quartz crystal until a stable and constant frequency change is obtained. The shutter can then be opened to deposit the desired material thickness on the substrate.

In principle, the more precisely Δf can be measured for a given resonant frequency, the smaller the mass changes that can be measured. Warner and Stockbridge[18] have developed a system in which frequency can be measured with a precision of one part in 10^{10}. All variables that affect the frequency are required to be carefully controlled. In particular the temperature of the quartz plate was held constant to 0.01°C. This system has a mass sensitivity of about 10^{-11} gm/cm^2 which is less than 1/1000 of a monolayer of most elements.

It should be emphasized that quartz crystal microbalances are readily adapted to monitor and control thin film deposition. Not only can the total thickness be controlled, but by differentiating the Δf versus time signal, the deposition rate can be monitored and controlled. Overall, the quartz crystal microbalance is one of the simplest techniques for making mass measurements inside of a vacuum system and is usable in ultrahigh vacuum and in bakeable systems. There have been many applications and refinements of quartz crystal microbalances and the reader is referred to the literature for some of the most recent advances.[19–21]

[17] M. T. Thomas and J. A. Dillon, Jr., *Trans. Intern. Vac. Congr., 3rd,* Vol. 2, p. 343. Pergammon Press, New York, 1966.

[18] A. W. Warner and C. D. Stockbridge, "Vacuum Microbalance Techniques," Vol. 2, p. 71, 93; Vol. 3, p. 55. Plenum Press, New York, 1962, 1963.

[19] E. P. Eer Nisse, *J. Vac. Sci. Technol.* **12,** 564 (1975).

[20] C. Lee, *J. Vac. Sci. Technol.* **12,** 578 (1975).

[21] F. Wehking, K. Hartig, and R. Niedermay, "Progress in Vacuum Microbalance Techniques" (T. Gast and E. Robens, eds.), p. 25. Heyden, London, 1972.

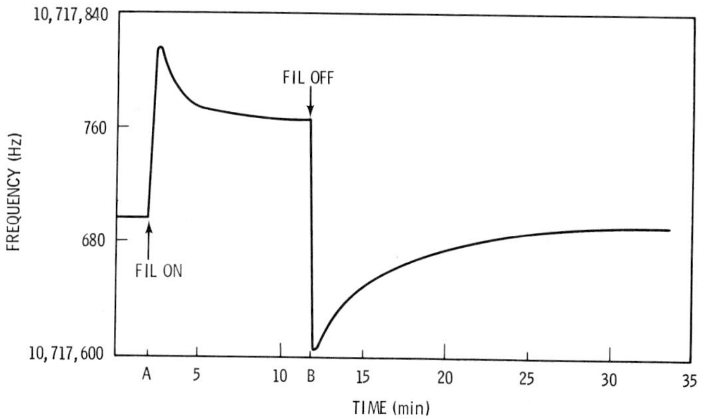

FIG. 13. Quartz crystal microbalance: thermal shock. Frequency of a vibrating quartz plate versus time showing the effect of thermal shock. At time A a tungsten filament was turned on, at time B it was turned off.

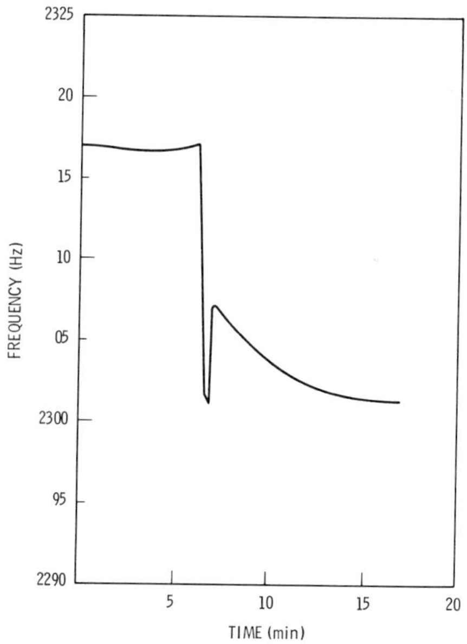

FIG. 14. Data from double quartz crystal microbalance. Frequency change recorded during film deposition when two quartz crystals were exposed to the same environment and thermal radiation. There is no frequency variation at the onset of deposition, see Fig. 13. It is possible to deposit films as thin as 0.1 monolayers reproducibly. The frequency excursion at the end of the deposit is due to the thermal energy carried to one of the quartz crystals by the evaporant.

16.2.5.3. Other Techniques. There are many other techniques for measuring film thickness and monitoring deposition rate. Glang,[4] Behrndt,[16] and Gillespie[22] give good descriptions and comparisons of various techniques. Table II lists some of the most widely used methods along with some of their properties.

16.2.6. Deposition Procedures

Although many different applications of evaporated thin film have their own film requirements, the actual procedures for producing a film are simple and, to a large extent, independent of the material or the type of source. However, compounds requiring reactive evaporation or flash evaporation are exceptions. The reader is referred to Glang[4] for details of procedures using these two techniques.

The first step in a film deposition procedure is to itemize the requirements of the film. The requirements are determined by the application and thus establish the deposition requirements. For example, the deposition parameters for a gold electrode on a transducer are quite different from those for a silver film that will be used for gas adsorption and work function experiments. The material to be evaporated determines to a large extent the type of source heater that must be used. In addition, if a filament basket, foil boat, crucible, etc., should be used, the film thickness and uniformity will help specify the source configuration.

After all important deposition parameters have been established, the vacuum system is loaded with the source material and the substrates. This step should be done under as clean and dust-free conditions as possible. Even though the substrates have been cleaned, degreased, and stored in a dust-free environment, the handling of the substrates during the loading process, the closing of the vacuum system, and the initial pumpdown can easily contaminate them with dust particles, and, of course, dust can produce pinholes in the film. Often the vacuum system that must be used for film deposition is not in a clean room but, rather, in the corner of some laboratory where the environment is difficult to control. In these cases, dry nitrogen flowing through the vacuum system or over the substrates will help to keep particulates from settling on them. It is also important to wear clean gloves and a headcover and to handle substrates and source material with degreased and clean tools. Before the system is closed all electrical connections should be checked to ensure electrical continuity.

Once the system has been evacuated and a suitable pressure

[22] D. J. Gillespie, "Measurement Techniques for Thin Films" (B. Schwartz and N. Schwartz, eds.), p. 102. The Electrochemical Society, New York, 1967.

TABLE II. Methods for Thickness and Deposition-Rate Measurements[a]

Method	Film material	Thickness or rate maximum	Sensitivity	Uhv			Remarks
				Precision	Application	Automation	
Ionization-gauge monitor	All	None	1 Å/sec	1–5%	Yes	Yes	Compensation for high residual gas pressure required
Current-meter movement	All	None	1 μg/cm^2	2%	No	Yes	Usable for rate or thickness measurement
Momentum meter	All	None	1.6 ± 0.2 Å/min deg		Yes	No	Slow response, but simple and robust device
Colors of films	Dielectric	~1 μm	100–200 Å	2–3%	Yes	No	Subjective method
Photometric	Metal	1000 Å	?	?	Possibly	No	Little experience with method.
	Dielectric	Many μm	Up to λ/300	1%	Possibly	No	Well suited for control of multilayers to λ/4
Polarimetric	Dielectric	Many μm	~10 Å	?	Possibly	No	Requires extensive calculations and preceding measurements
Interferometric	Dielectric	Several μm	λ/200	?	Possibly	No	Spacer layer in interference filter to 0.4%
Resistance	Metal	~1 μm	?	5%	Possibly	Yes	Can be precise, depending on film material and precautions
Capacitance	All	?	?	?	Possibly	Possibly	Methods rarely used, not enough information to evaluate them fully
Eddy-current damping	Metal	Several 1000 Å	?	?	?	?	
Crystal oscillator	All	A few μm	$\ll$1 Å/cm^2	1%	Yes	Yes	Easy to build and operate in spite of high sensitivity
Microbalance	All	None	<1 Å/cm^2	1%	Yes	Possibly	High-sensitivity balance frequently home built

[a] Reproduced from K. H. Behrndt in "Physics of Thin Films" (G. Hass and R. E. Thun, eds.), Vol. 3. Academic Press, New York, 1966.

attained—normally the base pressure of the system—the source is ready for processing. Processing consists of slowly heating the source so the evaporant is allowed to outgas. If there is a shutter in the system, it should be covering the substrates during source processing. Very gassy materials will cause the system pressure to rise rapidly. For clean films it is important to have base pressure before starting a deposition. The source temperature increases until the material in the source melts and wets the source supports. It is very important to use dark, eye protective glasses; when viewing a glowing filament, cobalt-colored glass is a good dark glass. Further increase in source temperature will cause the material to evaporate, and a layer will start to deposit on the substrate. Source bubbling and spattering can be minimized if the temperatures are increased slowly. If the system has a shutter, it can be opened after a few seconds of deposition to expose the substrates. In those situations where the substrate must be at some elevated temperature, it can be heated to the desired temperature during source processing. Once the film of the correct thickness is deposited, the source and substrates are allowed to cool to room temperature before the vacuum system is opened.

The actual process of depositing a film is very simple. Most of the labor occurs in deciding what film parameters are needed for a particular application and then discovering what deposition parameters in a given system will produce the desired results.

16.3. Vacuum Sputter Deposition

16.3.1. Simple Theory

Sputtering is the removal of atoms from a surface by the transfer of momentum and energy from an incident atom or ion to the atoms of the material being irradiated. Details of the slowing down processes of an incident particle on a material are extremely complex. Atoms in the material are scattered from their original location and sequences of collision take place which result in a momentum reversal so that surface atoms are ejected. In a crystalline or polycrystalline solid these slowing down processes cause lattice damage as well as implantation of the incident particles, which, in turn, cause the properties of the solid in the region to change.

No general or universal theory of sputtering from crystalline or polycrystalline solids exists. Sigmund[23] has published a theory of sputtering from amorphous solids based to some extent on Boltzmann's transport

[23] P. Sigmund, *Phys. Rev.* **184**, 383 (1969).

equation. A sputtering theory based on two body collisions has been developed by Thompson.[24] There also are many phenomenological theories or rules to describe the sputtering of a solid or group of solids by a given incident ion in a restricted energy range, many of which are based on the fact that the interaction between ions and atoms can be described by a hard sphere model for ion energies below about 50 keV.

The most significant parameter measured in sputtering experiments is the sputtering yield or sputtering rate defined as the number of atoms ejected per incident particle. It is the sputtering yield that a successful model must predict. For hard sphere collisions the maximum energy transferred in a collision is

$$E_T = [4M_1M_2/(M_1 + M_2)^2]E,$$

where E is the energy of the incident particles, and M_1 and M_2 are the masses of the colliding particles. Most particles ejected from a solid during sputtering come from the first few atomic layers of the solid. The sputtering yield S in the hard sphere model is proportional to the maximum energy transferred in a collision E_T, and inversely proportional to the mean free path λ of the incident particle in the solid. Therefore,

$$S = KE_T/\lambda,$$

where the mean free path is a function of the incident particle energy and the atomic density of the target, and where the parameter K is related to the binding energy of the surface target atoms. For a good review of sputtering theories, see Carter and Colligan,[25] Chapter 7 and Townsend *et al.*,[26] Chapter 6.

Figure 15 shows the variation of the sputtering yield of copper for argon ions of various energies, and the low energy sputtering yield for argon on copper is shown in Fig. 16. A number of important facts concerning sputtering are seen in these data. First, there exists a sputtering threshold which means that the impinging ions must have an energy greater than this threshold energy before target ions are emitted from the surface. Second, the sputtering yield increases more or less linearly until it reaches a broad maximum and then slowly decreases as the ion energy is further increased. The threshold energy and the sputtering yield maximum vary with material and with the mass of the bombarding ion. Table III lists the sputtering yields of different materials for argon ions at various energies.

[24] M. W. Thompson, *Phil. Mag.* **18,** 377 (1968).

[25] G. Carter and J. S. Colligan, "Ion Bombardment of Solids" Chapter 7. Elsevier, New York, 1968.

[26] P. D. Townsend, J. C. Kelly, and N. E. W. Hartley, "Ion Implantations, Sputtering and Their Applications," Chapter 6. Academic Press, London, 1976.

The variation of the sputtering yield with atomic number is shown in Fig. 17. This variation resembles the variation of the heat of sublimation with atomic number. The sputtering yield also varies as the angle the incident particles make with the surface normal is varied. Other variables affecting the sputtering yield are target crystallinity, target temperature, surface topography, electric field at the surface, ion current density, and background gas pressure, some of which are discussed in more detail below.

The sputtered or ejected atoms have much more energy than atoms that have been thermally evaporated. Figure 18 gives the energy distribution of copper atoms sputtered with 600 eV argon, helium, and mercury ions. The energy of the sputtered atoms is deposited onto the substrates and is one of the reasons that sputtered films adhere better than evaporated films. However, the main reason for better sputtered film adhesion is the chemically active species formed in the plasma.

Sputtered atoms are emitted over a range of angles. Atoms sputtered from single crystal surfaces are ejected preferentially along the close packed direction; this is seen by the patterns formed by the deposited film. Atoms sputtered from a polycrystalline surface have a cosine or near cosine angular distribution. There are a number of excellent review articles on sputtering which cover both theoretical and experimental studies and should be consulted by the reader desiring more details.[27–30]

16.3.2. Components for Vacuum Sputter Deposition

Systems used for sputter deposition consist of the following components: a vacuum system, a gas handling system, a target, substrates, and a plasma formation and ion acceleration system. Details of the vacuum system and the gas handling system are not discussed in this section. The most important vacuum parameter to consider in designing a sputter deposition experiment is the level of gas impurities that will be present during film formation. Which impurities are acceptable will, of course, depend on desired thin film parameters. In general, a low level of gas impurities is required, which implies that the vacuum system be free of real and virtual leaks and have a low outgassing rate when the walls are exposed to an energy flux. For many applications, a partial pressure of impurities between 0.01 and 0.1% of the pressure during film formation is necessary

[27] G. K. Wehner, "Advances in Electronics and Electron Physics" (L. Marton, ed.), Vol. 7, p. 239. Academic Press, New York, 1955.

[28] G. K. Wehner and G. S. Anderson, "Handbook of Thin Film Technology" (L. I. Maissel and R. Glang, eds.), Chapter 3. McGraw-Hill, New York, 1970.

[29] L. Maissel, "Handbook of Thin Film Technology" (L. I. Maissel and R. Gland, eds.), Chapter 4. McGraw-Hill, New York, 1970.

[30] W. D. Westwood, "Progress in Surface Science," p. 71. Pergamon, Oxford, 1976.

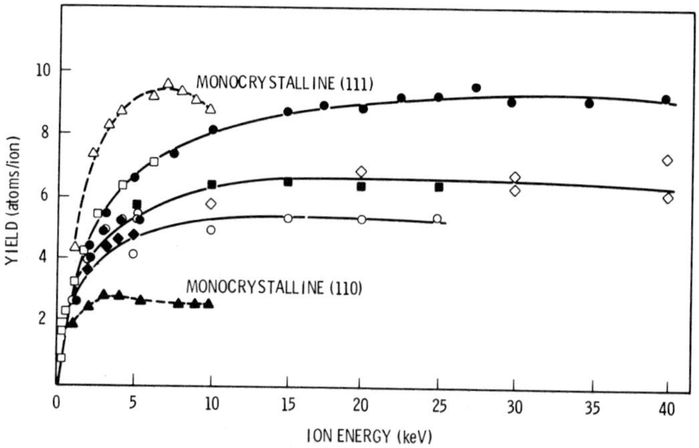

FIG. 15. Sputtering yield of Cu versus argon ion energy. Sputtering yields vary as a function of the ion energy. This figure is a compilation of data of the sputtering yield of Cu taken by various workers on different Cu samples as a function of argon ion energy. The high energy region is emphasized showing the broad maximum in the 10 keV region. The figure also shows the variation in sputtering yield with sample type. (Reproduced with permission from American Elsevier Publishing. See Ref. 25.)

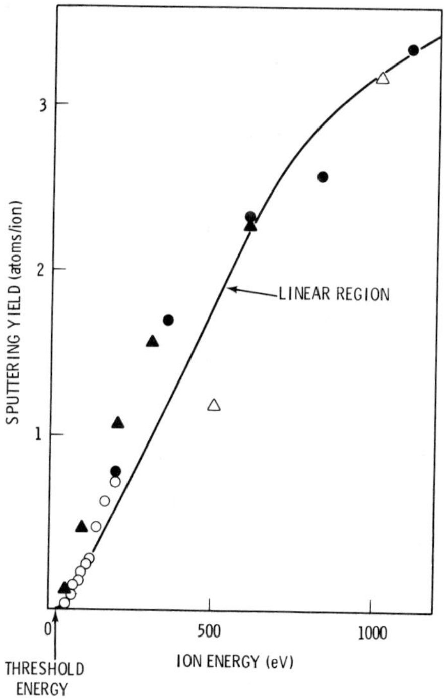

FIG. 16. Low energy sputtering yield of Cu versus argon ion energy. A compilation of data showing the sputtering yield of Cu as a function of argon ion energy in the energy region less than 1000 eV. This data shows a threshold energy and the linear increase of sputtering yield with ion energy. (Reproduced with permission from American Elsevier Publishing. See Ref. 25.)

TABLE III. Sputtering Yields of Various Materials in Argon*
(Bombarding Energy of Ar^+ in Volts)

Target	200	600	1,000	2,000	5,000	10,000	Reference
Ag	1.6	3.4	—	—	—	8.8	*a*, *b*
Al	0.35	1.2	—	—	2.0	—	*a*, *c*
Au	1.1	2.8	3.6	5.6	7.9	—	*a*, *d*
C	0.05†	0.2†	—	—	—	—	*e*
Co	0.6	1.4	—	—	—	—	*a*
Cr	0.7	1.3	—	—	—	—	*a*
Cu	1.1	2.3	3.2	4.3	5.5	6.6	*a*, *f*, *g*, *h*
Fe	0.5	1.3	1.4	2.0‡	2.5‡	—	*a*, *c*
Ge	0.5	1.2	1.5	2.0	3.0	—	*a*, *f*
Mo	0.4	0.9	1.1	—	1.5	2.2	*a*, *i*, *j*, *k*
Nb	0.25	0.65	—	—	—	—	*a*
Ni	0.7	1.5	2.1	—	—	—	*a*, *i*
Os	0.4	0.95	—	—	—	—	*a*
Pd	1.0	2.4	—	—	—	—	*a*
Pt	0.6	1.6	—	—	—	—	*a*
Re	0.4	0.9	—	—	—	—	*a*
Rh	0.55	1.5	—	—	—	—	*a*
Si	0.2	0.5	0.6	0.9	1.4	—	*a*, *f*
Ta	0.3	0.6	—	—	1.05	—	*a*, *c*
Th	0.3	0.7	—	—	—	—	*a*
Ti	0.2	0.6	—	1.1	1.7	2.1	*a*, *k*
U	0.35	1.0	—	—	—	—	*a*
W	0.3	0.6	—	—	1.1	—	*a*, *c*
Zr	0.3	0.75	—	—	—	—	*a*

*Reproduced from L. Maissel, "Handbook of Thin Film Technology" (L. I. Maissel and R. Glang, eds.), Chapter 4, Application of Sputtering to the Deposition of Films. McGraw-Hill, 1970. With permission from McGraw-Hill.

† Kr^+ ions.

‡ Type 304 Stainless steel.

[a] N. Laegreid and G. K. Wehner, *J. Appl. Phys.* **32**, 365 (1961).

[b] O. Almen and G. Bruce, *Nucl. Instr. Methods* **2**, 257 (1961).

[c] C. E. Carlston, G. D. Magnuson, A. Comeaux, and P. Mahadevan, *Phys. Rev.* **138**, A759 (1965).

[d] M. T. Robinson and A. L. Southern, *J. Appl. Phys.* **38**, 2969 (1967).

[e] D. Rosenberg and G. K. Wehner, *J. Appl. Phys.* **33**, 1842 (1962).

[f] A. L. Southern, W. R. Willis, and M. T. Robinson, *J. Appl. Phys.* **34**, 153 (1963).

[g] O. C. Yonts, C. E. Normand, and D. E. Harrison, *J. Appl. Phys.* **31**, 447 (1960).

[h] P. K. Rol, J. M. Fluit, and J. Kistemaker, "Electrostatic Propulsion" (D. B. Langmuir, E. Stuhlinger and J. M. Sellen, eds.). Academic Press, New York, 1961. p. 203.

[i] C. H. Weijsenfeld, A. Hoogendoorn, and M. Koedam, *Physica* **27**, 963 (1961).

[j] E. T. Pitkin in D. B. Langmuir, E. Stuhlinger, and J. M. Sellen (eds.), "Electrostatic Propulsion." Academic Press, New York, 1961. p. 195.

[k] O. K. Kurbatov, Soviet *Phys. Tech. Phys.* English Transl. **12**, 1328 (1968).

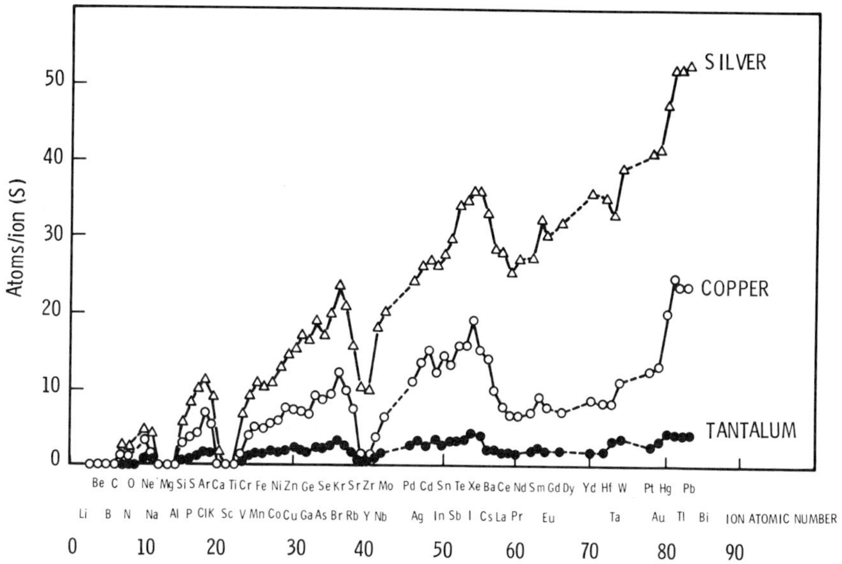

FIG. 17. Variation of sputtering yield with atomic number. The sputtering yield varies as a function of atomic number of the bombarding ion. The ion energy was 45 keV and the targets were copper, silver, and tantalum. (Reproduced with permission from American Elsevier Publishing. See Ref. 25.)

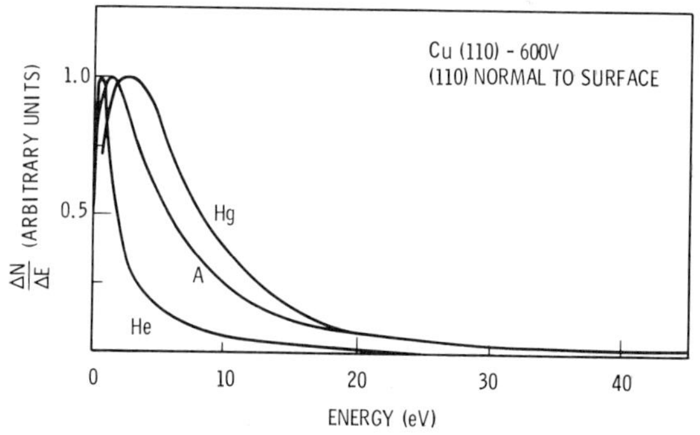

FIG. 18. Energy distribution of sputtered atoms. When sputtered atoms leave a surface they have considerable energy. This data shows the energy of sputtered Cu atoms when a single crystal Cu (110) was bombarded with 500 eV He, Ar, and Hg ions. The energy was measured in the ⟨110⟩ ejection direction, i.e., normal to the surface. (Reproduced with permission from American Elsevier. See Ref. 25.)

to control the film parameters. Inert gases are generally used for sputtering, so the vacuum pumps must be capable of handling large quantities of these gases. This is particularly true if the inert gas is continually flowing through the system to flush out the impurities desorbed from the walls during sputtering. Sputtering with the gas flowing is called dynamic sputtering, in contrast to static sputtering in which the sputter chamber is backfilled to a predetermined pressure with the pumps isolated. It is also important that the gas handling systems maintain the purity of the gases bled into the system. In fact, if the inert gas source is of questionable purity, or if the gas will pick up impurities in the tubulation leading to the vacuum chamber, it may be necessary to purify the gas by passing it through a titanium sublimation pump or over hot active metal chips. The need to control the impurities in a plasma is discussed in more detail below.

16.3.2.1. Target Materials. Any material can be sputter deposited if the appropriate sputter method is used. Section 16.3.3 will describe briefly the various types of sputtering techniques that can be used for film deposition. There are, however, some general characteristics of targets that are common to all types of sputtering. Almost any target shape can be used in sputtering, but a planar disk-shaped surface is generally used. Other geometries are chosen if the substrates to be coated have specific shapes such as cylindrical, spherical, or conical. The target shape is chosen to insure uniformity of film deposit and a high deposition efficiency. However, nonplanar targets may be difficult to fabricate and some other method to secure film uniformity such as moving the substrate may be used. Only one sputtering method does not use a plasma discharge (see Section 16.3.2.2) from which ions are accelerated to the target. To assure uniform sputtering, the field lines to the target must be uniform, thus requiring a flat smooth target and uniform electric fields around the edge of the target. The first of these is easy to ensure but the second requires a special ground shield arrangement around the target to shape the edge fields for uniform deposition. Figure 19 shows one such shield arrangement. With a shield placed a distance smaller than the dark space from the target (see Section 16.3.2.2), no discharge will exist and, hence, no sputtering. This is important in controlling unwanted sputtering from the target and target heating. In fact, target heating is one of the most severe problems in sputter deposition and is generally the limiting factor on the deposition rate. Maissel[29] has estimated that less than 1% of the total power in a discharge goes into ejecting atoms or electrons from the target. About three quarters of the power goes into heating the target and the rest is absorbed by the electrons in the plasma which collide with the substrate. For high deposition rates, the targets must be cooled

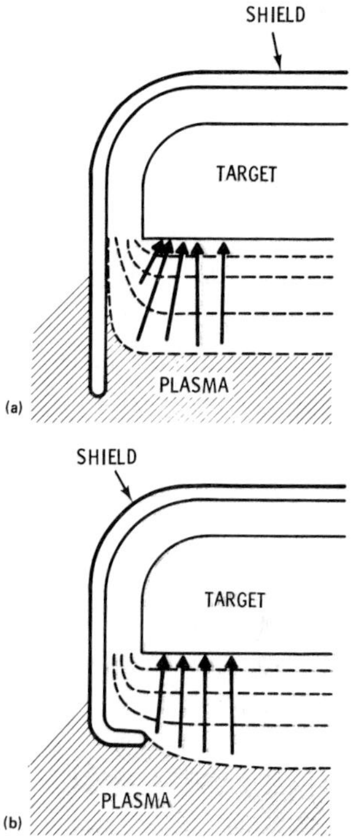

FIG. 19. Ground shields for sputtering targets. Two configurations of ground shields to reduce the fringe field edge effects to improve sputter film uniformity. (a) Extension of ground shield toward substrates, (b) bending the ground shield around the target. (Reproduced with permission from McGraw-Hill. See Ref. 29.)

to eliminate target failures such as melting. A number of good designs for water cooled targets exist in the literature.[29,31]

Ceramics, insulators, and organic compounds can be sputtered if they can be fabricated in a form consistent with target designs. Insulators must be bonded to a metal plate. Great care must be taken to assure good thermal contact and a low vapor pressure for the bonding agents. Many ceramics are formed from powders with a bonding agent. If the proper bonding agents are not selected, they will act as "impurities" in the target and contaminate the plasma, particularly the thin film. Many sputter target materials are available commercially; however, one should specify the target purity required for a particular application and insist on an elemental analysis from the manufacturer. It should be remembered that low Z impurities such as carbon, nitrogen, and oxygen are important in determining film properties but are often not reported as a target impurity because they are difficult to measure.

[31] D. M. Mattox, "Sputter Deposition and Ion Plating Technology" Sandia SLA-73-0619, 1973. Vacuum Science and Technology Short Course, Education Committee American Vacuum Society.

Metal alloy films can be deposited by sputtering; however, the composition of the resultant film may not be the same as the target. This difference may be due to different sputtering ratios of the constituents of the alloy target, different sticking coefficients of the constituent atoms on the substrate, or diffusion if the target is at an elevated temperature. For example, the constituent with the highest sputtering rate would be removed preferentially from the surface and, if bulk diffusion is rapid enough to keep the surface supplied with this element, the resulting film composition would be rich in this element. Selective sputtering and diffusion can be avoided by fabricating a target of different pure elements, i.e., make a composite target, and adjusting the areas of each material to produce a film of the desired composition. This usually is done by covering a fraction of an elemental target with another material. However, the target must have a uniform composition since the spatial distribution in the film of a particular element will depend on its geometrical arrangement on the target. The only way to be sure of the film composition is to analyze the film.

16.3.2.2. Ion Production. All sputtering techniques require energetic ions that are formed either in a self-sustained glow discharge or in a supported (thermionically or magnetically) glow discharge. Self-sustained glow discharges can be obtained by applying a dc voltage between two electrodes in a vacuum chamber when the gas pressure ranges from about 5×10^{-3} Torr to about 10 Torr. If the gas pressure in the system is between 0.01 and 0.1 Torr and the voltage across the electrodes is increased from zero, a current of approximately a few picoamperes will be measured because of ionization of the gas and electrodes from cosmic radiation. As the voltage is increased, gas collisional processes increase ionization and the current also increases slightly. This voltage–current regime is called a dark or Townsend discharge. If the voltage is increased further, a transition region is reached where the current increases very rapidly, the voltage drops, and a glow discharge is observed. A normal glow discharge has a characteristic appearance of bright regions and dark regions, shown schematically in Fig. 20. Also shown are the spatial distributions of positive and negative current densities and space charge densities, net charge density, electric field, voltage, and light intensity. In general, it is difficult to observe the Astron dark space, and the cathode layer (cathode spot) looks like it is sitting on the cathode. The light emitted from the cathode layer is characteristic of the cathode material since most of the light results from the deexcitation of sputtered material. The most important regions for sputtering are the cathode dark space, also called Crookes dark space, and the negative glow. Most of the volt-

FIG. 20. Schematic of a normal glow discharge. (a) Schematic of a normal glow discharge showing the spacial distribution of light and dark regions. Note that the cathode dark space is sometimes called Crookes dark space. (b) The spacial distribution of light intensity. (c) The spacial distribution of voltage. (d) The spacial variation of the voltage gradient. (e) The spacial variation of the net charge density. (f, g) The spacial distribution of positive and negative charge density. (h) The spacial distribution of electron current density j^-. Note: $*E = dV/dX$;**DS is dark space. (Reproduced with permission from Dover Publications from J. D. Cobine, "Gaseous Conductors," 1958.)

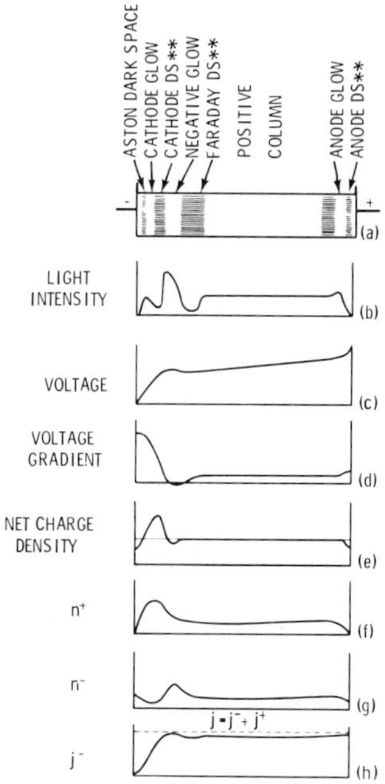

age drop occurs across the Crookes dark space [Fig. 20(c)] and most of the ionization occurs in the negative glow region.

Figure 20(e) shows the regions of maximum positive charge density. Size of the Crookes dark space is related to the electron mean path length before an ionization collision and thus is related to the type of gas composing the plasma and the gas pressure. If Λ is the width of the dark space and p the pressure of the gas, then

$$\Lambda p = C,$$

where C is a function of the gas composition and temperature. Table IV gives some typical values of C. Experimentally one observes Λ increase and the positive glow decrease as the pressure is reduced. If the pressure remains constant and the anode–cathode distance decreases, the Crookes dark space stays the same size, and the positive column decreases. If the anode is positioned inside the cathode dark space, the discharge is extinguished. This is used to control the discharge and keep the sputtering from occurring around the high voltage electrode by placing groundshields, as mentioned above.

Positive ions are produced by electron–ion scattering in the negative glow region and are accelerated by the cathode potential. It is possible

TABLE IV. Normal Cathode–Fall Thickness[a,b,c] $\Lambda p = C$ (cm Torr at Room Temperature)

Cathode	Air	A	H_2	He	Hg	N_2	Ne	O_2
Al	0.25	0.29	0.72	1.32	0.33	0.31	0.64	0.24
C	—	—	0.9	—	0.69	—	—	—
Cd	—	—	0.87	—	—	—	—	—
Cu	0.23	—	0.8	—	0.6	—	—	—
Fe	0.52	0.33	0.9	1.30	0.34	0.42	0.72	0.31
Hg	—	—	0.9	—	—	—	—	—
Mg	—	—	0.61	1.45	—	0.35	—	0.25
Ni	—	—	0.9	—	0.4	—	—	—
Pb	—	—	0.84	—	—	—	—	—
Pt	—	—	1.0	—	—	—	—	—
Zn	—	—	0.8	—	—	—	—	—

[a] A. V. Engel and M. Steenbeck, "Electrische Gasentladungen, ihre Physik u. Technik," Vol. 2, p. 104. Springer-Verlag, Berlin (1934).

[b] Reproduced from J. D. Cobine, "Gaseous Conductor." Dover Publications, New York, 1970, with permission from the publisher.

[c] Measured in cm/Torr at room temperature.

for ions to reach the cathode with the full cathode potential but most undergo atomic scattering and charge exchange that modify their energy and direction. Some of the ions become part of the plasma or positive column and others impinge on the chamber walls and substrate surfaces. Those ions that reach the cathode have an energy distribution shown in Fig. 21 and it can be seen that very few ions have energy corresponding to the cathode potential. Therefore, the actual sputtering rate or thin film growth rate is much smaller than would be calculated using the cathode potential and cathode ion current density. The ion energy incident on the target during low pressure, sustained plasma discharge sputtering (an example is triode sputtering) is close to that of the cathode potential, since there are fewer ion collisions. The average sputtering yield and the film deposition rate are considerably higher in the low pressure mode. However, the difficulty of maintaining a uniform high density plasma at pressures below ~10 mTorr is quite severe. Some of these problems have been solved, resulting in high sputter deposition rates for a number of materials.[32,33]

For any particular gas–cathode combination there is a minimum voltage at which a normal glow discharge will be produced. This voltage, called the "cathode drop," depends on the gas composition and cathode

[32] S. D. Dahlgren, E. D. McClanahan, J. W. Johnston, and A. G. Graybeal, *J. Vac. Sci. Technol.* **7,** 398 (1970).

[33] J. W. Patten and E. D. McClanahan, *J. Appl. Phys.* **43,** 4811 (1972).

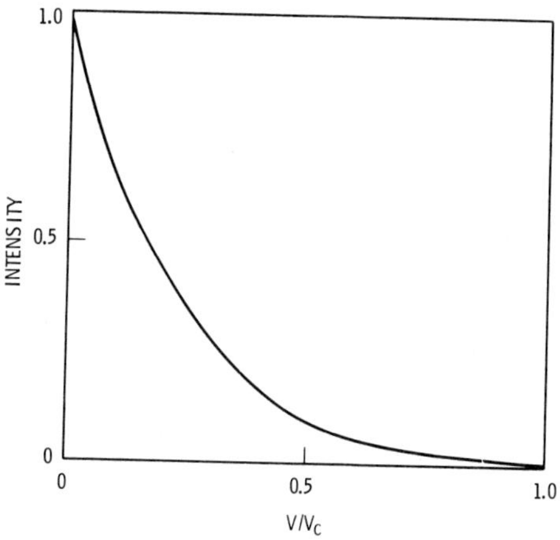

FIG. 21. Ion energy distribution at a cathode. Ion energy distribution at a cathode. Notice that very few ions have the energy equal to the full cathode potential V_c. (Reproduced with permission from D. Mattox, Sandia Laboratories. See Ref. 31.)

material. Measured cathode drop values for various gas–cathode combinations are given in Table V. At this voltage, if the pressure is high enough, the glow is self-sustained because enough secondary electrons are produced at the cathode to maintain the discharge. If more power is supplied to the cathode or if the pressure is increased, the cathode glow increases in size with the cathode current density remaining constant until the cathode glow covers the cathode. In this regime the cathode drop remains constant. Further increase in pressure or power will result in the cathode potential increasing along with the current density. This plasma mode is called "abnormal glow" and is the one usually used for film deposition.

The ions and atoms that impinge on the wall and substrate can have important effects on the sputtering and thin film parameters. In particular, gaseous impurities are desorbed from the walls because of the plasma–wall interaction, enter the plasma, and interact with the target cathode and the film. These impurities are generally water vapor, carbon monoxide, and hydrocarbons which are chemically active and which can affect the film characteristics. The gas composition and impurities concentration must be controlled in order to produce films with reproducible properties. Because of the energetic atoms and ions in the plasma, the plasma is very active chemically forming species which are thought to improve film–substrate bonding. General requirements for the vacuum system and techniques for processing the system will be discussed in Section 16.3.3.

In addition to sputtered atoms, high energy ions and atoms also produce secondary electrons when they collide with the target cathode. These electrons are essential for sustaining the plasma, but some are accelerated to the substrate resulting in substrate heating as will be discussed in Section 16.3.2.3. Another interaction that can occur between a

TABLE V. Normal Cathode Drop[a,b,c]

Cathode	Air	A	He	H_2	Hg	Ne	N_2	O_2	CO	CO_2	Cl
Al	229	100	140	170	245	120	180	311	—	—	—
Ag	280	130	162	216	318	150	233	—	—	—	—
Au	285	130	165	247	—	158	233	—	—	—	—
Ba	—	93	86	—	—	—	157	—	—	—	—
Bi	272	136	137	240	—	—	210	—	—	—	—
C	—	—	—	240	475	—	—	—	525	—	—
Ca	—	93	86	—	—	86	157	—	—	—	—
Cd	266	119	167	200	—	160	213	—	—	—	—
Co	380	—	—	—	—	—	—	—	—	—	—
Cu	370	130	177	214	447	220	208	—	484	460	—
Fe	269	165	150	250	298	150	215	290	—	—	—
Hg	—	—	142	—	340	—	226	—	—	—	—
Ir	380	—	—	—	—	—	—	—	—	—	—
K	180	64	59	94	—	68	170	—	484	460	—
Mo	—	—	—	—	353	115	—	—	—	—	—
Mg	224	119	125	153	—	94	188	310	—	—	—
Na	200	—	80	185	—	75	178	—	—	—	—
Ni	226	131	158	211	275	140	197	—	—	—	—
Pb	207	124	177	223	—	172	210	—	—	—	—
Pd	421	—	—	—	—	—	—	—	—	—	—
Pt	277	131	165	276	340	152	216	364	490	475	275
Sb	269	136	—	252	—	—	225	—	—	—	—
Sn	266	124	—	226	—	—	216	—	—	—	—
Sr	—	93	86	—	—	—	157	—	—	—	—
Th	—	—	—	—	—	125	—	—	—	—	—
W	—	—	—	—	305	125	—	—	—	—	—
Zn	277	119	143	184	—	—	216	354	480	410	—
CsO–Cs	—	—	—	—	—	37	—	—	—	—	—

[a] A. V. Engel and M. Steenbeck, "Elektrische Gasentladungen, ihre Physik u. Technik," Vol. 2, p. 103, Springer-Verlag, Berlin, 1934. J. J. Thomson and G. P. Thomson, "Conduction of Electricity Through Gases," Vol. 2, pp. 331–332. Cambridge Univ. Press, London, 1933.

[b] Reproduced from J. D. Cobine, "Gaseous Conductor." Dover Publications, New York, 1970, with permission from the publisher.

[c] Measured in volts.

high energy ion and the cathode surface is reflection and neutralization, whereby the depositing film is bombarded by high energy neutrals resulting in substrate heating, resputtering of the film, and gas incorporated or implanted into the film. The amount of gas implanted in the film is a function of the sputtering parameters, especially the discharge pressure, substrate temperature, and substrate potential. As much as a few atomic percent (by number) of inert gas can be incorporated in the film under the proper conditions.

16.3.2.3. Substrates and Substrate Holders. The substrate requirements for sputter film deposition are similar to those for vapor deposition. In principle, any vacuum-compatible material can be used as a substrate. Vacuum compatible is defined by the requirements of the process and material being deposited. A substrate that normally outgasses large quantities of water vapor, oxygen, and other chemically active molecules should be vacuumed processed before deposition of active materials. Substrate surface conditions such as cleanliness and smoothness are important in determining the final appearance and adhesion of the film. Dust on the surface results in pinholes in the sputtered film as it does in vapor deposited films. If the film is the first step in the formation of a device or is to be used for an experiment, the pinholes can seriously limit the usefulness of the film. Therefore, care must be exercised during substrate cleaning and loading in the vacuum chamber. Even though the deposition chamber is not in a clean room, it is advisable to use clean room techniques, such as wearing gloves and hair covers when loading and unloading substrate.

Requirements of the substrates are normally dictated by the experiment for which the film is being made. If the experiment is to study epitaxial growth as a function of sputtering parameters, then the substrate would probably be a single crystal, the lattice parameters of which would most closely match that of the film. Thermal coefficient of expansion and resistance to thermal shock are important for those experiments in which the film is deposited at high temperature or in which the film will undergo large thermal cycling, as in cryogenic experiments. Under these circumstances the substrate and film should have closely matching thermal coefficients of expansion. It is also important to consider the chemical properties (inertness) of the substrate if patterns must be etched in the film. For additional details concerning substrate materials the reader is referred to the article by Brown.[7]

Substrate temperature is an important parameter in sputter deposition as it is in vapor deposition. However, in sputter deposition a large fraction ($\sim 40\%$) of the power applied to the plasma can be dissipated in the substrate and substrate holder. This means that the substrates could increase in temperature a few hundred degrees centigrade if good thermal contact is not made to the substrate holder. A good bond must be made between the substrate and holder. Mechanical clamps might not always be sufficient and low vapor pressure bonding adhesives might have to be used. It may also be important to control the temperature, that is, maintain a substrate temperature that is either higher or lower than the equilibrium temperature obtained with a given set of sputtering parameters.

Heating the substrates and substrate holder is not too difficult since heater wire such as nichrome or other resistance wire can be incorporated in the holder design. Uniform temperature is important; so the location of the heaters is important as well as constructing the holder out of a good thermal conductor.

Cooling substrates during sputter deposition is much more difficult than heating. It is essential to provide good thermal contact with the substrate holder. Either water or liquid nitrogen cooling can be used depending on the temperatures required during film growth, and the desired temperature can be controlled by regulating the liquid flow. Design of hot or cold substrate holders is strongly dependent upon the requirements of the process and the geometry of the vacuum system. In general, substrate holders are constructed of materials with high thermal conductivity, low outgassing, and low sputtering yield. If substrate sputtering by high energy neutrals becomes a problem, then coating the holder with target or cathode material will minimize impurities from being incorporated into the film. If welds are necessary to introduce the cooling liquid to the substrate supports, then it is important that the proper vacuum fabrication techniques be employed.

To obtain thin films with given parameters from one deposition to another, not only must the substrate temperature be controlled but also the substrate potential. Often ceramic or glass substrates are used, and if they remain electrically isolated from the system, their surfaces will become charged and could attain some negative potential. Positive ions from the plasma are accelerated to the substrate and can resputter the film. In some situations (see Section 16.3.3) it is desirable to maintain the substrate at some electrical potential other than system ground. To do this, the substrate holder must be electrically isolated from the main system but connected to appropriate electrical vacuum feedthroughs. If the substrates are to be floated at some potential and also heated or cooled, proper isolation of the temperature controlling apparatus must be maintained. Whether this sophistication is necessary depends on what properties are required in the film and if the films must be made reproducible for some application or device.

16.3.3. Sputtering Arrangements

A number of different sputtering modes have been developed to solve specific application problems. These will be described briefly in this section.

16.3.3.1. Diode or Cathode Sputtering. Diode sputtering is the basic sputtering mode and is probably the method routinely used for many sputter film processes. All other sputtering methods are really modifications of diode sputtering. Figure 22 schematically illustrates the elements of a diode sputtering system. In this most simple form, the substrate holder is electrically connected to the vacuum system and any heating or cooling is integrated into the holder. Also shown in Fig. 22 is a shutter directly above the substrate which is used during presputter cleaning of the cathode and during processing the systems.

The procedure for producing a film by diode sputtering is: (1) clean and etch cathode and mount in vacuum system, (2) clean and load substrate and substrate holder, (3) evacuate system (a pressure of $\leq 10^{-6}$ Torr is a suitable ambient pressure), (4) backfill sputter chamber with argon gas to a pressure of 20–100 mTorr, (5) with the shutter closed, apply -2 to -5 kV to the cathode (at this point a glow discharge is observable) and presputter for 30–60 min, (6) open the shutter and continue sputtering until the desired film thickness is achieved, (7) allow the film and cathode to cool, (8) backfill with argon or nitrogen gas to atmospheric pressure and remove the film. This description assumes that the sputtering parameters to achieve desired film are known; that is, the proper argon pressure, cathode voltage, current density, and substrate temperature are known and obtained before the presputtering. Experimentation with a particular system may be necessary to achieve the desired results. The argon pressure can be obtained either by valving off the pump and backfill to the required pressure and sputter in a static system, or throttling the pump and adjusting the argon gas flow to obtain the sputtering pressure. There are

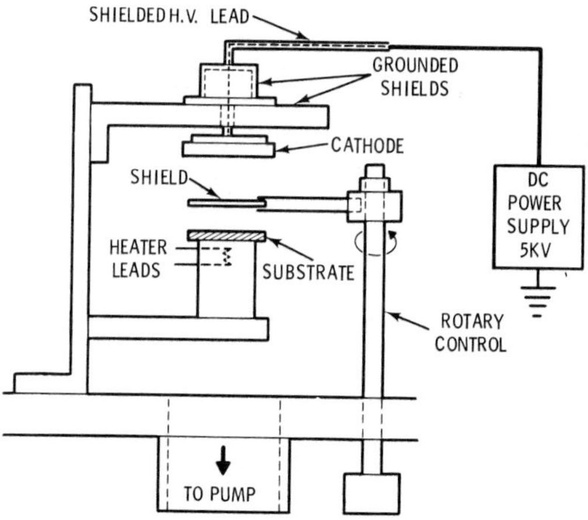

FIG. 22. Schematic diagram of a diode sputtering system. Schematic diagram of a diode sputtering system. This system is simple to construct since the only electrical feedthrough required is for a high voltage (1–5 kV), high current (100 mA) power supply. The substrate heater and the rotatable shield are optional and will depend on the application. (After M. H. Francombe in "The Use of Thin Films In Physical Investigations" (J. C. Anderson, ed.), p. 46. Academic Press, New York, 1966.)

some advantages of maintaining a gas flow during sputtering. First, the pressure remains constant since argon ions become implanted during sputtering so in a static system the pressure will decrease with time. Second and probably most important, the gas flow will help to flush out impurities removed from the wall by the plasma. In a static system these impurities can accumulate in the plasma, thus introducing impurities in the deposited film.

16.3.3.2. Triode Sputtering. In triode sputtering the plasma is produced and maintained thermionically, that is, a hot filament generates electrons that are used to sustain the plasma. The geometric and electrical configuration is shown in Fig. 23. Figure 23 does not show a shutter that is placed part way between the target and the substrate. Since the substrate is electrically separated from the vacuum system, it can be processed by ion sputter cleaning as indicated by the deposit-cleaned switch—a capability not unique to the triode system. During film deposition the substrate would be grounded. Because the hot filament electron source helps maintain the plasma, triode sputtering is done at gas pressures of ~1–10 mTorr, which is considerably lower than that used in diode sputtering systems. Lower pressures mean less ion scattering and so most of the ions have an energy close to the target potential. This results in a higher average sputtering yield and hence a higher film deposition rate. Also, lower pressures may assist in the control of impurities and are incorporated in the film during deposition. The main disadvantage of triode sputtering is the difficulty of maintaining a uniform discharge over a large area. If great care is not exercised in controlling system geometry, deposited film can have nonuniform thickness distribution.

16.3.3.3. Bias Sputtering. Substrates and the depositing film are continuously being bombarded by electrons and high energy ions. If the substrates are electrically floating, they will float to some negative potential, where they are also bombarded by ions. Ion and neutral bombardment cause the film to be resputtered. In bias sputtering the substrate is placed at a negative potential so the resputtering can be controlled. Figure 24 shows a schematic of the system when the substrate can be biased from 0 to −75 V dc. Of course, any sputtering technique can be adapted for bias sputtering. For certain metals such as niobium, bias sputtering can produce quite pure films which certainly affects film resistivity. Film adhesion, film epitaxi, and film microhardness also can be controlled to some extent by bias sputtering.

16.3.3.4. Rf Sputtering. Insulators, organic solids, or metals with insulating layers cannot be sputter deposited in a direct current plasma discharge. The front surfaces of these materials accumulate positive charges until a positive potential equal to the ion energy is reached which

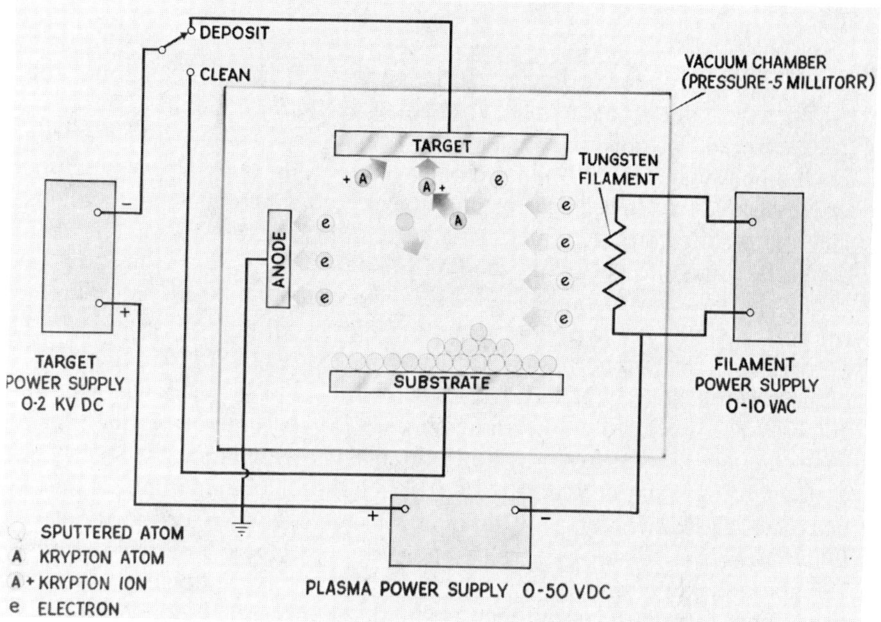

FIG. 23. Schematic of a triode sputtering system. A simple schematic of a triode sputtering system. The tungsten filament supplies electrons which are accelerated to the anode. These electrons help to maintain the glow discharge and enables sputtering to be done at pressures lower than for diode sputtering. The substrate can be sputter cleaned prior to deposition. Not shown in this figure are a shutter between the target and substrate, substrate heater, and target cooling details. A is a krypton atom, A^+ a krypton ion, and e an electron. (Reproduced with permission from E. McClanahan, Battelle Northwest Laboratories.)

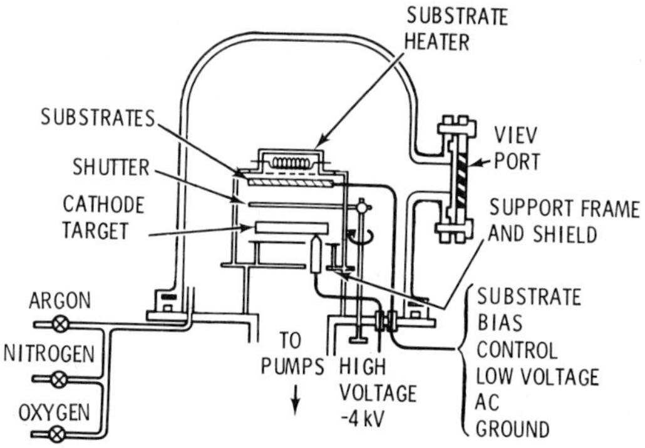

FIG. 24. Schematic of a bias sputtering system. Simple schematic of a bias sputtering system. The substrate bias control can apply a positive or negative low dc voltage (± 100 V dc), an ac voltage to the substrate, or it can ground the substrate. The cathode target requires the typical high voltage power supply. Also shown in the schematic are a rotatable shutter, substrate heater, and a gas manifold assembly. The gas manifold assembly is used during reactive gas sputtering.

stops further ion bombardment. If the positive charge can be neutralized by electron bombardment, then sputtering would be possible. This neutralization is accomplished by using an rf generated plasma. The electron has a much higher mobility in the plasma than ions and charges the insulating surface to a high negative potential which accelerates positive ions that sputter the surface. Sputtering occurs during every half cycle when the surface has the negative potential.

It is also possible to superimpose a dc potential on the rf to increase the sputtering yield. rf plasmas can be formed at quite low pressures, ~1 mTorr, so it is not necessary to use a triode configuration; the diode mode, shown in Fig. 22, is normally used. An rf generator is used in place of the dc power supply shown in Fig. 22. It is important that the rf generator impedance be matched to the plasma impedance. This can be done by using a simple *L* network (a capacitor and an inductor). If this is not done, a great deal of rf power can be dissipated uselessly and at potentially hazardous locations. The values required in the *L* network can be calculated using circuit theory once the impedances of the generator and the system (plasma) are known. It should also be pointed out that as in any rf work system grounds are important and should be carefully checked before operation. The rf mode of sputtering allows any vacuum-compatible material to be sputter deposited.

16.3.3.5. Reactive Sputtering. Freshly sputtered thin films of most elements are highly active chemically and will combine with non-noble gases in the vacuum chamber, which is why the partial pressure of active gases in the plasma should be a few orders of magnitude less than the sputtering pressure. On the other hand, it is possible to deposit stoichiometric compounds if the plasma has the correct partial pressures of the appropriate active gas. The most easily formed and widely studied compounds are oxides and nitrides, but others such as carbides and sulfides can be made. Various chemical states of a compound, nonstoichiometric layers, and mixed compound and metallic films can be grown depending on the partial pressure of the active gas in the plasma and on substrate temperature.

Figure 25 shows how the specific resistivity of tantalum films change with various partial pressures of oxygen, methane, and nitrogen in the plasma. The figure also shows the compounds formed on the film and the crystallographic data as determined by transmission electron microscopy. It is not possible to predict what type of film will be formed with a known target element and partial pressure of an active gas. It is necessary to determine empirically those parameters that will yield the type of film desired. When depositing from compounds, either binary or multielement, one of the species may be deficient in the film. This could be because the

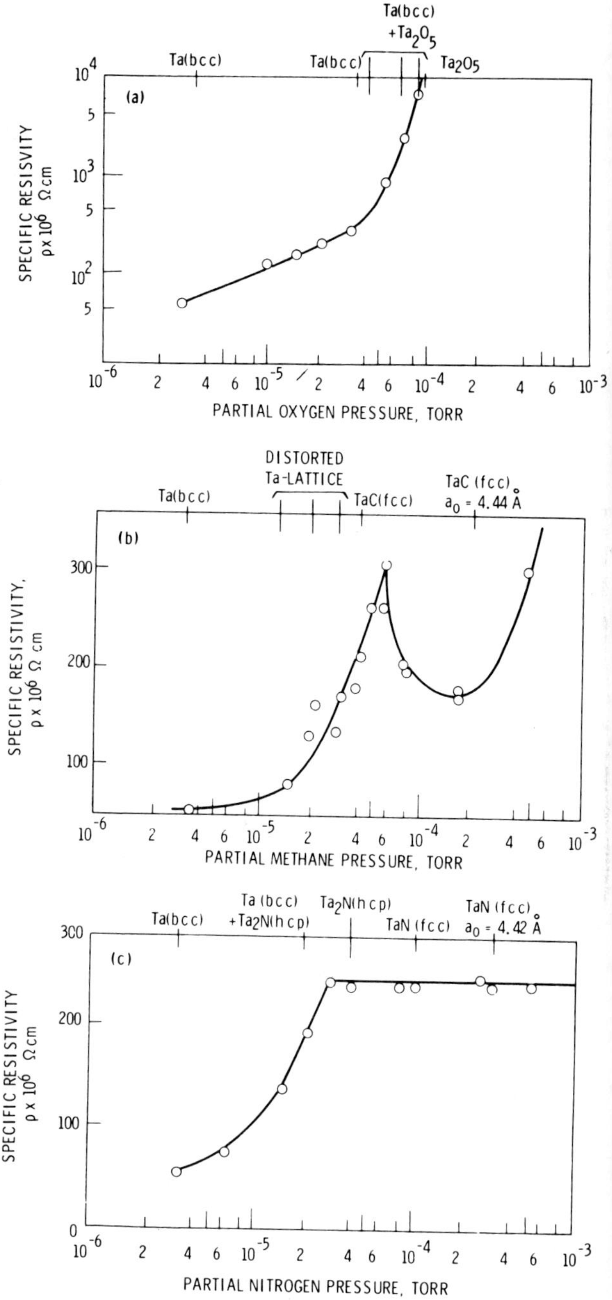

FIG. 25. Specific resistivity of tantalum films deposited during reactive sputtering. Specific resistivity of tantalum films sputtered for 10 min as a function of the partial pressure of (a) oxygen, (b) methane, and (c) nitrogen. Also indicated on each figure are the lattice changes and compounds formed. [After D. Gerstenberg and C. J. Calbick, *J. Appl. Phys.* **35,** 402 (1964).]

species has a high vapor pressure and evaporates from a hot substrate. In this case, vaporizing the deficient element in the plasma can help maintain the stoichiometric compound in the film. It is not known where in the sputtering chamber the compounds are formed. Under some sputtering conditions the compounds may be formed in the plasma; under others they may be formed at the target and then sputtered off. The fact that different mechanisms may be involved in compound formation could explain why different chemical species can be formed when the sputtering parameters are varied. In fact, the plasma characteristics and the gas mixture composing the plasma are strongly dependent on each other.

Reactive sputtering can be easily adopted in any of the sputtering modes described above. The only requirements are a gas manifold that can handle the number of gases needed for the experiment and a pressure gauge that will allow the partial pressure of active gas to be set. One procedure for reactive sputtering is as follows: (1) after the vacuum system has reached its base pressure, throttle pump and backfill with the active gas to a preselected pressure; (2) once this pressure has stabilized, backfill with argon or some other noble gas until the pressure needed for sputtering is reached; (3) the discharge can then be established and a presputter done to establish a dynamic equilibrium condition before the film is deposited on the substrate. This active gas must be kept flowing; otherwise the plasma would become deficient of this gas species. A freshly deposited metallic film acts like an efficient pump for most active gases.

16.3.3.6. Getter Sputtering. Freshly deposited thin films act as an effective pump for most active gases. In fact, this is the basis for getter pumps. In getter sputtering the geometry of the sputtering chamber is such that target material can be sputtered over a large area of the chamber as well as over the substrate. The chamber walls are often cooled to increase the pumping efficiency of the film. Also, the substrate area is small compared to the cathode target and a shutter above the substrate holder is essential. With this geometry, shown in Fig. 26, the gas in the plasma has a high probability of interacting with clean film before it is in the region to sputter deposit. In this way the partial pressures of any active gases can be kept very low in the plasma, and very pure films can be grown. This can be particularly important when epitaxial sputtered films are being studied.

16.3.3.7. Ion Beam Sputtering. Ion beam sputtering is quite different from any of the other methods of sputter deposition. A general plasma discharge between the cathode (target) and substrate is not used in this technique. Rather, ions are formed in an ion source of a particle acceler-

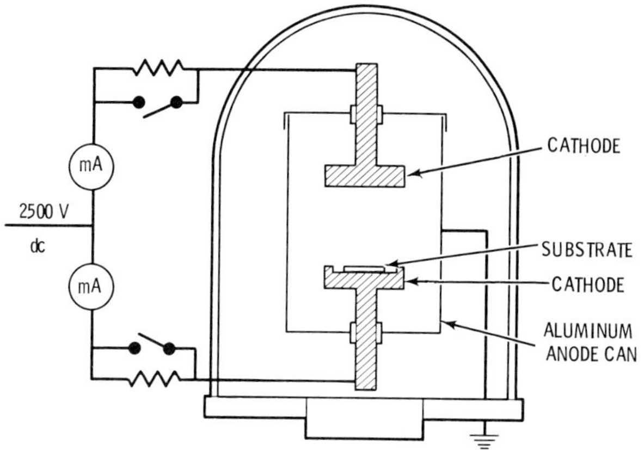

FIG. 26. Schematic for a getter sputtering system. Schematic for a dc getter sputtering system. The aluminum can acts as a large area substrate. Sputtered material from the cathode collects over the whole surface area of the aluminum can producing a surface that is very active chemically. This surface pumps the active gases in the plasma producing a pure discharge. The resulting film collected on the substrate has very few impurity molecules incorporated in it. This geometry works because the area of the aluminum can is large compared to the substrate. [Reproduced with permission from American Institute of Physics, see H. C. Theuerer and J. J. Hauser, *J. Appl. Phys.* **35,** 554 (1964).]

ator, ion gun.[34–36] These ions are extracted from the source, accelerated to source energy, and focused on a target. Atoms are ejected from the target and collected on a substrate. Figure 27 is a schematic of such an arrangement. This technique has the advantage of being able to sputter at a low ambient gas pressure and thus obtain purer films. Also, no electrons and few high energy neutrals bombard the substrates, so the substrate temperature can be controlled much easier than in diode sputtering. However, a serious disadvantage of ion beam sputtering is that only small substrates can be covered uniformly. Furthermore, an ion beam system is more complex to set up and operate and is normally not suited for routine film deposition.

16.3.4. Safety Considerations of Sputtering Techniques

All sputtering techniques require high voltage (0.5–10 keV) and high current power supplies. The current specification of the supply will depend not only on the size of the target but also on the target current den-

[34] F. Gaydou, *Vide* **126,** 454 (1966).
[35] F. Gaydou, *Vacuum Tech.* **15,** 161 (1966).
[36] K. L. Chopra and M. R. Randlett, *Rev. Sci. Instr.* **38,** 1147 (1967).

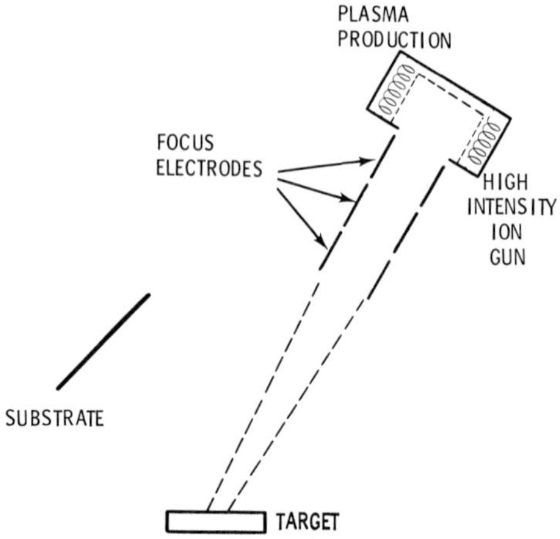

FIG. 27. Schematic of an ion beam sputtering system. A simple schematic of a high intensity ion gun that can be used for sputter deposition. The plasma is produced by electron bombardment and is confined. By placing the electrodes and shield of the plasma production region at a high positive potential (5–10 kV) the ions are accelerated toward the target. The focus electrodes increase the current density in the beam. It is possible to produce thin films at low pressures ($\sim 5 \times 10^{-5}$ Torr) and so control the substrate environment.

sity needed to achieve the necessary film growth rate. It is not uncommon for a target to draw 20–100 mA. These are lethal power supplies and it is essential to follow strict laboratory safety procedures. A number of good safety handbooks[37–39] including one on Vacuum Hazards[9] are available and should be used.

16.4. Comparison of Vacuum Evaporation and Sputtering Techniques

16.4.1. Advantages of Vacuum Evaporation

Vacuum evaporation techniques are easy to use and the equipment required to set up an evaporation station is readily available or inexpensive to obtain. This is particularly true for single film and small batch experi-

[37] N. V. Steere, "Safety in the Chemical Laboratory" Reprint from *J. Chem. Educ.*, American Chemical Society, 1967.

[38] "Guide for Safety in the Chemical Laboratory," 2nd ed. Manuf. Chem. Assoc., Van Nostrand–Reinhold, New York, 1972.

[39] "Handbook of Laboratory Safety" (N. V. Steere, ed.). The Chemical Rubber Co., Cleveland, 1971.

ments. A wide spectrum of materials, including most metals, many alloys, and a large number of compounds, can be vacuum deposited. Those compounds that decompose at high temperature or in a vacuum environment are the main class of materials that cannot be deposited with this technique.

The deposition parameters can be controlled independently so the film properties desired for a given application can be readily obtained. In general, deposition rates for vacuum evaporation are much larger than for sputtering and the rates can be varied over a wider range; in fact, using codeposition techniques makes it possible to produce alloy films over the complete range of compositions. Finally codeposition can be used to produce films of mutually insoluble materials.

16.4.2. Disadvantages of Vacuum Evaporation

Except for special geometries, it is difficult to coat large substrates uniformly with the vacuum evaporation technique. This is particularly true for large planar surfaces. Irregular shaped objects are also difficult to coat uniformly. These problems can be overcome or almost eliminated by certain techniques, the easiest of which is the use of a multiple source arranged to take advantage of the substrate symmetry. For small planar samples increasing the source–substrate distance helps to alleviate the problem although the deposition rate for a given source temperature will be decreased. Moving the substrate can be used to produce large uniform deposits, and rotating a cylindrical substrate allows all sides to be coated. However, adding motion to a deposition system increases its complexity and cost. Finally, introducing a high partial pressure (about 50 mTorr) of an inert gas such as neon or argon will produce gas–source vapor collisions. This scattering has the effect of making the source appear very large. The resulting deposit will cover a large area quite uniformly. There will, of course, be inert gas inclusions in the film.

Generally, the adhesion of evaporated films is not as good as that of sputtered films. Poor adhesion can be minimized or eliminated by raising the temperature of the substrate. Some deposited films adhere better on one type of substrate material than on another. So if a given substrate is not needed for the application, changing substrate material can also help an adhesion problem.

16.4.3. Advantages of Sputter Deposition

Sputter deposition produces film that covers large areas quite uniformly. This technique also makes it easier to cover complex surfaces uniformly without having to move the substrate. These advantages occur because the relatively high pressure of the sputtering gas produces many collisions between the gas and sputtering atoms, and because the cathodes used in the sputtering system are spatially much larger than evaporation sources. Sputtered films adhere much better to substrates than evaporated films. As discussed above, the combination of chemically active species formed in the plasma and the high energy atoms and electrons impinging on the substrate causing surface atom rearrangement result in better film adhesion.

With the proper choice of technique and sputtering parameters almost anything can be sputtered, including insulating materials and complex compounds. Refractory metals that are difficult to evaporate except by electron beam techniques can be easily sputtered. By adjusting the plasma gas composition, stoichiometric compounds can be sputter deposited, even if the cathode is a pure element. It is possible to obtain film properties that are quite different than the bulk material which can be beneficial in certain applications.

Once a set of sputtering parameters are chosen it is possible to maintain them for long periods of time. This is much more difficult in evaporation systems. For example, evaporation sources tend to change temperature as the evaporant charge decreases. By monitoring the deposition rate the source temperature can be changed to maintain a constant rate. For single or small batch film experiments this is not a problem, but it can be serious for larger batch or automated deposition systems. Sputtering techniques can be and have been set up for automated film production.

16.4.4. Disadvantages of Sputter Depositions

There are a number of disadvantages of sputter deposition, but like the disadvantages of evaporation, they can be overcome to a certain degree. The state of the art in sputtering has progressed far in the last few years, but the techniques that overcome some of the difficulties tend to be quite complex and expensive. For instance, deposition rates are much slower in sputtering than in evaporation, but techniques used to increase the deposition rates lead to other problems, such as increasing cathode and substrate temperatures. Of course, steps can also be taken to maintain constant cathode and substrate temperature.

It is somewhat more difficult to arrive at the sputtering parameters that will produce a film with a given set of properties, like bulk properties, for example. Also it is more difficult to monitor and change the deposition rate in sputter deposition. This is especially true for rf sputtering and bias sputtering.

Finally, it is difficult to produce sputtered films without large amounts of gas inclusions. This gas will cause high internal stresses in the film, which can cause the film to craze and flake. Substrate temperature and substrate bias can help control this problem.

Acknowledgments

This work was supported by many sponsors, one of which was the Department of Energy, Division of Materials Sciences, Office of Basic Energy Sciences.

SUBJECT INDEX

D

E

H

I

J

K

L

Q

R

T